# Lecture Notes in Nanoscale Science and Technology

Volume 22





For further volumes:
http://www.springer.com/series/7544

Anming Hu • Allen Apblett
Editors

# Nanotechnology for Water Treatment and Purification

Foreword by Mark R. Servos

Springer

*Editors*
Anming Hu
Department of Mechanical
Aerospace and Biomedical
Engineering
University of Tennessee
Knoxville, TN, USA

Allen Apblett
Department of Chemistry
Oklahoma State University
Stillwater, OK, USA

ISSN 2195-2159 ISSN 2195-2167 (electronic)
ISBN 978-3-319-37470-3 ISBN 978-3-319-06578-6 (eBook)
DOI 10.1007/978-3-319-06578-6
Springer Cham Heidelberg New York Dordrecht London



Printed on acid-free paper

Springer is part of Springer Science+Business Media (www.springer.com)

# About the Editors

**Allen Apblett** is an Oklahoma State University professor of chemistry, Councilor of the Oklahoma Section of the American Chemical Society, and President of XploSafe, LLC. His interests include industrial, materials, and environmental chemistry, catalysis, and metallo-organic chemistry applied to development of new chemical processes. Recently he is focused on explosive detection and neutralization, nanotechnology for water purification, sensors, and arsenic remediation.

**Anming Hu** is an assistant professor in the Department of Mechanical, Aerospace and Biomedical Engineering, University of Tennessee, USA, and former research assistant professor in the Department of Mechanical and Mechatronics Engineering at the University of Waterloo, Canada. He and his colleagues began studying the application of nanotechnology for water treatment and purification in 2009. The research is funded by the Canadian Water Network, the Natural Science and Engineering Research Council (NSERC), Canada. Anming Hu is also working on laser-based advanced manufacturing, nanophotonics, and ultrafast laser-materials interaction.

# Preface

Water is essential for life, but for many years it was generally taken for granted until increasing pollution and diminishment of potable water supplies delineated a need to protect our water supplies and develop new technologies to purify water for consumption. For example, it is imperative that pollutants in wastewater effluents from both industrial and domestic sources be removed or destroyed before discharge to the environment. These pollutants, produced as the by-products of industrialization and social development, are posing serious challenges to existing municipal water treatment technologies. This includes conventional and emergent contaminants such as harmful heavy metal ions, toxic dye molecules, radionuclides, pharmaceuticals, personal care products, and endocrine disruptor compounds. The public concern about the effectiveness of conventional water treatment processes for the removal of emergent environmental pollutants has generated a need for the development of novel water treatment methods. At the same time, as natural water resources continue to diminish due to overuse, waste, and pollution, the current water treatment capacities for securing clean water will be unable to meet the needs of our growing population. New treatment plants and facilities should be built to have high-energy efficiency, improved water treatment abilities, and a long service life. Conventional water treatment technologies fail to satisfy these requirements, prompting the need for new technologies. Furthermore, access to clean drinking water is a development issue faced by billions of people in developing and near-developed countries. Sustainable development in these regions requires innovative technologies to provide safe drinking water in a cost and energy efficient manner.

Nanotechnology offers practical solutions to the problems associated with procuring clean drinking water and protecting natural waters from being contaminated. Nanotechnology is one of the most significant technological achievements of the twentieth century. This importance results from the fact that at the nanoscale (1–100 nm) materials can display dramatically different physical and chemical properties from their bulk counterparts. Radically improved specific surface areas, significant increases in surface reactivity, and tunable energy gaps and electronic structures provide the means to achieve useful water treatment capabilities such as highly efficient surface adsorption, nanofitration, photocatalytic degradation, and solar disinfection. Nanotechnology can be utilized to safeguard our critically important water resources by providing a diverse range of tools and technologies for water treatment and purification. This involves harnessing the unique physicochemical, electronic, magnetic, and surface properties of nanoparticles along with the engineering applications of nanotechnologies to ensure a supply of safe drinking water for a growing population, generating freshwater from seawater, preventing contamination of the environment, and creating effective efficient methods for remediation of polluted waters.

*Nanotechnology for Water Treatment and Purification* is designed to serve as a desktop reference for the field at the boundary between nanotechnology, water science, analytical chemistry, and environmental engineering. Although this interdisciplinary topic is of current specialized research interest, the authors have also provided introductory materials and have utilized presentation styles and organization structures that make the book accessible to both students and scholars outside the field. To this end, the book consists of 11 chapters that reflect the progression from introductory to specialized topics.

Chapter 1 introduces the fundamentals of nanotechnology for water treatment. Starting from basic water treatment methods, the principles of membrane filtration, reverse osmosis, advanced oxidization, and photocatalytic degradation are reviewed. Chapter 2 presents the synthesis of $TiO_2$ nanowires, the fabrication methods for nanowire membranes, characterization of nanowire membranes, and then their application for water treatment through surface adsorption, photocatalytic degradation, disinfection of by-products and bio-pollutants, and nanofiltration. In Chap. 3, Zhang and Li show the application of other metal oxide semiconductor nanomaterials to decompose the persistent organic environment pollutant perfluorooctanoic acid (PFOA). The authors provide an overview of recent advances in the photocatalytic decomposition mechanism of PFOA by $In_2O_3$ and $Ga_2O_3$, nanostructures that are more effective than $TiO_2$ for the removal of PFOA. Potential applications in wastewater treatment are also discussed.

In Chap. 4, S. Tabe demonstrates that electrospun nanofiber membranes (ENMs) are a cutting-edge new generation of membranes that offer significantly higher flux at similar rejection rates in comparison to conventional membranes. Electrospinning has allowed for fabrication of highly porous ENMs with controllable pore size in the range required for microfiltration and ultrafiltration. These ENMs could replace conventional water treatment membranes with smaller systems that operate at lower pressures. In the next chapter (Chap. 5), Li and Seymour introduce fullerenes and carbon nano-onions as sorbents, photocatalysts, and membranes for wastewater treatment. These materials have emerged as promising candidates for environmental applications due to their unique properties.

In Chap. 6, S. Muley and N. Ravindra present graphene-based nanomaterials for water and air remediation via a variety of mechanisms. They extensively discuss the effects of adsorbents on the electronic properties, magnetic properties, and photoconductivity, along with the relevant mechanisms. They show that graphene nanosheets and graphene oxides, due to their ease of chemical modification, have emerged as ideal materials for adsorbents to eliminate heavy metal ions, inorganic ions, and organic dyes from water. Use of graphene as a photocatalyst in water remediation is also discussed. A brief overview of the latest scientific advances in the development of graphene-based devices to overcome the present shortcomings for fabricating sensors in the field of gas sensing, biosensing, and photonics for image sensing applications, along with their operating principles, is also presented.

Leshuk and Gu show the potential of magnetic separation of superparamagnetic nanoparticles from water for recycling and reuse as an efficient, practical, and low-cost slurry-type water treatment method. In this chapter they review the fundamental theory and concepts of magnetic nanoparticle separations and present both a synthesis process for size-tunable superparamagnetic iron oxide nanospheres and the application of these nanospheres as a core material for the immobilization of $TiO_2$, to be used in photocatalytic water treatment as magnetically recyclable composite particles.

In the following chapter, Trad and Apblett focus on the production of magnetic activated carbon materials that, when combined with magnetic filtration, provide efficient technology for removing arsenic and dyes from water. In Chap. 9, Susan Andrew's group reviews the progress of nanotechnologies for drinking water treatment. They demonstrate that various nanomaterials can provide concurrent disinfection, disinfection by-product (DBP) precursor reduction, and degradation of recalcitrant compounds. Research into DBP formation in water treatment processes employing $TiO_2$ photocatalysis is limited, but the studies that have been conducted to date suggest that these technologies create few DBPs directly and have site-specific effects on DBP precursor formation. UV/$TiO_2$ compares favorably to existing disinfection and DBP precursor removal technologies in terms of both DBP formation and DBP precursor removal. Hybrid systems incorporating UV/$TiO_2$ along with another DBP precursor removal technology such as coagulation, adsorption, or membrane filtration are also promising for both applications.

Last but not least, in Chap. 10, Oakes et al. describe how nanotechnologies could further allow economic utilization of mine water as a source of commercial viable products. They provide an overview of recent and future developments in nanotechnology that would benefit contemporary mine water treatment regimes, while investigating novel abate approaches possible through the use of nanotechnology. The chapter discusses candidate nanomaterials, their advantages and limitations relative to existing processes, and the unique properties and surface-active mechanisms that enable their adoption in mine water treatment applications.

The application of nanotechnology to drinking water purification is a rapidly expanding field with ever accelerating technical achievements and dramatic expansion of knowledge in this field. This book provides a snapshot of present-day technologies and provides a strong foundation for future developments. The rapid growth of the field means that the book cannot be all encompassing, but it does represent all of the important areas where nanotechnology is favorably impacting drinking water supplies. We owe special thanks to individual authors in each chapter.

It was our great honor to work with all those who have collaborated, shared, reviewed, helped, explained, and, in general, taught us about this interdisciplinary field. We thank Sara Kate Heukerott, who helped to organize, discuss, and edit this book.

Knoxville, TN, USA Anming Hu
Stillwater, OK, USA Allen Apblett

# Foreword

Provision of clean safe water is one of the major challenges facing societies around the globe. The continued urbanization of human populations, the escalating exploitation of natural resources, and the resulting pollution are putting tremendous pressure on water resources. Increasing demands for food, energy, and natural resources are expected to continue to accelerate into the near future in response to the demands of these changing human populations. In addition, the complexity of human activities is leading to a diversity of new chemical contaminants in the environment that represents a major concern for water managers. This will create increased pressure on both water quantity and quality, making it increasingly difficult to provide a sustainable supply of water for human welfare and activities. Although protection of the water resources is the best long-term solution, we will also need innovative new approaches and technologies to treat water to ensure an adequate high-quality supply to meet these needs.

Solving tomorrow's water issues is going to require innovative new approaches that incorporate emerging new technologies. Great advances have recently been made in material sciences especially in the area of nanotechnology. Nanoparticles can have distinctly different properties from their bulk counterparts creating the opportunity for new materials with a diversity of applications. Recent developments related to water treatment include the potential use of carbon nanotubes, nano-onions, nanospheres, nanofibers, and nanowires for the removal of a diversity of contaminants. By exploiting the properties and structure of these new materials, such as increased surface area, reactivity, and photocatalytic activity, it will be possible to create technologies that can be very efficient at removing and degrading environmental contaminants of concern. Understanding and using these unique properties should lead to innovative cost-effective applications for addressing the complexities of emerging needs for water treatment and protection. Although in early stages, research into the application of nanotechnology is showing great promise for solving some of these major global water issues.

Provision of clean water is becoming more complex and challenging. The traditional approaches are unlikely to be adequate to address the diversity of emerging issues and chemicals of concern. Solutions will require consideration of the complexity of human activities and their impacts on drinking water, water use, and ecosystems. Addressing this complexity will require interdisciplinary teams with complementary skills applied to the development of practical treatment options, approaches, and technologies.

Several of the researchers involved in this publication came together initially through a common interest in applying their research to emerging water issues. The collaboration was accelerated by a seed grant from the Canadian Water Network promoting innovative technologies for water treatment. Once this team came together they applied their combined expertise and resources to attract top students and maintain a focus on exploring the science and engineering aspects of creating innovative application of nanotechnology to water treatment. Funding agencies such as the National Science and Engineering Research Council of Canada and the Canada Research Chairs Program have played a critical role in supporting research and training making this research possible. We have been fortunate to have had the opportunity to work in an interdisciplinary group that has advanced the field by incorporating several disciplines including chemistry, biology, materials science, and water treatment. The advances and opportunities described in this book are the result of researchers actively seeking the expertise of other disciplines to enhance the understanding, application, and impact of their work. Although challenging at times because of the added complexity, the advantages and results of this type of interdisciplinary collaboration will be apparent in the papers presented. Together with a variety of other collaborators and researchers, this book presents the recent developments and opportunities for the application of nanotechnology to address the critical issue of water treatment and protection.

Waterloo, ON, Canada Mark R. Servos

# Contents

# Contributors

**Susan Andrews** Department of Civil Engineering, Drinking Water Research Group, University of Toronto, Toronto, ON, Canada

**Allen Apblett** Department of Chemistry, Oklahoma State University, Stillwater, OK, USA

**Stephanie Gora** Department of Civil Engineering, Drinking Water Research Group, University of Toronto, Toronto, ON, Canada

**Frank Gu** Department of Chemical Engineering, Waterloo Institute for Nanotechnology, University of Waterloo, Waterloo, ON, Canada

**Mélisa Hatat-Fraile** Department of Mechanical and Mechatronics Engineering, Centre for Advanced Materials Joining, University of Waterloo, Waterloo, ON, Canada

**Anming Hu** Department of Mechanical, Aerospace & Biomedical Engineering, University of Tennessee, Knoxville, TN, USA

**Rajendran Kaliaperumal** Verschuren Centre for Sustainability in Energy and the Environment, Cape Breton University, Sydney, NS, Canada

**Tim Leshuk** Department of Chemical Engineering, Waterloo Institute for Nanotechnology, University of Waterloo, Waterloo, ON, Canada

**Yusong Li** Department of Civil Engineering, University of Nebraska-Lincoln, Lincoln, NE, USA

**Zhenmin Li** Collaborative Innovation Center for Regional Environmental Quality, School of Environment, Tsinghua University, Beijing, China

**Robert Liang** Department of Mechanical and Mechatronics Engineering, Centre for Advanced Materials Joining, University of Waterloo, Waterloo, ON, Canada

Waterloo Institute for Nanotechnology, University of Waterloo, Waterloo, ON, Canada

**Martin Mkandawire** Verschuren Centre for Sustainability in Energy and the Environment, Cape Breton University, Sydney, NS, Canada

**Sarang V. Muley** Interdisciplinary Program in Materials Science & Engineering, New Jersey Institute of Technology, Newark, NJ, USA

**Ken Oakes** Verschuren Centre for Sustainability in Energy and the Environment, Cape Breton University, Sydney, NS, Canada

**Nuggehalli M. Ravindra** Interdisciplinary Program in Materials Science & Engineering, New Jersey Institute of Technology, Newark, NJ, USA

**Megan Seymour** HDR Inc., Omaha, NE, USA

**Zhi Shan** Verschuren Centre for Sustainability in Energy and the Environment, Cape Breton University, Sydney, NS, Canada

**Aleksandra Sokolowski** Department of Civil Engineering, Drinking Water Research Group, University of Toronto, Toronto, ON, Canada

**Shahram Tabe** Standards Development Branch, Ministry of the Environment, Toronto, ON, Canada

Department of Chemical Engineering and Applied Chemistry, University of Toronto, Toronto, ON, Canada

**Tarek Trad** Department of Chemistry and Environmental Sciences, University of Texas at Brownsville, Brownsville, TX, USA

**Pengyi Zhang** Collaborative Innovation Center for Regional Environmental Quality, School of Environment, Tsinghua University, Beijing, China

**Shine Xu Zhang** Verschuren Centre for Sustainability in Energy and the Environment, Cape Breton University, Sydney, NS, Canada

Department of Biology, Waterloo Institute of Nanotechnology, University of Waterloo, Waterloo, ON, Canada

**Norman Zhou** Department of Mechanical and Mechatronics Engineering, Centre for Advanced Materials Joining, University of Waterloo, Waterloo, ON, Canada

Waterloo Institute for Nanotechnology, University of Waterloo, Waterloo, ON, Canada

# Chapter 1
# Fundamentals on Adsorption, Membrane Filtration, and Advanced Oxidation Processes for Water Treatment

**Robert Liang, Anming Hu, Mélisa Hatat-Fraile, and Norman Zhou**

**Abstract** Water treatment is the processing of water to meet or achieve specified goals or standards set by regulatory agencies and end users. New water treatment technologies are being developed that need to be evaluated on a fundamental scientific and practical basis compared to traditional remediation processes. Recent advances in nanomaterial development for water treatment in the areas of filtration membranes, high surface area adsorbents, and efficient photocatalysts require approval for their effectiveness and safeness. Fundamental theories and concepts discussed in this chapter pertain to the areas of (i) adsorption and equilibrium isotherms (ii) pressure-driven membrane filtration and its rejection mechanisms for filtration and reverse osmosis processes; and (iii) advanced oxidation processes with a focus on semiconductor photocatalytic concepts.

## 1.1 Introduction

In the beginning of the twentieth century, many of the water treatment methods currently used today were already established [1]; these methods include mechanical separation, coagulation, chemical purification, disinfection, biological treatment,

R. Liang • M. Hatat-Fraile • N. Zhou
Department of Mechanical and Mechatronics Engineering, Centre for Advanced Materials Joining, University of Waterloo, 200 University Avenue West, Waterloo, ON N2L 3G1, Canada

Waterloo Institute for Nanotechnology, University of Waterloo, 200 University Avenue West, Waterloo, ON N2L 3G1, Canada

A. Hu (✉)
Department of Mechanical, Aerospace and Biomedical Engineering, University of Tennessee, 414 Dougherty Engineering Building, Knoxville, TN 37996-2210, USA
e-mail: ahu3@utk.edu

A. Hu and A. Apblett (eds.), *Nanotechnology for Water Treatment and Purification*, Lecture Notes in Nanoscale Science and Technology 22,
DOI 10.1007/978-3-319-06578-6_1, 

aeration, and boiling. The conventional drinking water treatment process, consisting of coagulation, flocculation, sedimentation, filtration, and chlorine disinfection, was developed and optimized throughout the twentieth century as a robust system to remove contamination [2]. Additionally, several technologies were developed over this time to meet complex end user goals which include aeration to control taste and odor, ion exchange and reverse osmosis for inorganic species removal, and adsorption of organic species using activated carbon. Towards the end of the twentieth century, the most significant additions to water treatment were the development of membrane technologies and advanced oxidation processes (AOPs). Membrane technologies allow for the potential to completely remove particulates and pathogens via size exclusion on a scale as required by the end user. AOPs are capable of the oxidation of many pollutants into intermediate species and, eventually, their mineralization into inorganic compounds.

With the growing demand for clean water sources, there is rising concern regarding the availability and strategies necessary for the deliverance of potable water [3–6]. To exacerbate the situation, there are also emerging pollutants in wastewater effluents that have potential adverse health effects; these include, but are not limited to, textile dyes, pharmaceuticals and personal care products (PPCPs), endocrine disrupting compounds (EDCs), and plasticizers [3, 7–10]. The increasing network between receiving waters for treated wastewater and source water withdrawal has created the possibility of exposure to these trace contaminants through the water supply. Recent research has found evidence of low concentrations of PPCPs and EDCs in source waters in many developed communities, which have no regulations in place to control these contaminants [11]. PPCPs, including medical products, cosmetics, and pesticides, enter the wastewater system via human excreta or urine, washing, improper disposal, lawn care, etc. They have unknown side effects and consequences for human life and water ecosystems. Endocrine disrupting chemicals may disrupt the human endocrine system, which controls metabolic processes. Wastewater effluents are also high in nutrients and may cause algal blooms that release toxins into source water. Addressing current and future problems requires new robust methods and technologies of purifying water at lower cost, energy, and environmental impact than current methods.

Nanomaterials are materials that are modified or synthesized in the nanoscale (1–100 nm) using a bottom-up approach and can exhibit properties that are not observed in the macro or micro scale. They are promising solutions for water supplies contaminated by organic and inorganic compounds, heavy metal ions, and microorganisms [12]. Specifically, nanomaterials can improve on some existing treatment technologies including adsorption of contaminants, membrane filtration, and AOPs.

The focus of this chapter is to introduce the reader to the fundamental theories and concepts behind pollutant removal processes that use nanomaterials for water treatment, which are required to understand and evaluate nanomaterials against current benchmarks. The topics included in this chapter are the fundamentals of adsorption and the related adsorption equilibrium and isotherms; membrane filtration and reverse osmosis processes; and an overview of AOPs with a focus on photocatalytic degradation processes.

## 1.2 Adsorption

### *1.2.1 Introduction to Adsorption*

Adsorption is the accumulation of substances at a surface interface through a mass transfer process. Lowitz, in 1785, first observed the phenomenon in solution and it was subsequently used in sugar refining processes for color removal [13]. The term itself was coined by Kayser (1881), but has been reported as early as 4000 BC in Sanskrit texts as a method to enhance the taste and odor of drinking water [14]. The mass transfer operation occurs when a substance, the adsorbate, present in the liquid phase diffuses and is accumulated and transferred onto a solid phase, the adsorbent. The reader is referred to several texts on adsorption, particularly on activated carbon, that underlie the concepts in this section [15–17].

### *1.2.2 Types of Adsorption*

#### 1.2.2.1 Chemical Adsorption

Chemical adsorption, also known as chemisorption, is a process that transpires when the adsorbate reacts with the adsorbent interface to form chemical or ionic bonds. The attractive forces of the adsorbent and adsorbate at the interface are akin to covalent bonds formed between atoms and obey Coulomb's law.

#### 1.2.2.2 Physical Adsorption

Adsorbates undergo physical adsorption if the attractive forces are through physical forces that exclude covalent bonding with the interface and coulombic attraction. Physical adsorption and chemisorption are sometimes difficult to differentiate. Physical adsorption is reversible due to weaker forces and energies of bonding compared to chemisorption. Physical forces operate over longer distances than chemisorption and have no preferential surface adsorption sites.

In water treatment, the adsorption of organic molecules from water, a polar solvent, to a nonpolar adsorbent, such as activated carbon, is dictated by van der Waals forces that occur between organic substances and graphitic carbon planes in activated carbon. In general, increasingly larger and nonpolar compounds adsorb more intensely to nonpolar adsorbents due to hydrophobic bonding [18]. The attraction, due to van der Waals forces, at an adsorbate-activated carbon interface increases with increasing polarizability and size.

### 1.2.3 *Adsorption Materials*

The most common commercial adsorbent used in water remediation is activated carbon, but there are others worth mentioning, such as zeolites and synthetic polymers. In order to classify the pore sizes of these adsorbents, the IUPAC defines pore classifications into three classes: macropores (>25 nm), mesopores (1–25 nm), and micropores (<1 nm). Zeolites (aluminosilicates) have very small pore sizes, which are capable of excluding some organic compounds. Synthetic polymer adsorbents usually have micropore distributions, which prevent them from adsorbing natural organic matter (NOM). Activated carbon has a range of pore sizes, depending on the manufacturing process, and can adsorb large organic molecules such as NOM and synthetic organic compounds (solvents, fuels, pesticides, and others). In addition to the commercial adsorbents, research has also been conducted on nanomaterials such as carbon nanotubes [19] and graphene [20] due to their high sorption capacity. Non-conventional low-cost adsorbents are continuously being sought after in order to improve performance to cost ratios [21].

#### 1.2.3.1 Activated Carbon

Activated carbon is typically made from organic materials (coconut husks, wood, peat, coal, and synthetic sources) that are heated to high temperatures (>700 °C) and pyrolized in an oxygen-deprived environment. In water treatment, it is used in either of two forms: granular-activated carbon (GAC) and powder-activated carbon (PAC). PAC (mean particle size: 20–50 μm) is usually added directly to water sources at various stages of the water treatment process and is removed through sedimentation or filtration. GAC (mean particle size: 0.5–3 mm) is typically employed as a fixed bed after filtration and before disinfection. GAC can also be impregnated with iron to remove arsenic, ammonia to increase adsorption capacity for anionic species, and silver for disinfection properties [22].

#### 1.2.3.2 Non-conventional Low-Cost Adsorbents

There is a need for non-conventional low-cost adsorbents because the thermal input required to manufacture activated carbon makes it relatively expensive. These low-cost adsorbents should be effective, but also abundant in nature, inexpensive, and require minimal processing. A list of non-conventional adsorbents is given in Table 1.1. These adsorbents are usually made from agriculture and industry by-products, natural materials, and biosorbents that contain high surface area and porosity [23].

**Table 1.1** List of nonconventional adsorbents

| Adsorbent | Example | References |
|---|---|---|
| Agriculture and industry waste | Teak wood bark | [24] |
| | Papaya seeds | [25] |
| | Sugar industry mud | [26] |
| | Grass waste | [27] |
| | Peels (pomelo, jackfruit, banana, garlic) | [28, 29] |
| | Rubber seed shell | [30] |
| | Fly ash | [31, 32] |
| | Coconut tree flower | [33] |
| Natural materials | Clay (montmorillonite, bentonite, fibrous clay, palygorskite, kaolin) | [34–38] |
| | Glass wool | [39] |
| Bioadsorbents | Biomass (algae, activated sludge) | [40–46] |
| | Fungi | [47, 48] |
| | Microbial | [49] |

Adapted from Crittenden et al. (2012), Ref. [22]

## *1.2.4 Important Factors Involved in Adsorption*

### 1.2.4.1 Surface Area and Pore size

Surface area and pore size are important factors that determine the number and availability of adsorption sites for adsorbates. Typically, there is an inverse relationship between surface area and pore size; in other words, the smaller the pore size of a given volume of absorbent, the greater the surface area that is available for the adsorption process. Additionally, depending on the size of the adsorbate and adsorbent pores, adsorption may be limited by steric effects. Adsorbents with large porosities are brittle and tend to break apart.

### 1.2.4.2 Background Matrix Effects

One factor that affects adsorption efficiency of a target compound is the influence of the water matrix. Increases in the concentrations of naturally occurring or anthropogenic compounds in the background matrix will decrease the adsorption capacity of the target compound due to competitive adsorption onto the adsorption sites.

## *1.2.5 Adsorption Kinetics*

The removal of a compound by adsorbents is a three-step process: external mass transfer (EMT), internal mass transfer (IMT), and adsorption (Fig. 1.1). Both EMT and IMT are considered to be diffusion-based processes and the driving force is the

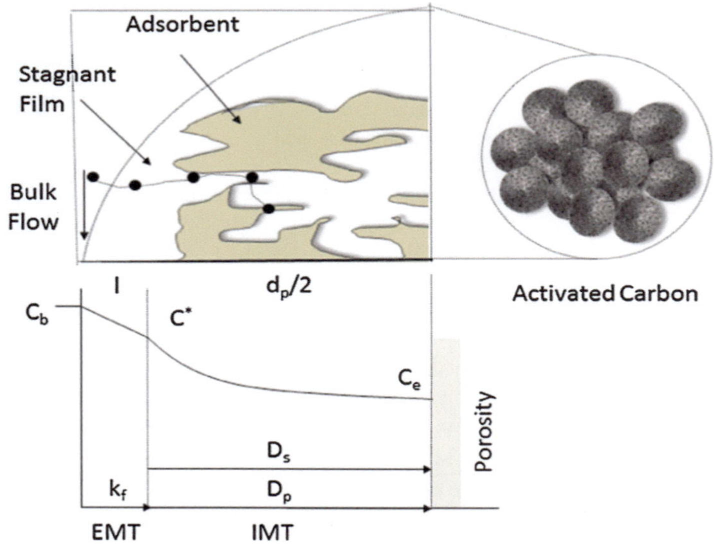


**Fig. 1.1** Illustration of mass transfer and adsorption processes, Ref. [17]

concentration gradient of the adsorbate; the difference between the bulk concentration ($C_b$) and the internal concentration at equilibrium ($C_e$). In the first step, EMT, or film diffusion, of the compound from the bulk liquid phase occurs through the hydrodynamic layer that surrounds the adsorbent. After EMT occurs and delivers the adsorbate to the external surface of the adsorbent, IMT delivers the compound to an adsorption site. IMT occurs either through the pore fluid or pore wall through pore diffusion ($D_p$) and surface diffusion ($D_s$). These diffusional processes are controlled by the pore structure or adsorbate properties. The IMT is often the slowest and thus controls the overall uptake of an adsorbate. The actual adsorption of the adsorbate, the final step, occurs quickly and is not considered rate-limiting.

### *1.2.6 Adsorption Equilibrium*

Adsorption isotherms are used to quantify the quantity of adsorbate that an adsorbent can adsorb at equilibrium conditions and constant temperature. From an experiment that varies the adsorbent dose and/or the initial adsorbate concentration, the equilibrium solid-phase adsorbate concentration can be calculated, and a relationship with the equilibrium liquid-phase concentration can be established.

This adsorption equilibrium relationship is termed the isotherm and allows the adsorption capacity of the adsorbent to be calculated at any given liquid-phase adsorbate concentration. The equilibrium capacity is calculated using a mass balance expression [22]:

$$q_e = \frac{V}{M}(C_o - C_e) \tag{1.1}$$

where $q_e$ is equilibrium concentration of adsorbate (mg g$^{-1}$), $C_o$ is initial concentration of adsorbate (mg L$^{-1}$), $C_e$ is equilibrium concentration of adsorbate (mg L$^{-1}$), $V$ is volume of aqueous solution added to bottle (L), and $M$ is mass of adsorbent (g).

Some isotherms commonly used in research are the Langmuir, Brauener-Emmett-Teller (BET), Freundlich, and Redlich-Peterson isotherms.

#### 1.2.6.1 Langmuir Isotherm

The Langmuir adsorption isotherm describes the reversible chemical equilibrium between species at the surface-solution interface [50]. The chemical reaction defines adsorbate species ($A$), which diffuses to vacant surface sites ($S_\mathrm{v}$), to form adsorbate species bound to surface sites (SA):

$$S_\mathrm{v} + A \leftrightarrow SA \tag{1.2}$$

where $S_\mathrm{v}$ is vacant surface sites (mmol m$^{-2}$), A is adsorbate species in solution, (mmol), and SA is adsorbate species bound to surface sites (mmol m$^{-2}$).

In Eq. (1.1), the reaction is assumed to have a constant value for free energy change ($\Delta G^o_\mathrm{ads}$) for all surface sites. Each site is capable of binding only one molecule or, in other words, the model only allows for accumulation of adsorbates up to a monolayer on the adsorbent. The equilibrium equation is written as:

$$K_\mathrm{ad} = \frac{SA}{S_\mathrm{v} C_A} = \mathrm{e}^{-\Delta G^o_\mathrm{ads}/RT} \tag{1.3}$$

where $K_\mathrm{ad}$ is adsorption adsorbent-phase equilibrium constant (L mg$^{-1}$), $C_A$ is equilibrium adsorbent-phase concentration of adsorbate A in solution (mg L$^{-1}$), $\Delta G^o_\mathrm{ads}$ is free-energy change for adsorption (J mol$^{-1}$), $R$ is universal gas constant, 8.314 J K$^{-1}$ mol$^{-1}$, and $T$ is absolute temperature (K).

Equation (1.3) can be expressed much more conveniently by fixing the total number sites and eliminating $S_\mathrm{V}$ as an unknown:

$$S_T = S_\mathrm{V} + SA = \frac{SA}{K_\mathrm{ad} C_A} + SA \tag{1.4}$$

where $S_T$ is total number of sites available or monolayer coverage.

Rearranging Eq. (1.4) for *SA* gives:

$$SA = \frac{S_T}{1 + \frac{1}{K_{\mathrm{ad}}C_A}} = \frac{K_{\mathrm{ad}}C_A S_T}{1 + K_{\mathrm{ad}}C_A} \tag{1.5}$$

Equation (1.5) can be expressed in terms of mass loading, $q_A$:

$$q_A = (SA)(A_{\mathrm{ad}})(\mathrm{MW}) = \frac{K_{\mathrm{ad}}C_A S_T A_{\mathrm{ad}}\mathrm{MW}}{1 + K_{\mathrm{ad}}C_A} = \frac{Q_M K_{\mathrm{ad}}C_A}{1 + K_{\mathrm{ad}}C_A} = \frac{Q_M b_A C_A}{1 + b_A C_A} \tag{1.6}$$

where $q_A$ is equilibrium adsorbent-phase concentration of adsorbate A, (mg adsorbate $g^{-1}$) adsorbent, $A_{\mathrm{ad}}$ is surface area per gram of adsorbent ($m^2$ $g^{-1}$), MW is molecular weight of adsorbate A (g $mol^{-1}$), $C_A$ is equilibrium adsorbent-phase concentration of adsorbate A in solution (mg $L^{-1}$), $Q_M$ is maximum adsorbent-phase concentration of adsorbate A when surface sites are saturated with adsorbate, (mg adsorbate $g^{-1}$ adsorbent), $b_A$ is Langmuir adsorption constant of adsorbate A, $K_{ad}$, (L $mg^{-1}$).

It is also convenient to rearrange Eq. (1.6) to a linear form:

$$\frac{C_A}{q_A} = \frac{1}{b_A Q_M} + \frac{C_A}{Q_M} \tag{1.7}$$

A plot of $C_A/q_A$ versus $C_A$ results in a straight line with slope of $1/Q_M$ and intercept $1/b_A Q_M$.

#### 1.2.6.2 Brunauer–Emmett–Teller Adsorption Isotherm

The BET adsorption isotherm [51] is an extension of the Langmuir model. Specifically, in order to accommodate multilayer adsorption at greater pressures, the BET isotherm expands on the assumption that a monolayer of molecules is formed during adsorption. Each additional layer of adsorbate molecules is in equilibrium with subsequent layers below it, which can vary in thickness. A consequence of the BET model is that it will always generate a smaller value for the predicted surface area compared to the Langmuir isotherm due to potential adsorbate layering of the material. The equation is:

$$\frac{q_A}{Q_M} = \frac{B_A C_A}{(C_{S,A} - C_A)\left[1 + (B_A - 1)\left(\frac{C_A}{C_{S,A}}\right)\right]} \tag{1.8}$$

$$B_A = \frac{K_{1,\mathrm{ad}}}{K_{i,\mathrm{ad}}} = \frac{\mathrm{e}^{-\Delta G^o_{\mathrm{ads}}}}{\mathrm{e}^{-\Delta G^o_{\mathrm{prec}}}} \tag{1.9}$$

where $q_A$ is equilibrium adsorbent-phase concentration of adsorbate A (mg $g^{-1}$), $Q_M$ is maximum adsorbent-phase concentration of adsorbate when surface sites are saturated with adsorbate (mg $g^{-1}$), $K_{1,\mathrm{ad}}$ is equilibrium constant for first layer

(L mg$^{-1}$), $K_{i,\text{ad}}$ is equilibrium constant for lower layers (L mg$^{-1}$), $B_A$ is ratio of $K_{1,\text{ad}}$ and $K_{i,\text{ad}}$, $C_A$ is equilibrium concentration of adsorbate A in solution (mg L$^{-1}$), $C_{S,A}$ is saturated solution concentration of A (mg L$^{-1}$), $\Delta G^o_{\text{ads}}$ is free energy of adsorption (J mol$^{-1}$), and $\Delta G^o_{\text{prec}}$ is free energy of precipitation (J mol$^{-1}$).

In the Langmuir model, the site energy for adsorption is equal for all surface sites and the largest capacity is one monolayer. These assumptions are not valid for many adsorbents, such as activated carbon. Although the BET isotherm allows for many layers to be modeled, it is assumed in Eq. (1.8) that site energy is equal to the first layer and to the free energy of precipitation for additional lower layers. In actuality, the site energy of adsorption is variable for most adsorbents because they are quite heterogeneous in nature. Often, in the cases of heterogeneous adsorbents with varying site energies, the Freundlich equation is used to describe isotherm data.

#### 1.2.6.3 Freundlich Isotherm

The Freundlich adsorption isotherm [52] equation can be used to model data for heterogeneous adsorbents and is a better fit for GAC isotherm data than the Langmuir equation [53]. The equation is:

$$q_a = K_A C_A^{\frac{1}{n}} \tag{1.10}$$

where $K_A$ is Freundlich adsorption capacity parameter, (mg/g) (L/mg)$^{1/n}$ and 1/n is Freundlich adsorption intensity parameter, unitless.

The Freundlich equation (Eq. (1.10)) is derived from the Langmuir equation (Eq. (1.5)) to model the adsorption of adsorbates onto surface sites of a given free energy with two additional assumptions: (i) the surface site energies for adsorption tend to a Boltzmann distribution with mean site energy $\Delta H^o_M$ and (ii) the change in surface site entropy is a linear function of site enthalpy $-\Delta H^o_{\text{ad}}$. Given these assumptions, the Freundlich adsorption intensity parameter, $n^{-1}$, is:

$$n = \frac{\Delta H^o_M}{RT} - \frac{r\Delta H^o_{\text{ad}}}{R} \tag{1.11}$$

where $\Delta H^o_M$ is mean site energy (J mol$^{-1}$), R is universal gas constant, 8.314 J K$^{-1}$ mol$^{-1}$, $\Delta H^o_{\text{ad}}$ is change in site enthalpy (J mol$^{-1}$), T is absolute temperature (K), and r is proportionality constant.

#### 1.2.6.4 Redlich-Peterson Isotherm

The Redlich-Peterson isotherm [54] combines features from the Langmuir and Freundlich equation. It is designated as a three parameter equation used to represent adsorption equilibria over a wider concentration range:

$$q_A = \frac{K_{R-P} C_A}{1 + \alpha_{R-P} C_A^{\beta}} \tag{1.12}$$

where $K_{R-P}$ is Redlich-Peterson parameter (L mg$^{-1}$), $\alpha_{R-P}$ is Redlich-Peterson parameter (L mg$^{-1}$), and $\beta$ is heterogeneity factor (values between 0 and 1).

Equation (1.12) approaches the Langmuir equation when $\beta$ approaches 1, and become's Henry's law equation when $\beta = 0$.

## 1.3 Membrane Filtration and Reverse Osmosis

### 1.3.1 Introduction

A membrane is a physical interface which separates two phases, forming a barrier to the transport of matter. Membranes have microscopic openings that allow water molecules to pass, but not compounds that are larger than the opening. Membrane filtration can be operated either as dead-end filtration or cross-flow filtration as shown in Fig. 1.2. In dead end filtration, the feed water flows perpendicular to the membrane surface; all solids will amass onto the membrane surface during filtration and are removed via backwashing. The accumulation of solids in dead-end filtration often results in a lower flux compared to cross-flow filtration.

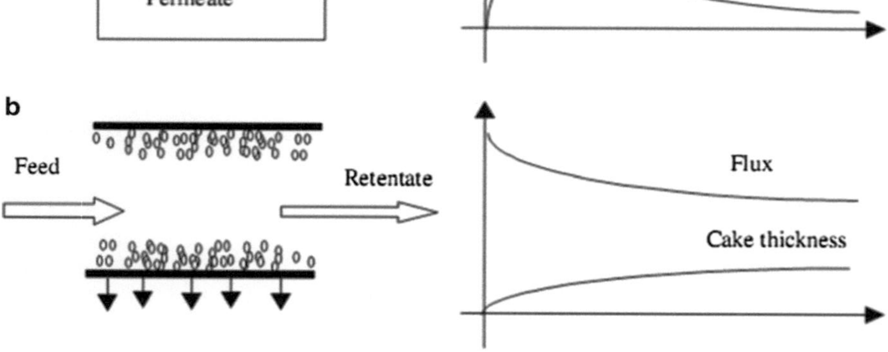


**Fig. 1.2** Schematic of (**a**) dead-end and (**b**) cross-flow filtration (*Source*: Li (2007), Ref. [56])

In cross-flow filtration, the feed water is parallel to the membrane surface. The flow velocity parallel to the surface of the membrane generates a shear force that reduces the growth of a filter cake [55]. Since the majority of solids pass with the retentate instead of collecting on the membrane surface, the system can function at higher flux.

## *1.3.2 Membrane Materials, Properties, and Geometries*

### 1.3.2.1 Membrane Materials

Most membranes are made from synthetic organic polymers, which can be either hydrophobic or hydrophilic. Hydrophilic polymer membranes include cellulose and its derivatives, polyacrylonitrile, hydrophilized polysulfone, hydrophilized polyethylene, and others. Hydrophobic membranes include polytetrafluroethylene (PTFE), polyethylene (PE), polyvinylidene fluoride (PVDF), and many more.

Additionally, membranes can be prepared from inorganic materials, including metals and ceramics. Inorganic membranes are used because of their high stability at high temperatures (over 100 °C) and at extreme pH, but are brittle due to their crystal structure. Ceramic materials, for the most part, are composite materials consisting of one or many different ceramic materials. They have a macroporous support, followed by a few layers of microporous top layers. The most commonly used materials for ceramic membranes are alumina ($Al_2O_3$), titanium dioxide ($TiO_2$), zirconium dioxide ($ZrO_2$), silicon dioxide ($SiO_2$), or a mixture of these materials.

### 1.3.2.2 Membrane Properties

Membranes have pore sizes ranging from 0.5 nm to 5 μm, which may be determined by bubble point analysis, microscopic analysis, or porosimetry. Membrane manufacturers also use the molecular weight cutoff (MWCO) to characterize the apparent size of particles that are retained. It is a parameter defined as the lower limit of a molecular weight of a solute for which the rejection is 95–98 %. The MWCO is an indicator of the membrane's actual ability to remove a given compound because factors other than membrane pore size (polarity, molecular shape, and surface interaction with the membrane) affect rejection characteristics [57].

Testing properties other than pore size are often necessary. Surface properties such as surface roughness, charge, hydrophobicity, and chemistry are necessary in determining the effectiveness of the membrane. From a commercial standpoint, a performance over cost ratio in terms of durability, stability, and effectiveness of membranes is necessary. Table 1.2 lists some of the properties that can be analyzed and the impact of each property on the performance of the membrane.

**Table 1.2** Important properties of membranes and their method of determination and impact on performance

| Property | Method of determination | Impact on membrane performance |
|---|---|---|
| Surface roughness | Atomic force microscopy | Rough materials will tend to foul more than smooth materials |
| Surface charge | Streaming potential | Repulsive forces between a charged membrane and ionic species can reduce fouling by minimizing contact between the membrane and foulant |
| Hydrophobicity | Contact angle | Hydrophobic materials are, generally, more prone to fouling than hydrophilic materials |
| Surface chemistry | ATR-FTIR, SIMS, XPS | The surface composition affects cleaning and fouling by influencing interactions between the membrane surfaces and constituents in the feed solution |
| Porosity | Thickness/weight measurements | Affects the head loss through a given membrane; higher porosity of the membrane will result in lower head loss |
| Mechanical durability | Mechanical tests | Affects the ability of the material to withstand stressors in the form of surges due to imperfect operation of mechanical pumps and valves |
| Cost | Material cost | Affects the cost of the membrane system |
| Chlorine/oxidant tolerance | | Affects the ability to disinfect the membrane equipment after use. Increasing the tolerance of the membrane to these chemicals increases the longevity of the membrane |
| Biological stability | Exposure to organisms | Affects the life span of the membrane; low biological stability can result in microorganism colonization and the degradation of the membrane by these microorganisms |
| Chemical and thermal stability | Exposure to chemical, pH, and temperature extremes | Affects the life span of the membrane; greater chemical and temperature tolerance allows for harsher, but more effective cleaning procedures with less degradation of the material |

Adapted from Crittenden et al. (2012), Ref. [22]

#### 1.3.2.3 Membrane Configurations

Membranes are fabricated into four configurations: (i) flat sheet and plate and frame, (ii) hollow fiber, (iii) tubular, and (iv) spiral wound. The flat sheet and plate and frame configuration is commonly used for laboratory separations. Hollow fiber and spiral wound membrane configurations are often used for NF and RO membranes. The tubular configuration is preferred for ceramic membranes that have low packing densities. These configurations are illustrated and described in Table 1.3.

**Table 1.3** Description and schematic of various membrane configurations

| Configuration | Schematic | Description |
|---|---|---|
| Flat sheet and plate and frame | 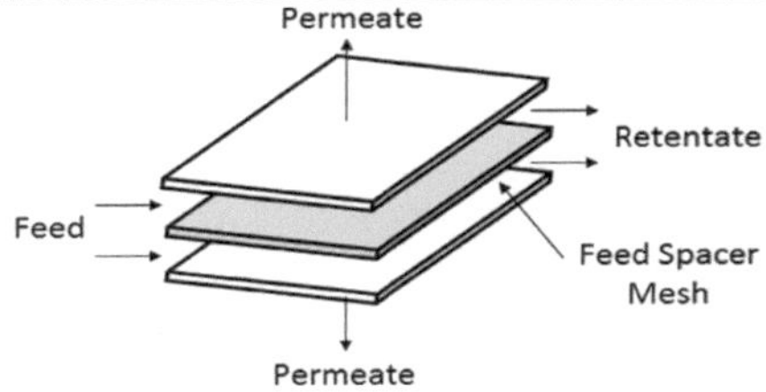 | Membranes are flat sheets and used as a layer or several layers interspersed with a frame in a filtration cell. They are common in laboratory separations, but difficult to implement at a large scale |
| Hollow fiber and hollow-fine fiber | 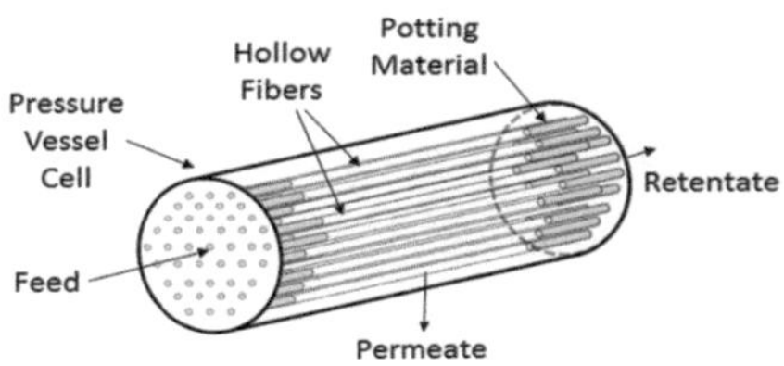 | Membranes are made as hollow tubes. They are the most common configuration in membrane filtration for water treatment |
| Tubular | 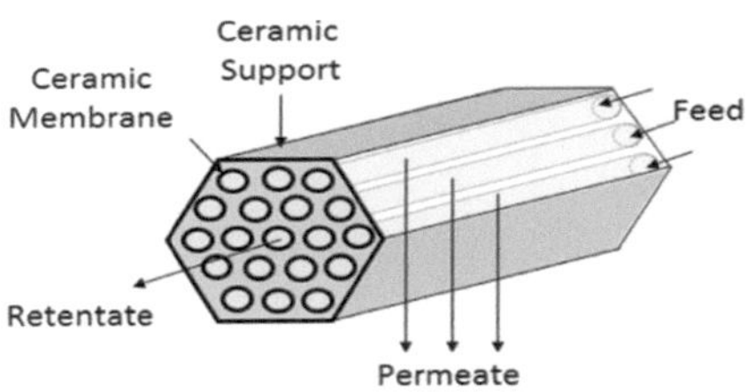 | Membranes are made into monoliths with channels through the structure. Tubular membranes are often constructed using ceramic membranes, which have low packing densities but can operate at high cross-flow velocity, where solute concentration is very high |
| Spiral wound | 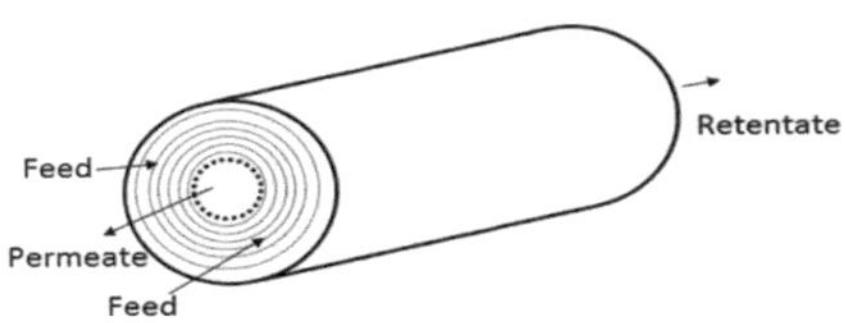 | These are many flat sheet membranes that are stacked and bundled around the center of a tube so that the permeate will travel via a spiral flow path toward the collection tube at the center. These are used often in NF and RO membranes, but not used in membrane filtration due to clogging of large particulates and backwashing problems |

## *1.3.3 Membrane Processes*

There are four distinguishable types of pressure-driven membrane processes: microfiltration (MF), ultrafiltration (UF), nanofiltration (NF), and reverse osmosis (RO). These membrane processes can also be separated into two classes: (i) membrane filtration (microfiltration and ultrafiltration) and (ii) reverse osmosis (and nanofiltration). Microfiltration and ultrafiltration are governed by Darcy's law, whereas nanofiltration and reverse osmosis are governed by Fick's law. These pressured-driven processes are illustrated in Table 1.4.

Table 1.4 Pressure-driven membrane processes

| Process | Separation concept | Pressure required (atm) | Materials passed | Governing equations |
|---|---|---|---|---|
| Microfiltration | 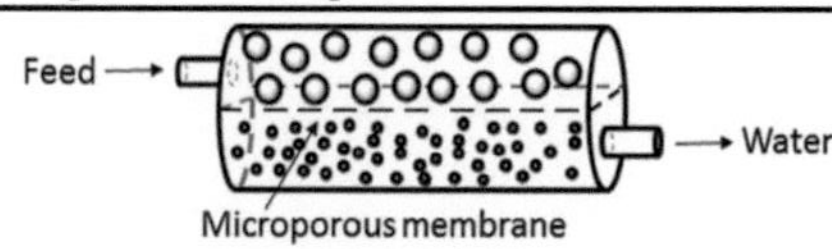  | 1–5 | Water and dissolved species | Darcy's law |
| Ultrafiltration | 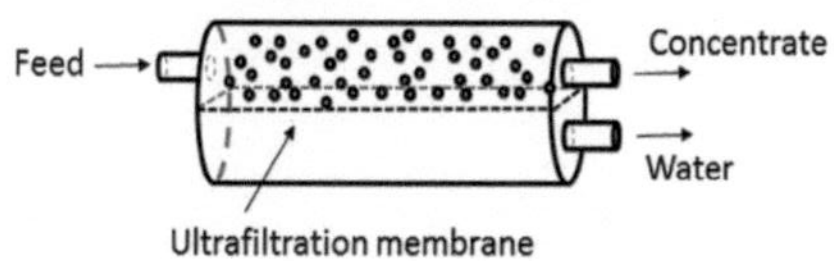  | 2–10 | Water and salts | Darcy's law |
| Nanofiltration | 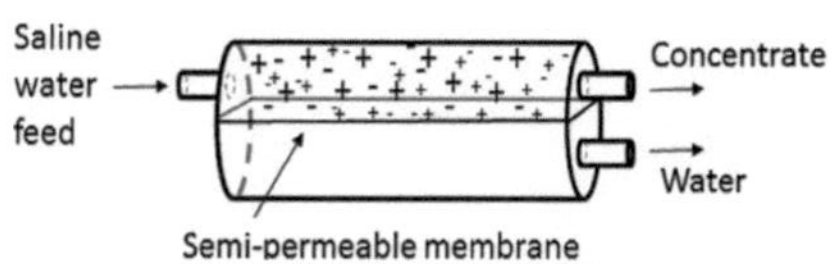  | 5–50 | Water and monovalent ions | Fick's law |
| Reverse Osmosis | | 10–100 | Water | Fick's law |

**Table 1.5** Description of sieving, adsorption, and cake formation processes

| Mechanism for rejection | Diagram | Description |
|---|---|---|
| Sieving | 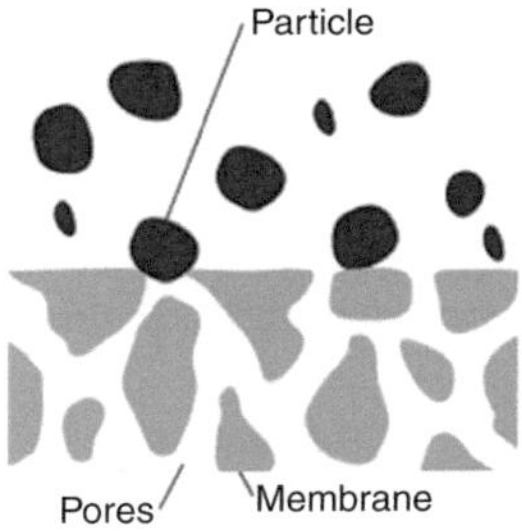 | Particles larger than the pores are retained |
| Adsorption | 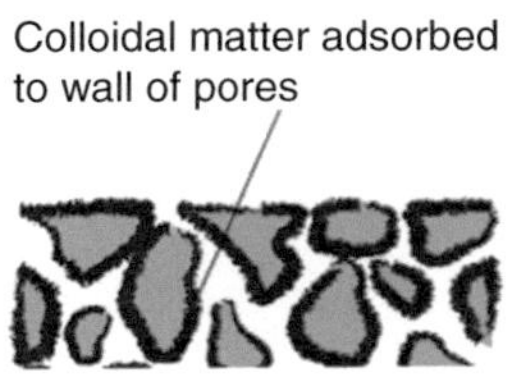 | Adsorption occurs when particulates are small enough to adsorb to the walls of pores |
| Cake formation | 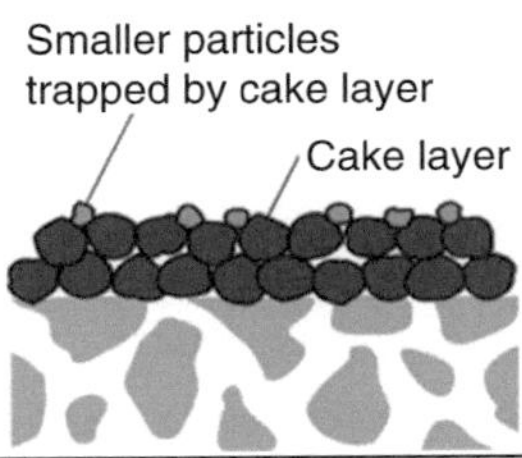 | Small particles are unable to pass through the membrane and are retained due to the accumulation of larger materials that amass at the membrane surface |

Adapted from Crittenden et al. (2012), Ref. [22]

MF membranes have large pore sizes and reject large particulates and microorganisms. UF membranes have smaller pores than MF membranes and can reject bacteria and soluble macromolecules. RO membranes are, for the most part, nonporous and reject particulates and numerous low molecular weight species such as salt ions and small organics molecules. NF membranes are thought of as a "loose" RO membrane exhibiting performance parameters between UF and RO membranes.

#### 1.3.3.1 Microfiltration and Ultrafiltration

Filtration in microfiltration and ultrafiltration is governed by Darcy's law and proceeds in three steps: (i) sieving, (ii) adsorption, (iii) and cake formation. Illustrated in Table 1.5, these mechanisms are critical in understanding and evaluating performance. Sieving, or straining, is the primary filtration mechanism in which particles larger than the MWCO collect at the surface of the membrane, while smaller particles in the water pass through. However, as mentioned before, particulate removal is not an ideal step function at the MWCO because of the properties of the membrane and particle-membrane interface interactions.

NOM is capable of adsorbing to membrane surfaces. Soluble substances may be excluded despite their physical dimensions being smaller than the MWCO. Adsorption plays a role in the primary stage of filtration using a clean membrane; however, the adsorption capacity is eventually exhausted and is not an effective mechanism for long-term filter use. Adsorption can have a significant impact on the operation of the membrane. Adsorbed material can decrease the MWCO by decreasing the size of pore cavities throughout the membrane, thus increasing the capability to retain smaller material via sieving. Furthermore, due to decreasing the MWCO, adsorption of NOM causes membrane fouling.

During filtration, a clean membrane will accumulate particles at the surface due to the sieving process. The accumulation of solids forms a particulate cake, which acts as a filtration medium that is dynamic in that its capability varies as a function of time from cake growth and backwashing processes.

The flow of feed water through MF and UF membranes follows Darcy's law for superficial fluid velocity:

$$v = \frac{k_P h_L}{L} \tag{1.13}$$

where $k_P$ is hydraulic permeability coefficient (m s$^{-1}$), $h_L$ is head loss across porous medium (m), and $L$ is thickness of porous medium (m).

The hydraulic permeability coefficient, $k_P$, in Eq. (1.13) is empirically obtained and is dependent on porosity and surface area of the membrane. The membrane flux equation incorporates the membrane thickness and the resistance coefficient; it is stated as:

$$J = \frac{Q}{a} = \frac{\Delta P}{\mu \kappa_m} \tag{1.14}$$

where $J$ is water flux through membrane (L m$^{-2}$ s$^{-1}$), $Q$ is flow rate (L s$^{-1}$), $a$ is membrane area (m$^2$), $\Delta P$ is differential pressure across membrane (bar), $\mu$ is dynamic viscosity of water (kg m$^{-1}$ s$^{-1}$), and $\kappa_m$ is membrane resistance coefficient (m$^{-1}$).

#### 1.3.3.2 Nanofiltration and Reverse Osmosis

Osmosis is a process of flow through a semipermeable membrane. Under isothermal ($\Delta T = 0$) and isobaric ($\Delta P = 0$) conditions between the two sides of the membrane, no water will flow through the membrane. However, when solutes are dissolved into the water on one side, a flow through the membrane from the pure water to the water containing salts will occur. In other words, the concentration of solutes will tend toward being equal on both sides.

Reverse osmosis is the process where pressure is applied on the side of the membrane where salts are added. The extra pressure will result in a flow of water through the membrane, but the salts do not flow through because of the membrane pore size.

In other words, reverse osmosis exploits two opposing forces: (i) the concentration gradient and (ii) the pressure gradient. Reverse osmosis is always operated under cross-flow conditions. The rejection of solutes of a RO and NF membranes is defined by salt rejection, given by:

$$\mathrm{Rej} = 1 - \frac{C_P}{C_F} \tag{1.15}$$

Rej is rejection (dimensionless), $C_P$ is concentration of permeate (mol $L^{-1}$), and $C_F$ is concentration in feed water (mol $L^{-1}$).

Equation (1.15) can be calculated from parameters such as total dissolved solids (TDS) or conductivity. Sodium chloride (NaCl) rejection is used as the standard for high pressure RO membranes, whereas magnesium sulfate ($MgSO_4$) is used as the standard for NF and low pressure RO membranes.

#### 1.3.3.2.1 Kinetic Equations

*Flux*

The flux, $J$, is the permeate flow through one square meter of membrane surface:

$$J = \frac{\Delta P_{\mathrm{trans}}}{\mu \kappa_m} \tag{1.16}$$

where $\Delta P_{\mathrm{trans}}$ is transmembrane pressure (Pa), $\mu$ is dynamic viscosity of water (N s $m^{-2}$), $\kappa_m$ is membrane resistance coefficient ($m^{-1}$), and $C_F$ is concentration in feed water (mol $L^{-1}$).

From Eq. (1.16), the net transmembrane pressure, $\Delta P_{\mathrm{trans_{net}}}$, is a driving force for RO and NF membranes and is given by:

$$\Delta P_{\mathrm{trans_{net}}} = \Delta P - \Delta \Pi = P_f - \frac{\Delta P_{\mathrm{hyd}}}{2} - P_p - \Delta \Pi \tag{1.17}$$

$$\Delta P_{\mathrm{hyd}} = P_f - P_c$$

where $P_f$ is pressure of feed (Pa), $\Delta P_{\mathrm{hyd}}$ is hydraulic pressure loss (Pa), $P_p$ is pressure of permeate (Pa), $\Delta \Pi$ is osmostic pressure difference (Pa), and $P_c$ is pressure of concentrate (Pa).

*Osmotic Pressure*

The reverse osmosis (including nanofiltration) driving force for diffusion is via a concentration gradient, or in thermodynamic terms, a gradient in Gibbs energy [58]. To describe osmosis, the general form of the Gibbs function is used:

$$\delta G = V\,\delta P - S\,\delta T + \Sigma_i \mu_i^o \delta n_i \tag{1.18}$$

$G$ is Gibbs energy (J), $V$ is volume (m$^3$), $P$ is pressure (Pa), $S$ is entropy (J K$^{-1}$), $T$ is absolute temperature (K), $\mu_i^o$ is chemical potential of solute I (J mol$^{-1}$), and $n_i$ is amount of solute I in solution (mol).

The chemical potential is defined as:

$$\mu_i^o = \left.\frac{\delta G}{\delta n_i}\right|_{P,T} \tag{1.19}$$

The last term in Eq. (1.18) describes the difference in Gibbs energy between the amount of solute in the feed and permeate sides. When volume is constant, the difference in the amount of solute equals the difference in concentrations. Under constant temperature ($\delta T = 0$), equilibrium occurs when the summation of the Gibbs energy gradient due to the chemical potential is counterbalanced by the pressure gradient between the feed and permeate sides:

$$V\delta P = -\Sigma_i \mu_i^o \delta n_i \quad \text{when } \delta G = 0 \tag{1.20}$$

To balance the difference in chemical potential of a solute, the pressure required is known as the osmotic pressure and denoted as Π. When the vessel reaches equilibrium conditions, the difference in hydrostatic pressure is equal to the difference in osmotic pressure between the two sides. The osmotic pressure can be derived using a thermodynamic treatment with the assumptions that the solution is incompressible and behaves ideally:

$$\Pi = -\frac{RT}{V_b}\ln(x_w) \tag{1.21}$$

where Π is osmotic pressure (bar), $V_b$ is molar volume of pure water (L mol$^{-1}$), $x_w$ is mole fraction of water (mol mol$^{-1}$), T is temperature (K), and R is universal gas constant, 0.083145 (L bar K$^{-1}$ mol$^{-1}$).

In dilute solution (i.e., $x_w = 1$), Eq. (1.21) can be approximated by the van't Hoff equation for osmotic pressure, which has the same form as the ideal gas law ($PV = nRT$):

$$\Pi = \frac{n_S}{V} RT = \mathrm{CRT} \tag{1.22}$$

where Π is osmotic pressure (bar), $n_s$ is total amount of all solutes in solution (mol), $C$ is concentration of all solutes (mol L$^{-1}$), and V is volume of solution (L).

To account for the assumption of diluteness, the nonideal behavior of concentrated solutions, and the compressibility of liquid at high pressure, a nonideality coefficient (osmotic coefficient $\phi$) must be incorporated into Eq. (1.22):

$$\Pi = \phi\,\mathrm{CRT} \tag{1.23}$$

where $\phi$ is osmotic coefficient (unitless).

*Osmotic Pressure Difference*

The osmotic pressure difference over a membrane, $\Delta\Pi$, is given by:

$$\Delta\Pi = \frac{\Pi_f + \Pi_c}{2} - \Pi_p \tag{1.24}$$

where $\Pi_f$ is osmotic pressure of feed (Pa), $\Pi_c$ is osmotic pressure of concentrate (Pa), and $\Pi_p$ is concentration of permeate (Pa).

Because the concentration of salts in the permeate feed is very low, the osmotic pressure in the permeate feed is almost always neglected. On the other hand, the osmotic pressure of the concentrate, $\Pi_c$, is higher than the osmotic pressure of the feed, $\Pi_f$. The osmotic pressure of the concentrate can be put in terms of the osmotic pressure of the feed:

$$\Pi_c = \Pi_f \left(\frac{1}{1-\gamma}\right) \tag{1.25}$$

$$\gamma = \frac{Q_p}{Q_f} \tag{1.26}$$

$\gamma$ is recovery, $Q_p$ is permeate flow ($m^3\ h^{-1}$), and $Q_f$ is feed flow ($m^3\ h^{-1}$).

Combining Eq. (1.24) with Eq. (1.25), the osmotic pressure difference over a membrane can be written as:

$$\Delta\Pi = \Pi_f \frac{2-\gamma}{2(1-\gamma)} \tag{1.27}$$

#### 1.3.3.2.2 Concentration Polarization

Concentration polarization is the accumulation of solutes close to the proximity of the surface of the membrane causing a decrease in performance. The exclusion mechanisms for solutes differ from particles because solutes stay in solution and form a concentration gradient with the highest concentration near the boundary layer of the adsorbent and liquid phase, causing polarization. The resulting effects due to concentration polarization are that: (i) water flux is lower due to higher osmotic pressure, (ii) rejection is lower due to the increase in concentration at the membrane and decrease in flux, and (iii) solubility limits of solute at or near the membrane may be exceeded and cause fouling, specifically precipitation or scaling (Fig. 1.3a).

Concentration polarization is reversible and will disappear as the driving force tends to zero. Concentration polarization can be limited with disturbance to the boundary layer via increasing the velocity along the membrane surface. The relationship between concentration close to the membrane surface and in the feed (Fig. 1.3b) is represented by the concentration polarization factor ($\beta$), which is given by:

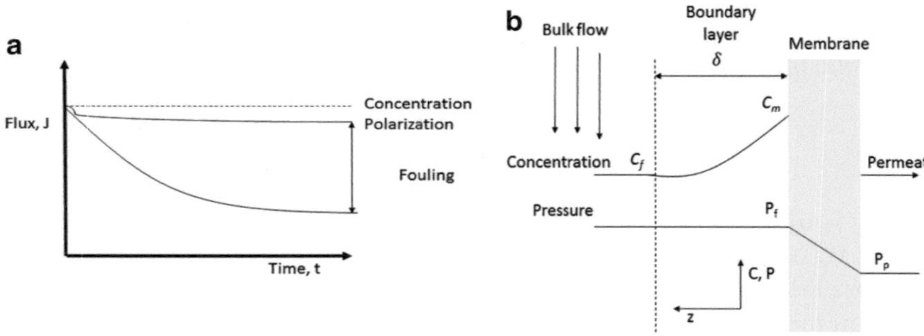


**Fig. 1.3** (**a**) Concentration polarization and fouling during cross-flow operation and (**b**) schematic of concentration polarization near the membrane

$$\beta = \frac{C_m - C_p}{C_f - C_p} = e^{\frac{J\delta}{D}} \tag{1.28}$$

where $C_p$ is concentration in permeate (mol $L^{-1}$), $C_m$ is concentration in membrane (mol $L^{-1}$), $C_f$ is concentration in feed (mol $L^{-1}$), $J$ is permeate flux ($ms^{-1}$), $\delta$ is thickness of boundary layer (m), and $D$ is diffusion coefficient ($m^2s^{-1}$).

$C_p$ can be neglected because $C_p \ll C_f < C_m$ and the coefficient $k$ is taken for the mass transfer and the following relation can be used:

$$k = \frac{D}{\delta} \tag{1.29}$$

$\beta$ can be rewritten as:

$$\beta = \frac{C_m}{C_f} = e^{\frac{J}{k}} \tag{1.30}$$

The high concentration polarization given by the factor $\beta$ leads to higher osmotic pressure difference across the membrane and therefore lowers the performance of the membrane.

### 1.3.4 Membrane Fouling

A major problem of membranes is fouling. Fouling occurs on the surface of the membrane and/or within the pores of the membrane and is detrimental in that it reduces flux. The types of fouling are shown in Fig. 1.4, which include biofouling, organic, colloidal, and scaling. Biofouling results from microorganism contamination of the feed water and produces a biofilm on the surface of the membranes. Organic fouling comes from substances such as hydrocarbons, dyes, and pesticides, which coat the surface and/or plug pores. Colloidal fouling mainly stems from

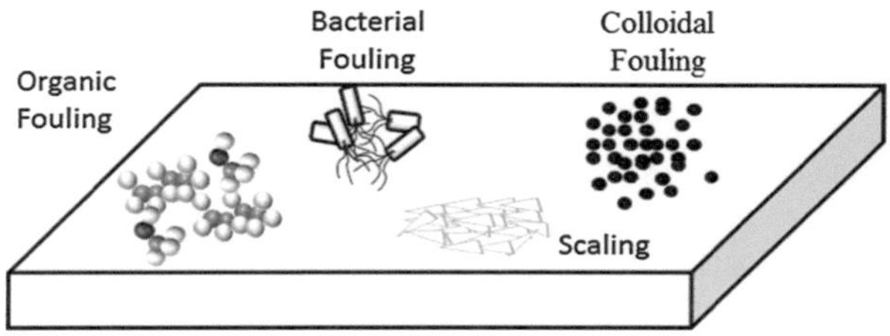


**Fig. 1.4** Fouling of membranes under organic, bacterial, colloidal, and precipitate (scaling) contaminants

particles, such as silica and clay, collecting on the surface of the membrane. Scaling arises from the deposition and precipitation of salts onto the membrane.

#### 1.3.4.1 Biofouling

Biofouling is defined as the loss of system performance due to the formation of a biofilm; a layer of microorganisms attached to the membrane surface [59]. The microorganisms can attach to the membrane quite strongly so that backwashing removal steps prove insufficient. To exacerbate the problem, they also produce extracellular material that can act as foulants. These biofoulants can typically be addressed using chlorine disinfectants in the feed water and/or backwashing.

#### 1.3.4.2 Organic Fouling

The most problematic membrane fouling is due to the adsorption of NOM at the membrane surface. The ability of dissolved organic matter (DOM) to adsorb to membranes has been demonstrated, with humic and fulvic solutions [60, 61]. Studies have also shown removing even a large fraction of DOM has little or no reduction of fouling [62, 63] and removing a small fraction of DOM can completely eliminate fouling [64]. Collectively, this research indicates that membrane fouling can be modeled by a power law. In other words, only a fraction of DOM causes the majority of fouling in membrane filtration, and high molecular weight and colloidal fractions cause the constriction of the membrane pores.

#### 1.3.4.3 Precipitation and Scaling

Precipitation of ions is an import factor that limits recovery in RO systems. To limit precipitation, the salt rejection rate, the recovery, and the degree of concentration polarization is important because precipitation occurs in close proximity to the membrane surface. Concentration polarization is reduced by achieving turbulence and maintaining minimum velocity in feed channels. The most common scales encountered are calcium carbonate ($CaCO_3$) and calcium sulfate ($CaSO_4$), but there are other scales that occur as well, as shown in Table 1.6. Pretreatment

**Table 1.6** Common salts, their solubility, and dissociated ions

| Salt | Equilibrium equation | Solubility of product ($pK_{sp}$, 25 °C) |
|---|---|---|
| Calcium carbonate | $CaCO_3(s) \leftrightarrow Ca^{2+} + CO_3^{2-}$ | 8.2 |
| Calcium fluoride | $CaF_2(s) \leftrightarrow Ca^{2+} + 2\,F^-$ | 10.3 |
| Calcium orthophosphate | $CaHPO_4(s) \leftrightarrow Ca^{2+} + HPO_4^{2-}$ | 6.6 |
| Calcium sulfate | $CaSO_4(s) \leftrightarrow Ca^{2+} + SO_4^{2-}$ | 4.6 |
| Strontium sulfate | $SrSO_4(s) \leftrightarrow Sr^{2+} + SO_4^{2-}$ | 6.2 |
| Barium sulfate | $BaSO_4(s) \leftrightarrow Ba^{2+} + SO_4^{2-}$ | 9.7 |
| Silica, amorphous | $SiO_2(s) + 2H_2O \leftrightarrow Si(OH)_4(aq)$ | 2.7 |

*Source*: Stumm and Morgan (1996), Ref. [65]

steps such as acidic pH adjustment of calcium carbonate can be used to produce bicarbonate and carbon dioxide and prevent precipitation. Additionally, antiscalent chemicals can be used to allow for supersaturation without precipitation by preventing crystal growth.

## 1.4 AOPs and Photocatalysis

### *1.4.1 Introduction*

The goal of advanced oxidation processes (AOPs) is to oxidize organic pollutants in water via hydroxyl radicals ($HO^{\cdot}$) thereby converting the constituents of the organic pollutants into relatively harmless organic or inorganic molecules. $HO^{\cdot}$ radicals are very powerful oxidants that are highly reactive, non-selective electrophilic oxidizing agents, and are ubiquitous in nature [9]. In addition, $HO^{\cdot}$ radicals react $10^6$–$10^{12}$ times more rapidly than alternative oxidants such as ozone ($O_3$) and have a high redox potential (2.80 V vs. normal hydrogen electrode, NHE) second only to fluorine, which is highly toxic [9, 66, 67]. A list of common oxidants is shown in Table 1.7.

### *1.4.2 Overview of AOP Processes*

AOPs can be subdivided into the following processes: photo-induced processes, ozonation, Fenton process, photo-Fenton process, photosensitized process, and ultrasound. These processes generate hydroxyl radicals coming from chemical, photochemical, and sonochemical methods. A description of the processes and their reaction mechanism, shown in Table 1.8, and their advantages and disadvantages, shown in Table 1.9, are covered.

**Table 1.7** Redox potentials of common oxidants

| Oxidant | Redox potential (V vs NHE, 25 °C) |
|---|---|
| $F_2$ | +3.03 |
| $HO^\bullet$ | +2.80 |
| $O^\bullet$ | +2.42 |
| $O_3$ | +2.07 |
| $H_2O_2$ | +1.78 |
| $HO_2^\bullet$ | +1.70 |
| $Cl_2$ | +1.36 |

*Source*: Belgiorno et al. (2011), Ref. [9]

#### 1.4.2.1 Photo-Induced Processes

Photo-induced processes can be thought of as either a photolysis process or a photochemical process. Photolysis is the use of radiant energy from UV-A, B, or C sources without the presence of a catalyst to irradiate a polluted aqueous solution, whereas a photochemical reaction is a chemical reaction induced or catalyzed by light. The primary step in UV photolysis is the electronic excitation of the substrate, S, followed by electron transfer from the excited state, $S^*$, to ground state molecular oxygen, $O_2$:

$$S + h\nu \rightarrow S^* \tag{1.31}$$

$$S^* + O_2 \rightarrow S\cdot^+ + O_2\cdot^- \tag{1.32}$$

Through homolytic scission, free radicals form and further reaction of these species with dissolved oxygen generates peroxide radicals. The formation of reactive oxygen species and radicals leads to the degradation of the parent compound. Photolysis of $H_2O_2$ and $O_3$ is used in water treatment.

#### 1.4.2.2 Ozonation Processes

Ozonation involves two species of interest: (i) ozone and (ii) hydroxyl radicals. Ozone is able to decompose into hydroxyl radicals. Disinfection occurs predominantly with ozone, which is a very selective oxidant compared to hydroxyl radicals which react nonselectively. NOM is converted into oxidized compounds and total degradation end-products through electrophilic attack or dipolar cyclo–addition [68]. Ozonation processes that are found in water treatment are $O_3$, UV/$O_3$, and $O_3$/$H_2O$.

#### 1.4.2.3 Fenton and Photo-Fenton Processes

The reaction of ferrous iron with $H_2O_2$ leads to the formation of ferric iron and hydroxyl radical [69]. Dark and photo-Fenton reactions take place under acidic

**Table 1.8** Overview of AOP processes and their reaction mechanisms

| AOP process | Description | Reaction mechanism |
|---|---|---|
| Photo-induced advanced oxidation processes | *UV/H$_2$O$_2$*<br>$H_2O_2$ is injected and mixed into the feed water and enters a reactor equipped with UV lights. The direct photolysis, at a wavelength of 254 nm, of hydrogen peroxide leads to the formation of $HO^{\cdot}$ radicals | $H_2O_2 + h\nu\ (\lambda < 254\ nm) \rightarrow 2HO^{\cdot}$<br>$H_2O_2 + HO^{\cdot} \rightarrow HO_2^{\cdot} + H_2O$<br>$H_2O_2 + HO_2^{\cdot} \rightarrow HO^{\cdot} + H_2O + O_2$<br>$HO_2^{\cdot} + HO_2^{\cdot} \rightarrow H_2O_2 + O_2$ |
| | *UV/O$_3$*<br>Ozone adsorbs UV radiation at 330 nm or less. $H_2O_2$ is produced as an intermediate, which later decomposes to $HO^{\cdot}$ | $O_3 + H_2O + h\nu\ (\lambda < 330\ nm) \rightarrow H_2O_2 + O_2$<br>$H_2O_2 + h\nu \rightarrow 2HO^{\cdot}$<br>$O_3 + H_2O_2 \rightarrow H_2O + 2O_2$ |
| | *VUV*<br>Extreme UV irradiation of $H_2O$ at wavelengths below 190 nm is able to lyse $H_2O^{\cdot}$ and generate HO radicals | $H_2O + h\nu\ (\lambda < 190\ nm) \rightarrow H^{\cdot} + HO^{\cdot}$ |
| Ozonation | *Ozone* is unstable in an aqueous medium, decomposing spontaneously and generating $HO^{\cdot}$ radicals | $O_3 + HO^{\cdot} \rightarrow O_3^{\cdot} - HO^{\cdot}$<br>$O_3^{\cdot -} \rightarrow O_2^{\cdot} + HO^{\cdot}$<br>$O^{\cdot -} + H^+ \rightarrow HO^{\cdot}$ |
| Fenton and Photo-Fenton processes | *Fenton process*<br>Fenton's reagent, a mixture of ferrous iron and $H_2O_2$ is used to catalytically decompose $H_2O_2$ into $OH^{\cdot}$radicals. The ferrous iron catalyst is oxidized into ferric iron and reduced back to ferrous iron by the same $H_2O_2$ source | $Fe^{2+} + H_2O_2 \rightarrow Fe^{3+} + [H_2O_2-] + HO^{\cdot} + OH-$<br>$Fe^{3+} + H_2O_2 \rightarrow Fe^{2+} + HO_2^{\cdot -} + H+$<br>$Fe^{2+} + HO^{\cdot} \rightarrow Fe^{3+} + OH-$ |
| | *Photo-Fenton Process*<br>Exposing UV light to the Fenton reagent can also generate $HO^{\cdot}$ radicals | $Fe^{III} + H_2O_2 + h\nu \rightarrow Fe^{II} + HO^{\cdot} + H^+$<br>$Fe^{II} + H_2O_2 \rightarrow Fe^{III} + HO^{\cdot} + OH^-$ |
| Sensitized processes | $TiO_2$ *semiconductor photocatalysis*<br>$TiO_2$ in aqueous solution acts as a photocell in which UV light with a wavelength of less than 390 nm can generate electron and hole pairs. These electron/hole pairs participate in redox reactions and produce $HO^{\cdot}$ radicals | $TiO_2 + h\nu\ (\lambda < 390\ nm) \rightarrow TiO_2(e^-_{CB} + h_{VB}-)$<br>$TiO_2$ electron donor reactions<br>$TiO_2(e_{CB^-}) + O_{2ads} \rightarrow TiO_2 + O_2{\bullet}^-$<br>$TiO_2(e_{CB^-}) + HO_2{\bullet} + H^+ \rightarrow TiO_2 + H_2O$<br>$2HO_2^{\cdot} \rightarrow H_2O_2 + O_2$<br>$H_2O_2 + O_2^{\cdot} \rightarrow HO^{\cdot} + O_2 + OH-$<br>$H_2O_2 + TiO_2(e_{CB^-}) \rightarrow TiO_2 + HO^{\bullet} + OH^-$<br>$O_2^{\cdot} + H^+ \rightarrow HO_2^{\cdot}$<br>$HO_2^{\cdot} \leftrightarrow O_2^{\cdot -} + H^+$<br>$TiO_2$ electron acceptor reaction<br>$TiO_2(h^+_{VB}) + H_2O_{ads} \rightarrow TiO_2 + HO^{\cdot}_{ads} + H^+$<br>$TiO_2(h^+_{VB}) + HO^-_{ads} \rightarrow TiO_2 + HO^{\cdot}_{ads}$<br>$h_{VB}{}^+ + RX_{ads} \rightarrow RX^{\cdot -}_{ads} \rightarrow RX^{\cdot -}_{ads}$<br>$h_{VB}{}^+ + H_2O_{ads} \rightarrow HO^{\cdot}{}_{ads} + H^+$ |

(continued)

**Table 1.8** (continued)

| AOP process | Description | Reaction mechanism |
|---|---|---|
| Sonolysis | Ultrasonic irradiation generates high local temperatures and pressures as a result of acoustic cavitation resulting in HO˙ production | $R \xrightarrow{\Delta+\text{ultrasonic}} \text{pyrolysis product}$<br>$H_2O \xrightarrow{\Delta+\text{ultrasonic}} H^{\bullet} + HO^{\bullet}$<br>$HO\cdot + R \xrightarrow{\Delta+\text{ultrasonic}} \text{product}$ |

Adapted from Belgiorno et al. (2011), Ref. [9]; Gottschalk and Libra (2000), Ref. [71]

conditions, and irradiation of the Fenton process by visible or near-UV radiation considerably improves the rate of removal of pollutants.

#### 1.4.2.4 Sensitized Processes

Sensitized AOPs can be classified as either dye-sensitized and semiconductor-sensitized processes. In a dye-sensitized process, the dye molecules absorbing visible light are excited to a higher energy state. The excited dye transfers its excess energy to other molecules in the reaction medium producing a chemical reaction. When dissolved oxygen is present, it accepts energy from the excited dye molecule and the dissolved oxygen is converted to singlet oxygen, which is a strong oxidant. Semiconductor-sensitized processes are concerned with the effect of light on interacting molecules and solid semiconductor surfaces like $TiO_2$. The process depends on the formation of electron–hole pairs upon irradiation of light energy exceeding the energy of the band gap of the semiconductor.

#### 1.4.2.5 Sonolysis

Ultrasonic irradiation, or sonolysis, can be employed for the degradation of many organic compounds. Ultrasound irradiation alone has not been able to provide sufficient degradation rates; especially for hydrophilic compounds. The degradation rate by sonolysis alone is much slower than other AOPs. Ultrasound destroys organics by cavitation of small bubbles. Extremely high temperatures and pressures occur during bubble cavitation at a localized region. This causes pyrolysis of organic molecules and produces reactive chemical radicals. The estimated transient temperature is about 4,000–10,000 K and the pressure is in the range of 30–98 MPa [70]. As a result of cavitation, two processes occur that degrade organics: (i) pyrolysis due to high temperature and pressures and (ii) the production of radical species.

**Table 1.9** Common AOPs and their advantages and disadvantages

| AOP | Advantages | Disadvantages |
|---|---|---|
| $H_2O_2$–UV | • $H_2O_2$ is relatively stable and can be stored for long periods<br>• UV irradiation can serve as disinfectant<br>• Capable of full-scale drinking water treatment | • Turbidity affects UV light penetration<br>• Special reactors for UV illumination are required<br>• Residual $H_2O_2$ must be addressed |
| $O_3$ | • Very strong oxidant with short reaction time<br>• Can eliminate wide variety of organic and inorganic contaminants<br>• The treatment does not add chemicals to the water | • High capital and maintenance costs<br>• Energy intensive<br>• Potential fire hazard and toxicity<br>• pH adjustment required and not practical |
| $O_3$–$H_2O_2$ | • High efficiency<br>• Waters with poor UV transmission can be treated and custom reactors for UV are not required<br>• Disinfection<br>• Established technology for remediation | • Potential for bromate formation<br>• Energy and cost-intensive<br>• May require treatment of excess $H_2O_2$ due to potential for microbial growth<br>• Must maintain proper $O_3/H_2O_2$ dosages<br>• Low pH is detrimental to the process |
| $O_3$–UV | • Residual oxidant will degrade rapidly<br>• Ozone adsorbs more UV light than hydrogen peroxide (more efficient that $H_2O_2$/UV)<br>• UV irradiation can serve as disinfectant | • Energy and cost-intensive<br>• Custom reactors for UV irradiation are required<br>• Volatile compounds are stripped from the process<br>• Turbidity affects UV light penetration<br>• Ozone diffusion can result in mass transfer limitations<br>• Potential increase in trihalomethanes (THM) and haloaceticacids (HAA) formation when combined with chlorine disinfection |
| $TiO_2$–UV | • Activated at near UV light (300–380 nm) compared to other UV oxidation processes that require lower wavelengths | • Fouling of catalyst may occur<br>• When used as a slurry, $TiO_2$ must be recovered |
| Sonolysis | • Useful with waters with low light penetration<br>• Economical treatment of small volumes | • Increases water turbidity<br>• High energy consumption<br>• High maintenance costs of probes<br>• Design criteria is still being developed |
| Fenton | • Some groundwaters have sufficient Fe to drive Fenton's reaction<br>• Not energy intensive like $O_3$ , UV, or sonolysis<br>• Commercial processes are available | • Low pH (<2.5) is required so that iron does not precipitate<br>• Requires iron extraction system<br>• Operational costs associated with pH adjustment |

Adapted from Belgiorno et al. (2011), Ref. [9]; Crittenden et al. (2012), Ref. [22]

#### 1.4.2.6 Major Factors that Impact AOPs

The major factors that impact AOPs are pH, carbonate, NOM, and reduced metal ions. The performance of AOPs is affected by pH in three ways: (i) pH affects the charge on organic compounds that are slightly acidic or basic, (ii) pH affects the concentration of $HCO_3^-$ and $CO_3^{2-}$, and (iii) pH affects the concentration of $HO_2^-$, which is important in UV/$H_2O_2$, UV/$O_3$, and $H_2O_2/O_3$ AOPs. For these AOPs, low pH (<5.0) reduces the rate of production of $HO^{\cdot}$ [72]. Bicarbonate and carbonate ions are $HO^{\cdot}$radical scavangers and are often found in high concentrations compared to the concentration of organic pollutant [73]. NOM reacts with $HO^{\cdot}$ radicals and quenches the reaction [74, 75]. Metal ions in reduced oxidation states, such as Fe (I) and Mn (II), can consume a significant amount of chemical oxidants, and can scavenge $HO^{\cdot}$ radicals.

### *1.4.3 Semiconductor Photocatalytic Degradation*

#### 1.4.3.1 Semiconductor Basics

##### 1.4.3.1.1 General Characteristics of Semiconductors

The de Broglie relation associates a wavelength with the electron as follows:

$$\lambda = \frac{h}{p} = \frac{h}{mv} \tag{1.33}$$

where $h$ is Planck constant and $p$ is momentum.

The wavelength is a function of the inverse of the momentum of the particle and the frequency is a function of the particle's kinetic energy:

$$f = \frac{E}{h} \tag{1.34}$$

The wave number corresponds to the number of repeating units per unit of space:

$$\overline{\nu} = \frac{1}{\lambda} \left( \overline{\nu} = \frac{f}{v} \text{ for non-dispersive waves and } \overline{\nu} = \frac{f}{c} \text{ for electromagnetic waves} \right) \tag{1.35}$$

where $v$ is propagation velocity of the wave (m $s^{-1}$) and $c$ is speed of light (m $s^{-1}$).

The wave vector is a vector related to a wave, and its amplitude is equal to the wave number while its direction is that of the propagation of the wave:

$$\overline{k} = \frac{2\pi}{\lambda} \tag{1.36}$$

Therefore the electron momentum can be expressed as:

$$\overline{p} = \frac{\hbar}{\lambda} = \hbar\overline{k} \tag{1.37}$$

where $\overline{h} = \frac{h}{2\pi}$ is reduced Plank constant.

In the free electron model, the electron energy is therefore:

$$E = \frac{p^2}{2m} = \frac{\hbar^2\overline{k}^2}{2m} \tag{1.38}$$

where $m$ is inertial mass of the wave-particle.

As m varies with the wave vector, it is called effective mass ($m^*$), defined as:

$$m^* = \frac{\hbar^2}{\partial^2 E/\partial k^2} \tag{1.39}$$

In a free electron the wave function are plane waves, defined as:

$$\Psi_k(r) = \mathrm{e}^{\mathrm{ikr}} \tag{1.40}$$

When an electron is placed in a periodic potential such as a crystal lattice with a lattice distance of $a$, the wave function of the electron is known as the Bloch wave function [76]:

$$\Psi_{nk}(r) = \mathrm{e}^{\mathrm{ikr}} U_{nk}(x) \tag{1.41}$$

where $U(x)$ is a periodic function that matches the periodic potential ($U(x+a) = U(x)$).

In the propagation of a wave through a crystal lattice, the frequency is a periodic function of some wave vector $k$, in which multiple solutions to the Schrödinger equation exist, labeled by n, the band index that numbers the energy bands. Each of these energy levels, with changes in k, forms a smooth band of states. For each band, a function $E_n(k)$ can be defined. The wave vector takes on any value inside what is known as the Brillouin zone [77], which is a polyhedron in reciprocal space of the crystal lattice.

Energy band gaps can be classified as either direct or indirect band gaps (Fig. 1.5). A semiconductor is called a direct band gap semiconductor if the energy of the top valence band lies below the minimum energy of the conduction band without a change in momentum. On the other hand, a semiconductor is called an indirect band gap semiconductor if the minimum energy in the conduction band is shifted by a change in momentum $\Delta p$.

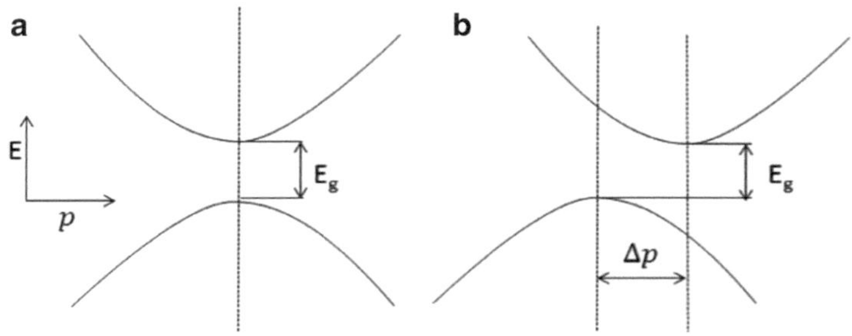


**Fig. 1.5** Energy vs. Momentum diagram for a (**a**) direct bandgap semiconductor and (**b**) an indirect band gap semiconductor

The probability *f*(*E*), that an energetic level of a solid is occupied by electrons, is determined by the Fermi-Dirac distribution function [78]:

$$f(E) = \frac{1}{1 + \exp\frac{\left(E - E_F^o\right)}{k_B T}} \tag{1.42}$$

where $E$ is energy (J), $E_F^o$ is Fermi level (J), $k_B$ is Boltzmann constant (1.3806 $10^{-23}$ $m^2$ kg $s^{-2}$ $K^{-1}$), and $T$ is temperature (K).

The Fermi level ($E_F^o$) represents the 50 % probability of finding an electron in this level. For intrinsic semiconductors, $E_F^o$ falls inside the energetic gap and depends on the mass of electrons at the end of the conduction band ($m_e^*$), on the effective mass of electrons at the beginning of the valence band ($m_h^*$), and on the amplitude of the band gap ($E_g$). The equation of the Fermi level is as follows:

$$E_F^o = \frac{1}{2}E_G + \frac{3}{4}kT\ln\left(\frac{m_h^*}{m_e^*}\right) \tag{1.43}$$

The value of $E_F^o$ is equivalent to the electrochemical potential of an electron.

#### 1.4.3.1.2 Semiconductor Doping

Some types of impurities and imperfections to the crystal lattice may drastically affect the electrical properties of a semiconductor. The conductivity of a semiconductor can be increased by doping, a technique that introduces foreign atoms into the lattice. This makes electrons available in the conduction band and holes available in the valence band. For example, in the case of the Si lattice, with each Si atom having four covalent bonds with four nearby Si atoms, the addition of atoms—arsenic, phosphorous, or antimony—having one more valence electron

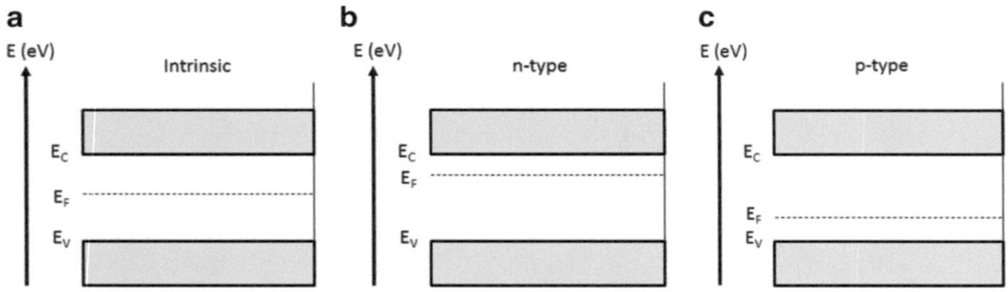


**Fig. 1.6** Three types of semiconductors: (**a**) intrinsic, (**b**) *n*-type, and (**c**) *p*-type

compared to Si will lead to an excess positive charge due to the transfer of an electron from the foreign atom to the conduction band. This is called donor doping and creates an n-type semiconductor, where the Fermi level will be close to the conduction band. On the other hand, if the foreign atom is boron, gallium, and indium, which have one valence electron less than Si, it can accept one electron from the valence band. This is called acceptor doping and creates a p-type semiconductor, where the Fermi level will reside closer to the valence band. The Fermi levels of intrinsic, n-type, and p-type semiconductors are shown in Fig. 1.6.

#### 1.4.3.1.3 Semiconductor in Electrolyte

The energy levels of electrons in solids can be extended to the case of an electrolytic solution containing a redox system [79]. The occupied electronic levels correspond to energetic states of reduced species, whereas unoccupied states correspond to energetic states of oxidized species. The Fermi level of the redox couple, $E_{\text{f, redox}}$, corresponds to the electrochemical potential of electrons in the redox system and is equivalent of the reduction potential $V_o$. To correlate energetic levels of the semiconductor and the redox couple in an electrolyte, two different scales are used (in eV and V). There are two scales because in solid-state physics zero is the level of an electronic vacuum and in electrochemistry the reference is the potential of the normal hydrogen electron (NHE). The two scales are correlated using the potential of NHE, which is equal to −4.5 eV and is referred to as that of the electron in a vacuum [80].

If a semiconductor is placed in contact with a solution containing a redox couple, equilibrium is reached when the Fermi levels of both phases become equal. This occurs by means of an electron exchange from solid and electrolyte, which leads to the generation of charge inside the semiconductor. This charge is distributed in a spatial charge region near the surface, in which the value of holes and electron concentrations also differ considerably from those inside the semiconductor. Figure 1.7a shows schematically the energetic levels of an *n*-type semiconductor and a redox electrolyte before contact and at equilibrium after contract. In particular, as the energy of the Fermi level is higher than that of the electrolyte,

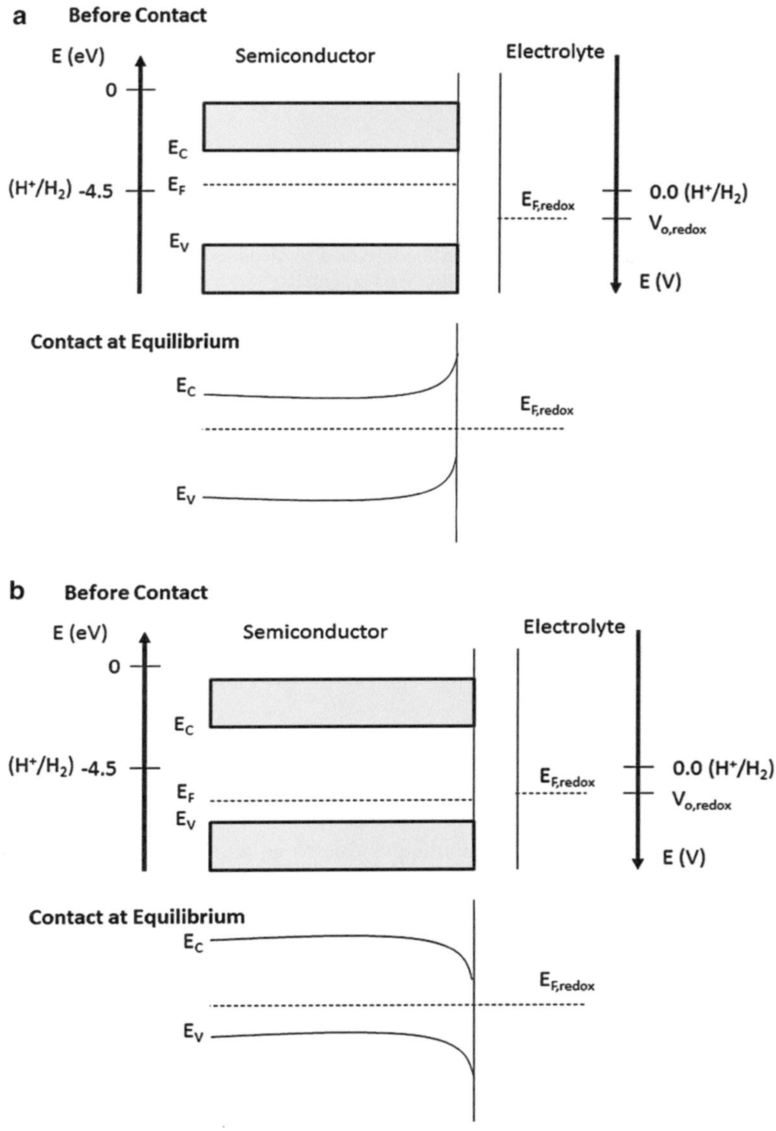


**Fig. 1.7** Formation of a junction between an (**a**) *n*-type semiconductor and (**b**) *p*-type semiconductor in an electrolyte solution before contact and at equilibrium after contact

equilibrium is reached by electron transfer from the semiconductor r to the solution. An electric field is produced by this electron transfer, which is represented by upward band bending. Owing to the presence of the field, holes in excess generated in the spatial charge region move toward the semiconductor surface, whereas electrons in excess migrate from the surface to the bulk of the solid. Figure 1.7b

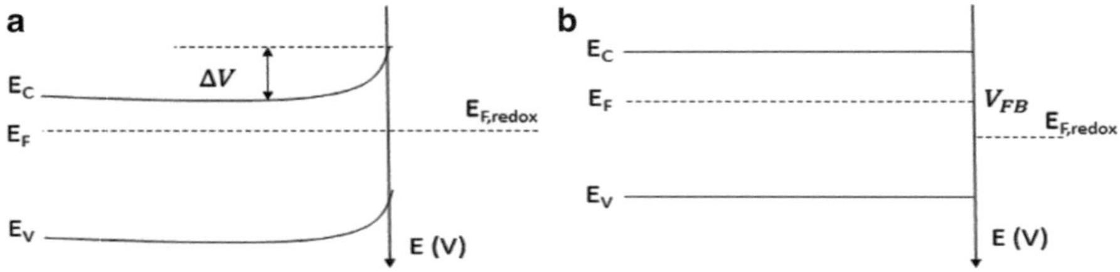


**Fig. 1.8** Scheme of the energetic levels at the interface semiconductor-electrolyte for an *n*-type semiconductor at (**a**) equilibrium and (**b**) flat band potential

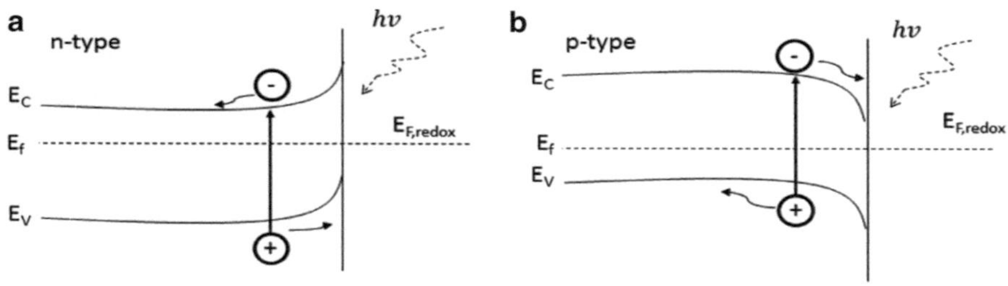


**Fig. 1.9** Generation of an electron–hole pair after irradiation of (**a**) n-type semiconductor and (**b**) *p*-type semiconductor

shows the contact between a redox electrolyte and a p-type semiconductor. In this case transfer of electrons occurs from the electrolyte to the semiconductor and band bending is downward.

If the potential of the electrode changes due to an anodic or cathodic polarization, a shift of the Fermi level of the semiconductor with respect to that of the solution occurs with an opposite curvature of the bands as seen in Fig. 1.8. For a particular value of the electrode potential, the charge excess disappears and the bands become flat from the bulk to the surface of the semiconductor. The corresponding potential is called the flat band potential, $V_{FB}$, and the determination of this potential allows for the calculation of the energy of the conduction and the valence bands [81].

When a semiconductor is irradiated by radiation of suitable energy equal to or higher than that of the band gap, $E_g$, electrons can be promoted from the valence band to the conduction band. Figure 1.9 shows the scheme of electron–hole pair formation due to the adsorption of a photon by a semiconductor.

The existence of an electric field in the spatial charge region allows for the separation of the photogenerated pairs. In the case of n-type semiconductor, electrons migrate toward the bulk whereas holes move to the surface. In the case of p-type semiconductors, holes move toward the bulk whereas electrons move to the surface.

Photoproduced holes and electrons, during their migration in opposite directions, can (i) recombine and dissipate their energy as either electro-magnetic

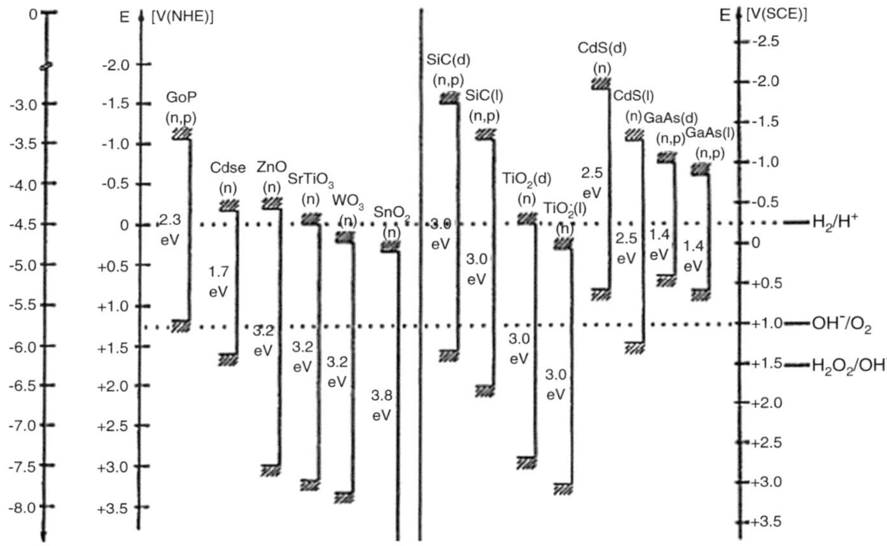


**Fig. 1.10** Positions of band edges for some semiconductors in contact with aqueous electrolyte at pH 0. *Source*: Memming (1994), Ref. [82]

radiation (photon emission) or simply as heat, or (ii) react with electron-acceptor or electron-donor species present at the semiconductor electrolyte interface, thereby reducing or oxidizing them, respectively.

The energy of the conduction band edge ($E_c$) corresponds to the potential of the photogenerated electrons, whereas the energy of the valence band gap corresponds to the potential of the holes. If $E_c$ is more negative than the potential of a species present in solution, electrons reaching the interface can reduce the oxidized form of the redox couple. Conversely, if the potential of $E_v$ is more positive than that of the redox couple, photoproduced holes can oxidize its reduced form. Knowledge of the relative edge positions of the bands and of the energetic levels of the redox couples is essential to establish if thermodynamics allow the occurrence of oxidation and/or reduction of the species in solution. The band gaps and the positions of the valence band and conduction band edges for various semiconductors are given in Fig. 1.10.

#### 1.4.3.2 Photocatalysis Mechanism

Photocatalytic properties of a semiconductor depend on (i) the position of the energetic levels; (ii) the mobility and mean lifetime of the photogenerated electrons and holes, (iii) the light absorption coefficient, and (iv) the nature of the interface. The photoactivity depends on the method of preparation of the catalyst, where the physicochemical properties of the semiconductor are altered.

Each particle of semiconductor in a photocatalytic system is akin to a photoelectrochemical cell using a semiconductor electrode in contact with an electrode of an inert metal [83]. In a photoelectrochemical cell an oxidation or reduction reaction may occur on the semiconductor electrode, whereas in a semiconductor particle immersed in an electrolyte solution both reactions occur simultaneously via hole transfer from the valence band and by electron transfer from the conduction band.

A semiconductor powder suspension is a mixture of many small photocells. Separation of electron–hole pairs increases with increasing thickness of the spatial charge region, which depends on the doping of the semiconductor. As the volume of the semiconductor particle decreases, charge separation becomes a minimum when the particle size is smaller than the thickness of the spatial charge region. The adsorption of radiation of a suitable wavelength by the semiconductor allows the transformation of photons into chemical energy; this phenomenon is termed photocatalysis. When aqueous solutions of semiconductor powders are irradiated at the solid–liquid interface of the photocatalyst, photo-induced chemical reactions occur that are able to degrade organic and inorganic molecules. Degradation occurs directly by the electron–hole pairs, or via formation of reactive radical species that are generated in the presence of $O_2$ and $H_2O$. The photocatalytic process also eliminates many inorganic ionic pollutants present in water by reduction into elemental form in the vicinity of the catalyst particle or transformation into less noxious species.

#### 1.4.3.3 Kinetics of Photocatalysis

Heterogeneous catalysis involves systems in which reactants and catalyst are physically separated phases. Photocatalysis in particular is a change in the rate of a chemical reaction or its generation when a semiconductor, the photocatalyst, absorbs ultraviolet-visible-infrared radiation. Photocatalysis taking place at the boundaries between two phases can be expressed as:

$$A \xrightarrow{h\nu,\text{ catalyst}} B \tag{1.44}$$

where $A$ is reactant and $B$ is product.

Photocatalysis of a heterogeneous system can be thought of as two steps: (i) photoadsorption/desorption processes and (ii) photoexcitation. Photosorption can be represented as:

$$A \xrightarrow{h\nu,\text{ catalyst}} A_{\text{ads}} \tag{1.45}$$

where $A$ is reactant and $A_{\text{ads}}$ photoadsorbed species.

The photoadsorption of oxygen is focused on $TiO_2$ surface [84–88]. Photoadsorbed species can act as surface-hole trappings and photoelectrons can be trapped in the bulk

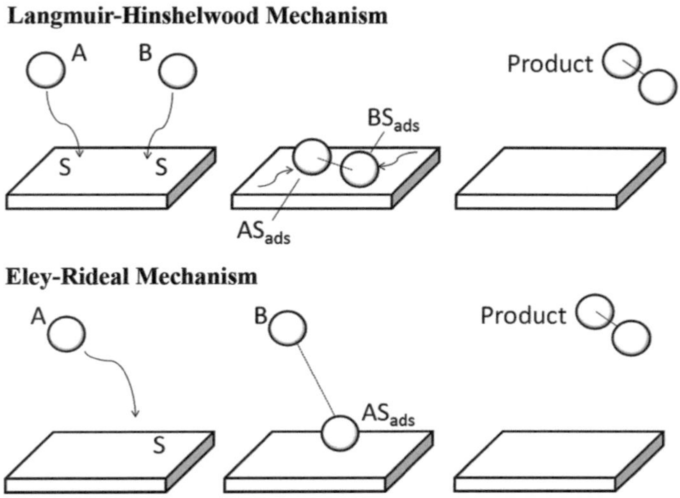


**Fig. 1.11** Top model: Langmuir-Hinshelwood Mechanism—two molecules adsorb onto the surface and diffuse and interact with each other until a product is formed and desorbs from the surface; Bottom model: Eley-Rideal Mechanism—a molecule adsorbs onto the surface and another molecule interacts with the adsorbed one until a product is formed and desorbs from the surface

of the solid or as surface electron trappings. It will depend on the chemical nature of the molecule adsorbed and on the type of the solid adsorbent. These changes are generally fast and reversible, such that once irradiation has stopped the surface recovers its previous features under equal initial conditions.

Photoexcitation of a semiconductor can be separated into four simplified types of electronic excitations induced by light adsorption. In a perfect crystal lattice, absorption can produce only intrinsic photoexcitations with (i) the promotion of electrons from the valence band to the conduction band forming free electron-hole pairs and (ii) the formation of free bulk excitons (the combination of an electron and a positive hole that is free to move through a non-metallic crystal as a unit). In an imperfect lattice, the presence of defects causes extrinsic absorption of light, in particular (iii) photon absorption due to the generation of electronically excited defects and bound and /or self-trapped excitons and (iv) photon absorption due to ionization of defect transitions between localized and delocalized electron states [89, 90].

The adsorption of molecules by the semiconductor in aqueous solution can be described by two mechanisms (Fig. 1.11): (i) the Langmuir-Hinshelwood mechanism and (ii) the Eley-Rideal mechanism. The Langmuir-Hinshelwood mechanism is used to explain the interaction of surface charge carriers and excitons with adsorbed molecules that can promote surface chemical processes, whereas the Eley-Rideal mechanism is used to explain the interaction of molecules with surface active centers that can initiate surface chemical processes.

In the Eley-Rideal mechanism [91] proposed by Eley and Rideal in 1938, only one of the reactant molecules adsorbs and the other reacts without adsorbing:

$$A + S_{\mathrm{ads}} \leftrightarrow AS_{\mathrm{ads}} \tag{1.46}$$

$$AS_{\mathrm{ads}} + B_{\mathrm{g}} \leftrightarrow \text{Product} \tag{1.47}$$

In the Langmuir-Hinshelwood mechanism both molecules adsorb and then undergo a bimolecular reaction:

$$A + S \leftrightarrow AS_{\mathrm{ads}} \tag{1.48}$$

$$B + S \leftrightarrow BS_{\mathrm{ads}} \tag{1.49}$$

$$AS_{\mathrm{ads}} + BS_{\mathrm{ads}} \leftrightarrow \text{Product} \tag{1.50}$$

The Langmuir-Hinshelwood model (LH) is widely applied to liquid and gas phase systems for the degradation of organic substrates on $TiO_2$ surfaces in the presence of oxygen [92–95]. It explains the kinetics of reactions that occur between two adsorbed species. The two main assumptions of the LH model are (i) the adsorption equilibrium is established at all times and the reaction rate is less than the rate of adsorption or desorption and (ii) the reaction is assumed to occur between adsorbed species whose coverage, on the catalyst surface, is in equilibrium with the concentration of the species in the fluid phase, so that the rate-determining step of the photocatalytic process is a surface reaction.

The decrease of the amount of species in a liquid–solid phase photocatalytic system is the combination of photoadsorption and photoconversion processes. To describe this system, a molar balance can be applied to the species at any time [94]:

$$n_T = n_L + n_S \tag{1.51}$$

where $n_T$ is total number of moles present in a photoreactor (mol), $n_L$ is number of moles in the fluid phase (mol), and $n_S$ is number of moles photoadsorbed by the solid (mol).

The molar balance can be rearranged in terms of the total concentration of the species ($C_T$) (mol $L^{-1}$), by dividing by the volume ($V$) of the liquid phase ($L$), to obtain:

$$C_T = C_L + \frac{n_S}{V} \tag{1.52}$$

where $C_L$ is concentration in the liquid phase (mol $L^{-1}$).

Both substrate and oxygen must be present for the occurrence of the photoreaction, then it is assumed that the total disappearance rate of substrate per unit surface area, $r_T$, relies on second-order kinetics (or a first order model with respect to the substrate coverage and oxygen coverage):

$$r_T = -\frac{1}{S}\frac{dn_T}{dt} = k''\theta_{\mathrm{sub}}\theta_{\mathrm{Ox}} \tag{1.53}$$

$$\theta_{\text{sub}} = \frac{n_S}{WN_S^*} \tag{1.54}$$

$$\theta_{Ox} = \frac{n_{S,Ox}}{WN_{S,Ox}^o} \tag{1.55}$$

where $S$ is catalyst surface area (mg g$^{-1}$), $t$ is time (s), $k''$ is second order rate constant, $\theta_{Sub}$ is substrate fractional coverage of the surface, $\theta_{Ox}$ is oxygen fractional coverage of the surface, $n_{S,Ox}$ is the number of moles of oxygen photoadsorbed on the solid on the unit mass of irradiated solid (mol), $N_S^*$ is maximum capacity of photoadsorbed moles of substrate on the unit mass of irradiated solid (mol g$^{-1}$), $N_{S,Ox}^*$ is maximum capacity of photoadsorbed moles of oxygen (mol g$^{-1}$), and $W$ is the mass of catalyst (g).

If oxygen is continuously bubbled into the dispersion its concentration in the liquid phase does not change and it is always in excess. The $\theta_{Ox}$ term of Eq. (1.53) is then constant so we can define $k = k''\theta_{Ox}$ and Eq. (1.53) turns to a pseudo first-order rate equation:

$$r_T = -\frac{1}{S}\frac{dn_T}{dt} = -\frac{1}{S_S W}V\frac{dC_T}{dt} = k\theta_{\text{sub}} \tag{1.56}$$

where $S_S$ surface area per unit mass of catalyst and $k$ is pseudo first-order rate constant.

The kinetic information on a photoprocess consists of knowledge of substrate concentration values in the liquid phase, $C_L$, as a function of irradiation time. $r_T$ and $\theta_{sub}$ can be formally written as a function of $C_L$, where $\theta_{\text{sub}}$ and $C_L$ relations can be obtained from an appropriate isotherm (See Sect. 2.6):

$$r_T = -\frac{1}{S_S W}V\frac{dC_T(C_L)}{dt} = k\theta_{\text{sub}}(C_L) \tag{1.57}$$

In a batch photocatalytic experiment, the substrate concentration values measured in the liquid phase represent the substrate concentration in equilibrium with an unknown substrate concentration on the catalytic surface. This is apparent to all the measured values of substrate concentration except for the initial one. The substrate concentration measured at the start of a photodegradation experiment is a system without irradiation, where the initial substrate concentration is under dark conditions.

#### 1.4.3.4 Photocatalytic Materials

$TiO_2$ is the most widely used semiconductor photocatalyst in water treatment due to its low cost, chemical stability, and abundance. Current research focuses on increasing photocatalytic reaction kinetics and photoactivity range to provide either high performance UV-activated photocatalytic reactors or energy-efficient solar/visible light-activated photocatalytic reactors. The optimization approaches for enhancing the reaction kinetics under UV include increasing surface area [96, 97] using 1D nano morphology [98, 99], noble metal doping [100], and producing reactive crystal facets [101, 102]. Researchers have attempted to alter the photoactivity range to solar/visible light by, for example, doping via metal impurities, anions, dyes, and narrow band-gap semiconductors [96, 100].

Other than $TiO_2$, $WO_3$ and fullerene derivatives have potential to be used in photocatalytic water treatment. $WO_3$ has a narrower band gap compared to $TiO_2$, allowing it to be activated by visible light (<450 nm) [103]. Fullerene derivatives such as aminofullerenes [104], fullerol, and $C_{60}$ [105] have been proposed to generate $^1O_2$ under visible light irradiation, despite it currently being more expensive and less available than $TiO_2$.

#### 1.4.3.5 Physical Parameters that Affect Photo-Activity

There are five physical parameters that affect photocatalytic activity: (i) mass of catalyst, (ii) wavelength, (iii) initial concentration of reactant, (iv) temperature, and (v) radiant flux. These are shown in Fig. 1.12.

#### 1.4.3.6 Radiation Sources

The radiation source used in a photocatalytic application largely determines the performance. There are six irradiation sources that can be used: (i) arc lamps, (ii) fluorescent lamps, (iii) incandescent lamps, (iv) lasers, (v) light-emitting diodes (LEDs), and (vi) solar light. The six processes are listed in Table 1.10 below.

Arc lamps are made from neon, argon, xenon, krypton, sodium, metal halide, and mercury. For mercury lamps, there are four classifications that have been made:

1. Low pressure: the lamp contains Hg vapor at 0.1 Pa at 298 K (emission wavelengths: 253.7 and 184.9 nm).
2. Medium pressure: the lamp contains Hg vapor from 100 kPa to several hundred kPa (emission wavelengths: 313, 366, 436, 576, and 578 nm).
3. High pressure: the lamp contains Hg vapor greater or equal to 10 MPa (emission: continuous background from 200 to 1,000 nm with broad lines).
4. Hg-Xe: the lamp contains a mixture of Hg and Xe vapors at high pressure and is used to simulate solar radiation.

Fluorescent lamps are filled with gas typically containing a mixture of low-pressure mercury vapor and argon (or xenon). The inner surface of the lamp is coated with a fluorescent coating made of vary blends of metallic and rare earth phosphor salts. The UV light produced by the cathode emits light at 253.7 and 185 nm. The UV light is absorbed by the bulb's fluorescent coating, which reradiates the energy at two intense lines of 440 and 546 nm in the visible light region.

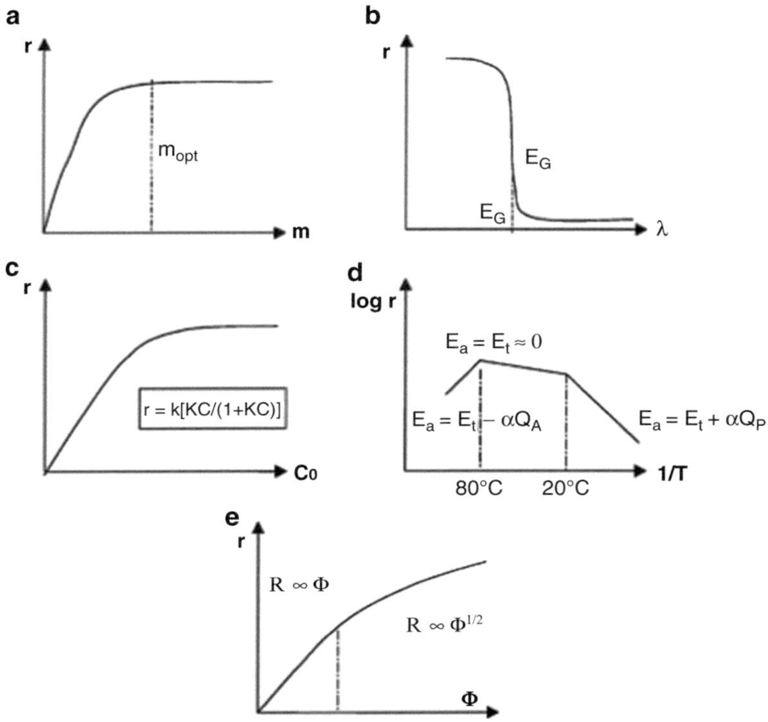


**Fig. 1.12** Influence of physical parameters that affect kinetics of photocatalysis: (**a**) mass of catalyst; (**b**) wavelength $\lambda$; (**c**) initial concentration of the reactant; (**d**) temperature; (**e**) radiant flux Φ. (*Source*: Herrmann (2010), Ref. [106]

**Table 1.10** List of irradiation sources

| Type of radiation source | Description of emission process |
|---|---|
| Arc lamp | Emission is obtained by activation of a gas by collision with accelerated electrons generated by an electric discharge between two electrodes, typically made from tungsten |
| Fluorescent lamp | The cathode is heated sufficiently to emit electrons which collide with a gas, and ionize it to form a plasma, known as impact ionization. As the conductivity of the ionized gas rises, higher currents ionize mercury causing it to emit light in the UV region of the spectrum. The UV light is adsorbed by the bulb's fluorescent coating, which re-radiates energy to lower frequencies to emit visible light. |
| Incandescent lamp | Emission is obtained by heating filaments of various substances to very high temperatures via current circulation |
| Lasers | Emission occurs via a quantum-mechanical and thermodynamic process. Light of a specific wavelength is amplified through a gain medium by a process known as pumping |
| Light-emitting diodes | Electrons and holes are injected in a recombination zone through two parts of the diode doped in different ways; n-type impurities for electrons and p-type for holes. When these electron-hole pairs recombined light is emitted |
| Solar light | Light emission from sun |

Adapted from Augugliaro et al. (2010), Ref. [91]

An incandescent lamp is a halogen lamp, where the tungsten filament is sealed into a small envelop filled with a halogen such as iodine or bromine. In applications that required UV radiation, the lamp envelope is made of quartz and is a source of UV-B radiation. The lamp is designed to run about 2,000 h.

#### 1.4.3.7 Quantum Yield

The quantum yield is the fundamental parameter in heterogeneous photocatalysis and is described as:

$$\Phi_\lambda = \frac{\text{moles of reactant or product}}{\text{moles of photons (einsteins)adsorbed}} \tag{1.58}$$

This parameter is useful as a way to compare efficiencies between various photocatalysts; however, there are two experimental issues: (i) most common light sources are not monochromatic and (ii) the inherent difficulty of measuring the photons adsorbed by the photocatalyst. For photocatalytic reactions in aqueous solutions, a standard protocol has been proposed that uses a standard reactant, such as phenol, and a standard photocatalyst (Degussa P25) to compare results from any laboratory or experimental arrangement in terms of quantum yield. The apparent quantum yield for most photocatalytic reaction range from 0.1 to 3 % depending on the following properties: (i) reactant and its concentration, (ii) light intensity, and (iii) the constituents in the water matrix. There are examples of increasing the quantum yield, such as by reducing catalyst electron–hole recombination. This is an area where ongoing research is being conducted [91].

## References


1. A. Hazen, *Clean Water and How to Get It* (Wiley, New York, 1909)
2. G.W. Fuller, Progress in water purification. J. AWWA **25**(11), 1566–1576 (1933)
3. S. Malato, P. Fernández-Ibáñez, M.I. Maldonado, J. Blanco, W. Gernjak, Decontamination and disinfection of water by solar photocatalysis: recent overview and trends. Catal. Today **147**(1), 1–59 (2009)
4. S.D. Richardson, A.D. Thurston, T.W. Collette, K.S. Patterson, B.W. Lykins, J.C. Ireland, Identification of TiO2/UV disinfection byproducts in drinking water. Environ. Sci. Technol. **30**, 3327–3334 (1996)
5. S. Suárez, M. Carballa, F. Omil, J.M. Lema, How are pharmaceutical and personal care products (PPCPs) removed from urban wastewaters? Rev. Environ. Sci. Biotechnol. **7**, 125–138 (2008)
6. T. Wintgens, F. Salehi, R. Hochstrat, T. Melin, Emerging contaminants and treatment options in water recycling for indirect potable use. Water Sci. Technol. **57**(1), 99–107 (2008)
7. L. Bousselmi, S.U. Geissen, H. Schroeder, Textile wastewater treatment and reuse by solar catalysis: results from a pilot plant in Tunisia. Water Sci. Technol. **49**(4), 331–337 (2004)

8. S. Mozia, M. Tomaszewska, A.W. Morawski, Photocatalytic membrane reactor (PMR) coupling photocatalysis and membrane distillation—effectiveness of removal of three azo dyes from water. Catal. Today **129**(1–2), 3–8 (2007)
9. V. Belgiorno, V. Naddeo, L. Rizzo, *Water, Wastewater and Soil Treatment by Advance Oxidation Processes* (Lulu, Raleigh, NC, 2011)
10. L. Rizzo, S. Meric, M. Guida, D. Kassinos, V. Belgiorno, Heterogeneous photocatalytic degradation kinetics and detoxification of an urban wastewater treatment plant effluent contaminated with pharmaceuticals. Water Res. **43**(16), 4070–4078 (2009)
11. S.A. Snyder, P. Westerhoff, Y. Yoon, D.L. Sedlak, Pharmaceuticals, personal care products, and endocrine disruptors in water: implications for the water industry. Environ. Eng. Sci. **20**(5), 449–469 (2003)
12. J. Theron, J.A. Walker, T.E. Cloete, Nanotechnology and water treatment: applications and emerging opportunities. Crit. Rev. Microbiol. **34**(1), 43–69 (2008)
13. J.W. Hassler, *Activated Carbon* (Chemical Publishing, New York, 1974)
14. M.N. Baker, *The Quest for Pure Water, vol. 1*, 2nd edn. (American Water Works Association, Denver, CO, 1981)
15. H. Sontheimer, J.C. Critenden, R.S. Summers, *Activated Carbon for Water Treatment*, 2nd edn. (DVGW-Forschumgsstelle, University of Karlsruhe, Germany, 1988). Distributed in the US by the American Water Works Association
16. R.S. Summers, D.R.U. Knappe, V.L. Snoeyink, Adsorption of organic compounds by activated carbon, in *Water Quality and Treatment*, ed. by J.K. Edzwald, 5th edn. (McGraw-Hill and American Water Works Association, New York, 2010)
17. Z.K. Chowdhury, R.S. Summers, G.P. Westerhoff, B.J. Leto, K.O. Nowack, C.J. Corwin, L.B. Passantino, *Activated Carbon: Solutions for Improving Water Quality* (American Water Works Association, Denver, CO, 2013)
18. G. Nemethy, H.A. Scherage, Structure of water and hydrophobic bonding in proteins. I. A model for the thermodynamic properties of liquid water. J. Chem. Phys. **36**, 3382–3401 (1962)
19. M. Terrones, Science and technology of the twenty-first century: synthesis, properties and applications of carbon nanotubes. Ann. Rev. Mater. Res. **33**, 419–501 (2003)
20. C. Kemp, H. Seema, M. Saleh, H. Le, K. Mahesh, V. Chandra, K.S. Kim, Environmental applications using grapheme composites: water remediation and gas adsorption. Nanoscale **5**, 3149–3171 (2013)
21. T.T. Teng, L.W. Low, Removal of dyes and pigments from industrial effluents, in *Advances in Water Treatment and Pollution Prevention*, ed. by S.K. Sharma, R. Sanghi (Springer, The Netherlands, 2012)
22. J.C. Crittenden, R.R. Trussel, D.W. Hand, K.J. Howe, G. Tchobanoglous, *MWH's Water Treatment—Principles and Design*, 3rd edn. (John Wiley & Sons, 2012)
23. G. Crini, Non-conventional low-cost adsorbents for dye removal: a review. Bioresour. Technol. **97**, 1061–1085 (2006)
24. G. McKay, J.F. Porter, G.R. Prasad, The removal of dye colours from aqueous solutions by adsorption on low-cost materials. Water Air Soil Pollut. **114**, 423–438 (1999)
25. B.H. Hameed, Evaluation of papaya seeds as a novel non-conventional low-cost adsorbent for removal of methylene blue. J. Hazard. Mater. **162**, 939–944 (2009)
26. Y.H. Magdy, A.A.M. Daifullah, Adsorption of a basic dye from aqueous solutions onto sugar-industry-mud. Waste Manag. **18**, 219–226 (1998)
27. B.H. Hameed, Grass waste: a novel sorbent for the removal of basic dye form aqueous solution. J. Hazard. Mater. **166**, 233–238 (2009)
28. B.H. Hameed, D.K. Mahmoud, A.L. Ahmad, Sorption of basic dye from aqueous solution by pomelo (Citrus grandis) peel in a batch system. Colloids Surf. A **316**, 78–84 (2008)
29. B.H. Hameed, A.A. Ahmad, Batch adsorption of methylene blue from aqueous solution by garlic peel, an agricultural waste biomass. J. Hazard. Mater. **164**, 870–875 (2009)
30. N.A. Oladoja, I.O. Asia, C.O. Aboluwoye, Y.B. Oladimeji, A.O. Ashogbon, Studies on the sorption of basic dye by rubber (Hevea brasiliensis) seed shell. Turk. J. Eng. Environ. Sci. **32**, 143–152 (2008)

31. D.S. Sun, X.D. Zhang, Y.D. Wu, X. Liu, Adsorption of anionic dyes from aqueous solution on fly ash. J. Hazard. Mater. **181**, 335–342 (2010)
32. T.C. Hsu, Adsorption of an acid dye onto coal fly ash. Fuel **87**, 3040–3045 (2008)
33. S. Senthilkumaar, P. Kalaamani, K. Porkodi, P.R. Varadarajan, C.V. Subburaam, Adsorption of dissolved reactivered dye from aqueous phase onto activated carbon prepared from agricultural waste. Bio. Resour. Technol. **97**, 1618–1625 (2006)
34. C.A.P. Almeida, N.A. Debacher, A.J. Downs, L. Cottet, C.A.D. Mello, Removal of methylene blue from colored effluents by adsorption on momtmorillonite clay. J. Colloid Interface Sci. **332**, 46–53 (2009)
35. S. Hong, C. Wen, J. He, F. Gan, Y.S. Ho, Adsorption thermodynamics of methylene blue onto bentonite. J. Hazard. Mater. **167**, 630–633 (2009)
36. M. Hajjaji, A. Alami, A. El-Bouadili, Removal of methylene blue from aqueous solution by fibrous clay minerals. J. Hazard. Mater. **135**, 188–192 (2006)
37. A. Al-Futaisi, A. Jamrah, R. Al-Hanai, Aspects of cationic dye molecule adsorption to palygorskite. Desalination **214**, 327–342 (2007)
38. S. Chakrabarti, B.K. Dutta, Note on the adsorption and diffusion of methylene blue in glass fibers. J. Colloid Interface Sci. **286**, 807–811 (2005)
39. V. Vimonses, S.M. Lei, B. Jin, C.W.K. Chow, C. Saint, Adsorption of Congo red by three Australian kaolins. Appl. Clay. Sci. **43**, 465–472 (2009)
40. K. Marungrueng, P. Pavasant, High performance biosorbent (Caulerpa lentillifera) for basic dye removal. Bioresour. Technol. **98**, 1567–1572 (2007)
41. O. Gulnaz, A. Kaya, F. Matyar, B. Arikan, Sorption of basic dyes from aqueous solution by activated sludge. J. Hazard. Mater. **108**, 183–188 (2004)
42. E. Rubin, P. Rodriguez, R. Herrero, J. Cremades, I. Barbara, M.E. Sastre de Vicente, Removal of methylene blue from aqueous solutions using as biosorbent Sargassum muticum: an invasive macroalga in Europe. J. Chem. Technol. Biotechnol. **80**, 291–298 (2005)
43. A. El-Sikaily, A. Khaled, A. El-Nemr, O. Abdelwahab, Removal of methylene blue from aqueous solution by marine green alga Ulva lactuca. Chem. Ecol. **22**, 149–157 (2006)
44. D. Caparkaya, L. Cavas, Biosorption of methylene blue by a brown alga Cystoseira barbatula Kutzing. Acta. Chim. Solv. **55**, 547–553 (2008)
45. M.C. Ncibi, A.M.D. Hamissa, A. Fathallah, M.H. Kortas, T. Baklouti, B. Mahjoub, M. Seffen, Biosorptive uptake of methylene blue using Mediterranean green alga Enteromorpha spp. J. Hazard. Mater. **170**, 1050–1055 (2009)
46. T. Akar, A.S. Ozcan, S. Tunali, A. Ozcan, Biosorption of a textile dye (Acid Blue 40) by cone biomass of Thuja orientalis: estimation of equilibrium, thermodynamic and kinetic parameters. Bioresour. Technol. **99**, 3057–3065 (2008)
47. N.S. Maurya, A.K. Mittal, P. Cornel, E. Rother, Biosorption of dyes using dead macro fungi: effect of dye structure, ionic strength and pH. Bioresour. Technol. **97**, 512–521 (2006)
48. K. Kumari, T.E. Abraham, Biosorption on anionic textile dyes by nonviable biomass of fungi and yeast. Bioresour. Technol. **98**, 1704–1710 (2007)
49. M.P. Elizalde-González, L.E. Fuentes-Ramírez, M.R.G. Guevara-Villa, Degradation of immobilized azo dyes by Klebsiella sp. UAP-b5 isolated from maize bioadsorbent. J. Hazard. Mater. **161**, 769–774 (2009)
50. I. Langmuir, The adsorption of gases on plane surfaces of glass, mica, and platinum. J. Am. Chem. Soc. **40**, 1361–1402 (1918)
51. S. Brunauer, P.H. Emmett, E. Teller, Adsorption of gases in multimolecular layers. J. Am. Chem. Soc. **60**, 309–319 (1938)
52. H. Freundlich, *Groundwater Pollution* (Elsevier Scientific, Amsterdam, 1975)
53. G.D. Halsey, H.S. Taylor, Adsorption of hydrogen on tungsten powders. J. Chem. Phys. **15**, 624–630 (1947)
54. O. Redlich, O. Peterson, A useful adsorption isotherm. J. Phys. Chem. **63**, 1024 (1959)
55. M.R. Wiesner, S. Chellam, Mass transport considerations for pressure-driven membrane processes. J. AWWA **84**(1), 88–95 (1992)

56. K. Li, *Ceramic Membranes for Separation and Reaction* (Wiley, West Sussex, England, 2007)
57. M. Mulder, *Basic Principles of Membrane Technology* (Kluwer, The Netherlands, 1996)
58. K.J. Laidler, J.H. Meiser, *Physical Chemistry* (Houghton Mifflin, Boston, MA, 1999)
59. H.F. Ridgway, H.F. Flemming, Membrane biofouling, in *Water Treatment Membrane Processes*, ed. by J. Mallevialle, P.E. Odendaal, M.R. Wiesner (McGraw-Hill, New York, 1996)
60. G. Crozes, C. Anselme, J. Mallevialle, Effect of adsorption of organic matter on fouling of ultra filtration membranes. J. Memb. Sci. **84**(1–2), 61–77 (1993)
61. V. Lahoussine-Turcaud, M.R. Wiesner, J.Y. Bottero, Fouling in tangential-flow ultra filtration: the effect of colloid size and coagulation pretreatment. J. Memb. Sci. **52**(2), 173–190 (1990)
62. C.F. Lin, T.Y. Lin, O.J. Hao, Effects of humic substance characteristics on UF performance. Water Res. **34**(4), 1097–1106 (2000)
63. T. Carroll, S. King, S.R. Gray, B.A. Bolto, N.A. Booker, Fouling of microfiltration membranes by NOM after coagulation treatment. Water Res. **34**(11), 2861–2868 (2000)
64. K.J. Howe, Effect of Coagulation Pretreatment on Membrane Filtration Performance, Ph.D. Thesis (University of Illonois at Urbana-Champaign, Urbana, IL, 2001)
65. W. Stumm, J.J. Morgan, *Aquatic Chemistry: Chemical Equilibria and Rates in Natural Waters*, 3rd edn. (Wiley, New York, 1996)
66. J.J. Pignatello, E. Oliveros, A. Mackay, Advanced oxidation processes for organic contaminant destruction based on the Fenton reaction and related chemistry. Crit. Rev. Environ. Sci. Technol. **36**(1), 1–84 (2006)
67. Solarchem Environmental Systems, *The UV/ Oxidation Handbook* (Solarchem Environmental Systems, Markham, ON, 1994)
68. F.J. Beltran, *Ozone Reaction Kinetics for Water and Wastewater Systems* (Lewis, Boca Raton, FL, 2004)
69. H.J.H. Fenton, Oxidation of tartaric acid in the presence of iron. Chem. Soc. H. Lond. **65**, 899–910 (1894)
70. M.R. Hoffmann, I. Hua, R. Höchemer, Application of ultrasonic irradiation for degradation of chemical contaminants. Water. Ultrason. Sonochem. **3**, S163–S172 (1996)
71. C. Gottschalk, J.A. Libra, A. Saupe, *Ozonation of Water and Waste Water* (Wiley-VCH, New York, 2000)
72. J. Hoigné, H. Bader, The role of hydroxyl radical reactions in ozonation process in aqueous solutions. Water Res. **10**(5), 377–386 (1976)
73. W.H. Glaze, J. Kang, Advanced oxidation processes for treating groundwater contaminated with TCE and PCE: laboratory studies. J. AWWA **81**(5), 57–63 (1988)
74. P. Westerhoff, G. Aiken, G. Amy, J. Debroux, Relationships between the structure of natural organic matter and its reactivity towards molecular ozone and hydroxyl radicals. Water Res. **33**(10), 2265–2276 (1999)
75. P. Westerhoff, S.P. Mazyk, W.J. Cooper, D. Minakata, Electron pulse radiolysis determination of hydroxyl radical rate constants with Suwannee river fulvic acid and other dissolved organic matter isolates. Environ. Sci. Technol. **41**, 4640–4646 (2007)
76. F. Bloch, Über die Quantenmechanik der Elektronen in Kristallgittern. Z. Physik **52**, 555–600 (1929)
77. L. Brillouin, Les électrons dans les métaux et le classement des ondes de de Broglie correspondantes. C. R. Hebd. Seances Acad. Sci. **191**, 292 (1930)
78. A.J. Dekker, *Solid State Physics* (Prentice-Hall, Englewood Cliffs, NJ, 1957)
79. H. Gerischer, An advanced treatise, in *Physical Chemistry, vol. IX*, ed. by H. Eyring, D. Henderson, W. Host (Academic Press, New York, 1970)
80. F. Lohmann, Fermi-Niveau und flachbandpotential von molekülkristallen aromatischer kohlenwasserstoffe. Z. Naturforsch. Teil. A **22**, 843–844 (1967)
81. Y.V. Pleskov, Y.Y. Gurevich, *Semiconductor Photoelectrochemistry* (Consultants Bureau, New York, 1986)

82. R. Memming, Photoinduced charge transfer processes at semiconductor electrodes and particles. Top. Curr. Chem. **169**, 105–181 (1994)
83. A.J. Bard, Photoelectrochemistry and heterogeneous photo-catalysis at semiconductors. J. Photochem. **10**, 59–75 (1979)
84. P. Meriaudeau, J.C. Vedrine, Electron paramagnetic resonance investigation of oxygen photoadsorption and its reactivity with carbon monoxide on titanium dioxide: the O3-3 species. J. Chem. Soc. Faraday Trans. **72**, 472–480 (1976)
85. W.R. Murphy, T.H. Veerkamp, T.W. Leland, Effect of ultraviolet radiation on zinc oxide catalysts. J. Catal. **43**, 304–321 (1976)
86. N. Sakai, R. Wang, A. Fujishima, T. Watanabe, K. Hashimoto, Effect of ultrasonic treatment on highly hydrophilic $TiO_2$ surfaces. Langmuir **14**, 5918–5920 (1998)
87. R. Wang, N. Wakai, A. Fujishima, T. Watanabe, K. Hashimoto, Studies of surface wettability conversion on $TiO_2$ single-crystal surfaces. J. Phys. Chem. B **103**, 2188–2194 (1999)
88. W.C. Wu, L.F. Liao, J.S. Shiu, J.L. Lin, FTIR study of interactions of ethyl iodide with powdered $TiO_2$. Phys. Chem. Chem. Phys. **2**, 4441–4446 (2000)
89. A. Emeline, A. Salinaro, V.K. Ryabchuk, N. Serpone, Photoinduced processes in heterogeneous nanosystems. From photoexcitation to interfacial chemical transformations. Int. J. Photoenergy **3**, 1–16 (2001)
90. V. Ryabchuk, Photophysical processes related to photoadsorption and photocatalysis on wide band gap solids: a review. Int. J. Photoenergy **6**, 95–113 (2004)
91. V. Augugliaro, L. Vittorio, M. Pagliaro, G. Palmisano, L. Palmisano, *Clean by Light Irradiation: Practical Applications of Supported $TiO_2$* (RSC Publishing, Cambridge, UK, 2010)
92. G. Palmisano, V. Loddo, S. Yurdakal, V. Auguliaro, L. Palmisano, Reaction pathways and kinetics of photocatalytic oxidation of nitrobenzene and phenylamine in aqueous $TiO_2$ suspensions. AIChE J. **53**, 961–968 (2007)
93. A. Gora, B. Toepfer, V. Puddu, G. Li Puma, Photocatalytic oxidation of herbicides in single-component and multicomponent systems: reaction kinetics analysis. Appl. Catal. B **65**, 1–10 (2006)
94. H. De Lasa, B. Serrano, M. Salaices, *Photocatalytic Reaction Engineering* (Springer, New York, 2005)
95. H. Ibrahim, H. De Lasa, Kinetic modeling of the photocatalytic degradation of air-bourne pollutants. AIChE J. **50**, 1017–1027 (2004)
96. A. Fujishima, X.T. Zhang, D.A. Tryk, TiO(2) photocatalysis and related surface phenomena. Surf. Sci. Rep. **63**(12), 515–582 (2008)
97. Z.B. Zhang, C.C. Wang, R. Zakaria, J.Y. Ying, Role of particle size in nanocrystalline $TiO_2$-based photocatalysts. J. Phys. Chem. B **102**(52), 10871–10878 (2012)
98. J.M. Macak, M. Zlamal, J. Krysa, P. Schmuki, Self-organized $TiO_2$ nanotube layers as highly efficient photocatalysts. Small **3**(2), 300–304 (2007)
99. A. Hu, R. Liang, X. Zhang, S. Kurdi, D. Luong, H. Huang, P. Peng, E. Marzbanrad, K.D. Oakes, Y. Zhou, M.R. Servos, Enhanced photocatalytic degradation of dyes by TiO2 nanobelts with hierarchical structures. J. Photochem. Photobiol. A: Chem. **256**(15), 7–15 (2013)
100. M. Ni, M.K.H. Leung, D.Y.C. Leung, K. Sumathy, A review and recent developments in photocatalytic water-splitting using $TiO_2$ for hydrogen production. Renew. Sust. Energ. Rev. **11**(3), 401–425 (2007)
101. N. Murakami, Y. Kurihara, T. Tsubota, T. Ohno, Shape-controlled anatase titanium (IV) oxide particles prepared by hydrothermal treatment of peroxo titanic acid in the presence of polyvinyl alcohol. J. Phys. Chem. C **113**(8), 3062–3069 (2009)
102. X.G. Han, Q. Kuang, M.S. Jin, Z.X. Xie, L.S. Zheng, Synthesis of titania nanosheets with a high percentage of exposed (001) facets and related photocatalytic properties. J. Am. Chem. Soc. **131**(9), 3152 (2009)

103. H. Kominami, K. Yabutani, T. Yamamoto, Y. Kara, B. Ohtani, Synthesis of highly active tungsten(VI) oxide photocatalysts for oxygen evolution by hydrothermal treatment of aqueous tungstic acid solutions. J. Mater. Chem. **11**(12), 3222–3227 (2001)
104. R. Lof, M. Van Veenendaal, H. Jonkman, G. Sawatzky, Band gap, excitons and Coulomb interactions of solid C 60. J. Electron Spectrosc. Related Phenomena **72**, 83–87 (1995)
105. L. Brunet, D.Y. Lyon, E.M. Hotze, P.J.J. Alvarez, M.R. Wiesner, Comparative photoactivity and antibacterial properties of C-60 fullerenes and titanium dioxide nanoparticles. Environ. Sci. Tech. **43**, 4355–4360 (2009)
106. J.M. Herrmann, Photocatalysis fundamentals revisited to avoid several misconceptions. App. Catal. B: Environ. **99**, 461–469 (2010)

# Chapter 2
# Development of $TiO_2$ Nanowires for Membrane Filtration Applications

**Robert Liang, Anming Hu, Mélisa Hatat-Fraile, and Norman Zhou**

**Abstract** $TiO_2$ membranes are multifunctional in that they provide liquid separation of both contaminated and treated water and offer photocatalytic degradation of pollutants. The use of $TiO_2$ nanowires for membranes exhibit greater photocatalytic efficiency compared to conventional bulk materials due to the higher surface area and size effects on the quantum scale. $TiO_2$ nanowires can be synthesized from hydrothermal and sol-gel methods. This chapter focuses on the synthesis, characterization, and applications of $TiO_2$ nanowire membrane. The functional features of $TiO_2$ membranes include filtration, self-cleaning, photocatalytic degradation, and disinfection.

R. Liang • N. Zhou
Centre for Advanced Materials Joining, Department of Mechanical and Mechatronics Engineering, University of Waterloo, 200 University Avenue West, Waterloo, ON, Canada N2L 3G1

Waterloo Institute for Nanotechnology, University of Waterloo, 200 University Avenue West, Waterloo, ON, Canada N2L 3G1

A. Hu (✉)
Centre for Advanced Materials Joining, Department of Mechanical and Mechatronics Engineering, University of Waterloo, 200 University Avenue West, Waterloo, ON, Canada N2L 3G1

Department of Mechanical, Aerospace, Biomedical Engineering, University of Tennessee, Knoxville, TN 37996, USA
e-mail: ahu3@utk.edu

M. Hatat-Fraile
Centre for Advanced Materials Joining, Department of Mechanical and Mechatronics Engineering, University of Waterloo, 200 University Avenue West, Waterloo, ON, Canada N2L 3G1

A. Hu and A. Apblett (eds.), *Nanotechnology for Water Treatment and Purification*, Lecture Notes in Nanoscale Science and Technology 22,
DOI 10.1007/978-3-319-06578-6_2, 

## 2.1 Introduction

Drinking water contaminated with microbiological organisms and viruses is a major cause of diseases in developing countries. There are roughly 3.8 billion human beings who have limited or no access to potable water [1]. Chemical disinfectants such as chlorine and membrane filtration systems which are deployed to control microbial pathogens are continually being questioned on their ability to remove these pathogens. Chemical disinfectants may produce harmful disinfection by-products (DBPs), and pathogens may be resistant to these disinfectants [2]. Additionally, membranes used in filtration suffer from a variety of problems, including biofouling and virus penetration [3].

Membrane filtration provides a physical barrier for undesired pollutants based on a size exclusion process, allowing one to filter various types of water sources, including wastewater, groundwater, and surface water. As an important key water treatment process, membrane filtration systems provide automation, require very little land and chemicals to setup and use, and allows for flexible modular designs to be implemented. An inherent trade-off in membrane filtration is between membrane selectivity and permeability. There exists a very high energy consumption associated with this process, which limits the application of these pressure-driven processes. Additionally, membrane fouling adds to the complexity of the process and reduces the lifetime of the membranes, along with increasing the energy consumption.

The performance of membranes for water treatment in pressure-driven processes largely depends on the membrane material. Incorporation of nanomaterials, such as nanowire and nanofibers, into membranes offers an opportunity to improve membrane properties, such as permeability, fouling resistance, and mechanical properties. Filtration systems may be improved using membranes containing nanowires or nanofibers. The high surface-to-volume ratio, small pore size, and enhanced permeability allow the nanowire membrane to remove unwanted submicron particles [3]. Furthermore, nanomaterials, such as $TiO_2$, can impart functionalities, which include contaminant degradation and self-cleaning through photocatalysis.

The focus of this chapter is on the development of $TiO_2$ nanowire membranes, including synthesis methods and the multifunctional capabilities of the technology, which include photocatalytic degradation of pollutants, disinfection, antifouling, and filtration characteristics. The $TiO_2$ nanomaterial synthesis methods that this chapter focuses on include hydrothermal and sol-gel processing methods.

## 2.2 $TiO_2$ Properties

### *2.2.1 The Lattice and Electronic Structure of $TiO_2$*

$TiO_2$ occurs naturally in the form of rutile, anatase, and brookite crystal structures as shown in Fig. 2.1. The rutile and anatase phase have a tetragonal crystal structure, whereas brookite has an orthorhombic structure with the lattice

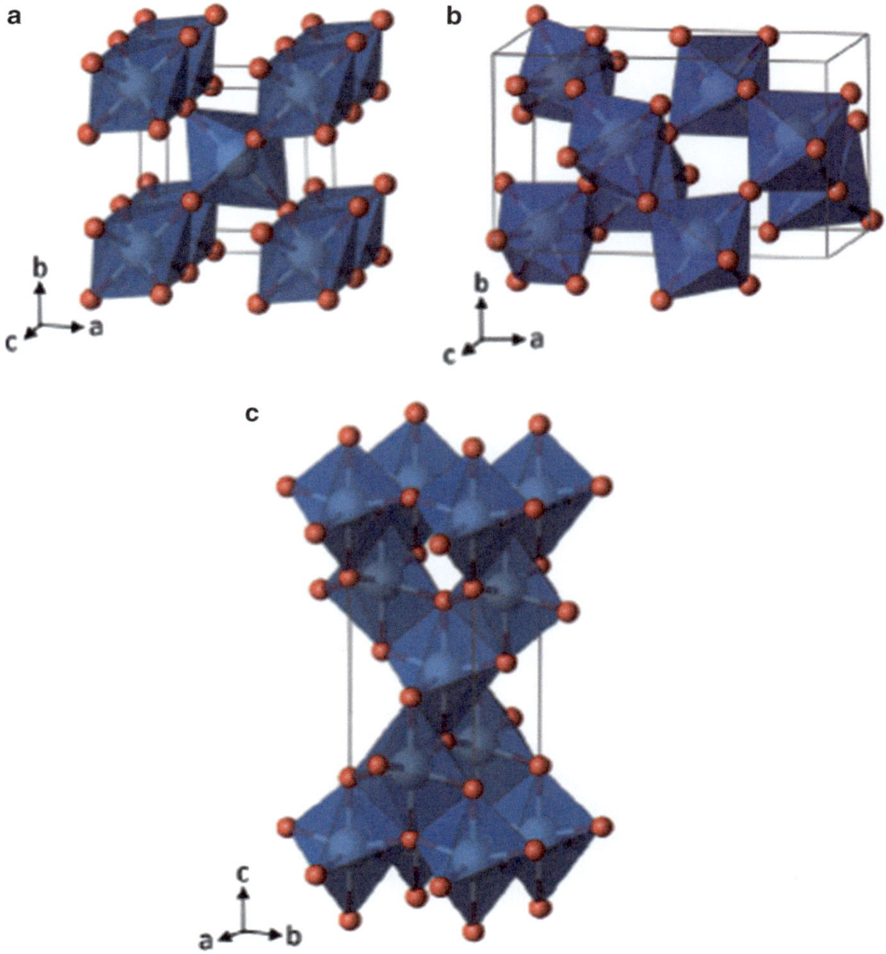


**Fig. 2.1** $TiO_6$ polyhedra for $TiO_2$ phases rutile (**a**), brookite (**b**), and anatase (**c**). (Reproduced with permission from [4])

parameters given in Table 2.1. For general nanostructured samples, anatase and brookite transformation are dependent on the initial particle sizes, which determine the thermodynamic phase stability [8, 9].

### 2.2.2 *Photocatalytic Mechanism of Pure $TiO_2$ and Doped $TiO_2$*

Photocatalysis occurs through the absorption of photon energy, $hv_1$, equal to or greater than the bandgap energy of $TiO_2$ (~3.2 eV for the anatase phase—see Sect. 2.5.4), producing an electron–hole pair on the surface of $TiO_2$ nanomaterial

**Table 2.1** Select $TiO_2$ crystal phases and their lattice parameters

| | Lattice parameter | | |
|---|---|---|---|
| Crystal phase | a (Å) | b (Å) | c (Å) |
| Rutile[a] | 4.5937 | 4.5937 | 2.9581 |
| Brookite[b] | 9.16 | 5.43 | 5.13 |
| Anatase[c] | 3.7842 | 3.7842 | 9.5146 |

[a][5]
[b][6]
[c][7]

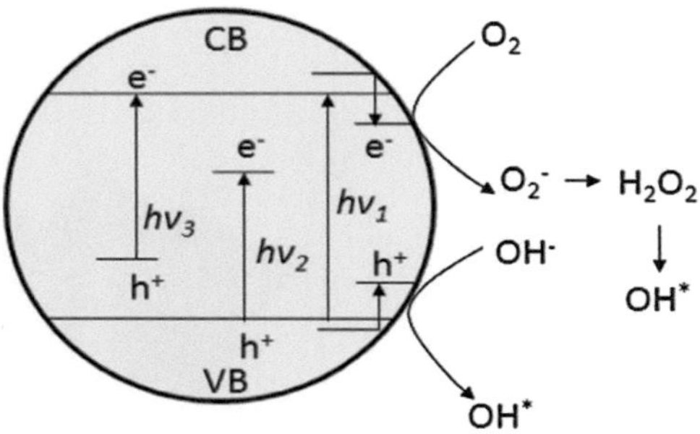


**Fig. 2.2** $TiO_2$ photocatalytic mechanisms: $h\nu_1$ pure $TiO_2$, $h\nu_2$ metal-doped $TiO_2$, $h\nu_3$ nonmetal-doped $TiO_2$

as shown in Fig. 2.2. Typically, for pure $TiO_2$, UV illumination is the primary source of photoactivity. An electron from the valence band is promoted to the conduction band (CB), leaving a hole formed in the valence band (VB). Excited electron–hole pairs have the probability of the following processes occurring: (1) electron–hole recombination and dissipation as heat, (2) electron and hole trapping in metastable surface states, or (3) reaction of electrons and holes with electron donors and acceptors adsorbed in the electrochemical double layer of the nanomaterials. After reaction with water, holes on the surface of the $TiO_2$ nanomaterial can produce hydroxyl radicals, which are very powerful oxidants. Depending on the environmental conditions, oxidants produced in this process can play a role in the photocatalytic mechanism [10, 11].

The visible light photocatalytic activity of metal-doped $TiO_2$ is due to the metal nanoparticles deposited into the $TiO_2$ matrix. As shown in Fig. 2.2, electrons are excited from the defect state, imparted by the metal doping, to the CB by a photon with energy equals $h\nu_2$. Additionally, transition metal doping improves the trapping of electrons and decreases the occurrence of electron–hole recombination during irradiation, which results in enhanced photoactivity.

In nonmetal-doped $TiO_2$, there are various perspectives regarding the mechanism in which doping imparts visible light photoactivity, which include: (1) bandgap narrowing, (2) impurity energy levels, and (3) oxygen vacancies. Taking nitrogen doping as an example, for bandgap narrowing, in nitrogen-doped anatase $TiO_2$ (N-$TiO_2$), the N 2p state overlap O 2p states since their energies are very close, so the bandgap energy of N-$TiO_2$ is narrowed allowing for visible light absorbance [12]. By introducing nitrogen atoms into the $TiO_2$ matrix, the oxygen sites in $TiO_2$ are substituted with nitrogen atoms in the form of impurity energy levels above the valence band. Under UV irradiation, electrons are excited in the VB and impurity energy levels, $h\nu_3$; but under visible irradiation, electrons are only excited from impurity energy levels [13]. Another perspective on the mechanism of $TiO_2$ visible light photoactivity is that oxygen vacancies or deficient sites are formed at grain boundaries and are important in imparting visible photoactivity in nitrogen-doped $TiO_2$ as a blocker for re-oxidation [14].

## 2.3 Preparation of $TiO_2$ Powder Nanomaterials

### *2.3.1 Hydrothermal Synthesis*

Hydrothermal synthesis of $TiO_2$ nanoparticles [15–17] or nanorods [18] are carried out in water or in mixtures water/ethanol and, generally, pressure autoclaves are used to control the temperature and pressure. The technique allows for the crystallization of precursors from the use of aqueous solutions at an elevated temperature and high vapor pressures. The crystal growth synthesis occurs in an autoclave, in which $TiO_2$ precursor is used in a growth solution whereby nanoparticles, nanorods, nanotubes, and nanowires of $TiO_2$ can be prepared, as seen in Table 2.2, via various synthesis protocols.

#### 2.3.1.1 Chemical Process for Formation of 1D $TiO_2$ Nanomaterials

One-dimensional nanomaterials (nanowires, nanobelts, and nanotubes) produced from hydrothermal synthesis [20, 21] occur due to a phase transformation. All the 1D nanostructures originate from an alkaline hydrothermal process. $Na_2Ti_3O_7$ nanobelts and nanotubes are synthesized through the following chemical reaction:

$$3TiO_2 + 2NaOH \rightarrow Na_2Ti_3O_7 + H_2O \tag{2.1}$$

The $TiO_2$ precursor of anatase phase is a $TiO_6$ octahedral structure. During the hydrothermal process in Eq. (2.1), some of the Ti–O–Ti bonds break from the $TiO_2$

**Table 2.2** Preparation of $TiO_2$ nanomaterials

| Nanomaterial | Preparation technique | References |
|---|---|---|
| Nanoparticles | Titanium tetraisopropoxide in water/ethanol solution at pH 0.7 using nitric acid at 353 K for 4 h. | [15–17] |
| Nanorods | $TiCl_4$ in water at 606–696 K for 12 h in acids or inorganic salts | [18] |
| Nanotubes | P25 precursor in 403 K for 48 h in 10 M NaOH at 423 K | [19] |
| Nanowires/nanobelts | P25 precursor in 453 K for 72 h in 10 M NaOH | [20, 21] |

precursor and a six-coordinated monomer, $[Ti(OH)_6]^{2-}$, is formed [22]. During the hydrothermal process, the alkaline solution is saturated and the $[Ti(OH)_6]^{2-}$ monomer is unstable and combines via oxolation or olation, forming a nuclei. As these nuclei grow, they become thermodynamically stable when the size of the nuclei exceeds the critical nuclei size. Thin nanosheets, composed of layer unit cells, can form during the growth process.

The growth of these nanosheets is anisotropic, with the growth along the *b*-axis being the fastest, which leads to the formation of 1D $Na_2Ti_3O_7$ nanostructures. The crystal structure of $Na_2Ti_3O_7$ is a monoclinic structure with layers of $[TiO_6]$ octahedral sites with shared edges and vertices. The $Na^+$ cations of these nanosheets are located between the $[TiO_6]$ layers.

Hydrogen deficiencies exist on the surface of the nanosheets, which increases surface tension and the propensity for the surface layer to bend. The surface strain energy becomes larger due to increasing hydrogen-deficient sites to the point when the surface layer can overcome coupling between layers and peel off from the layer underneath. The morphologies of the nanostructures depend on various parameters. The synthesis of $Na_2Ti_3O_7$ nanotubes occurs at lower synthesis temperatures; this is due to the hydrogen deficiency in the surface layer. On the other hand, autoclaving at higher (180 °C) temperatures leads to an increased growth rate of nanosheets, which form long micron length nanobelts. This process may require larger strain energy than at lower temperatures, but the strain energy is not sufficient to overcome layer-by-layer coupling.

The $Na_2Ti_3O_7$ conversion to $H_2Ti_3O_7$, which is a related structure, requires an ion exchange process. When $Na_2Ti_3O_7$ is immersed and washed with dilute hydrochloric acid solution, the $Na^+$ ions in the $TiO_2$ matrix are replaced by $H_3O^+$ ions to form $H_2Ti_3O_7$:

$$Na_2Ti_3O_7 + 2HCl \rightarrow H_2Ti_3O_7 + 2NaCl \tag{2.2}$$

Following the conversion of $Na_2Ti_3O_7$ to $H_2Ti_3O_7$ in Eq. (2.2), anatase may be obtained by annealing the $H_2Ti_3O_7$ at 700 °C for 2 h [20] through crystal lattice rearrangement as given by Eq. (2.3):

$$H_2Ti_3O_7 \rightarrow 3TiO_2 + H_2O \tag{2.3}$$

Despite the heat treatment, anatase retains the 1D structure of $H_2Ti_3O_7$. The lattice distortion during heat treatment is due to rotation and rearrangement of the $[TiO_6]$ octahedral; this occurs during the $H_2Ti_3O_7$ to anatase conversion process.

### 2.3.2 *Sol-Gel Method*

The sol-gel technique is a method to create colloidal suspensions, or sol, obtained from liquid solutions of organometallic or inorganic precursors, which result in highly pure and homogeneous materials. The stability of a sol suspension is dependent on the van der Waals attraction and Coulombic repulsive forces among very small particles. The transformation of a sol into another phase (the gel) can be induced under various experimental conditions. The gel is a viscous solid that surrounds molecules of the chosen solvent. By drying the gel, it is possible to obtain porous solids and films.

In a typical sol-gel process to produce $TiO_2$, the sol is the result of hydrolysis and polymerization reactions of $TiO_2$ precursors, generally an organic titanium alkoxide (tetra-*n*-butoxide and tetraisopropoxide). The precursor is dissolved in an alcohol solvent (ethanol, 2-propanol, 1-pentanol, etc.) and then added to water, in Eq. (2.4), to induce hydrolysis to titanium hydroxide:

$$\mathrm{Ti(iso\text{-}OC_3H_7)_4 + 4H_2O \rightarrow Ti(OH)_4 + 4C_3H_7OH} \tag{2.4}$$

The size for the $TiO_2$ particles depends on the experimental and process parameters used, such as the molar ratio between precursor and water, initial pH, reaction time, presence of external ionic species, and temperature. The polymerization reaction and the loss of solvent transforms the liquid sol into a solid gel phase culminating in the formation of Ti–O–Ti chains. These chains are facilitated by the low content of water, low hydrolysis rate, and excess of precursor in the reaction.

The sol-gel method is a versatile approach to preparing $TiO_2$ with certain morphological properties and removal of organics from $TiO_2$ gels plays an important role in the preparation of $TiO_2$ samples on substrates, as residual organic moieties can affect photocatalytic efficiency [23, 24]. The gelling process can be carried out by treating the reactant mixtures by reducing pressure or adding various additives, such as chelating agents (HCl or $SnCl_2$) which allow one to control hydrolysis and condensation reaction rates [25].

### 2.3.3 Chemical Doping

To increase the photocatalytic rate in the visible light region, there are many approaches proposed to dope impurities into $TiO_2$, including ion implantation, chemical vapor deposition, plasma deposition, and dopant incorporation in sol-gel synthesis methods. Table 2.3 gives a list of selected publications of metal and nonmetal doping into the $TiO_2$ matrix.

The addition of noble metals via various methods such as solution-based processing and sputtering into the $TiO_2$ matrix can change the surface properties, thereby changing the efficiency of photocatalysis. Metals (Ag, Au, Pt, Pd, V, and Fe) can enhance the rate of the photocatalytic reaction; this was observed for the photoconversion of $H_2O$ to $H_2$ and $O_2$ using the $Pt/TiO_2$ system [36]. After excitation of the electron via visible light illumination, the electron travels to the metal where it is trapped, and thus electron–hole recombination is suppressed [34]. When the semiconductor and metal come in contact, the Fermi levels of these two materials align causing electrons to flow from the semiconductor to the metal. The decrease in

**Table 2.3** Chemical doping of $TiO_2$ via nonmetal and metal additions

| Type of dopant | Doped element | Preparation method | Reference |
|---|---|---|---|
| *Non metal* | B | *Mechanical mixing*: anatase $TiO_2$ powder with boric acid triethyl ester was grounded and calcined in air at 723 K | [26] |
| | C | *Solution-based processing (sol-gel)*: titanium (IV) isopropoxide was dissolved in alcohol and hydrochloric acid solution. The sol gel was aged for several days and calcined in air for 3 h at 338 K and 3 h at 523 K and grounded | [27] |
| | P | *Solution-based processing (sol-gel)*: titanium (IV) isopropoxide was hydrolyzed with isopropanol and water. Phosphoric acid was added after hydrolysis and the dispersion was stirred for 2 h, centrifuged, and dried at 373 K. The powder was calcined at 573 K | [28] |
| | N | *Chemical vapor deposition*: anatase $TiO_2$ powder was treated in ammonia (67 %) in argon at a temperature of 873 K for 3 h | [29] |
| | S | *Thermal process*: oxidation annealing of titanium disulphide at 573–873 K | [30] |
| *Metal* | Ag | *Solution-based processing (sol-gel)*: silver nitrate was mixed in sodium citrate tribasic dehydrate (reducing agent) at 353 K under continuous stirring. Titanium(IV) isopropoxide and nitric acid was added and the reaction maintained at 323 K for 24 h. The prepared sol was dried at 378 K for 24 h and calcined at 573 K | [31] |
| | Au | *Solution-based processing (sol-gel)*: titanium(IV) butoxide in ethanol was added to a solution of tetrachloroauric acid, acetic acid, and ethanol. The resulting suspension was aged for 2 days and dried, ground, and calcined at 923 K | [32] |
| | Fe | *Reactive Magnetron Sputtering*: a titanium target and iron were placed in a reaction chamber where argon and oxygen were introduced into the chamber during discharge | [33] |
| | Pt | *Photoreduction processing*: $TiO_2$ was suspended in methanol containing hexachloroplatinic acid. The suspension was irradiated with a mercury lamp (125 W) for an hour. Pt-$TiO_2$ was separated via filtration, washed with distilled water, and dried at 373 K for 24 h | [34] |
| | V | *Solution-based processing (sol-gel)*: a solution containing vanadyl acetylacetonate dissolved in *n*-butanol and mixed with a solution containing acetic acid in titanium butoxide. The mixture was hydrolyzed for 24 h by water generated via the esterification of acetic and butanol groups. The suspension was dried at 523 K, ground, and calcined at 673 K for 0.5 h | [35] |

electron concentrations in the semiconductor increases the hydroxyl group acidity and affects the photocatalytic activity.

Additionally, $TiO_2$ can also be doped with nonmetals to increase visible light photoactivity. Nonmetals such as boron, carbon, phosphorus, nitrogen, and sulphur have been shown to enhance the degradation rate compared to pure $TiO_2$ anatase under visible light irradiation due to impurity trapping and bandgap narrowing as discussed in Sect. 2.2.2. These nonmetals can be incorporated via solution-based processing, mechanical mixing, and chemical vapor deposition.

## 2.4 $TiO_2$ Membrane Fabrication Methods

### 2.4.1 *Dip Coating on Substrates*

Dip coating is carried out by immersing a support in a liquid in which the precursor of a material of interest is present, and then withdrawing the support at a controlled speed and temperature under atmospheric conditions. The process has generally been developed for small surface for lab bench and research purposes. The process has been used to coat $TiO_2$ nanomaterials for plate glass, solar energy systems, anti-reflecting coatings on windows, and photocatalytic films [37].

The process of dip coating can be used to coat substrates with $TiO_2$ films, which are a few nm to 1 μm in thickness (see Sect. 2.3.2). The thickness can be obtained with high precision by selecting a suitable viscosity of the coating liquid and withdrawal rate according to the Landau–Levich equation [38]:

$$t = 0.94 \frac{(\mu \cdot U)^{2/3}}{\gamma_{LV}^{1/6} (\rho g)^{1/2}} \tag{2.5}$$

where $t$ is the coating thickness, $\mu$ is the viscosity, $\gamma_{LV}$ is the liquid–vapor surface tension, $\rho$ is the solution density.

Equation (2.5) can be applied under the following conditions: (1) the withdrawal rate must be $U > 1\ \text{mm s}^{-1}$ and (2) particles cannot repel each other. Additionally, the ambient environment controls the evaporation of both the solvent and the subsequent transformation of the sol into the film. The thickness of the layer also depends on the dipping angle between the support and the liquid surface and different thickness can be produced at the top and bottom sides of the surface that is coated.

### 2.4.2 *Electrospinning on Substrates*

Electrospinning is a production technique to produce continuous ultrafine fibers (with diameters of 10 nm to 10 μm) based on forcing a polymer melt or organic-based solution through a spinneret with an electrical driving force. The technique is comprised of three components: a high voltage source, a metallic collecting plate, and a capillary tube equipped with a needle of small diameter. In this process, an electrically charged jet of the spinning solution is produced out of the nozzle or needle when a high voltage is applied between two electrodes connected to the outlet. The spinning solution is ejected from the nozzle forming a charged jet that continually deposits onto the collector electrode until no spinning solution, supplied from an infusion pump, remains.

Electrospinning can be used to create $TiO_2$ nanofibers for physical separation applications. It is possible to create nanofibrous membranes using this technique in order to generate higher porosities and interconnected porous structures that are more water permeable than conventional counterparts [39, 40]. Nanofibers have a much higher surface-to-volume ratio than conventional microfibers, providing efficient separation of particulates in a solution. Electrospinning also offers opportunities to fine tune surface functionality through polymer chemistry. Electrospun nanofibers can be tailored for many applications where the diameter, composition, morphology, and spatial alignment can be changed [40]. Even though this technique is a simple method, the design of functional nanofibers for filter membranes requires specific and controlled parameters for reproducible and controlled electrospinning.

Although nanofiber membranes have been previously used for air filtration applications, there is potential in using these membranes in water treatment for the removal of micro-sized particles from the liquid phase at a high rejection rate without substantial fouling [41]. These nanofiber membranes have been proposed to be used as pretreatment step prior to ultrafiltration and reverse osmosis. Functional nanomaterials can be created from spinning solutions with dopants to produce impregnated nanofibers or formed in situ. The tunable process allows electrospun nanofibers to be an effective research and development platform for constructing multifunctional membrane filters by either using multifunctional materials, such as $TiO_2$, or by introducing functional materials onto the nanofibers network through a deposition process. For instance, by incorporating specific chelating agents on a $TiO_2$ nanofiber matrix, $TiO_2$ nanofiber membranes can be designed to selectively remove, or have an affinity towards, heavy metals and organic pollutants during filtration.

## 2.4.3 Mechanical Pressing

### 2.4.3.1 Self-standing $TiO_2$ Nanowire Membranes

$TiO_2$ nanobelt membranes can be fabricated through filtration and hot-press method [42]. To obtain these membranes, a $H_2Ti_3O_7$ nanowire suspension is deposited on filter paper using vacuum filtration. The deposited $H_2Ti_3O_7$ nanowire cake is sandwiched between alumina disks and pressed under uniaxial pressure at a temperature of 200 °C. The resulting $TiO_2$ filter cakes are fired at 700 °C for 1 h to anneal the $H_2Ti_3O_7$ monoclinic phase to a $TiO_2$ anatase phase.

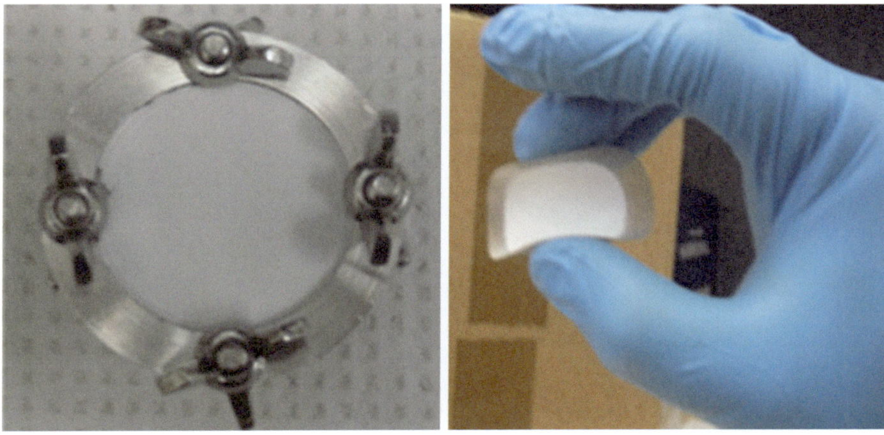

**Fig. 2.3** $TiO_2$ nanowire membrane on 316 stainless steel holder (*left*) and $TiO_2$ nanowire supported on 316 stainless steel mesh (*right*)

#### 2.4.3.2 $TiO_2$ Nanowire Membranes Immobilized on Stainless Steel Supports

$TiO_2$ nanowires can be immobilized on a stainless steel mesh forming a porous $TiO_2$ active layer on top of an inactive support. A photocatalytic membrane in a reactor system is an attractive solution to particle separation because it can serve multiple functions, such as the photocatalytic degradation of organic compounds and physical separation of target contaminants and reaction intermediates. The $TiO_2$ membranes shown in Fig. 2.3 are created via filtration of a 100 mg $TiO_2$ nanowire layer on a stainless steel 304 (200 × 200 mesh) support and mechanical hot pressing under uniaxial pressure at 200 °C. The fabrication method has advantages of (1) creating a homogeneous distribution of $TiO_2$ nanowires, (2) controlling the membrane thickness by suspension concentration and volume of the suspension filter, and (3) obtaining membrane flexibility.

## 2.5 Characterization of 1D $TiO_2$ Nanowires

### *2.5.1 Field Emission Scanning Electron Microscopy*

The field emission scanning electron microscopy (FESEM) images (Fig. 2.4) were obtained from a ZEISS LEO 1550 FESEM at an accelerating voltage of 10 kV. Figure 2.4 show hierarchical $TiO_2$ nanowires with widths ranging from 30 to 100 nm and lengths in the range of tens of μm. The hierarchical nanowires overlap with each other, and contain nanoparticles and truncated rods fused onto the nanobelt structure, thereby promoting the integrity of the porous structure.

**Fig. 2.4** FESEM images of hierarchal $TiO_2$ nanowires: (**a**) low magnification of $TiO_2$ nanobelts and (**b**) high magnification. (Reproduced with permission from [20])

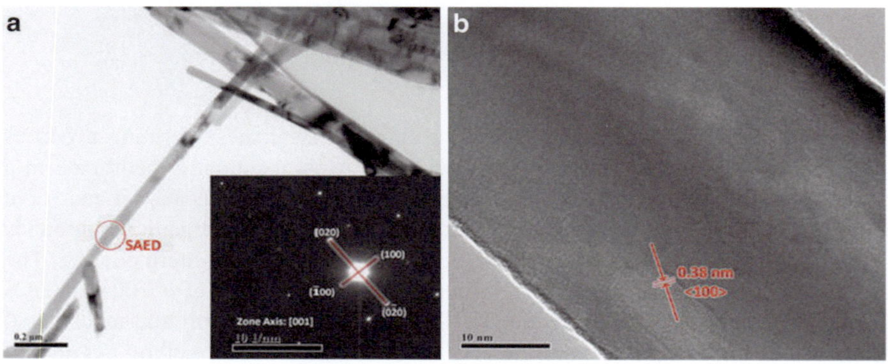


**Fig. 2.5** HRTEM images of hierarchical $TiO_2$ nanobelts: (**a**) low magnification of single nanobelt with indexed SAED pattern and (**b**) high resolution of nanobelts with crystal d-spacing of 0.38 nm. (Reproduced with permission from [20])

### *2.5.2 High-Resolution Transmission Electron Microscopy*

Using high-resolution transmission electron microscopy (HRTEM) from a JEOL 2010F instrument, the detailed lattice structure of $TiO_2$ nanowires is shown in Fig. 2.5. Figure 2.5a shows a nanowire with an indexed selected area electron diffraction (SAED) pattern obtained using a zone axis of [001]. The indexed SAED pattern indicates that the crystal structure can be attributed to the anatase phase. Additionally, the growth direction of the nanobelts is in the {100} family of directions. Figure 2.5b depicts that the d-spacing of the crystal lattice is 3.8 Å, corresponding to the <100> family of planes in the crystal structure.

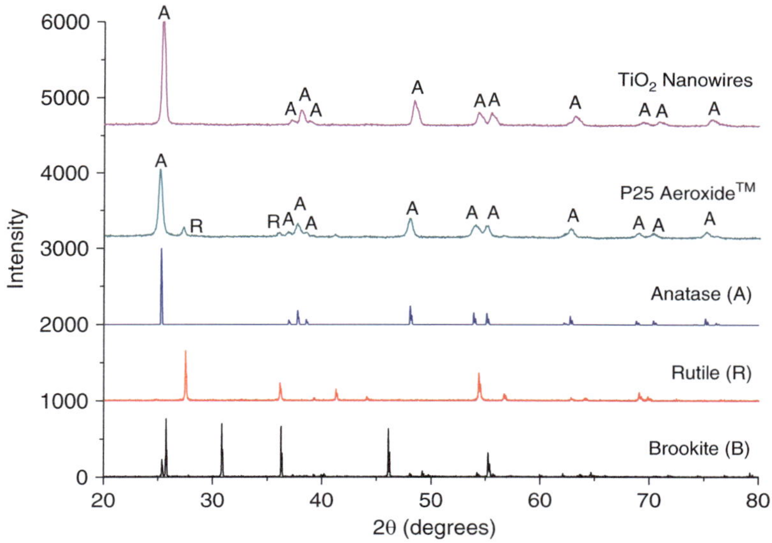


**Fig. 2.6** XRD patterns of $TiO_2$ nanowires annealed at 700 °C, P25 Aeroxide™, anatase, rutile, and brookite

## 2.5.3 X-ray Diffraction

The X-ray diffraction (XRD) technique was used to verify the crystal structure contained in both $TiO_2$ nanowires and P25 Aeroxide™ in Fig. 2.6. Powder XRD was conducted on a Rigaku SA-HF3 X-ray diffractometer using Cu Kα radiation (1.54 Å) X-ray source, operating at a voltage of 50 kV. There are several characteristic anatase diffraction peaks that are seen in both samples, which come from (101), (004), (200), (105), (211), (204), (116), (220), and (215) planes. However, P25 Aeroxide™ contains peaks associated with the rutile phase, as seen in the XRD spectra, indicated by the (110), (101), and (111) rutile diffraction peaks.

## 2.5.4 Bandgap Energy Determination

The bandgap energy of a semiconductor can be calculated from Tauc plots of $(\alpha h v)^n$ vs. photon energy ($hv$), where $\alpha$ is the absorption coefficient, $h$ is the Planck constant, $v$ is the light frequency, and $n$ depends on the characteristic of the transition in the semiconductor, where $n = 2$ for direct transitions and $n = 0.5$ for indirect transitions. The absorption coefficient is estimated from Eq. (2.6):

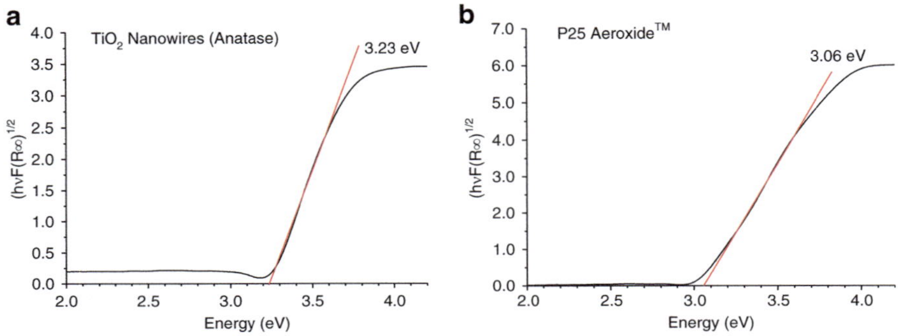


**Fig. 2.7** Tauc plots of (**a**) $TiO_2$ nanowires and (**b**) P25 Aeroxide™

$$\alpha = F(R_\infty) = \frac{(1 - R_\infty)^n}{2R_\infty} \tag{2.6}$$

where $R_\infty$ is the reflectance of an infinitely thick sample with respect to a reference at each wavelength.

$TiO_2$ semiconductors have bandgaps that are indirect ($n = 0.5$). The bandgap energy ($E_g$) can be estimated from the Tauc plot by determining the intercept of the tangent at the inflection point.

The bandgap energy in Fig. 2.7 is obtained from a diffuse reflectance spectra using a Shimdazu UV-2501 PC spectrophotometer equipped with an integrating sphere. Figure 2.7 depicts Tauc plots of $TiO_2$ nanowires and its precursor material, P25 Aeroxide™. The bandgap of the $TiO_2$ nanowires is 3.23 eV because of its pure anatase structure, whereas the bandgap for P25 is narrower at 3.06 eV because it contains rutile phase [43] as shown by XRD in Sect. 2.4.3.

### 2.5.5 *Surface Photovoltage*

The surface photovoltage (SPV) and corresponding phase spectra measurements are carried out on a measurement system that consists of a sample chamber, a lock-in amplifier with a light chopper, a source of monochromatic light, and a monochromator. The monochromator and the lock-in amplifier are connected to a computer for data acquisition. The monochromatic light is chopped at a frequency in the range of tens of Hz, and its intensity depends on the energy distribution of the lamp spectrum. The configuration of the electrode in the SPV measurement is a sandwich-like structure consisting of indium tin oxide (ITO)-doped glass sandwiching the sample layer [44].

Generally, SPV spectra can be measured by a lock-in amplifier as the co-phase and phase photovoltage signals shifted by 90° ($U_x$ and $U_y$ signals, respectively),

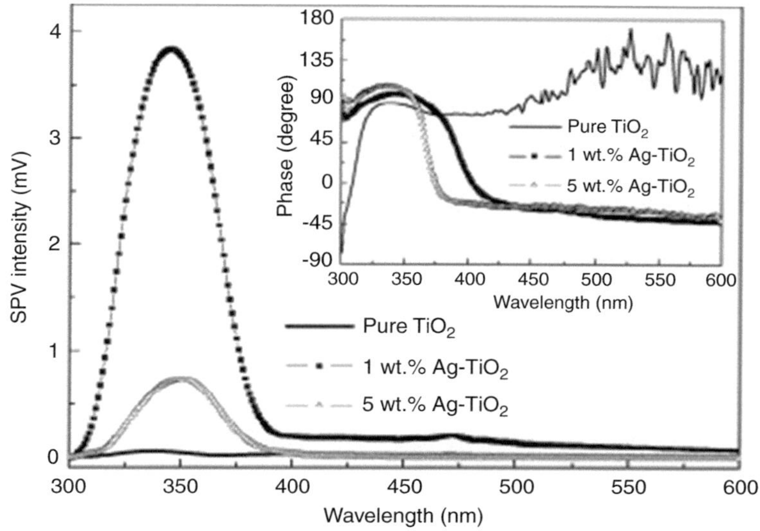


**Fig. 2.8** SPV spectra of pure $TiO_2$, 1 wt% Ag-$TiO_2$, and 5 wt% Ag-$TiO_2$ for different wavelength from 300 to 600 nm (inset: SPV phase spectra of the three samples). (Reproduced with permission from [44])

with respect to the chopped light. The signals can also be presented as the SPV amplitude ($U_R$), Eq. (2.7), and phase ($\theta$), Eq. (2.8), represented as:

$$U_R = U_x^2 + U_y^2 \tag{2.7}$$

$$\tan\theta = \frac{U_y}{U_x} \tag{2.8}$$

A positive $U_x$ signal represents the surface being positively charged under illumination. As $U_x = U_R \cos\theta$, this means the phase angles will be in the range $-90°$ to $90°$. In a single SPV process, the phase angles are in the range $-90°$ to $0°$ for an n-type semiconductor and 90–180° for a p-type semiconductor. When $U_x$ is used in the SPV experiments, only the net result of charge carrier transfer properties is recorded, but usually at the same there may be more than one process that can affect the SPV response.

The lock-in-based SPV measurements characterizes the light absorption, charge separation, and charge transport properties of the photoactive material. In a study by Ma et al. [44]. The SPV spectra of three $TiO_2$ samples are compared to address the behaviors of photogenerated charges upon illumination, as shown in Fig. 2.8. The sample containing 1 wt% Ag/$TiO_2$ possesses the highest SPV intensity response, which indicates an improved separation of photogenerated electrons and holes, or less recombination. Additionally, the UV response below 400 nm shows broad SPV tailing towards the visible region for the 1 wt% Ag/$TiO_2$, which is attributed to the plasmonic effect of Ag nanoparticles adsorbed at the surface of the $TiO_2$.

**Table 2.4** Separation factors of the $TiO_2$ nanowire membrane using polystyrene microspheres

| Polystyrene diameter (μm) | Separation factor (%) | Error value (%) |
|---|---|---|
| 0.05 | 89.5 | 1.9 |
| 0.1 | 96.3 | 1.3 |
| 0.2 | 98.5 | 0.4 |
| 0.5 | 99.2 | 0.1 |
| 1 | 99.7 | 0.1 |
| 2 | 99.9 | 0.1 |

Reproduced with permission from [48]

The corresponding inset figure indicates that the net charge transfer direction during illumination [45] accumulates towards the ITO electrode below 360 nm. From the inset figure, at the Ag absorption band, a positive SPV is obtained. Due to plasmon resonance, the Ag nanoparticles can be photoexcited by wavelengths of light from 390 to 600 nm, and the photoexcited electrons from the Ag nanoparticles are partially transferred to the $TiO_2$ conduction band [46], where the electrons will further diffuse towards the $TiO_2$ bulk, increasing the surface potential. The holes remaining in the Ag nanoparticles are expected to positively contribute in photocatalytic activity. However, excess metal loading on the semiconductor surface will detrimentally affect photoactivity due to an increased probability of charge recombination [47] as seen with 5 wt% Ag/$TiO_2$ sample.

### 2.5.6 *$TiO_2$ Nanowire Pore Size Distribution*

To identify the pore size distribution of $TiO_2$ nanowire membranes, aqueous suspensions of polystyrene (PS) microspheres (0.05, 0.1, 0.2, 0.5, and 1 μm in diameter, respectively) can be used. The separation factors of the $TiO_2$ nanowire membrane were determined by Zhang et al. [48]. The concentration of PS microspheres in the feed and permeate was detected via total organic carbon (TOC) analyzer. The separation factor is given by Eq. (2.9):

$$\text{S.F.} = \left(1 - \frac{C_{\text{permeate}}}{C_{\text{feed}}}\right) \times 100\% \tag{2.9}$$

where $C_{\text{permeate}}$ is the polystyrene solution concentration of the permeate collected, $C_{\text{feed}}$ is the polystyrene concentration of the original feed.

The results in Table 2.4 indicate that the separation factors for PS microspheres of different sizes (0.2, 0.5, 1 and 2 μm) were more than 99 %, indicating that the $TiO_2$ nanowire membrane is able to reject these microspheres from the permeate. Decreasing the PS microsphere diameter decreased the $TiO_2$ membrane separation factor. The pore size of a $TiO_2$ nanowire membrane can be defined as the diameter of latex microspheres which are 90 % retained by the $TiO_2$ membrane and, in this case, the pore size is approximately 0.05 μm.

### 2.5.7 $TiO_2$ Adsorption and Photocatalytic Models

#### 2.5.7.1 Adsorption Model

To evaluate the adsorption mechanism, kinetic data for adsorption can be modeled using a pseudo second-order rate equation given by Eqs. (2.10) and (2.11):

$$\frac{t}{q_t} = \frac{1}{kq_e^2} + \frac{1}{q_e}t \tag{2.10}$$

$$q_t = \frac{C_0 - C_t}{C_0} \tag{2.11}$$

where $t$ is the time (min), $q_t$ is the adsorption capacities at time $t$ (g g$^{-1}$), $q_e$ is the equilibrium adsorption capacity (g g$^{-1}$), $k$ is the initial adsorption rate constant (g g$^{-1}$ min$^{-1}$), $C_0$ is the initial concentration (mol/L), $C$ is a concentration (mol/L), $k_{ap}$ is the apparent rate constant (min$^{-1}$), and $t$ is time (min).

The values of $k$ and $q_e$ are obtained from the linear plot of $t/q_t$ vs. t, and if the fit of the data is linear, it suggests that chemisorption takes places [49, 50].

#### 2.5.7.2 Photocatalytic Degradation Model

The photocatalytic degradation can be described using a pseudo first-order kinetic model (Eq. (2.12)) and its integrated form (Eq. (2.13)):

$$-\frac{dC}{dt} = k_{ap}C \tag{2.12}$$

$$\ln(C/C_0) = k_{ap}t \tag{2.13}$$

where $C_0$ is the initial concentration (mol/L), $C$ is a concentration (mol/L), $k_{ap}$ is the apparent rate constant (min$^{-1}$), and $t$ is time (min).

A $\ln(C/C_0)$ vs. $t$ plot and a line of best fit yields $k_{ap}$.

## 2.6 Applications in Water Treatment

### 2.6.1 Photocatalytic Degradation of Emerging Pollutants

Pharmaceuticals and personal care products (PPCPs) are a group of emerging pollutants that are being release into the environment by industrial and urban waste, without having been regulated due to a lack of research regarding their

**Table 2.5** List of pharmaceuticals and personal care products

| Emerging contaminants | Examples |
|---|---|
| *Pharmaceuticals* | |
| Antibiotics | Sulfamethoxazole and trimethoprim |
| Analgesics | Aspirin, ibuprofen, and acetaminophen |
| Lipid regulators | Atorvastatin |
| Mood regulators | Fluoxetine |
| Antiepileptics | Carbamazepine |
| *Personal care products* | |
| Detergents | sodium dodecyl sulphate |
| Insect repellents | *N,N*-diethyl-*m*-toluamide (DEET) |
| Fragrances | Nitro and polycyclic compounds |
| Antimicrobial topical agents | Triclosan and triclocarban |

occurrence and environmental effects [21, 51]. These pharmaceuticals vary in their removability in wastewater treatment effluents based on their properties. Conventional drinking water treatment techniques, such as coagulation/flocculation, filtration, and sedimentation have been shown to be ineffective against treating some pharmaceuticals below detection limits [52–54].

The increasing interconnectedness between water treatment operations and source waters also raises concern as to whether trace contaminants can escape treatment and are able to enter the water supply. PPCPs are not regulated in many nations, but there is a broad concern of their presence in water, as there is evidence of low concentrations of PPCPs and endocrine disrupting compounds (EDCs) in community source waters in developed nations [55–57].

PPCPs, as seen in Table 2.5, can be found in wastewater effluents due to ingested pharmaceuticals excreted via human waste and entering the wastewater system. Personal care products can be washed from individuals during hygienic activities. The effluent discharge from wastewater treatment plants (WWTPs) is a major source of many PPCPs in the environment. Additionally, other sources of PPCPs include untreated animal waste, pesticides, agricultural runoffs, and manufacturing waste.

Additionally, another class of emerging contaminants includes EDCs, which have the capability of stimulating or repressing hormonal function in the human endocrine system. EDCs can interfere with sex hormones, hormones that control metabolism, or other systems in the body. EDCs can come from natural hormones that are excreted, ingested hormones (i.e., birth control pills), or synthetic compounds that function as hormones (e.g., bisphenol A).

Section 2.6.1 describes the adsorption (Sect. 2.6.1.1) and photocatalytic degradation (Sect. 2.6.1.2) by $TiO_2$ nanowires on selected PPCPs. Additionally, DBPs may be removed using the $TiO_2$/UV process (Sect. 2.6.1.3) because of concerns relating to the carcinogenic potential of these compounds.

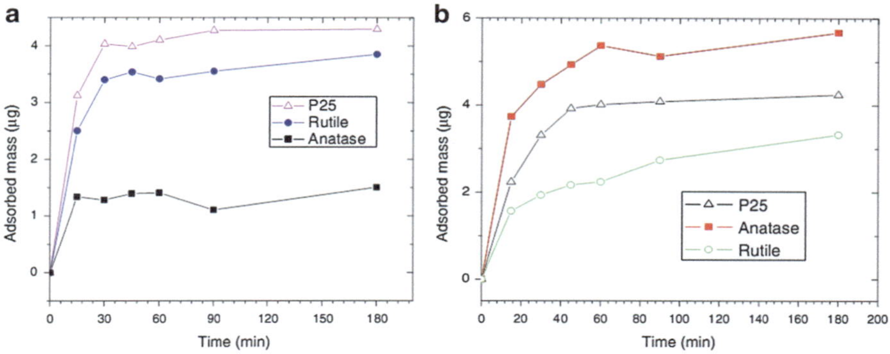


**Fig. 2.9** Adsorption kinetics of (**a**) fluoxetine and (**b**) naproxen at 100 mL solution dispersed with 100 mg of the following nanomaterials: P25, anatase nanowires, and rutile nanowires. (Reproduced with permission from [21])

#### 2.6.1.1 Adsorption Kinetics of Selected Pharmaceuticals on $TiO_2$ Nanowires

The surface adsorption response kinetics are illustrated in Fig. 2.9 for (1) fluoxetine and (2) naproxen on three kinds of $TiO_2$ nanomaterials: (a) P25Aeroxide™, (b) rutile nanowires, and (c) anatase nanowires. The adsorption reaches equilibrium after 20–40 min using 20 µg of pharmaceutical in 100 mL. Additionally, from the adsorption kinetics P25 strongly adsorbs fluoxetine, while anatase nanowires are mostly effective to adsorb naproxen. According to the specific area determined by BET method, rutile is most effective adsorbent of fluoxetine per unit of surface area. The saturated adsorption of fluoxetine is given as 2.034 µg $m^{-2}$ for rutile, 0.918 µg $m^{-2}$ for anatase, and 0.806 µg $m^{-2}$ for P25. For naproxen the saturate adsorption per unit of surface area is 1.296 µg $m^{-2}$ for rutile, 3.518 µg $m^{-2}$ for anatase, and 0.784 µg $m^{-2}$ for P25. There is a significant enhancement of surface adsorption, normalized by the surface area, by nanowires relative to commercial $TiO_2$ nanoparticles (e.g., P25).

#### 2.6.1.2 Photocatalytic Degradation of Pharmaceuticals

The degradation responses, using first-order kinetics, have been previously observed for PPCPs and bioactive environmental analytes with an initial concentration of 100 ppb given in Table 2.6. From the analysis of 13 pharmaceuticals, a fairly benign degradation rate was observed for carbamazepine and ibuprofen, which are often, passed through WWTPs near surface waters of urban centers [59, 60]. Despite the low performance of carbamazepine and ibuprofens, the photocatalytic degradation associated with $TiO_2$ nanowires can still effectively remove many PPCP pollutants from drinking water sources. To evaluate the photocatalytic activities of these

**Table 2.6** Photocatalytic degradation using $TiO_2$ nanowires analyzed by pseudo first-order kinetics for individual pharmaceuticals with an initial concentration of 100 ppb

| Compound | $k_{ap}$ | $r^2$ |
|---|---|---|
| Norfluoxetine | 0.1239 | 0.999 |
| Atorvastatin | 0.0688 | 0.999 |
| Lincomycin | 0.043 | 0.999 |
| Fluoxetine | 0.0408 | 0.995 |
| Venlafaxin | 0.0319 | 0.997 |
| Sulfamethoxazole | 0.0422 | 0.989 |
| Diclofenac | 0.0398 | 0.999 |
| Trimethoprim | 0.0269 | 0.997 |
| Bisphenol A | 0.0227 | 0.988 |
| Gemfibrozil | 0.0159 | 0.993 |
| Atrazine | 0.0155 | 0.999 |
| Carbamazepine | 0.0008 | 0.971 |
| Ibuprofen | 0.0005 | 0.945 |

Reproduced with permission from [58]

compounds to the full extent, the extent of mineralization must be obtained through ion chromatography or TOC analysis. Additionally, other studies can be conducted to determine the effect of process parameters—temperature, concentration of pollutant, irradiation intensity, pH, and wavelength of light—as well as the intermediate species during the photocatalytic degradation process, which may form DBPs.

#### 2.6.1.3 Disinfection By-product Precursor Removal

Water purveyors need to be aware of DBPs, which are a cause for health concern due to the absence of regulations or standards in place. The formation of these DBPs has undergone significant scrutiny due to their carcinogenetic nature and the fact that these emerging contaminants can be difficult to remove using current infrastructure. Techniques such as coagulation/flocculation, chlorination, and ozonation are commonly used in water treatment plants for the removal of particulates found within wastewater. However, this technology is unable to completely remove small molecular weight natural organic matter (NOM). These small molecular weight NOMs (i.e., fulvic acids and humic acids) have been shown to react with the major disinfectants found in conventional oxidation processes—chlorine, chlorine dioxide, chloramines, and ozone—to produce DBPs which include trihalomethanes (THMs), haloacetic acids (HAAs), bromoform ($CHBr_3$), dibromoacetic acid (DBAA), dibromophenol (2,4-DBP), and others [61].

The different classes of DBP precursors have varying degrees of removal through UV/$TiO_2$ processes [62]. In a study by Kent et al. [62], trihalomethanes (THMs) were difficult to remove even with UV irradiation of $TiO_2$; however HAA removal from this process is substantial. Figure 2.10 shows the Total Trihalomethanes Formation Potential (TTHMFP) and Total Haloacetic Acids Formation Potential ($THAA^9FP$) values for the raw water and the degree of formation potential changes after three different UV/$TiO_2$ treatments were applied. The three UV/$TiO_2$

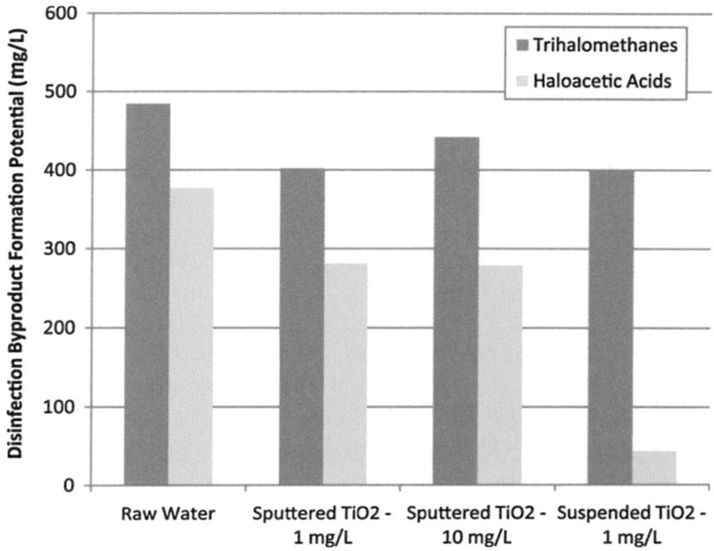


**Fig. 2.10** DBP formation potential of raw and treated wastewaters. (Reproduced with permission from [62])

treatments exhibited less than 20 % removal of the formation potential of THM compounds. Studies conducted by other researchers using higher $TiO_2$ doses have shown a higher removal rate of THM precursor compounds [63, 64]. In contrast, the THAA[9]FP was reduced by 88 % with UV/$TiO_2$ treatment despite the low concentration of $TiO_2$ (1 mg $L^{-1}$).

### 2.6.2 *Superhydrophilicity and Self-cleaning Membranes*

One primary characteristic of $TiO_2$-based materials is their superhydrophilicity [37]. When water drops are deposited on a $TiO_2$ functionalized surface, the contact angle is of several tens of degrees. Upon UV irradiation, however, water begins to exhibit a decrease in contact angle and starts to spread flat instead of beading up. After a certain time the water contact angle reaches 0°, meaning that the surface is non-water repellent, or superhydrophilic [65]. If the material is kept in the dark, a very low contact angle is retained for a period of time until the angle starts increasing and the surface becomes hydrophobic again. Exposure of UV light can impart the hydrophilic properties again to the surface. This property is permanent and is retained provided that the film is supported and does not delaminate.

In terms of the chemical mechanism during UV illumination of $TiO_2$ and water, the electrons tend to reduce Ti(IV) cations to the Ti(III) state, and the holes oxidize $O_2^-$ anions. During this process, oxygen atoms are ejected, creating oxygen vacancies. This process can coexist with photocatalysis, generated by electron–hole pairs, which give rise to redox reactions that target pollutants rather than Ti cations. Depending on the film morphology one phenomenon can prevail over another. When the $TiO_2$ surface is irradiated with UV light, the hydrogen bond distribution in the $H_2O$ molecules and surface tension of $H_2O$ clusters decreases. Furthermore, the decrease in the amount of $H_2O$ adsorbed on the $TiO_2$ surface implies a decrease in the outer surface areas of the $H_2O$ clusters. It can be said that the surface relaxation energies of $H_2O$ clusters are stabilized and the surface areas of the $H_2O$ clusters increase spontaneously. A strict consequence of this behavior is the lower amount of adsorbed $H_2O$ molecules due to the thermodynamically driven process of $H_2O$ clusters spreading out on the $TiO_2$ surfaces.

The hydrophilicity of a $TiO_2$ surface hampers the adhesion of organic compounds, making such surface easy to clean. However, a surface that is superhydrophilic cannot always guarantee a self-cleaning effect. Nevertheless, the photocatalytic effect of $TiO_2$ allows for both direct and indirect cleaning effects, which include direct self-cleaning via photocatalytic decomposition of organic compounds and recoverable hydrophilicity due to photocatalytic behavior, which gives it an indirect cleaning effect.

### *2.6.3 UV and Solar Disinfection*

Upon UV light or solar irradiation, $TiO_2$ exhibits an oxidizing effect that leads to the destruction of many harmful bacteria, viruses, and protozoa that can inflict human disease as seen in Table 2.7. Since the reaction proceeds through heterogeneous catalysis, physical contact between the target organism and the catalyst is necessary. Due to pathogenic bacteria and viruses having exponential growth rates that can significantly compromise water supplies, the bactericidal effect of $TiO_2$ is important to inhibit growth of these pathogens and disinfect water treatment effluents.

The bactericidal effect of $TiO_2$ has been studied on *Escherichia coli* (*E. coli*) cells. When irradiation of $TiO_2$ occurs with microbes on its surface, they are the primary target of the initial oxidative attack. Polyunsaturated phospholipids are important components of a bacterial cell membrane, which are susceptible to attack by reactive oxygen species produced upon irradiation of $TiO_2$. Several functions and features of a cell, such as cellular respiration, semipermeability, and oxidative phosphorylation reactions rely on the integrity of the cell membrane; however when lipid peroxidation occurs due to UV irradiation of the $TiO_2$ surface, the bacterial cell membrane is compromised, as seen in Fig. 2.11, and these functions are impaired leading to the destruction of viable cells.

An estimation of membrane damage can also be made by monitoring the production of malondialdehyde (MDA), a product of lipid peroxide. Figure 2.12 shows the results of studies conducted by Maness et al. [66], in which *E. coli* cells were incubated in suspension with $TiO_2$ (Degussa P25) and irradiated with UV light ($8\ W\ m^{-2}$) for 30 min with continuous stirring. The addition of light or $TiO_2$

**Table 2.7** Common waterborne pathogenic organisms

| Common waterborne pathogenic organisms | Disease |
|---|---|
| *Bacteria* | |
| *Bacillus anthracia* | Antrax |
| *Brucella species* | Malta fever |
| *Chiamydia trachomatis* | Conjunctivitis |
| *Clostridium botulinum* | Botuliamus |
| *Escherichia coli* | Enteriticen and enterotoxamien |
| *Listeria monocytogenas* | Listeriose |
| *Leptospira species* | Weil'sche disease |
| *Salmonella typhy* | Thyphoid |
| *Salmonella paratyphy A, B, C* | Parathyphoid |
| *Shigella species* | Ruhr |
| *Vibrio cholerae* | Cholera |
| *Virus* | |
| *Coxsackievirus A,V* | Meningitis, eczema |
| *Polovirus* | Meningitis, polio |
| *Hepatitis A* | Epidemic hepatitis |
| *Protozoa* | |
| *Cryptosporidium parvum* | Cryptosporidiosis |
| *Entamoeba histolytica* | Amoebaeruhr |
| *Giardia lamblia* | Lambliaeruhr |

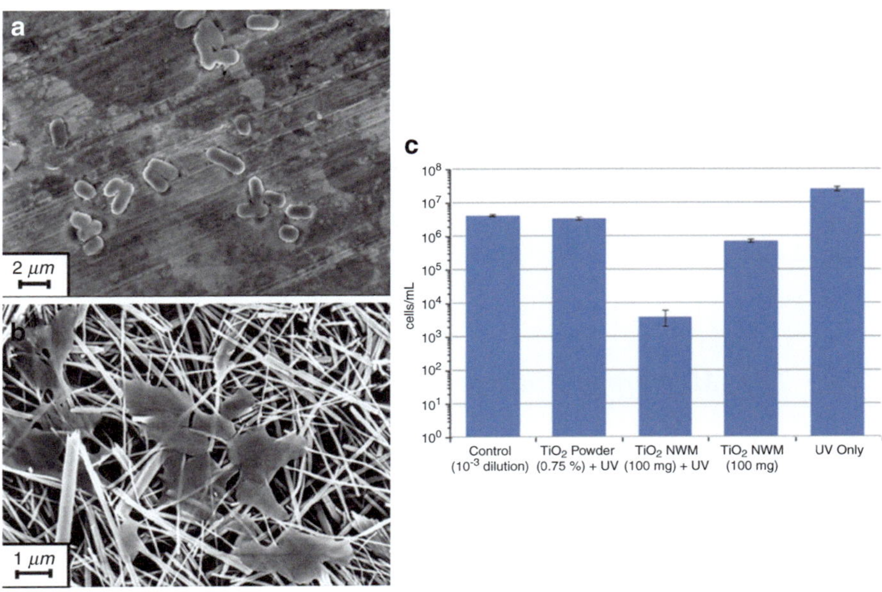


**Fig. 2.11** (**a**) SEM image of *E. coli* on substrate. (**b**) SEM image of *E. coli* on $TiO_2$ after UV irradiation. (**c**) The number of *E. coli* cells per volume of PBS under various conditions

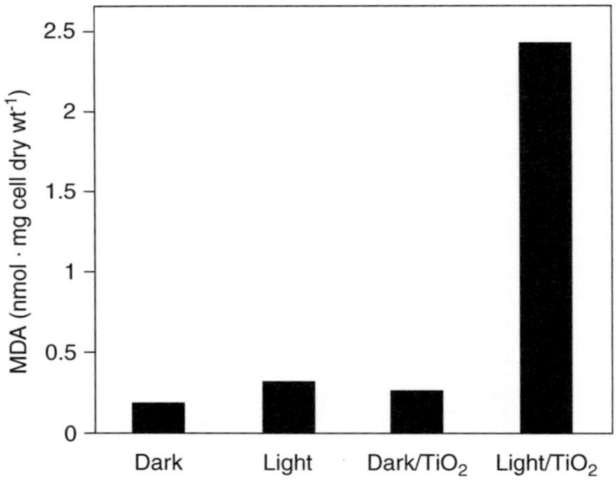


**Fig. 2.12** Effects of light and $TiO_2$ on MDA concentration produced by lipid peroxidation of *E. coli*. *Cells source*: (Reproduced with permission from [66])

slightly promote cell destruction, whereas the coupled presence of light and $TiO_2$ significantly increases cell destruction.

MDA, a product of lipid peroxidation, is produced by *E. coli* cells under various conditions and can be used to estimate the membrane damage in the cell due to oxidation. In the experiment, *E. coli* cells ($2.5 \times 10^8$ CFU $mL^{-1}$) were incubated with 0.1 mg of $TiO_2$ $mL^{-1}$ in a stirred slurry solution under UV illumination, and 2.4 nmol of MDA per mg of dry cell mass was extracted. However, under no UV illumination, 0.28 nmol of MDA per mg was detected. When no $TiO_2$ was present, control *E. coli* cells in the dark and under UV illumination conditions produced similarly low levels of MDA, indicating that UV light alone did not result in a significant level of lipid peroxidation. Hence, the lipid peroxidation process depends on the presence of light and $TiO_2$ as confirmed by the *E. coli* cell count in Fig. 2.11c.

### 2.6.4 $TiO_2$ Membrane Filtration

$TiO_2$ nanopowders in suspension can be utilized in to increase the probability of contact between $TiO_2$ particles and target pollutants in water, and hence improve photocatalytic efficiency. However, the separation of these powders from treated wastewater prior to discharge is a time and resource consuming process, which limits the application of $TiO_2$ photocatalysis in water treatment. This drawback can be avoided by immobilizing $TiO_2$ on a substrate. However, the result of immobilization has its own drawbacks such as decreased photocatalytic ability,

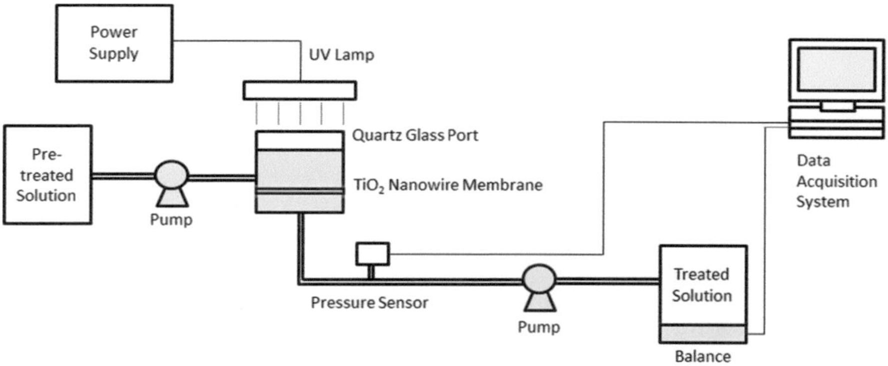


**Fig. 2.13** Schematic of a photocatalytic dead-end filtration setup

as immobilized $TiO_2$ is no longer dispersed in water to have full contact with target pollutants. Instead of focusing on the degradation performance of immobilized $TiO_2$, one can use immobilized $TiO_2$ as a membrane that can be used as a filtration apparatus, with a photocatalytic self-cleaning ability used to increase the lifespan of the membrane that would otherwise be ineffective due to organic and biofouling.

There are many $TiO_2$ membrane substrates that can be used in filtration applications and can be classified as (1) non-adsorbent and (2) adsorbent substrates. Non-adsorbent substrates, such as glass [67] and stainless steel, have been used in the past. Additionally, adsorbent substrates with large surface area, such as silica [68], zeolite [69], activated carbon [70], and activated carbon fibers [71] are increasingly being researched because the adsorption of unwanted contaminants on substrates can rectify some of the loss of photocatalytic ability caused by lower surface area to volume caused by immobilization.

To test the $TiO_2$ nanowires membranes for fouling and filtration efficiency, a setup as seen in Fig. 2.13, which measures the permeate flux and transmembrane pressures (TMP) over time can be tested. In a typical setup for filtration tests, the devices required include the following: (1) photoreactor, (2) UV source, (3) pumps, (4) pressure sensors, (5) weighing balance, and (6) data acquisition system.

In a study by Liu et al. [72], the flux performance of an Ag/$TiO_2$ nanofiber membrane, pure $TiO_2$ nanofiber membrane and P25 membrane were investigated using a dead-end membrane setup. The permeate flux was calculated by dividing the permeate mass by the filtration time and the effective surface area ($L\,m^{-2}\,min^{-1}$). As seen in Fig. 2.14, the trends for both the pure $TiO_2$ nanofiber membrane and Ag/$TiO_2$ nanofiber membrane follow a linear regression ($R^2 > 0.99$), suggesting that $TiO_2$ nanofiber and Ag/$TiO_2$ nanofiber membrane behave well in terms of permeate flux. On the other hand, the permeate flux as a function of TMP did not fit a linear regression ($R^2 = 0.77$) for commercial P25 deposited on a glass filter membrane. At the same TMP, the P25 deposited membrane showed a decrease in flux compared to the $TiO_2$ nanofiber membrane and Ag/$TiO_2$ nanofiber membrane.

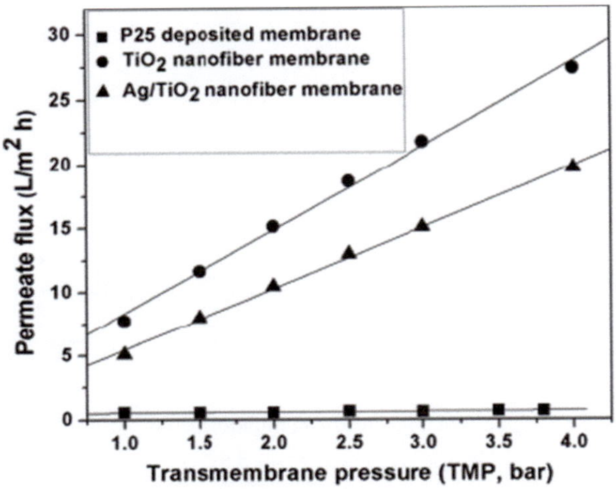


**Fig. 2.14** Change in permeate flux of P25, $TiO_2$ nanofiber and Ag-$TiO_2$ nanofiber membranes under different *TMP source*: (Reproduced with permission from [72])

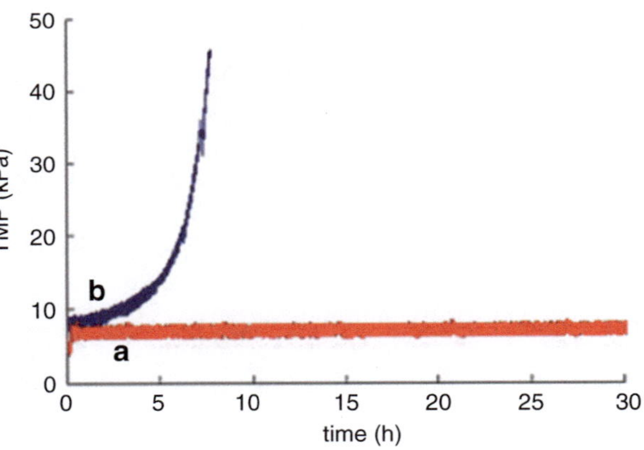


**Fig. 2.15** TMP changes of the $TiO_2$ nanowire membrane during filtration (**a**) with UV irradiation, (**b**) without UV irradiation. (Reproduced with permission from [48])

These results indicate that (1) 1D $TiO_2$ nanomaterials are favorable to maintain membrane flux compared to commercial P25 and (2) the Ag/$TiO_2$ nanofiber membrane demonstrates a slight decrease of flux compared with pure $TiO_2$ nanofiber membrane due to the addition of Ag deposited on the membrane.

In another study by Zhang et al. [48], the TMP as a function of time was evaluated for their nanowire membranes with and without UV irradiation to evaluate the existence of fouling. Figure 2.15 exhibits the outcome of TMP changes during filtration, monitored via a pressure sensor, with UV irradiation and without

UV irradiation. The TMP of the $TiO_2$ nanowire membrane as a function of filtration time increases rapidly after 7 h without UV irradiation. However, the TMP of the $TiO_2$ nanowire membrane under UV irradiation stays constant after 30 h of filtration time, after accounting for the initial resistance of the membrane in the initial stage of the experiment. A constant TMP throughout the duration experimental filtration time (30 h) implicitly reveals that fouling of $TiO_2$ nanowire membrane does not occur in this case because the formation of a fouling cake, which would increase the TMP, is inhibited.

## 2.7 Concluding Remarks

$TiO_2$ nanowire membranes can be applied to large scale water and WWTPs and overcome existing problems of conventional membrane fouling, resulting in lower maintenance costs. These membranes have several functions that make it an interesting option for water treatment applications, including (1) solar disinfection and detoxification, (2) superhydrophilicity and antifouling, (3) degradation of organic pollutants, and (4) filtration. In order for $TiO_2$ photocatalysis to become a competitive process for full-scale applications many efforts and research studies should be focused on easy scale-up, effective reactor design, UV fouling reduction, and $TiO_2$ fouling prevention. It is believed that the catalyst activity should be increased at least one order of magnitude for it to be viable. The synthesis of more efficient $TiO_2$, which can be activated under visible light irradiation, has a tremendous impact on the development of sustainable solar-energy-based treatment systems, and the move to membrane systems can reduce the systemic environmental risks of $TiO_2$ slurry systems and remediation costs.

Even though high performance is a factor of the adoption of new technologies, it mostly depends on cost effectiveness and managing the risks associated with the technology. $TiO_2$ nanomaterial costs are increasing due to the increasing demand of $TiO_2$ for many industries. One approach to mitigate the cost of $TiO_2$ nanomaterials is to use nanomaterials that are low in purity, in order to maximize efficiency over the production cost associated with purification [73]. Alternatively, cost-effectiveness of these nanomaterials can be realized by having the ability to retain and reuse nanomaterials via filtration and magnetic recovery technologies (see Chap. 7). Although $TiO_2$ nanomaterials can be quite effective in the removal of pollutants, these nanomaterials present their own unique challenges for environmental risk management as they contain small nanostructures that have not been studied extensively compared to molecules or ions for which risk assessment frameworks and protocols are already in place. $TiO_2$ nanomaterials require better understanding to mitigate potential environmental hazards in water and wastewater treatment and for broad public acceptance in using these new technologies.

The compatibility between $TiO_2$ membrane technologies and current water treatment infrastructure also need to be addressed because in many developed countries these systems are expected to remain in place for many years. Therefore, it is important to be able to implement novel membrane technologies with insignificant changes to existing infrastructure; however, in the interim, feasible nanotechnologies can be used in treatment processes where infrastructure does not exist or as point-of-use devices.

## References


1. S. Malato, P. Fernandez-Ibanez, M.I. Maldonando, J. Blanco, W. Gernjak, Decontamination and disinfection of water by solar photocatalysis: recent overview and trends. Catal. Today **147** (1), 1–59 (2009)
2. Q.L. Li, S. Mahendra, D.Y. Lyon, L. Brunet, M.B. Liga, D. Li, Antimicrobial nanomaterials for water disinfection and microbial control: potential applications and implications. Water Res. **42**, 4591–4602 (2008)
3. R.S. Barhate, S. Ramakrishna, Nanofibrous filtering media: filtration problems and solutions from tiny materials. J. Membr. Sci. **296**, 1–8 (2007)
4. M. Landmann, E. Rauls, W.G. Schmidt, The electronic structure and optical response of rutile, anatase, and brookite $TiO_2$. J. Phys. Condens. Matter **24**, 195503 (2012)
5. R.W.G. Wycoff, *Crystal Structures*, vol. 1 (Wiley, New York, 1963)
6. C. Hermann, O. Lohrmann, H. Philipp, *Strukturbericht Band II 1928-1932* (Akademische Verlagsgesellschaft MBH, Leipzig, 1937)
7. M. Horn, C.F. Schwerdfeger, E.P. Meagher, Refinement of the structure of anatase at several temperatures. Z. Kristallogr **136**, 273–281 (1972)
8. H. Zhang, H.F. Banfield, Understanding polymorphic phase transformation behaviour during growth of nanocrystalline aggregates: insights from $TiO_2$. J. Phys. Chem. B **104**, 3481–3487 (2000)
9. A. Fujishima, X. Zhang, D.A. Tryk, $TiO_2$ photocatalysis and related surface phenomena. Surf. Sci. Rep. **63**, 515–582 (2008)
10. M.R. Hoffman, S.T. Martin, W. Choi, D.W. Bahnemann, Environmental applications of semiconductor photocatalysis. Chem. Rev. **95**, 69–94 (1995)
11. A. Fujishima, T.N. Rao, D.A. Tryk, Titanium dioxide photocatalysis. J. Photochem. Photobiol. C Photochem. Rev. **1**, 1–21 (2000)
12. R. Asahi, T. Morikawa, T. Ohwaki, K. Aoki, Y. Taga, Visible light photocatalysis in nitrogen-doped titanium dioxide. Science **293**, 269–271 (2001)
13. H. Irie, Y. Watanabe, K. Hashimoto, Nitrogen-concentration dependence on photocatalytic activity of $TiO_{2-x}N_x$ powders. J. Phys. Chem. B **107**(23), 5483–5486 (2003)
14. T. Ihara, M. Miyoshi, Y. Triyama, O. Marsumato, S. Sugihara, Visible-light-active titanium oxide photocatalyst realized by an oxygen-deficient structure and by nitrogen doping. Appl. Catal. B **42**, 403–409 (2003)
15. Z. Zhang, C.C. Wang, R. Zakaria, J.Y. Ying, Role of particle size in nanocrystalline $TiO_2$-based photocatalysts. J. Phys. Chem. B **102**, 10871–10878 (1998)
16. K. Hashimoto, K. Wasada, N. Toukai, H. Kominami, Y. Kera, Photocatalytic oxidation of nitrogen monoxide over titanium(IV) oxide nanocrystals large size areas. J. Photochem. Photobiol. A Chem. **136**, 103–109 (2000)
17. H. Kominami, H. Kumamoto, Y. Kera, B. Ohtani, Immobilization of highly active titanium (IV) oxide particles: a novel strategy of preparation of transparent photocatalytic coatings. Appl. Catal. B Environ. **30**, 329–335 (2001)
18. Q. Zhang, L. Gao, Preparation of oxide nanocrystals with tubable morphologies by the moderate hydrothermal method: insights from rutile $TiO_2$. Langmuir **19**, 967–971 (2003)

19. S.H. Lim, J. Luo, Z. Zhong, W. Ji, J. Lin, Room-temperature hydrogen uptake by $TiO_2$ nanotubes. Inorg. Chem. **44**, 4124–4126 (2005)
20. R. Liang, A. Hu, W. Li, Y.N. Zhou, Enhanced degradation of persistent pharmaceuticals found in wastewater treatment effluents using $TiO_2$ nanobelt photocatalysts. J. Nanopart. Res. **15**, 1990 (2013)
21. A. Hu, X. Zhang, D. Luong, K.D. Oakes, M.R. Servos, R. Liang, S. Kurdi, P. Peng, Y. Zhou, Adsorption and photocatalytic degradation kinetics of pharmaceuticals by $TiO_2$ nanowires during water treatment. Waste Biomass Valoriz. **3**, 443–449 (2013)
22. Y.Q. Zheng, E.W. Shi, Z.Z. Chen, W.J. Li, X.F. Hu, Influence of solution concentration on the hydrothermal preparation of titania crystallites. J. Mater. Chem. **11**, 1547–1551 (2001)
23. D. Robert, J.V. Weber, Titanium dioxide synthesis by sol gel methods and evaluation of their photocatalytic activity. J. Mater. Sci. Lett. **18**, 97–98 (1999)
24. R. Campostrini, G. Carturan, L. Palmisano, M. Schiavello, A. Sclafani, Sol-gel derived anatase $TiO_2$: morphology and photoactivity. Mater. Chem. Phys. **38**, 277–283 (1994)
25. M. Ischia, R. Campostrini, L. Lutterotti, E. Garcia-Lopez, L. Palmisano, M. Schiavello, F. Pirillo, R. Molinari, Synthesis, characterization, and photocatalytic activity of $TiO_2$ powders prepared under different gelling and pressure conditions. J. Sol-Gel Sci. Technol. **33**, 201–213 (2008)
26. A. Zaleska, J.W. Sobczak, E. Grabowska, J. Hupka, Preparation and photocatalytic activity of boron-modified $TiO_2$ under UV and visible light. Appl. Catal. B **78**, 92–100 (2007)
27. S.Y. Treschev, P.W. Chou, T.H. Tseng, J.B. Wang, E.V. Perevedentseva, C.L. Cheng, Photoactivities of the visible light-activated mixed-phase carbon-containing titanium dioxide: the effect of carbon incorporation. Appl. Catal. B **79**, 8–16 (2008)
28. L. Korosi, I. Dekany, Preparation and investigation of structural and photocatalytic properties of phosphate modified titanium dioxide. Colloids Surf. A **280**, 146–154 (2006)
29. Z. Wu, F. Dong, W. Zhao, S. Guo, Visible light induced electron transfer process over nitrogen doped $TiO_2$ nanocrystals prepared by oxidation of titanium nitride. J. Hazard. Mater. **157**(1), 57–63 (2008)
30. K. Takeshita, A. Yamakata, T. Ishibashi, H. Onishu, K. Nishijima, T. Ohno, Transient IR absorption study of charge carriers photogenerated in sulphur-doped $TiO_2$. J. Photochem. Photobiol. **177**, 269–275 (2006)
31. M.S. Lee, S.S. Hong, M. Mohseni, Synthesis of photocatalytic nanosized $TiO_2$-Ag particles with sol-gel method using reduction agent. J. Mol. Catal. A **2442**, 135–140 (2005)
32. X.Z. Li, F.B. Li, Study of $Au/Au^{3+}$-$TiO_2$ photocatalysts towards visible photooxidation for water and wastewater treatment. Environ. Sci. Technol. **35**, 2381–2387 (2001)
33. J.O. Carneiro, V. Teixeira, A. Portinha, L. Dupak, A. Magalhaes, P. Coutinho, Study of the deposition parameters and Fe-dopant effect in the photocatalytic activity of $TiO_2$ films prepared by dc reactive magnetron sputtering. Vacuum **78**, 37–46 (2005)
34. X.Z. Li, F.B. Li, The enhancement of photodegradation efficiency using Pt-$TiO_2$ catalyst. Chemosphere **48**, 1103–1111 (2002)
35. J.C.S. Wu, C.H. Chen, A visible-light response vanadium-doped titania nanocatalyst by sol-gel method. J. Photochem. Photobiol. A **163**, 509–515 (2004)
36. S. Sato, J.M. White, Photodecomposition of water over platinum/titanium. Chem. Phys. Lett. **72**, 83–86 (1980)
37. V. Augugliaro, L. Vittorio, M. Pagliaro, G. Palmisano, L. Palmisano, *Clean by Light Irradiation: Practical Applications of Supported* $TiO_2$ (RSC Publishing, Cambridge, 2010)
38. L.D. Landau, B.G. Levich, Dragging of a liquid by a moving plate. Acta Physicochim. U.R.S.S. **17**, 42–54 (1942)
39. T.E. Cloete, M.D. Kwaadsteniet, M. Botes, J.M. Lopez-Romero, *Nanotechnology in Water Treatment Applications* (Caister Academic Press, Norfolk, 2010)
40. D. Li, Y.N. Xia, Electrospinning of nanofibers: reinventing the wheel? Adv. Mater. **16**(14), 1151–1170 (2004)
41. S. Ramakrishna, K. Fujihara, W.E. Teo, T. Yong, Z.W. Ma, R. Ramaseshan, Electrospun nanofibers: solving global issues. Mater. Today **9**(3), 40–50 (2006)

42. A. Hu, R. Liang, X. Zhang, S. Kurdi, D. Luong, H. Huang, P. Peng, E. Marzbanrad, K.D. Oakes, Y. Zhou, M.R. Servos, Enhanced photocatalytic degradation of dyes by $TiO_2$ nanobelts with hierarchical structures. J. Photochem. Photobiol. A **256**, 7–15 (2013)
43. A.D. Paola, M. Bellardita, L. Palmisano, Brookite, the least known $TiO_2$ photocatalyst. Catalysts **3**(1), 36–73 (2013)
44. N. Ma, X. Fan, X. Quan, Y. Zhang, Ag-$TiO_2$/HAP/$Al_2O_3$ bioceramic composite membrane: fabrication, characterization and bactericidal activity. J. Membr. Sci. **336**, 109–117 (2009)
45. V. Donchev, K. Kirilov, T. Ivanov, K. Germanova, Surface photovoltage phase spectroscopy—a handy tool for characterisation of bulk semiconductors and nanostructures. Mater. Sci. Eng. B **129**, 186–192 (2006)
46. Y. Tian, T. Tatsuma, Mechanisms and applications of plasmin-induced charge separation at $TiO_2$ films loaded with gold nanoparticles. J. Am. Chem. Soc. **127**, 7632–7637 (2005)
47. G. Li, K.A. Gray, The solid-solid interface: explaining the high and unique photocatalytic reactivity of $TiO_2$-based nanocomposite materials. Chem. Phys. **339**, 173–187 (2007)
48. X. Zhang, A.J. Du, P. Lee, D.D. Sun, J.O. Leckie, $TiO_2$ nanowire membrane for concurrent filtration and photocatalytic oxidation of humic acid in water. J. Membr. Sci. **313**(1–2), 44–51 (2008)
49. Y. Ho, G. McKey, The kinetics of sorption of basic dyes from aqueous solution by sphagnum moss peat. Can. J. Chem. Eng. **76**(4), 822–827 (1998)
50. K.V. Kumar, V. Ramamurthi, S. Sivanesan, Modeling the mechanism involved during the sorption of methylene blue onto fly ash. J. Colloid Interface Sci. **284**, 114–121 (2005)
51. V. Naddeo, L. Rizzo, V. Belgiorno, *Water, wastewater and soil treatment by advance oxidation processes* (Lulu, Raleigh, 2011)
52. M. Klavarioti, D. Mantzavinos, D. Kassinos, Removal of residual pharmaceuticals from aqueous systems by advanced oxidation processes. Environ. Int. **35**(2), 402–417 (2009)
53. L. Rizzo, S. Meric, M. Guida, D. Kassinos, V. Belgiorno, Hetergeneous photocatalytic degradation kinetics and detoxification of an urban wastewater treatment plant effluent contaminated with pharmaceuticals. Water Res. **43**(16), 4070–4078 (2009)
54. R. Rosal, A. Rodriguez, J.A. Perdigon-Melon, A. Petre, E. Garcia-Calvo, M.J. Gomez, A. Aguera, A.R. Fernandez-Alba, Occurrence of emerging pollutants in urban wastewater and their removal through biological treatment followed by ozonation. Water Res. **44**(2), 578–588 (2010)
55. O.P. Togunde, K.D. Oakes, M.R. Servos, J. Pawliszyn, Determination of pharmaceutical residues in fish bile by solid-phase microextraction with LC/MS/MS. Environ. Sci. Technol. **36**(10), 5302–5309 (2012)
56. J.S. Ings, M.M. Vijayan, M.R. Servos, Tissue-specific metabolic changes in response to an acute handling disturbance in juvenile rainbow trout exposed to municipal wastewater effluent. Aquat. Toxicol. **108**, 53–59 (2012)
57. G.R. Tetreault, C.J. Bennett, C. Cheng, M.R. Servos, M.E. McMaster, Reproductive and histopathological effects in wild fish inhabiting an effluent-dominated stream, Wascana Creek, SK, Canada. Aquat. Toxicol. **110**(111), 149–161 (2012)
58. A. Hu, X. Zhang, K.D. Oakes, P. Peng, Y.N. Zhou, M.R. Servos, Hydrothermal growth of free standing $TiO_2$ nanowire membranes for photocatalytic degradation of pharmaceuticals. J. Hazard. Mater. **189**(1–2), 278–285 (2011)
59. C.D. Metcalfe, C.D. Koenig, X.S. Miao, B.G. Koenig, J. Struger, Distribution of acidic and neutral drugs in surface waters near sewage treatment plants in the lower great lakes, Canada. Environ. Toxicol. Chem. **22**, 2872–2889 (2003)
60. M.J. Benotti, B.J. Vanderford, J.C. Holady, B.D. Stanford, S.A. Snyder, Pharmaceuticals and endocrine disrupting compounds in U.S. drinking water. Environ. Sci. Technol. **40**, 1427–1431 (2009)
61. R.J. Bull, Health effects of drinking water disinfectants and disinfection by-products. Environ. Sci. Technol. **18**, 554A–559A (1982)
62. F.C. Kent, K.R. Montreuil, R.M. Brookman, R. Sanderson, J.R. Dahn, G.A. Gagnon, Photocatalytic oxidation of DBP precursors using UV with suspended and fixed $TiO_2$. Water Res. **45**(18), 6173–6180 (2011)

63. S. Liu, M. Lim, R. Fabris, C. Chow, M. Drikas, R. Amal, $TiO_2$ photocatalysis of natural organic matter in surface water: impact on trihalomethane and haloacetic acid formation potential. Environ. Sci. Technol. **42**(16), 6218–6223 (2008)
64. K.K. Philippe, C. Hans, J. MacAdam, B. Jefferson, J. Hart, S.A. Parsons, Photocatalytic oxidation of natural organic matter surrogates and the impact on trihalomethane formation potential. Chemosphere **81**(11), 1509–1516 (2010)
65. M. Shimohigoshi, Y. Saeki, in *Research and Applications of Photocatalyst Tiles, Photocatalysts, Environment and Construction Materials*, ed. by P. Baglioni, L. Cassar (RILEM Publications, 2007)
66. P.C. Maness, S. Smolinski, D.M. Blake, Z. Huang, E.J. Wolfrum, W.A. Jacoby, Bactericidal activity of photocatalytic $TiO_2$ reaction: toward an understanding of its killing mechanism. Appl. Environ. Microbiol. **65**(9), 4094–4098 (1999)
67. H.D. Mansilla, A. Mora, C. Pincheira, M.A. Mondaca, P.D. Marcato, N. Duran, J. Freer, New photocatalytic reactor with $TiO_2$ coating on sintered glass cylinders. Appl. Catal. B Environ. **76**(1–2), 57–63 (2007)
68. P. Pucher, M. Benmami, R. Azouani, G. Krammer, K. Chhor, J.F. Bocquet, A.V. Kanaev, Nano-$TiO_2$ sols immobilized on porous silica as new efficient photocatalyst. Appl. Catal. A Gen. **332**(2), 297–303 (2007)
69. M. Huang, C. Xu, Z. Wu, Y. Huang, J. Lin, J. Wu, Photocatalytic discolorization of methyl orange solution by Pt modified $TiO_2$ loaded on natural zeolite. Dyes Pigments **77**, 327–334 (2008)
70. G.L. Puma, A. Bono, D. Krishnaiah, J.G. Collin, Preparation of titanium dioxide photocatalyst loaded onto carbon support using chemical vapour deposition: a review. J. Hazard. Mater. **157** (2–3), 208–219 (2008)
71. J. Shi, H. Cui, J. Chen, M. Fu, B. Xu, H. Luo, Z. Ye, $TiO_2$/activated carbon fibers photocatalyst: effects of coating procedures on the microstructure, adhesion property, and photocatalytic ability. J. Colloid Interface Sci. **388**(1), 201–208 (2012)
72. L. Liu, Z. Liu, H. Bai, D.D. Sun, Concurrent filtration and solar photocatalytic disinfection/ degradation using high-performance Ag/$TiO_2$ nanofiber membrane. Water Res. **46**(4), 1101–1112 (2012)
73. X.L. Qu, J. Brame, Q. Li, J.J.P. Alvarez, Nanotechnology for a safe and sustainable water supply: enabling integrate water treatment and reuse. Acc. Chem. Res. **46**(3), 834–843 (2013)

# Chapter 3
# Photocatalytic Degradation of Perfluorooctanoic Acid

**Pengyi Zhang and Zhenmin Li**

**Abstract** Perfluorooctanoic acid (PFOA) is an emerging persistent organic pollutant receiving increasing attention due to its global occurrence and resistance to most conventional degradation methods. Some special techniques such as ultrasonication, aqueous electron reduction, persulfate photolysis, and phosphotungstic-acid photocatalysis have been developed to decompose PFOA. However, these methods are not satisfactory due to their harsh reaction conditions or/and high energy consumption. Heterogeneous photocatalysis based on $TiO_2$, an effective, mild method for the treatment of most waters contaminated with organic compounds, is ineffective in degrading PFOA. Recently, we found other semiconductor materials possess higher PFOA degradation activities than $TiO_2$, such as $\beta$-$Ga_2O_3$ and $In_2O_3$. This paper provides an overview of some recent advances in the photocatalytic degradation of PFOA: (1) mechanism for photocatalytic degradation of PFOA by $In_2O_3$, (2) the photocatalytic performance of different $In_2O_3$ nanostructures, (3) photocatalytic degradation of PFOA by $\beta$-$Ga_2O_3$ nanomaterials, (4) potential applications in wastewater treatment.

## 3.1 Introduction

Perfluorocarboxylic acids (PFCAs, $C_nF_{2n+1}COOH$) are widely used as industrial surfactants, additives, firefighting foams, coatings, and lubricants owing to their versatile surface activities, and high chemical and thermal stability. Among them,

---

P. Zhang (✉) • Z. Li
Collaborative Innovation Center for Regional Environmental Quality,
Tsinghua University, Beijing 100084, China

School of Environment, Tsinghua University, Beijing 100084, China
e-mail: zpy@tsinghua.edu.cn

A. Hu and A. Apblett (eds.), *Nanotechnology for Water Treatment and Purification*, Lecture Notes in Nanoscale Science and Technology 22,
DOI 10.1007/978-3-319-06578-6_3, 

perfluorooctanoic acid (PFOA, $C_7F_{15}COOH$) is the most commonly used industrial PFCAs. A large amount of PFOA is manufactured through the electrochemical fluorination (ECF) and telomerization process. PFOA may also be the breakdown product of other perfluoroalkyl substances (PFASs) [1–4]. Direct or indirect emissions of PFOA during manufacture, use, and disposal have resulted in its widespread occurrence in the environment. Over the past decade, PFOA has been detected in various environmental matrices such as water, sediments, domestic sludge, dust, and biotic samples such as fish, human blood, and liver tissue [5–9].

As an emerging persistent organic pollutant, PFOA is the subject of increasing regulatory interest because of its bioaccumulation, environmental persistence, and growing evidence of its toxicity to humans. Significant levels of PFOA have been detected in serum samples in USA [10], China [11], and Japan [12]. The PFOA concentrations detected in 2,094 human serum samples from 2003 to 2004 in USA National Health and Nutrition Examination Survey (NHANES) ranged from 0.1 to 77.2 ng/mL with a median of 4.0 ng/mL [13]. In general, the longer the carbon chain length, the longer they can stay in the body. For example, perfluorobutane sulfonate, which has four carbons, has a half-life for elimination in a little over 1 month in humans, while PFOA, so called C8 compounds, has a half-life of 3.8 years [14].

Various in vivo and in vitro models have been used to assess the potential bioeffects of PFOA over the past decades. Distribution of PFOA in human tissues is unclear, but in vivo studies revealed the high likelihood of its accumulation in, primarily, the liver, kidney, and blood [15]. It was also found that exposure to PFOA could arrest cell cycle replication [16], alter peroxisomal and MAPK-related signaling pathways [17], and induce oxidative DNA damage in mammalian cells [18, 19]. The U.S. Environmental Protection Agency (EPA) has considered PFOA to be a "likely carcinogen."

Perfluorinated compounds (PFCs) have unique physiochemical properties gained by replacing the majority of the hydrogens in an organic molecule with fluorine. Organic poly- and perfluorination yields a more thermally stable (i.e., $C_2H_5$–H of 101 kcal/mol vs. $C_2F_5$–F of 127 kcal/mol, and $CF_3$–$CF_3$ of 99 kcal/mol vs. $CH_3$–$CH_3$ of 89 kcal/mol), more resistant to oxidation (i.e., $F + e^- \rightarrow F^-$, $E^0 = 3.6$ V), and weakly polarizable organic compound [20].

Unlike most persistent and bioaccumulative organic pollutants, PFOA is water-soluble and does not strongly adsorb to soil or sediments; thus, it has a high migration potential and poses a serious threat to ground and surface water resources. PFOA discharged or emitted from fluorochemical plants could pose potential threats on surrounding ecosystems due to contamination of food webs [21]. The consumption of contaminated drinking water and groundwater has been determined to be an important human exposure route in PFCs-contaminated areas. In January 2009, the U.S. EPA issued provisional health advisories for PFOA and perfluorooctanesulfonate (PFOS) in drinking water at 400 and 200 ng/L respectively after several revisions [22].

PFCs are particularly recalcitrant and environmentally persistent. The physiochemical properties of PFCs make them difficult to treat by most conventional methods. Due to its inherent resistance to chemical and microbiological treatment, many technologies have been developed to decompose PFOA, such as ultrasonication, aqueous electron reduction, direct photolysis, persulfate photolysis, and phosphotungstic-acid photocatalysis.

PFOA has strong absorption from deep UV region to 200 nm, thus it can be directly decomposed by UV with wavelength <200 nm. Phosphotungstic acid, $H_3PW_{12}O_{40}$, has been reported to be utilized for photocatalytic degradation of PFOA [23]. $H_3PW_{12}O_{40}$ is a heteropolyacid or polyoxometalate that has been used as an electron shuttle. Hori et al. proposed that $H_3PW_{12}O_{40}$ photocatalysis of PFOA involves a photo-Kolbe type mechanism. Hori et al. also reported that sulfate radical anions ($SO_4^{\bullet -}$) produced by $S_2O_8^{2-}$ photolysis can decompose PFOA and other shorter-chain PFCAs via an electron transfer from PFOA to $SO_4^{\bullet -}$ radical [24, 25]. In other oxidative processes, a similar decomposition mechanism was proposed, i.e., a PFOA molecule first loses an electron and then it is decarboxylated to form a perfluoroheptyl radical [23, 26, 27]. On the other hand, PFOA can be reductively decomposed by aqueous electrons ($e_{aq}^{-}$) via a defluorination step [28].

However, these methods are not satisfactory due to their harsh reaction conditions or/and high energy consumption. Heterogeneous photocatalysis is an effective, mild method for the treatment of various organic compound-contaminated waters. The advantages of the photocatalytic processes (operation under ambient temperature and pressure, high stability and low cost of the catalyst, complete mineralization without selectivity restrictions, and possibility of using solar light as energy source) open a wide range of environmentally friendly applications that are either under investigation or already in the market. Both the technological and economic relevance of photocatalysis has considerably increased over the past decade. $TiO_2$, $In_2O_3$ and β-$Ga_2O_3$ have been investigated as photocatalysts for PFOA decomposition.

## 3.2 Photocatalytic Degradation of PFOA by $In_2O_3$ and $TiO_2$

The applications of semiconductor photocatalysis are usually based on its properties of oxidation, reduction, sterilization, and super-hydrophilicity or a combination thereof. Figure 3.1 schematically shows oxidizing species which are usually involved in the photocatalytic process. The oxidation of surface $OH^-$ or $H_2O$ to hydroxyl radicals ($OH^\bullet$) takes place in the photogenerated holes; on the other hand, the photogenerated electrons reduce adsorbed oxygen to superoxide radical anion ($O_2^{\bullet -}$), which can be further transformed into hydroperoxyl radical ($^\bullet OOH$), hydrogen peroxide ($H_2O_2$), and hydroxyl radicals ($OH^\bullet$). In the presence of organic compounds, parent compounds and subsequent intermediates will be oxidized through several steps before final mineralization to carbon dioxide, water, and inorganic acids by the highly oxidizing species or by direct reaction with the photogenerated holes.

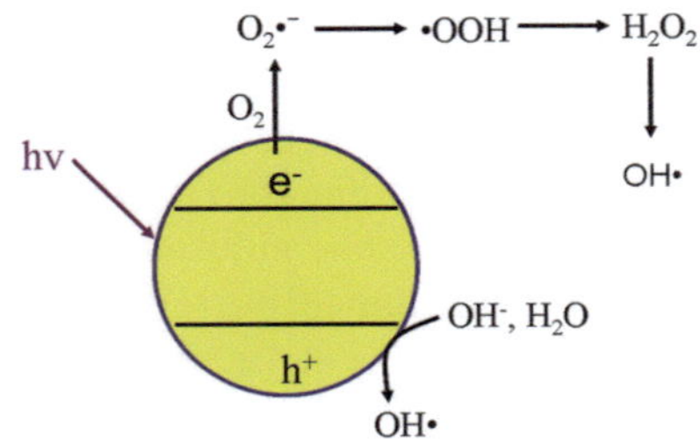


**Fig. 3.1** Active oxygen species generated in the photocatalytic process

Because of high energy of carbon–fluorine bonds (C–F, 116 kcal/mol) and high reduction potential of 3.6 V (Eq. (3.1)), the PFCs are chemically stable and resistant to oxidation, reduction, and biodegradation. Hence, the cleavage of C–F bonds was the most important criteria for PFCs detoxification and persistence elimination [29].

$$F + e^- = F^- \left(E_0 = 3.6\,\mathrm{V}\right) \tag{3.1}$$

Hydroxyl radicals normally react with saturated organics through an H-atom abstraction to form water (Eq. (3.2)) and react with unsaturated organics primarily via an addition reaction. The hydroxyl radical reacts with most aliphatic and aromatic organics at near diffusion-controlled rates at environmentally relevant pH values. PFOA contains no hydrogen atoms for abstraction by $OH^{\bullet}$ and perfluorination reduces electron density of $–COO^-$ group in PFOA, thus the direct electron transfer between $OH^{\bullet}$ and –COO– group in PFOA is not favorable. Therefore, the hydroxyl radical must act through a direct electron transfer to form the less thermodynamically favored hydroxyl ion (Eq. (3.3)) [20].

$$OH^{\bullet} + H \rightarrow H_2O\left(E^0 = 2.7\,\mathrm{V}\right) \tag{3.2}$$

$$OH^{\bullet} + e^- \rightarrow OH^-\left(E^0 = 1.9\,\mathrm{V}\right) \tag{3.3}$$

Thus the oxygen-containing radicals generated in the photocatalytic process cannot decompose PFOA. It is reported that the first step of the decomposition of PFCA in a $TiO_2$-mediated heterogeneous photocatalysis is initiated by an electron transfer from the adsorbed perfluorocarboxylate to the holes in the valence band (Eq. (3.4)).

$$F - (CF_2)_n CF_2COO^- \xrightarrow{h^+} F\text{-}(CF_2)_n CF_2COO^{\bullet} \tag{3.4}$$

In advanced oxidation processes, hydroxyl radicals generated through hydrogen peroxide photolysis, ozonation, photo-Fenton process, sonolysis, and peroxone chemistry. Therefore, some conventional advanced oxidation methods based on active oxygen radicals are not effective for PFOA decomposition.

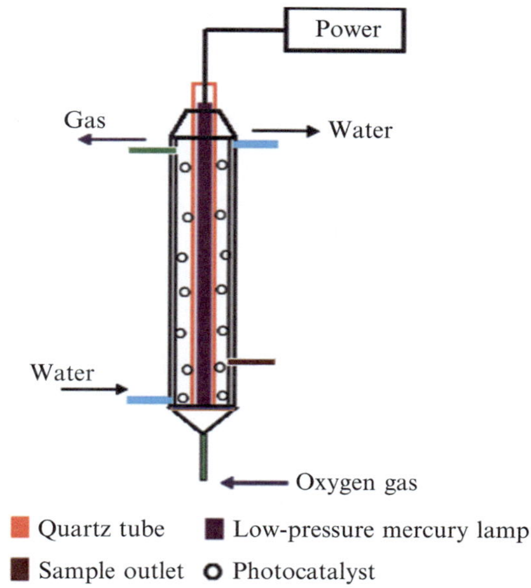


**Fig. 3.2** Schematic illustration of the photocatalytic system

### *3.2.1 The Photocatalytic Performance of $In_2O_3$ and $TiO_2$*

It was reported that the most common photocatalyst, $TiO_2$, is ineffective in degrading PFOA. Thus, we investigated the activity of other semiconductors for PFOA decomposition. Here we report the photocatalytic activity of commercial $In_2O_3$ relative to $TiO_2$. The commercial $In_2O_3$ nanoparticles with Brunauer–Emmett–Teller (BET) surface area of 12.6 $m^2/g$ and Degussa P25 $TiO_2$ with BET surface area ~50 $m^2/g$ were used as photocatalysts for degradation of PFOA.

A tubular quartz vessel reactor was used for photocatalytic decomposition of PFOA (Fig. 3.2). As shown in Fig. 3.2, a low-pressure mercury lamp (23 W) emitting 254 nm was placed in the center of the reactor with a quartz tube protection. The reaction temperature was kept at ~25 °C with a cooling water jacket around the reactor. Oxygen gas was continuously bubbled into the reactor through a porous glass plate during the whole reaction. The photocatalyst was suspended in the PFOA aqueous solution by the bubbling of oxygen gas. The initial concentration of PFOA aqueous solution was ~100 μmol/L (41.4 mg/L) and the dosage of photocatalysts in suspension was ~0.5 g/L.

Concentrations of PFOA and degradation intermediates formed were measured on a Waters Acquity UPLC system, coupled with a Micromass Quattro Premier tandem quadrupole mass spectrometric system (Waters, Milford, USA), i.e., UPLC-MS/MS system. PFOA weakly absorbs UV light with wavelengths longer than 220 nm, thus the direct photolysis by 254 nm UV light is very slow and considered negligible in these experiments.

Indium oxide ($In_2O_3$) possesses much higher photocatalytic activity than $TiO_2$ [30]. The photocatalytic decomposition of PFOA followed pseudo-first-order kinetics. The rate constant by $In_2O_3$ is about 0.378 $h^{-1}$, and it is 0.045 $h^{-1}$ by $TiO_2$ and 0.026 $h^{-1}$ for the direct photolysis, respectively, which means the decomposition rate constant of PFOA by $In_2O_3$ is 8.4 times higher than that by $TiO_2$.

The main degradation intermediates were shorter-chain PFCAs, including perfluoroheptanoic acid (PFHpA, $C_6F_{13}COOH$), perfluorohexanoic acid (PFHxA, $C_5F_{11}COOH$), perfluoropentanoic acid (PFPeA, $C_4F_9COOH$), perfluorobutanoic acid (PFBA, $C_3F_7COOH$), pentafluoropropionic acid (PFPrA, $C_2F_5COOH$), and trifluoroacetic acid (TFA, $CF_3COOH$).

Formations of fluoride ions (i.e., $F^-$) during PFOA decomposition were monitored by ion chromatography (Dionex ICS-2000, USA). TOC (total organic carbon) measurements are usually used to assess mineralization of the substrate. However, in the study of photocatalytic decomposition of PFOA, TOC data from the commercial TOC analyzer could not be obtained reliably. PFOA and its degradation intermediates are fully fluorinated hydrocarbons, which are resistant to most conventional treatment processes, and they are very stable under conditions under which their hydrocarbon analogues are degraded. To thermally decompose them, high temperatures (~1,200 °C) are required. And two commercial methods to measure TOC (i.e., catalytic oxidation at 680 ºC or UV-persulfate oxidation) cannot completely oxidize PFOA and its degradation intermediates. Other indices, i.e., defluorination ratio (formed fluoride ion/total amount of fluorine contained in initial PFOA), can be used to reflect the mineralization of PFOA.

During the photocatalysis, most fluorine atoms (F) stay in aqueous solution, although others may be transformed into gaseous products. Total F content in aqueous solution consists of four parts; remaining PFOA, shorter-chain PFCAs, $F^-$ and PFCAs adsorbed on catalyst surface. The F content on the catalyst surface could be estimated in terms of the adsorption amount and the XPS quantitative result. The gaseous products were analyzed by ATD/GC-MS. A small amount of HCOOH and $C_4$–$C_6$ alkane such as $C_nF_{2n+2}$ and $C_nHF_{2n+1}$ were detected in the gas phase.

### 3.2.2 *Coordination of PFOA to $In_2O_3$ and $TiO_2$*

To elucidate why $In_2O_3$ is a better photocatalyst than $TiO_2$, we investigated the adsorption status of PFOA on $In_2O_3$ and $TiO_2$. Adsorption of reactants to the photocatalyst surface is a critical step in the photocatalytic process. The diffuse reflectance infrared Fourier transform spectroscopy (DRIFTS) is a highly sensitive method to characterize structural changes of adsorbed species. Figure 3.3 shows the spectra of PFOA/$In_2O_3$, PFOA/$TiO_2$, and PFOA/KBr. An absorbance at 1,769 $cm^{-1}$ in IR spectrum of PFOA/KBr is typical of C = O vibration for carboxylic acid. The strong bands in the range of 1,300–1,100 $cm^{-1}$ are assigned to C–F stretching. The C–OH vibration is overlaid with an absorbance peak at 1,210 $cm^{-1}$ [31].

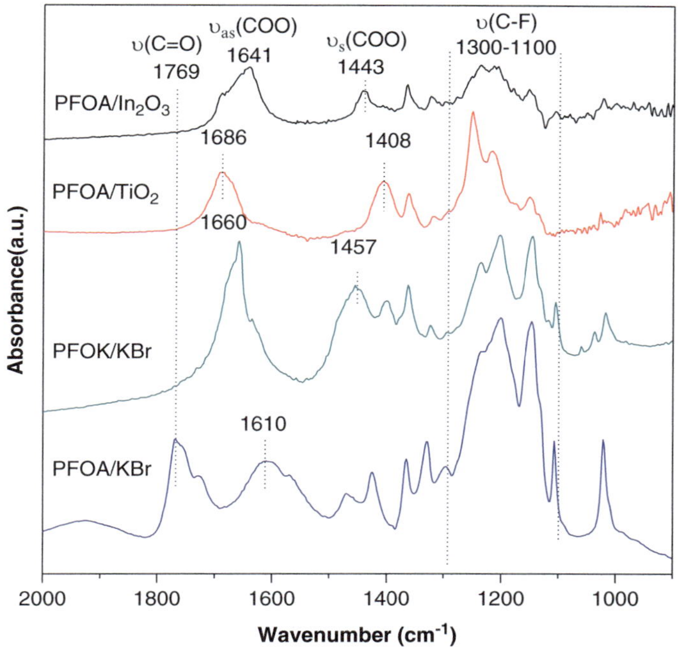


**Fig. 3.3** DRIFT spectra of PFOA mixed with KBr and adsorption-equilibrium on $In_2O_3$ and $TiO_2$ photocatalysts at room temperature. The samples of PFOA/photocatalyst were filtered and dried at room temperature for 24 h before measurement. Reproduced with permission from [30]

The broadbands at 1,640–1,600 $cm^{-1}$ are assigned to adsorbed water molecules. As for potassium perfluorooctanoate (PFOK), the $\nu$(C=O) vibration peak is replaced with the asymmetric ($\nu_{as}(COO^-)$) and symmetric stretches ($\nu_s(COO^-)$) of carboxylate, appearing at 1,660 and 1457 $cm^{-1}$ respectively. Upon adsorption on $In_2O_3$, $\nu$(C = O) stretch of PFOA vanishes, and two new peaks appear at 1,641 and 1,443 $cm^{-1}$, which are assigned to the asymmetric and symmetric stretching modes of the $-COO^-$ group [31–34]. Similarly, upon adsorption on $TiO_2$, $\nu$(C = O) vibration peak disappears, and $\nu_{as}(COO^-)$ and $\nu_s(COO^-)$ appear at 1,686 and 1,408 $cm^{-1}$ respectively. These results indicate that PFOA coordinates with both $In_2O_3$ and $TiO_2$ via its carboxylate group.

According to the stretching frequencies of the carboxylate, Deacon et al. concluded an empirical relationship between the frequency difference [35], $\Delta\nu = \nu_{as}(COO^-) - \nu_s(COO^-)$, and the types of bonding of carboxylate to cations. The $\Delta\nu$ value which is substantially greater than the ionic $\Delta\nu$ indicates a monodentate coordination, while the $\Delta\nu$ value which is significantly less than the ionic $\Delta\nu$ indicates a bidentate or bridging coordination. When the $\Delta\nu$ value is close to the ionic $\Delta\nu$, chelating and/or bridging carboxylates cannot be excluded. The $\Delta\nu$ value observed for PFOA/$TiO_2$ (278 $cm^{-1}$) is substantially greater than that of $K^+$ salt (203 $cm^{-1}$), which indicates that PFOA coordinates to $TiO_2$ in a monodentate mode. While the $\Delta\nu$ value for PFOA/$In_2O_3$ (198 $cm^{-1}$) is somewhat smaller than

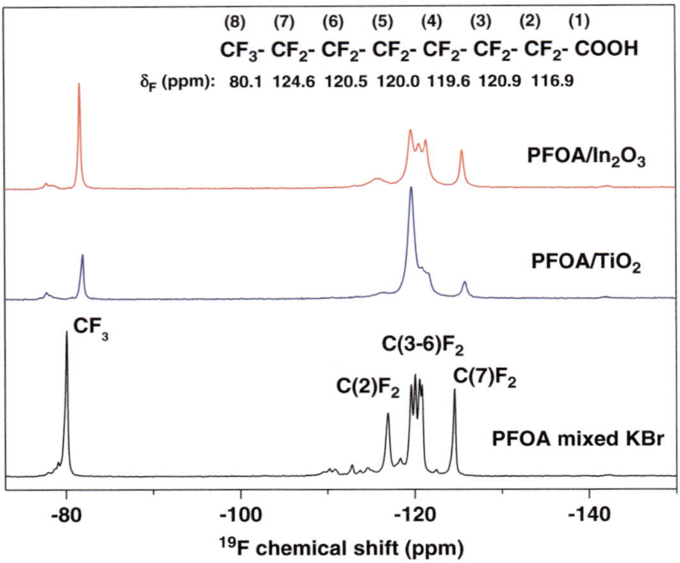


**Fig. 3.4** $^{19}F$ MAS NMR spectra of PFOA mixed with KBr and adsorption-equilibrium on $In_2O_3$ and $TiO_2$ photocatalysts at room temperature. The samples of PFOA/photocatalyst were filtered and dried at room temperature for 24 h before measurement. Reproduced with permission from [30]

the ionic $\Delta\nu$, considering the stronger adsorption of PFOA on $In_2O_3$ ($In_2O_3$, 30 μmol/g; $TiO_2$, 14.2 μmol/g), it can be concluded that PFOA is bound to $In_2O_3$ in a bidentate or bridging configuration.

Figure 3.4 shows the $^{19}F$ MAS NMR spectra of PFOA and samples adsorbed on $In_2O_3$ or $TiO_2$. For the sample of PFOA/KBr, peaks at −80.1, −116.9, and −124.6 ppm are assigned to the terminal $CF_3$, the $CF_2$ group next to the carboxylic group, and the $CF_2$ group adjacent to the terminal $CF_3$ respectively. Peaks in the range from −119.6 to −120.9 ppm are assigned to other $CF_2$ groups. The $^{19}F$ chemical shifts of PFOA in solution was also confirmed by Buchanan et al. [36, 37], who used $^{19}F$–$^{19}F$ correlation spectrometry technique. After PFOA adsorbed, changes of the chemical shifts and line widths were observed. The terminal $CF_3$ group showed a significant shift to the high field about 1.5 and 1.8 ppm for $In_2O_3$ and $TiO_2$ respectively, and the $CF_2$ group adjacent to $CF_3$ group also shifted upfield. The shift of the $CF_3$ and its adjacent $CF_2$ group can be attributed to these groups located at the air/monolayer interface as Pawsey et al. [38] explained for PFCAs adsorbed on $ZrO_2$. In addition, the line width of the $CF_2$ group next to the carboxylic group broadened, which is attributed to the deprotonation of carboxylic acid group and its coordination to the photocatalyst. Notably, there is a great difference in $^{19}F$ line width of other inner $CF_2$ groups between $In_2O_3$ and $TiO_2$. For the PFOA/$TiO_2$, the peaks heavily overlapped and lumped into a big peak, while those for PFOA/$In_2O_3$ were less changed and remained a shape similar to

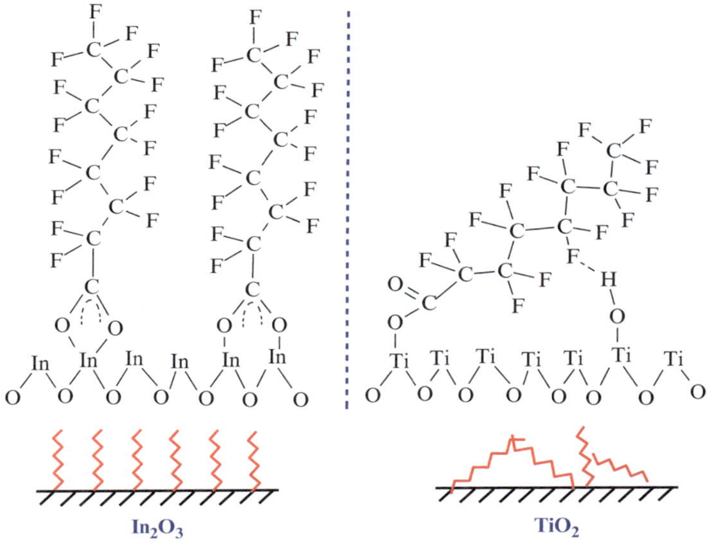


**Fig. 3.5** Schematic diagram of PFOA configurations adsorbed on $In_2O_3$ and $TiO_2$. Reproduced with permission from [30]

those of the bulk acid. These observations indicate that the inner $CF_2$ groups of PFOA may interact with $TiO_2$ surfaces, but have little interaction with $In_2O_3$.

According to the above results, the schematic configurations of PFOA adsorbed on $In_2O_3$ and $TiO_2$ are shown in Fig. 3.5. As for PFOA/$In_2O_3$, PFOA closely coordinates to $In_2O_3$ in a bidentate or bridging mode, resulting in a vertical and ordered configuration of PFOA chain on $In_2O_3$ surface. While in the case of PFOA/$TiO_2$, PFOA binds on $TiO_2$ surface in a monodentate mode with its carboxylate group, resulting in a tilted configuration of PFOA on $TiO_2$ surface. Consequently, the inner $CF_2$ group of PFOA may interact with $TiO_2$ surface OH group via hydrogen bonds. The formation of hydrogen bonds at the organic–inorganic interface have been widely reported and observed by solid-state NMR.

## *3.2.3 Mechanism for Photocatalytic Degradation of PFOA*

Photocatalytic oxidation can proceed via direct hole oxidation or via indirect $^{\bullet}OH$ radicals, with the tight adsorption of the electron donor an essential requisite for direct hole oxidation [39]. PFOA closely coordinates to $In_2O_3$ in a bidentate or bridging configuration, and it is reasonable that the PFOA decomposition may be induced by direct hole oxidation. We detected $^{\bullet}OH$ radicals formed in the presence of different photocatalysts by ESR using DMPO as the spin-trap reagent.

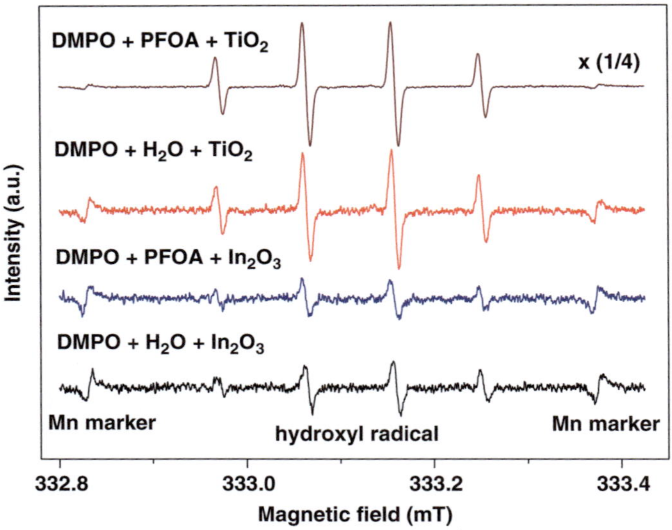


**Fig. 3.6** DMPO spin-trapping ESR spectra under UV irradiation for 4 min at room temperature of water and PFOA solution in the presence of $In_2O_3$ or $TiO_2$. Reproduced with permission from [30]

The DMPO-$^{\bullet}$OH generation was characterized by four peaks with intensity 1:2:2:1 appearing in the ESR spectra. Figure 3.6 shows that the intensities of $^{\bullet}$OH generated in $TiO_2$ suspension were much greater than those in $In_2O_3$ suspension. This indicates that the photogenerated holes in the valence band of $TiO_2$ are largely transformed into $^{\bullet}$OH radicals, while those of $In_2O_3$ react slowly with surface-bound water or hydroxyl group (–OH). After addition of PFOA, the signals of $^{\bullet}$OH peaks increased greatly in the $TiO_2$ system. This indicates that the production of $^{\bullet}$OH radicals was enhanced by PFOA. It was reported that $F^-$ and $CF_3COO^-$ adsorbed on $TiO_2$ have strong electron-withdrawing ability to reduce the recombination of electrons and holes [40–42], PFOA may also have this ability. As photogenerated holes of $TiO_2$ are quickly and mostly transformed into $^{\bullet}$OH radicals after addition of PFOA, and $^{\bullet}$OH radicals are not effective to degrade PFOA [43, 44], it is understandable that $TiO_2$ shows a low activity for the PFOA decomposition. However, intensities of $^{\bullet}$OH radicals in the $In_2O_3$ system were slightly lowered after addition of PFOA, which implied that photogenerated holes could directly react with PFOA and thus fewer holes were transformed into OH radicals. The possible mechanisms of PFOA decomposition on $In_2O_3$ and $TiO_2$ are summarized in Fig. 3.7. Compared with $TiO_2$ material, $In_2O_3$ has a higher adsorption capacity and tightly coordinates with PFOA in a bidentate or bridging configuration, which is beneficial for PFOA to be decomposed via the direct hole oxidation. Thus, $In_2O_3$ possesses much higher activity than $TiO_2$ to decompose PFOA under UV irradiation.

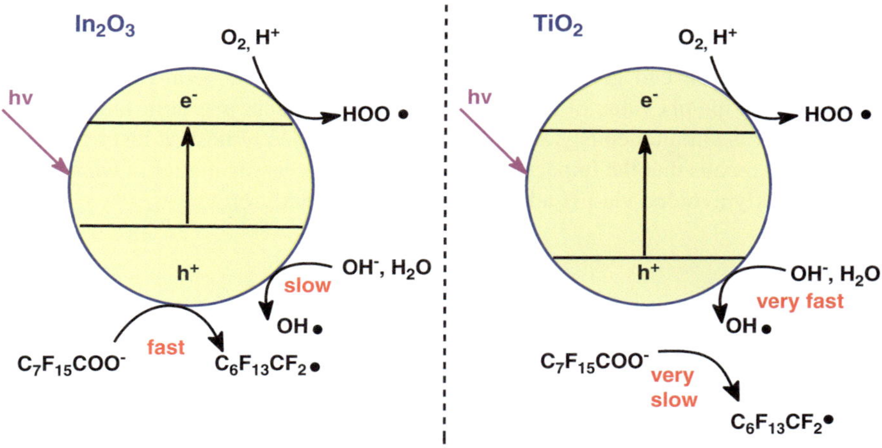


**Fig. 3.7** Possible mechanisms of the photocatalytic decomposition of PFOA by $In_2O_3$ and $TiO_2$. Reproduced with permission from [30]

**Table 3.1** Characteristics of the secondary-treated effluent taken from a municipal wastewater plant in Beijing, China

| Parameter | Value | Parameter | Value |
|---|---|---|---|
| TOC | 18.9 mg/L | pH | 7.80 |
| Bicarbonate | 4.76 mmol/L | $Na^+$ | 81.8 mg/L |
| TDS | 452 mg/L | $K^+$ | 14.6 mg/L |
| $F^-$ | 0.28 mg/L | $Mg^{2+}$ | 30.0 mg/L |
| $Cl^-$ | 107 mg/L | $Ca^{2+}$ | 85.7 mg/L |
| $SO_4^{2-}$ | 108 mg/L | Mn | 0.08 mg/L |
| $NO_3^-$ | 40.2 mg/L | Fe | <0.03 mg/L |

## 3.2.4 Potential Applications in Wastewater Treatment

In real wastewater, PFOA generally coexists with other chemical compounds, such as organic pollutants, natural organic matter, and bicarbonate, which may reduce the PFOA decomposition efficiency [45, 46]. The characteristics of the secondary-treated effluent were shown in Table 3.1. The decomposition of PFOA by $In_2O_3$ photocatalysis in the original secondary effluent was almost inhibited.

Moreover, the adsorption of PFOA on $In_2O_3$ became insignificant, which can be attributed to the competitive adsorption of bicarbonate anion on $In_2O_3$ carrying positive charges at pH 7.8 (the $pH_{zpc}$ of $In_2O_3$ is 8.7). Bicarbonate ($HOCOO^-$) has the same carboxyl group as PFOA, and its concentration usually much higher than PFOA concentration added in the secondary-treated effluent, the surface of $In_2O_3$ was mostly occupied by bicarbonate. As a result, little PFOA was adsorbed on $In_2O_3$ and its decomposition was accordingly inhibited. The adsorption of PFOA on $In_2O_3$ increased, to some extent, when the secondary-treated effluent was adjusted to acidic conditions. However, this has little influence on the PFOA decomposition in secondary effluent, which means it is necessary to degrade coexisting organic matters to recover the photocatalytic activity of $In_2O_3$ for PFOA decomposition.

Ozone addition can greatly accelerate the photocatalytic removal of organic matter. When the pH value of the secondary-treated effluent was adjusted to 4, and ozone gas was simultaneously added, PFOA decomposed almost as fast as in pure water. This means that the impacts of bicarbonate and organic matter in wastewater can be mostly avoided via pH adjustment and ozone addition.

### 3.2.5 The Enhancement of $TiO_2$ Photocatalysis of PFOA

The photocatalytic degradation of PFOA by using $TiO_2$ is very slow. And some efforts also have been made to explore the effectiveness of $TiO_2$ composites in degrading PFOA. $TiO_2$ photocatalytic degradation of PFOA is slightly enhanced by the addition of phosphotungstic ($H_3PW_{12}O_{40}$) acid, which acts as an electron shuttle from $TiO_2$ to $O_2$ [43]. Photoelectrocatalysis utilizing $TiO_2$ with a submonolayer Ni–Cu coating was also developed to enhance the rate of PFOA decomposition [47]. A rough analysis yields a power density-normalized absolute rate of $TiO_2$/Ni–Cu to be 7.4 times faster than $TiO_2$ alone. A reductive degradation process was proposed.

A combination of treatment techniques have been studied largely for the treatment of wastewater with organic pollutants. The efficiency of PFOA photocatalysis by $TiO_2$ is enhanced when coupled with sonolysis. The sonication enhances the photocatalytic decomposition of PFOA from 22 to 45 % after 7 h with $TiO_2$ as photocatalyst [48]. Sonication improves the photocatalytic performance through physical dispersion of $TiO_2$ and easing the mass transfer which keeps regenerating the $TiO_2$ surface.

The photocatalytic process is influenced by pH and surface chemistry. $TiO_2$ has a pH of zero point of charge ($pH_{zpc}$) of 6.25, which indicates that $TiO_2$ has a positive surface charge at $pH < 6.25$ and a negative charge at $pH > 6.25$. This charge-specific behavior of $TiO_2$ favors anionic electron donors at $pH < pH_{zpc}$ in photocatalysis. Therefore positive holes are considered to be the predominant oxidizing species at lower pHs while hydroxyl radicals play an important role at higher pH values. Owning to the involvement of electron transfer from the adsorbed perfluorocarboxylate to the valence hole, it was thought that an acidic solution might favor the decomposition of PFCAs more effectively. Perchloric acid ($HClO_4$) has been reportedly used as an additive to enhance the $TiO_2$ photocatalysis [49], the decomposition of 99 % and defluorination of 38 % were achieved in 7 h under highly acidic conditions ($pH = 0.9$–$1.2$).

Reducing the combination of $e^-$–$h^-$ pairs is an important approach to improve photocatalytic efficiency. Carbon nanotubes (CNTs) are an effective support for $TiO_2$ due to their electron accepting and transport capacity, and the possibility to modify their porosity and surface. They provide a convenient way to direct the flow of photogenerated charges and increase the lifetime of $e^-$–$h^-$ pairs generated by semiconductor photocatalysts. The porosity of CNTs favors adsorption of pollutants on the catalyst surface. The composite $TiO_2$ with multiple wall CNTs ($TiO_2$–MWCNT) has been used as a photocatalyst to degrade PFOA in water [50]. $TiO_2$–MWCNT displayed much higher photocatalytic ability towards PFOA relative to pure $TiO_2$ under UV irradiation.

It has been reported that the photocatalytic decomposition of PFOA in aqueous solution using Fe and Nb co-doped $TiO_2$ (Fe:Nb–$TiO_2$) [51] prepared by the sol-gel method showed higher activity compared to both undoped and commercially available $TiO_2$. The enhanced activity was attributed to the effects of co-doping on both the physicochemical properties and surface interfacial charge transfer.

Although some advancement has been made, the photocatalytic activity of $TiO_2$ on the decomposition of PFOA is still lower than that of $In_2O_3$.

## 3.3 $In_2O_3$ Nanomaterials for Photocatalytic Degradation of PFOA

$In_2O_3$ is an indirect band semiconductor with a direct band gap of 3.6 eV and an indirect bandgap of 2.8 eV [52]. As a wide bandgap semiconductor, $In_2O_3$ has been applied to improve the photocatalytic efficiency of other semiconductors. $In_2O_3$ and its composite oxides have also been investigated as water-splitting photocatalysts. As mentioned above, $In_2O_3$ showed higher photocatalytic activity for PFOA decomposition than $TiO_2$. We have found that a high adsorption capacity of $In_2O_3$ and tight coordination to PFOA is beneficial for PFOA degradation via direct hole oxidation. The following discussion introduces the preparation of $In_2O_3$ nanomaterials and photocatalytic performance of nanostructured $In_2O_3$ in detail. The photocatalytic decomposition of PFOA was conducted in a tubular quartz reactor under 254 UV light irradiation as described above. The initial concentration of PFOA was ~30 mg/L and the dosage of photocatalyst was ~0.5 g/L.

### *3.3.1 $In_2O_3$-Graphene Nanocomposites*

Graphene, a monolayer of two-dimensional (2D) carbon atomic sheets, exhibits excellent mobility of charge carriers (200,000 $cm^2/V\ s$), large surface area (calculated value is as high as 2,630 $m^2/g$), optical transparency, and chemical stability [53]. The decoration of photocatalysts such as $TiO_2$, ZnO, CdS, and $Sr_2Ta_2O_7$ with graphene can enhance their activity because the introduction of graphene can reduce the recombination of photogenerated electron–hole pairs. Because the decomposition of PFOA is initiated by an electron transfer from adsorbed PFOA to a valence band hole, the coverage with graphene on $In_2O_3$ nanoparticles influences the photocatalytic activity of $In_2O_3$.

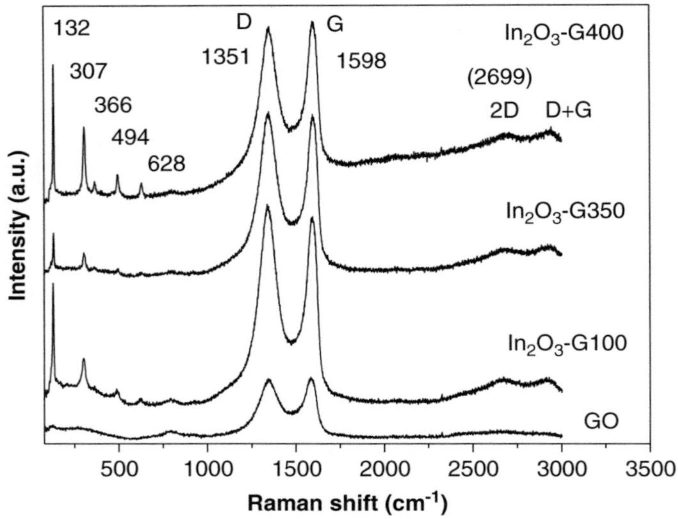


**Fig. 3.8** Raman spectra of graphene oxide and $In_2O_3$-graphene composites. Reproduced with permission from [54]

The $In_2O_3$-graphene composites were synthesized via a sonication-assisted solution route [54]. $In_2O_3$ nanoparticles and graphene oxide (GO) were dispersed in water by ultrasonication to form two uniform suspensions, which were then mixed and continuously sonicated to form the GO–$In_2O_3$ nanocomposite. The GO–$In_2O_3$ was reduced by hydrazine and ammonia at 95 °C. The $In_2O_3$-graphene composite ($In_2O_3$–G) obtained was calcined under nitrogen atmosphere. $In_2O_3$-graphene composite treated at 100, 350, and 400 °C were named as $In_2O_3$-G100, $In_2O_3$-G350, and $In_2O_3$-G400, respectively.

The coverage of graphene on the surface of $In_2O_3$ nanoparticles was identified by Raman spectrum. Figure 3.8 shows the Raman spectra of graphene oxide and $In_2O_3$-graphene composite. The Raman-active $E_{2g}$ mode at about 1,598 cm$^{-1}$ is characteristic of the presence of sp$^2$ carbon-type structures within the $In_2O_3$-graphene composites. The D band around 1,351 cm$^{-1}$ is associated with the presence of defects in the hexagonal graphical layers [55]. The 2D band around 2,679 cm$^{-1}$ provides evidence for the formation of graphene. The shape of Raman spectrum changes depending on graphene structure and layer.

The coverage of graphene on $In_2O_3$ was influenced by the calcination temperature, which was confirmed by HRTEM. Figure 3.9 shows the HRTEM images of $In_2O_3$ nanoparticles and $In_2O_3$-graphene composite. It can be seen that the carbon layers surrounding $In_2O_3$ particles cracked to some extent with increased heat. Therefore, the graphene-wrapped area of $In_2O_3$-graphene composite decreases with increases in heat treatment temperature. It has been reported that the absorption intensity of the graphene-semiconductor composites in the visible-light region is enhanced by increased graphene amounts [56, 57]. We found that the

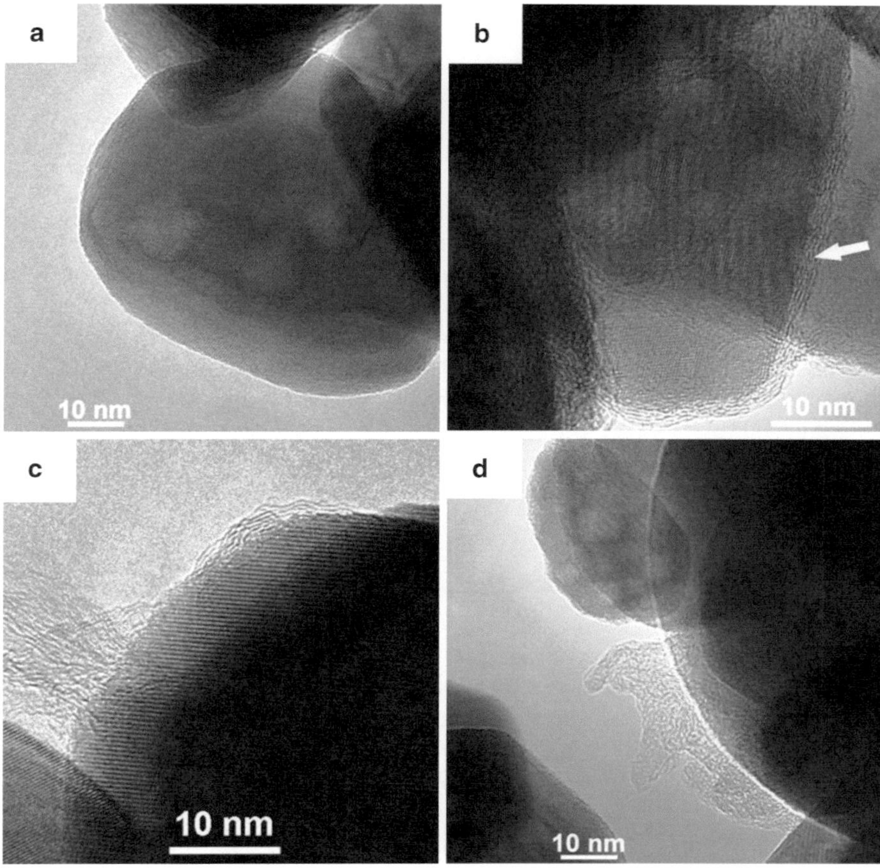


**Fig. 3.9** HRTEM images of samples. (**a**) $In_2O_3$ nanoparticles, (**b**) graphene-wrapped $In_2O_3$ nanocomposite ($In_2O_3$-G100), (**c**) $In_2O_3$-graphene nanocomposite with heat treatment at 350 °C ($In_2O_3$-G350), and (**d**) $In_2O_3$-graphene nanocomposite with heat treatment at 400 °C ($In_2O_3$-G400). The *white arrow* in (**b**) indicates the graphene. Reproduced with permission from [54]

absorption intensity of $In_2O_3$-graphene composites in the visible-light region also decreases with the decrease of graphene-wrapped area, which was caused by heat treatment.

The photocatalytic performance of PFOA decomposition depends on the coverage of graphene on $In_2O_3$. Compared with $In_2O_3$ nanoparticles and other composite photocatalysts, $In_2O_3$-G400 exhibits the highest photocatalytic activity. The PFOA decomposition appears to follow pseudo-first-order kinetics. The reaction rate constants ($k$) and half-life of PFOA ($\tau_{1/2}$) by different photocatalysts are listed in Table 3.2. The defluorination ratio of PFOA by $In_2O_3$-G400, $In_2O_3$-G350, $In_2O_3$-G100, $In_2O_3$ nanoparticles reached 60.9 %, 37.7 %, 12.9 %, and 29.7 % within 3 h, respectively.

**Table 3.2** Reaction rate constants and half-life of PFOA by using different photocatalysts

| Photocatalyst | $k$ ($h^{-1}$) | $\tau_{1/2}$ (min) |
|---|---|---|
| $In_2O_3$ | 0.66 | 63.0 |
| $In_2O_3$-G100 | 0.12 | 341.5 |
| $In_2O_3$-G350 | 0.68 | 61.2 |
| $In_2O_3$-G400 | 1.1 | 38.4 |

The photocatalytic activity of $In_2O_3$-graphene nanocomposites depends on two factors: exposed surface area of $In_2O_3$ as active adsorption sites, and supported graphene for separation of photogenerated carriers. The $In_2O_3$-graphene nanocomposites have the following advantages: (1) the exposed $In_2O_3$ surface can provide more active sites for the adsorption of PFOA, which favors the electron transfer from the adsorbed perfluorocarboxylate to valence band holes; (2) the supported graphene can act as an electron shuttle to carry excited electrons from the semiconductor to the electron acceptor, which is beneficial for efficient separation of photogenerated hole–electron pairs.

### 3.3.2 *$In_2O_3$ Porous Nanostructures*

Porous materials offer many advantages. Large surface area and porous structures are conducive to mass transfer, which are helpful for the photocatalytic reaction. Moreover, for photocatalysts with a porous structure, recombination of photogenerated carriers ($e^-/h^+$) in the bulk is reduced and a fast surface $e^-/h^+$ separation is achieved. As mentioned above, high adsorption of $In_2O_3$ is beneficial for PFOA decomposition via direct hole oxidation. Accordingly, $In_2O_3$ photocatalysts with porous nanostructure and high specific surface area are expected to show high photocatalytic activity for PFOA decomposition.

$In_2O_3$ nanospheres with nanoporous structures and a uniform size at ~100 nm were synthesized by a solvothermal method followed by calcinations [58]. For preparation of precursor of $In_2O_3$ nanospheres, i.e., $In(OH)_3$ porous nanospheres, ethylenediamine—ethanol was used as a mixed solvent. $In(NO_3)\cdot 4.5H_2O$ was dissolved in ethanol, and the obtained solution was mixed with ethylenediamine. The mixture was transferred into a Teflon-lined stainless steel autoclave, and maintained at 180 °C for 16 h. The resulting white precipitate (precursor) was thoroughly washed with deionized water and ethanol. The dried precursor was calcined for 2 h in air to form $In_2O_3$ nanospheres.

$In_2O_3$ has two crystal phases, i.e., cubic $In_2O_3$ (c–$In_2O_3$) and hexagonal $In_2O_3$ (h–$In_2O_3$). The $In_2O_3$ nanosphere that we prepared is cubic phase, as confirmed by X-ray diffraction.

Figure 3.10a–c show the TEM and SEM images of $In_2O_3$ nanospheres. The nanospheres were built of numerous nanoplates, which connected each other to form a nanoporous structure. The ring-like SAED pattern (inset of Fig. 3.10a) indicates a polycrystalline structure of as-synthesized $In_2O_3$ nanospheres, and the

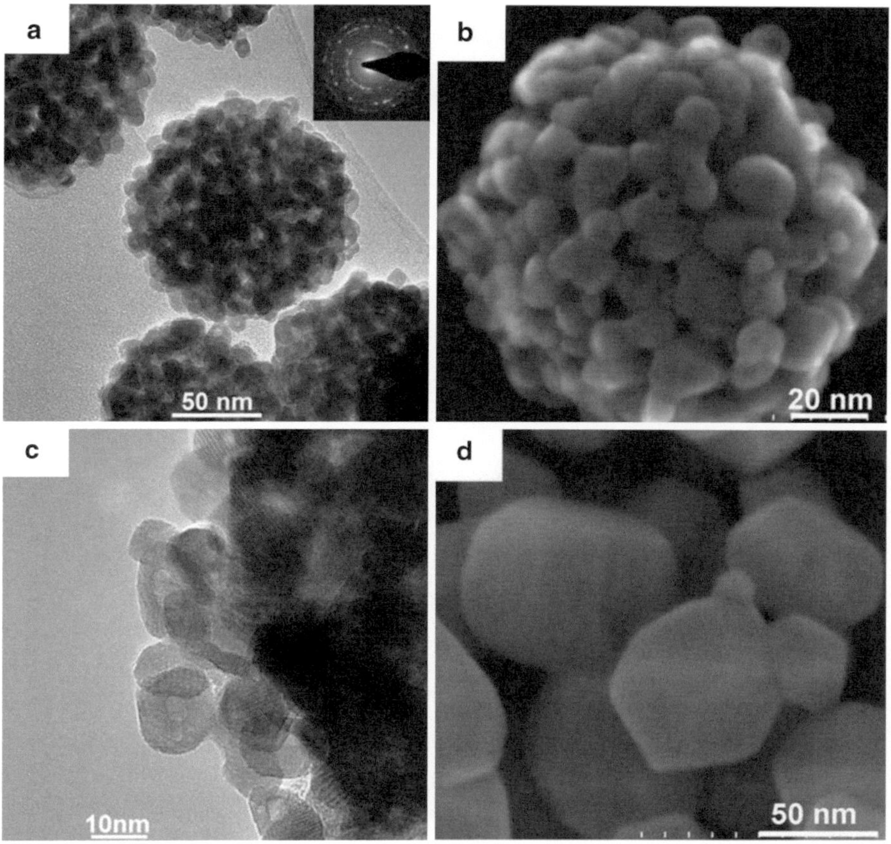


**Fig. 3.10** (**a**) Low-magnification TEM images, (**b**) FESEM image, and (**c**) high-magnification TEM image of $In_2O_3$ nanospheres; (**d**) FESEM image of commercial $In_2O_3$ nanoparticles. The inset of (**a**) is SAED pattern. Reproduced with permission from [58]

presence of obvious discrete spots suggests that the materials were well crystallized. The high-magnification TEM image in Fig. 3.10c shows $In_2O_3$ nanospheres were constructed by the interlocking stack of ultrathin nanoplates and confirms the single crystalline property of each nanoplate. The TEM image in Fig. 3.10c also reveals that the plate-like nanoparticles are irregular polygonal in shape, with size of about 5–12 nm. Commercial $In_2O_3$ nanoparticles (Beijing Chemical Co., Ltd., >99.99 %) and $TiO_2$ (Degussa P25) were adopted as the reference with which to compare the photocatalytic activity under the same experimental conditions. The diameters of most commercial $In_2O_3$ nanoparticles were in the range of 40–90 nm (Fig. 3.10d).

$In_2O_3$ nanospheres have characteristics of mesoporous materials, as confirmed by BET gas sorptometry measurements. The existence of a number of ~6 nm pores was identified, and those pores presumably arose from the stacks of nanoplates.

**Table 3.3** The BET surface area of different photocatalysts and reaction rate constants and half-life of photocatalytic decomposition of PFOA

| Photocatalyst | BET surface area ($m^2/g$) | $k$ ($h^{-1}$) | $\tau_{1/2}$ (min) |
|---|---|---|---|
| $In_2O_3$ nanospheres | 39.0 | 5.89 | 7.1 |
| $In_2O_3$ nanoparticles | 12.7 | 0.66 | 63.0 |
| P25 $TiO_2$ | 50 | 0.11 | 389.4 |

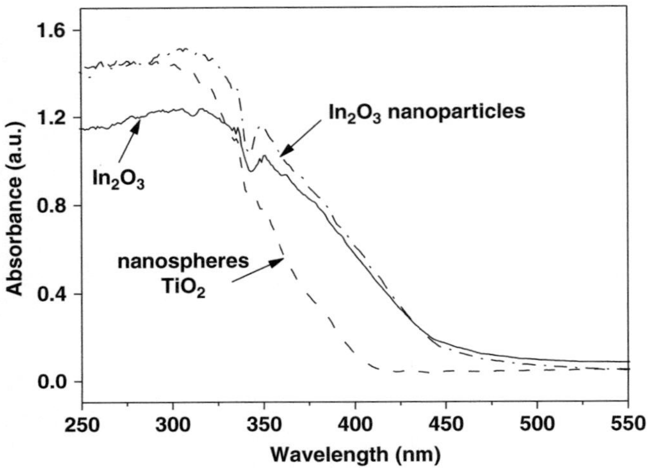


**Fig. 3.11** UV-vis absorption spectra of $In_2O_3$ nanoparticles, $In_2O_3$ nanospheres and $TiO_2$. Reproduced with permission from [58]

$In_2O_3$ nanospheres had a high BET surface area about 39.0 $m^2/g$ (Table 3.3), while that of commercial $In_2O_3$ nanoparticles was only 12.7 $m^2/g$. The BET-specific surface area of $TiO_2$ (P25) was ~50 $m^2/g$. The difference of the BET surface area between $In_2O_3$ nanospheres and $In_2O_3$ nanoparticles can be attributed to the fact that the former have nanoporous architecture.

The absorption of $In_2O_3$ nanospheres showed a slight blue-shift compared with those of $In_2O_3$ nanoparticles (Fig. 3.11). $In_2O_3$ has a small Bohr exciton radius of 2.14 nm. The slight blue-shift of UV-vis absorption may be resulted from the quantum confinement effect.

Comparative experiments were conducted to investigate the photocatalytic activity of $In_2O_3$ nanospheres, $In_2O_3$ nanoparticles and $TiO_2$ for PFOA decomposition. The PFOA decomposition appears to follow pseudo-first-order kinetics. The reaction rate constants ($k$) and half-life of PFOA ($\tau_{1/2}$) by different photocatalysts are listed in Table 3.3. The rate constant of $In_2O_3$ nanospheres was 5.89 $h^{-1}$, which was nearly 9 and 54.6 times higher than that of $In_2O_3$ nanoparticles and $TiO_2$ respectively. $In_2O_3$ nanospheres also showed a higher defluorination ratio. The defluorination ratio of PFOA by $In_2O_3$ nanospheres, $In_2O_3$ nanoparticles and $TiO_2$ reached 71.0 %, 29.7 %, and 5.1 % after 3 h, respectively.

The much higher activity of $In_2O_3$ nanospheres can be attributed to larger specific surface area and nanoporous structure. Large surface area can offer more adsorption and reaction sites, and accordingly, has a beneficial effect on the photocatalytic activity. The adsorption percentages of PFOA in the dark on $In_2O_3$ nanospheres, $In_2O_3$ nanoparticles, and $TiO_2$ were 27.4 %, 19.4 %, and 2.23 %, respectively. The nanoporous structure is beneficial for the diffusion of reactants and intermediates during reaction, which is one of the important steps determining the reaction rate. Besides, the nano-size of ultrathin plates constructing $In_2O_3$ nanospheres shortens the diffusion distance of photogenerated carriers and reduces recombination of electron–hole pairs [59], which accordingly favors the photocatalysis of PFOA.

### 3.3.3 Effects of $In_2O_3$ Morphology on Photocatalytic Degradation of PFOA

The catalytic performance of nanomaterials is determined either by the composition in terms of the atomic structure [60], or by the morphologies that affect surface atomic arrangements, coordination, and specific surface area [61, 62]. Earlier discussions have demonstrated that the photocatalytic decomposition of PFOA involves electron transfer from PFOA to photocatalyst [30]. The morphology of photocatalysts is thus thought to play an important role in improving the decomposition rate. $In_2O_3$ nanostructures with different morphologies have been developed and exhibit novel shape-dependent properties, such as improved optical properties and excellent gas-sensing characteristics. In the case of photocatalytic decomposition of PFOA, it is very important to study the relationship of photocatalytic performance and morphologies of $In_2O_3$ [63]. Thus three $In_2O_3$ nanostructures with different morphologies were synthesized to study the correlation between activity and morphology.

$In_2O_3$ microspheres were synthesized as follows. The preparation of precursors was carried out in a 100-mL Teflon-lined stainless steel autoclave. $In(NO_3)\cdot 4.5H_2O$ was used as the source material. $In(OH)_3$ microspheres were synthesized by a solvothermal process and a mixed solvent of ethanol/1,2-propane-diamine at 180 °C for 16 h. $In_2O_3$ porous microspheres were obtained by calcination of the precursor $In(OH)_3$ microsphere at 500 °C for 2 h. $In_2O_3$ nanocubes and nanoplates were synthesized as follows. $In_2O_3$ nanoplates and nanocubes were synthesized by a similar procedure, except that the solvents were changed to $H_2O$/1,2-propane-diamine for nanocubes, and $H_2O$/1,3-propane-diamine for nanoplates, respectively.

The crystal structure and the phase purity of the as-synthesized samples with different solvents were verified by XRD. All three synthesized products, microspheres, nanocubes, and nanoplates, are pure body-centered cubic (*bcc*) phase of In $(OH)_3$ (JCPDS No. 01-073-1810) and no other impurities were detected.

**Fig. 3.12** FESEM images of as-obtained $In_2O_3$ products: (**a**) microspheres, (**b**) nanocubes and (**c**) nanoplates. Reproduced with permission from [63]

**Table 3.4** The BET surface area of different $In_2O_3$ nanostructures, reaction rate constants, and half-life of PFOA with photocatalytic decomposition

| Photocatalyst | BET surface area ($m^2/g$) | $k$ ($h^{-1}$) | $\tau_{1/2}$ (min) |
|---|---|---|---|
| $In_2O_3$ microspheres | 42.3 | 7.94 | 5.3 |
| $In_2O_3$ nanoplates | 18.9 | 4.45 | 9.4 |
| $In_2O_3$ nanocubes | 13.6 | 1.83 | 22.7 |

Figure 3.12 shows the FESEM images of the calcined samples, i.e., as-obtained $In_2O_3$ products. Figure 3.12a shows the $In_2O_3$ microspheres have a diameter of about 180 nm and a porous structure. The porous microsphere is composed of several nanoparticles with diameters of 5–15 nm. As shown in Fig. 3.12b, the synthesized $In_2O_3$ nanocubes have a side length of about 40–150 nm. Although some $In_2O_3$ samples agglomerated and their morphology distorted, many particles still had a cube-like shape as indicated by white arrows in Fig. 3.12b. Figure 3.12c shows that the surface of nanoplates displayed some defects. The morphologies and structures of $In(OH)_3$ precursors usually play an important role in the transformation from indium hydroxide to indium oxide during calcination.

$In_2O_3$ microspheres have a characteristic of mesoporous materials, as confirmed by BET gas sorptometry measurements. $In_2O_3$ microspheres have a narrow pore-size distribution centered at 6 nm. The BET surface area values of $In_2O_3$ microspheres, $In_2O_3$ nanoplates and $In_2O_3$ nanocubes were 42.3, 18.9, and 13.6 $m^2/g$, respectively (Table 3.4). The relatively higher BET surface area of the microspheres confirms that the $In_2O_3$ microspheres have a porous structure.

Comparative experiments were carried out to investigate the photocatalytic activities of $In_2O_3$ products with different morphologies for PFOA decomposition. $In_2O_3$ microspheres showed the highest activity for PFOA among the three samples. The photocatalytic decomposition of PFOA by different photocatalysts followed pseudo-first-order kinetics. The rate constants of PFOA decomposition by $In_2O_3$ microspheres, nanoplates, and nanocubes were 7.94, 4.45, and 1.83 $h^{-1}$, respectively (Table 3.4). The corresponding half-life of PFOA decomposition with $In_2O_3$ microspheres was only 5.3 min, while 9.4 and 22.7 min were the experimentally-determined half-lives for $In_2O_3$ nanoplates and $In_2O_3$ nanocubes, respectively.

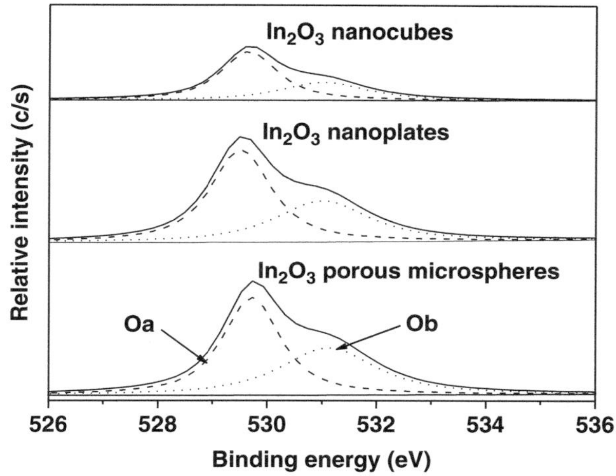


**Fig. 3.13** XPS spectra of O 1s of $In_2O_3$ samples with different morphologies. Reproduced with permission from [63]

Differing photocatalytic activities possessed by $In_2O_3$ in various nanostructures can be attributed to differences in their surface area and oxygen vacancies as discussed below. Since the photocatalytic reaction takes place on the surface of a photocatalyst, normally a high specific surface area has a beneficial effect on the activity of catalysts. As shown in Table 3.4, the values of BET surface area of $In_2O_3$ microspheres, $In_2O_3$ nanoplates, and $In_2O_3$ nanocubes correspond linearly with their corresponding reaction rates.

Due to PFOA decomposition involving the direct charge transfer between $In_2O_3$ and PFOA, the chemical states of surface In and O elements may have significant effects on PFOA adsorption and subsequent decomposition. Figure 3.13 shows the high-resolution XPS spectra of three $In_2O_3$ photocatalysts with different morphologies. As indicated in Fig. 3.13, all the O 1s core-level spectra are asymmetrical with a hump on the higher binding energy (BE) side, which can be fitted to two components with peaks around 529.7 and 531.1 eV, respectively. The two peaks of O 1s are distinguished as $O_a$ and $O_b$ [64–66]. The $O_b$ has a BE about 1.4 eV higher than that of the $O_a$. Fan [64] and Donley [66] suggested that the $O_a$ peak is assigned to $O^{2-}$ ions surrounded by In atoms with their full complement of six nearest-neighbor $O^{2-}$ ions and the $O_b$ peak is assigned to the oxygen atom adjacent to oxygen deficiency sites (i.e., they do not have neighbor In atoms with their full complement of six nearest-neighbor $O^{2-}$ ions). It is hypothesized that oxygen atoms adjacent oxygen defect sites donate some of their electron density towards In atoms that are no longer fully coordinated, which appears to cause the O 1s peak to shift towards the higher binding energy [60]. The $O_b/O_a$ ratio therefore magnifies and provides a sensitive indicator of the level of oxygen vacancy in the material. The $O_b/O_a$ ratios of three as-obtained $In_2O_3$ materials, i.e., microspheres, nanoplates, and nanocubes were determined as 0.80, 0.69, and 0.56, respectively. The high $O_b/O_a$ ratio of $In_2O_3$ microspheres may be ascribed to the ethanol solvent used during their solvothermal synthesis. It is well known that ethanol is a weak

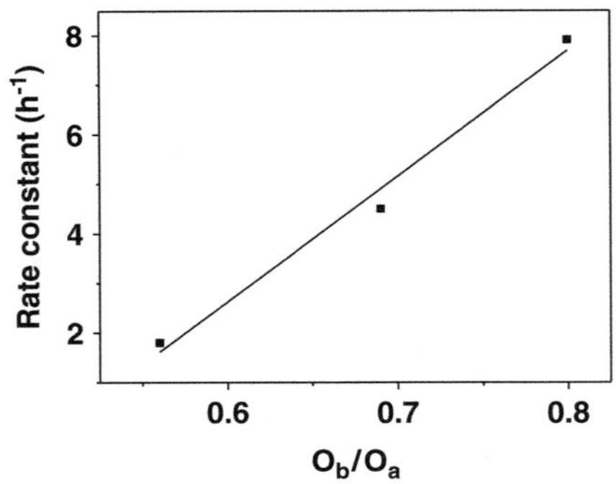


**Fig. 3.14** Photocatalytic activity of three nanostructured $In_2O_3$ materials vs. their $O_b/O_a$ ratio. Reproduced with permission from [63]

reducing agent under high temperature and pressure. When the precursor is treated solvothermally (ethanol as the solvent) at 180 °C for 16 h, the oxygen vacancy should be generated on the surface of $In_2O_3$ nanocrystals.

The $O_b/O_a$ ratio of three $In_2O_3$ materials linearly corresponds to their photocatalytic activity towards PFOA, as shown in Fig. 3.14, which is similar to those reported in the literature [67–69]. It was reported that the presence of oxygen vacancy on the surface of ZnO nanorods promotes the separation of photogenerated electron–hole pairs, thus enhancing photocatalytic activity [67]. Meng et al. [68] reported that organic dye can bind around the O vacancy by inserting an O atom from –COOH group into the vacant position on $TiO_2$ surfaces; the oxygen vacancy defects stabilize dye adsorption and facilitate charge injection. Furthermore, STM observations revealed that formic acid can also bond to $TiO_2$ (110) surface by inserting an O atom of formate into an oxygen vacancy site to form bridge and monodentate configurations, and oxygen vacancies are essential for the catalytic dehydration reaction process of formic acid on the $TiO_2$ surface [69]. Considering that PFOA molecules own a terminal –COOH group, it is suggested that PFOA molecules can also insert an O atom from its –COOH group into an oxygen vacancy site on the $In_2O_3$ surface to form tight and close contact with $In_2O_3$, which is beneficial for direct charge transfer and subsequent photocatalytic decomposition under UV irradiation. The $In_2O_3$ nanomaterials, especially porous microspheres, have high oxygen vacancy defects, thus demonstrating efficient photocatalytic activity to degrade PFOA.

### *3.3.4 Quantum Efficiency of Photocatalytic Degradation of PFOA*

The quantum efficiency (QE) of PFOA decomposition was estimated as follows (Eq. (3.5)):

**Table 3.5** The BET surface area and reaction rate constants and half-life of PFOA by different $In_2O_3$ nanostructures

| Photocatalyst | QE (%) |
|---|---|
| $In_2O_3$-graphene | 0.88 |
| $In_2O_3$ nanospheres | 5.28 |
| $In_2O_3$ microspheres | 7.92 |
| $In_2O_3$ nanoplates | 3.96 |
| $In_2O_3$ nanocubes | 0.77 |
| $In_2O_3$ nanoparticles | 0.25 |

$$\mathrm{QE}\ [\%] = \frac{\text{number of decomposed PFOA molecules}}{\text{number of incident photons}} \times 100 \tag{3.5}$$

The quantum efficiency (QE) of PFOA decomposition using different photocatalysts is shown in Table 3.5. The incident UV intensity (the main wavelength is 254 nm) was ca. 3.2 W/cm$^2$, and the UV radiation area was ca. 113 cm$^2$, thus the estimated UV power was 0.36 W.

The photocatalytic decomposition of PFOA proceeds via sequential steps, i.e., one $CF_2$ unit is removed, and shorter-chain PFCAs are generated in each step. Shorter-chain PFCAs can also be photocatalytically decomposed by $In_2O_3$ photocatalysts. Moreover, the $^{\bullet}OH$ generated cannot decompose PFCAs including PFOA, so it is difficult to calculate the QE of the reaction accurately. Equation (3.5) only estimates QE of the PFOA decomposition reaction. The QE of the entire reaction using $In_2O_3$ photocatalysts is actually higher than estimated here.

## 3.4 $Ga_2O_3$ Nanomaterials for Photocatalytic Decomposition of PFOA

Gallium oxide ($Ga_2O_3$) is an important semiconductor with a wide bandgap ($E_g = 4.9$ eV) and excellent chemical and thermal stability, unique conduction and luminescence properties. $Ga_2O_3$ has five polymorphs (α, β, δ, γ, and ε phases). Among these polymorphs, β-$Ga_2O_3$ is a thermodynamically stable phase with monoclinic structure, while the others are metastable and the ε-$Ga_2O_3$ exhibits the lowest symmetry. All the phases can be obtained from orthorhombic gallium oxide hydroxide (α-GaOOH) by annealing at a suitable temperature with the transformation dependent on the type of gallium precursors and the methodology used. For photocatalytic applications, β-$Ga_2O_3$ has been reported to show high activity and stability towards benzene degradation under UV irradiation [70]. As reported in the literature [70], the high activity and long-term stability of β-$Ga_2O_3$ is ascribed to its stronger oxidative capability and higher specific surface area.

We prepared the needle-like β-$Ga_2O_3$ and sheaf-like β-$Ga_2O_3$ nanostructures, and investigated their photocatalytic activity towards the decomposition of PFOA in pure water and wastewater.

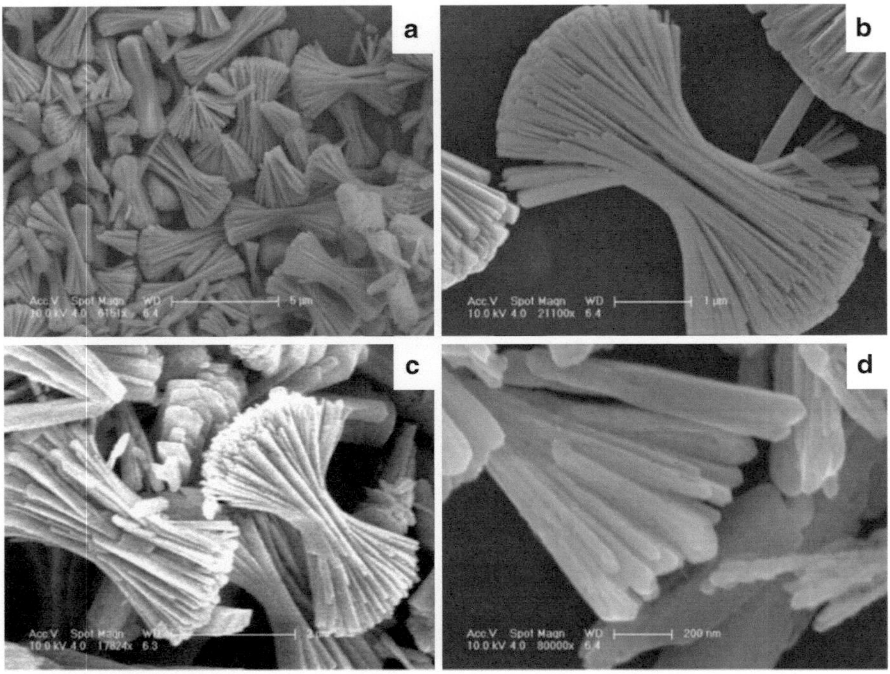


**Fig. 3.15** (**a**) Low- and (**b**) high-magnification FESEM images of as-synthesized β-GaOOH and (**c**, **d**) β-$Ga_2O_3$. Reproduced with permission from [71]

## *3.4.1 Sheaf-Like β-$Ga_2O_3$ and Needle-Like β-$Ga_2O_3$*

The sheaf-like β-$Ga_2O_3$ was synthesized by a hydrothermal method followed by calcination. In a typical procedure, $Ga(NO_3)_3 \cdot xH_2O$ and PVA (MW = 22,000) were dissolved in pure water [71]. The mixture was transferred into a Teflon-lined stainless steel autoclave and maintained at 200 °C for 8 h. The white precipitates (precursor of $Ga_2O_3$) were collected by centrifugation, and then washed with pure water and ethanol. β-$Ga_2O_3$ powder was obtained from the precursor via calcination at 700 °C for 2 h under a nitrogen atmosphere. The needle-like β-$Ga_2O_3$ was synthesized with a similar procedure as above, except that the pH value of the aqueous solution of $Ga(NO_3)_3 \cdot xH_2O$ and PVA was adjusted to 6.4 using NaOH solution [72]. The sheaf-like and needle-like products were all monoclinic phases of β-$Ga_2O_3$ and no other impurity was found, as confirmed by X-ray diffraction.

Figure 3.15 shows the FESEM images of β-GaOOH precursor and β-$Ga_2O_3$ sheaf-like nanostructure. Figure 3.15a, b indicates that the individual sheaf (β-GaOOH) has a length in the range of 2–3 µm and an average diameter in the range of 0.5–1 µm. The individual nanoplates have average widths of 100 nm and thicknesses of 10 nm. Upon calcination at 700 °C for 2 h under a nitrogen atmosphere, as-obtained $Ga_2O_3$ product largely retains the morphology and

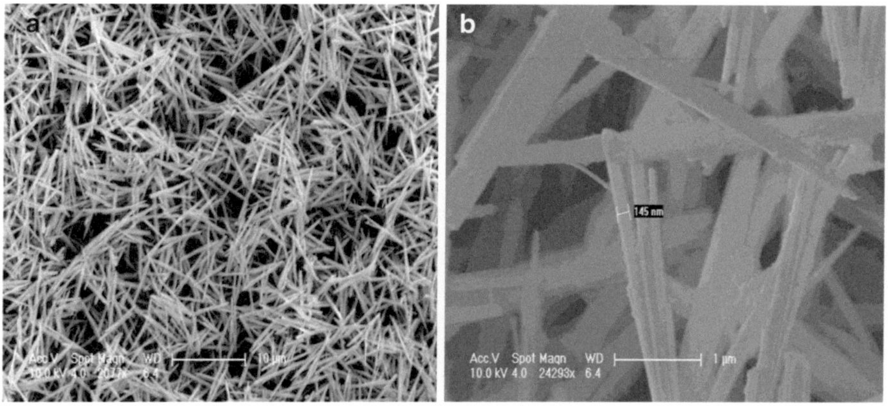


**Fig. 3.16** (**a**) Low- and (**b**) high-magnification SEM images of the β-$Ga_2O_3$ nanostructure. Reproduced with permission from [72]

architecture of its precursor, as show in Fig. 3.15c. Figure 3.15d shows the surface of material has become rough due to the dehydration shrinkage during calcination.

Figure 3.16 shows the FESEM images of the β-$Ga_2O_3$ needle-like nanostructure. The low-magnification image of Fig. 3.16a shows a needle-like morphology with a fairly uniform shape and size. Fig. 3.16b shows that individual needles have a length of 3–6 µm, and a width of 100–200 nm. The surfaces of the needles were relatively rough due to dehydration shrinkage during calcination.

The sheaf-like β-$Ga_2O_3$ and needle-like β-$Ga_2O_3$ contained mesopores in the structure, which were identified by BET gas sorptometry measurements. The pore-size distributions of these two products were both in the range of 2–4 nm. These small pores may arise from the loss of adsorbed PVA molecules, and dehydration shrinkage during the conversion process from β-GaOOH to β-$Ga_2O_3$. The BET surface area values of the sheaf-like β-$Ga_2O_3$ and needle-like β-$Ga_2O_3$ were 36.1 and 26.0 $m^2$/g, respectively (Table 3.6). The BET-specific surface area of commercial β-$Ga_2O_3$ was 11.5 $m^2$/g. The commercial β-$Ga_2O_3$ was adopted as the reference with which to compare the photocatalytic activity under the same experimental conditions.

### *3.4.2 UV Photocatalysis of PFOA in Pure Water by β-$Ga_2O_3$*

The photocatalytic decomposition of PFOA was conducted in a tubular quartz reactor vessel under ultraviolet irradiation (254 nm) as described above. The initial concentration of aqueous PFOA solution was ~500 µg/L, and the dosage of photocatalyst was ~0.5 g/L.

**Table 3.6** The BET surface area of different β-$Ga_2O_3$ photocatalysts and reaction rate constants and PFOA half-lives during photocatalytic decomposition

| Photocatalyst | BET surface area ($m^2/g$) | $k$ ($h^{-1}$) | $\tau_{1/2}$ (min) |
|---|---|---|---|
| Sheaf-like β-$Ga_2O_3$ | 36.1 | 4.85 | 8.7 |
| Needle-like β-$Ga_2O_3$ | 26.0 | 2.28 | 18.2 |
| Commercial β-$Ga_2O_3$ | 11.5 | 0.3 | 137.3 |
| P25 $TiO_2$ | 50.0 | 0.135 | 308.1 |

**Table 3.7** Decomposition rate and half-life of PFOA in sewage effluent under UV or VUV irradiation

| | | Blank | P25 $TiO_2$ | Commercial $Ga_2O_3$ | Sheaf-like $Ga_2O_3$ | |
|---|---|---|---|---|---|---|
| Light source | | (pH 4.3) | (pH 4.3) | (pH 4.3) | (pH 4.3) | (pH 7.8) |
| UV | $k$ ($h^{-1}$) | – | 0.09 | 0.10 | 1.43 | 1.00 |
| | $t_{1/2}$ (h) | – | 8.06 | 6.59 | 0.483 | 0.69 |
| VUV | $k$ ($h^{-1}$) | 1.21 | 0.98 | 1.39 | 4.29 | 1.95 |
| | $t_{1/2}$ (h) | 0.57 | 0.70 | 0.50 | 0.16 | 0.35 |

The reaction rate constants and half-lives are shown in Table 3.6. The sheaf-like β-$Ga_2O_3$ showed remarkably high activity in comparison to the needle-like β-$Ga_2O_3$ and commercial β-$Ga_2O_3$ as the larger surface area of the sheaf-like β-$Ga_2O_3$ provides more adsorption and reaction centers. Meanwhile, compared to $TiO_2$, all three β-$Ga_2O_3$ samples exhibited better photocatalytic activity for PFOA decomposition. The excellent performance of the synthesized β-$Ga_2O_3$ for PFOA decomposition can be attributed to their unique bonding to PFOA, which is similar to that between $In_2O_3$ and PFOA. This is beneficial for PFOA decomposition by holes of the photocatalyst generated under UV irradiation.

### 3.4.3 UV Photocatalysis of PFOA in Sewage Water

To validate the feasibility of the sheaf-like $Ga_2O_3$ photocatalysis to decompose PFOA in wastewater in which coexisting compounds may reduce decomposition efficiency, we investigated the decomposition of PFOA added to a secondarily-treated effluent (its composition is listed in Table 3.1) taken from a municipal wastewater plant in Beijing, China. The experiment conditions were the same as those described for pure water.

In the original secondarily-treated effluent with pH of 7.8, the decomposition of PFOA was obviously retarded, the rate constant was reduced to 1.00 $h^{-1}$ (Table 3.7). Similarly, the decomposition rate of PFOA in the presence of commercial $Ga_2O_3$ or P25 $TiO_2$ was also reduced. The lower decomposition rate of PFOA in the secondarily-treated effluent can be attributed to the influence of bicarbonate and organic material. Bicarbonate ($HCO_3^-$) has the same carboxyl group as PFOA, and its concentration (4.76 mmol/L) was nearly 3,000 times higher than that of PFOA added to the effluent, the competitive adsorption of bicarbonate on the surface of $Ga_2O_3$ inhibited the adsorption of PFOA, thus reducing its decomposition efficiency. After the pH value of the secondary effluent was adjusted to 4.3, transforming bicarbonate into carbonic acid, the rate constant increased from 1.00 to 1.43 $h^{-1}$. However, PFOA decomposition was still inhibited compared to pure water, reflecting the impact of organic materials in the effluent, which may also competitively adsorb and attenuate UV penetration in water.

### 3.4.4 *Vacuum Ultraviolet Photocatalysis of PFOA in Wastewater*

Considering the synergistic effects of vacuum ultraviolet (VUV) irradiation on a photocatalyst and its ability to degrade natural organic materials [73, 74], the combination of $Ga_2O_3$ with VUV irradiation was evaluated for degradation of PFOA in wastewater. The solution of PFOA (500 μg/L) in wastewater was irradiated with a 185 nm VUV lamp in the presence of the sheaf-like $Ga_2O_3$, commercial $Ga_2O_3$, or P25 $TiO_2$. To avoid the competitive influence of bicarbonate, pH was adjusted to 4.3 with HCl. A negative control (i.e., direct VUV photolysis), was carried out under the same conditions but without a photocatalyst.

Since PFOA strongly absorbs from the deep UV region to 200 nm [13], VUV irradiation itself can degrade PFOA. As shown in Table 3.7, the degradation rate constants of PFOA in wastewater for sheaf-like $Ga_2O_3$/VUV was 4.29 $h^{-1}$, which was nearly as effective as sheaf-like $Ga_2O_3$/UV in pure water (4.85 $h^{-1}$). While that was 1.21 $h^{-1}$ for direct VUV irradiation. The high efficiency and stability of the sheaf-like $Ga_2O_3$/VUV process for PFOA removal from wastewater is not only enhanced by VUV irradiation by eliminating adverse impacts of coexisting organic materials, but also attributed to the unique role of nanostructured $Ga_2O_3$ to decompose PFOA. When P25 $TiO_2$ or commercial $Ga_2O_3$ was used to replace sheaf-like $Ga_2O_3$ under VUV irradiation, as shown in Table 3.7, no significant improvement occurred compared with the VUV direct photolysis.

## 3.5 Summary

Heterogeneous photocatalysis based $In_2O_3$ or $Ga_2O_3$ is a promising technique for degradation of environmentally persistent PFOA. We first demonstrated that $In_2O_3$ and β-$Ga_2O_3$ exhibit higher photocatalytic activity for PFOA degradation relative to $TiO_2$. The terminal carboxylate group of PFOA molecule tightly coordinates to the $In_2O_3$ (β-$Ga_2O_3$) surface in a bidentate or bridging configuration, which is beneficial for PFOA to be directly decomposed by photogenerated holes of $In_2O_3$ under

UV irradiation. While PFOA coordinates to $TiO_2$ in a monodentate mode, and photogenerated holes of $TiO_2$ preferentially transform into hydroxyl radicals, which are inert to react with PFOA. The surface modification of $In_2O_3$ by graphene can enhance its photocatalytic activity for PFOA degradation, which was influenced by the coverage ratio of graphene on the surface of $In_2O_3$ nanoparticles.

In addition, several nanostructured $In_2O_3$ including nanospheres, nanoplates, nanocubes, and microspheres have been synthesized to obtain more efficient photocatalysts for PFOA degradation. The $In_2O_3$ nanospheres show excellent activity under mild conditions due to its porous nanostructures, with the first-order rate constant ~9 and 54.6 times higher than that by $In_2O_3$ nanoparticles and $TiO_2$ respectively. $In_2O_3$ microspheres, $In_2O_3$ nanoplates, and $In_2O_3$ nanocubes exhibit different photocatalytic activities, the corresponding first-order rate constants were of 7.94, 4.45, and 1.83 $h^{-1}$. The magnitude of oxygen vacancies owned by the different nanostructured $In_2O_3$ was found to be linear to their photocatalytic activity. Similarly, two kinds of nanostructured $\beta$-$Ga_2O_3$, i.e., sheaf-like $\beta$-$Ga_2O_3$ and needle-like $\beta$-$Ga_2O_3$, were also synthesized for PFOA degradation. The morphology of $\beta$-$Ga_2O_3$ also influences its activity for PFOA degradation.

Finally, the feasibility of nanostructured $In_2O_3$ and $\beta$-$Ga_2O_3$ for photocatalytic degradation of PFOA from real wastewater was investigated. The adverse impacts of bicarbonate and coexisting organic matters in wastewater can be mostly eliminated via pH adjustment and 185 nm VUV irradiation. For the real application of $In_2O_3$ and $\beta$-$Ga_2O_3$, further studies are necessary, such as coating of these nanomaterials and their activity for other pollutants.

## References

1. J.P. Benskin, L.W.Y. Yeung, N. Yamashita, S. Taniyasu, P.K.S. Lam, J.W. Martin, Perfluorinated acid isomer profiling in water and quantitative assessment of manufacturing source. Environ. Sci. Technol. **44**, 9049–9054 (2010)
2. H.M. Shin, V.M. Vieira, P.B. Ryan, R. Detwiler, B. Sanders, K. Steenland, S.M. Bartell, Environmental fate and transport modeling for perfluorooctanoic acid emitted from the Washington roliferat works facility in west Virginia. Environ. Sci. Technol. **45**, 1435–1442 (2011)
3. G.B. Post, J.B. Louis, K.R. Cooper, B.J. Boros-Russo, R.L. Lippincott, Occurrence and potential significance of perfluorooctanoic acid (PFOA) detected in New Jersey public drinking water systems. Environ. Sci. Technol. **43**, 4547–4554 (2009)
4. T. Zhang, Q. Wu, H.W. Sun, X.Z. Zhang, S.H. Yun, K. Kannan, Perfluorinated compounds in whole blood samples from infants, children, and adults in China. Environ. Sci. Technol. **44**, 4341–4347 (2010)
5. B.G. Loganathan, K.S. Sajwan, E. Sinclair, K.S. Kumar, K. Kannan, Perfluoroalkyl sulfonates and perfluorocarboxylates in two wastewater treatment facilities in Kentucky and Georgia. Water Res. **41**, 4611–4620 (2007)
6. R. Guo, W.J. Sim, E.S. Lee, J.H. Lee, J.E. Oh, Evaluation of the fate of perfluoroalkyl compounds in wastewater treatment plants. Water Res. **44**, 3476–3486 (2010)

7. R. Loos, G. Locoro, S. Comero, S. Contini, D. Schwesig, F. Werres, P. Balsaa, O. Gans, S. Weiss, L. Blaha, M. Bolchi, B.M. Gawlik, Pan-European survey on the occurrence of selected polar organic persistent pollutants in ground water. Water Res. **44**, 4115–4126 (2010)
8. J.Y. Hu, J. Yu, S.H. Tanaka, S.G. Fujii, Perfluorooctane sulfonate (PFOS) and perfluorooctanoic acid (PFOA) in water environment of Singapore. Water Air Soil Pollut. **216**, 179–191 (2011)
9. H.W. Sun, F.S. Li, T. Zhang, X.Z. Zhang, N. He, Q. Song, L.J. Zhao, L.N. Sun, T.H. Sun, Perfluorinated compounds in surface waters and WWTPs in Shenyang, China: mass flows and source analysis. Water Res. **45**, 4483–4490 (2011)
10. G.W. Olsen, T.R. Church, E.B. Larson, G. van Belle, J.K. Lundberg, K.J. Hansen, J.M. Burris, J.H. Mandel, L.R. Zobel, Serum concentrations of perfluorooctanesulfonate and other fluorochemicals in an elderly population from Seattle, Washington. Chemosphere **54**, 1599–1611 (2004)
11. L.W. Yeung, M.K. So, G. Jiang, S. Taniyasu, N. Yamashita, M. Song, Y. Wu, J. Li, J.P. Giesy, K.S. Guruge, P.K. Lam, Perfluorooctanesulfonate and related fluorochemicals in human blood samples from China. Environ. Sci. Technol. **40**, 715–720 (2006)
12. K. Harada, A. Koizumi, N. Saito, K. Inoue, T. Yoshinaga, C. Date, S. Fujii, N. Hachiya, I. Hirosawa, S. Koda et al., Historical and geographical aspects of the increasing perfluorooctanoate and perfluorooctane sulfonate contamination in human serum in Japan. Chemosphere **66**, 293–301 (2007)
13. A.M. Calafat, L.Y. Wong, Z. Kuklenyik, J.A. Reidy, L.L. Needham, Polyfluoroalkyl chemicals in the US population: data from the National Health and Nutrition Examination Survey (NHANES) 2003–2004 and comparisons with NHANES 1999–2000. Environ. Health Perspect. **115**, 1596–1602 (2007)
14. K.S. Betts, Perfluoroalkyl acids: what is the evidence telling us? Environ. Health Perspect. **115**, A250–A256 (2007)
15. C. Lau, K. Anitole, C. Hodes, D. Lai, A. Pfahles-Hutchens, J. Seed, Perfluoroalkyl acids: a review of monitoring and toxicological findings. Toxicol. Sci. **99**, 366–394 (2007)
16. I.G. Shabalina, T. Panaretakis, A. Bergstrand, J.W. DePierre, Effects of the rodent peroxisome proliferator and hepatocarcinogen, perfluorooctanoic acid, on apoptosis in human hepatoma HepG2 cells. Carcinogenesis **20**, 2237–2246 (1999)
17. X.M. Fang, Y.X. Feng, Z.M. Shi, J.Y. Dai, Alterations of cytokines and MAPK signaling pathways are related to the immunotoxic effect of perfluorononanoic acid. Toxicol. Sci. **108**, 367–376 (2009)
18. X. Yao, L. Zhong, Genotoxic risk and oxidative DNA damage in HepG2 cells exposed to perfluorooctanoic acid. Mutat. Res. **587**, 38–44 (2005)
19. A. Takagi, K. Sai, T. Umemura, R. Hasegawa, Y. Kurokawa, Short-term exposure to the peroxisome proliferators, perfluorooctanoic acid and perfluorodecanoic acid, causes significant increase of 8-hydroxydeoxyguanosine in liver DNA of rats. Cancer Lett. **57**, 55–60 (1991)
20. C.D. Vecitis, H. Park, J. Cheng, B.T. Mader, M.R. Hoffmann, Treatment technologies for aqueous perfluorooctanesulfonate (PFOS) and perfluorooctanoate (PFOA). Front. Environ. Sci. Eng. China **3**, 129–151 (2009)
21. J.W. Martin, D.M. Whittle, D.C.G. Muir, S.A. Mabury, Perfluoroalkyl contaminants in a food web from Lake Ontario. Environ. Sci. Technol. **38**, 5379–5385 (2004)
22. U.S. Environmental Protection Agency, Provisional Health Advisories for perfluorooctanoic acid (PFOA) and perfluorooctane sulfonate (PFOS), (2009), http://www.epa.gov/waterscience/criteria/drinking/pha-PFOA_PFOS.pdf
23. H. Hori, E. Hayakawa, H. Einaga, S. Kutsuna, K. Koike, T. Ibusuki, H. Kiatagawa, R. Arakawa, Decomposition of environmentally persistent perfluorooctanoic acid in water by photochemical approaches. Environ. Sci. Technol. **38**, 6118–6124 (2004)
24. H. Hori, A. Yamamoto, E. Hayakawa, S. Taniyasu, N. Yamashita, S. Kutsuna, Efficient decomposition of environmentally persistent perfluorocarboxylic acids by use of persulfate as a photochemical oxidant. Environ. Sci. Technol. **39**, 2383–2388 (2005)

25. S. Kutsuna, H. Hori, Rate constants for aqueous-phase reactions of $SO_4^-$ with $C_2F_5C(O)O^-$ and $C_3F_7C(O)O^-$ at 298 K. Int. J. Chem. Kinet. **39**, 276–288 (2007)
26. H. Hori, A. Yamamoto, K. Koike, S. Kutsuna, I. Osaka, R. Arakawa, Photochemical decomposition of environmentally persistent short-chain perfluorocarboxylic acids in water mediated by iron(II)/(III) redox reactions. Chemosphere **68**, 572–578 (2007)
27. Y. Wang, P. Zhang, G. Pan, H. Chen, Ferric ion mediated photochemical decomposition of perfluorooctanoic acid (PFOA) by 254 nm UV light. J. Hazard. Mater. **160**, 181–186 (2008)
28. H. Park, C.D. Vecitis, J. Cheng, W. Choi, B.T. Mader, M.R. Hoffmann, Reductive defluorination of aqueous perfluorinated alkyl surfactants: effects of ionic headgroup and chain length. J. Phys. Chem. A **113**, 690–696 (2009)
29. K. Zhang, J. Huang, G. Yu, Q. Zhang, S. Deng, B. Wang, Destruction of perfluorooctane sulfonate (PFOS) and perfluorooctanoic acid (PFOA) by ball milling. Environ. Sci. Technol. **47**, 6471–6477 (2013)
30. X. Li, P. Zhang, L. Jin, T. Shao, Z. Li, J. Cao, Efficient photocatalytic decomposition of perfluorooctanoic acid (PFOA) by indium oxide and its mechanism. Environ. Sci. Technol. **46**, 5528–5534 (2012)
31. F.P. Rotzinger, J.M. Kesselman-Truttmann, S.J. Hug, V. Shklover, M. Gratzel, Structure and vibrational spectrum of formate and acetate adsorbed from aqueous solution onto the $TiO_2$ rutile (110) surface. J. Phys. Chem. B **108**, 5004–5017 (2004)
32. M.I. Franch, J.A. Ayllon, J. Peral, X. Domenech, Enhanced photocatalytic degradation of maleic acid by Fe(III) adsorption onto the $TiO_2$ surface. Catal. Today **101**, 245–252 (2005)
33. T. Meulen, A. Mattson, L. Osterlund, A comparative study of the photocatalytic oxidation of propane on anatase, rutile, and mixed-phase anatase-rutile $TiO_2$ nanoparticles: role of surface intermediates. J. Catal. **251**, 131–144 (2007)
34. K. Nakamoto, *Infrared and Raman spectra of inorganic and coordination compounds*, 4th edn. (Wiley, New York, 2009)
35. G.B. Deacon, R.J. Phillips, Relationships between the carbon-oxygen stretching frequencies of carboxylato complexes and the type of carboxylate coordination. Coord. Chem. Rev. **33**, 227–250 (1980)
36. G.W. Buchanan, E. Munteanu, B.A. Dawson, D. Hodgson, Concerning the origin of $^{19}F$-$^{19}F$ NMR COSY and NOESY connections in the spectra of perfluorooctanoic acid, $R_F$-palmitic acid-$F_{13}$ and diethyl perfluorosuberate. Magn. Reson. Chem. **43**, 528–534 (2005)
37. W.R. Dolbier, *Guide to fluorine NMR for organic chemists* (Wiley, New Jersey, 2009)
38. S. Pawsey, L. Reven, $^{19}F$ fast magic-angle spinning NMR studies of perfluoroalkanoic acid self-assembled monolayers. Langmuir **22**, 1055–1062 (2006)
39. M.R. Hoffmann, S.T. Martin, W. Choi, D.W. Bahnemann, Environmental applications of semiconductor photocatalysis. Chem. Rev. **95**, 69–96 (1995)
40. J.C. Yu, W.K. Ho, J.G. Yu, S.K. Hark, K. Iu, Effects of trifluoroacetic acid modification on the surface microstructures and photocatalytic activity of mesoporous $TiO_2$ thin films. Langmuir **19**, 3889–3896 (2003)
41. J.G. Yu, W.G. Wang, B. Cheng, B.L. Su, Enhancement of photocatalytic activity of mesoporous $TiO_2$ powders by hydrothermal surface fluorination treatment. J. Phys. Chem. C **113**, 6743–6750 (2009)
42. S. Yang, C. Sun, X. Li, Z. Gong, X. Quan, Enhanced photocatalytic activity for titanium dioxide by co-modifying with silica and fluorine. J. Hazard. Mater. **175**, 258–266 (2010)
43. S. Kutsuna, Y. Nagaoka, K. Takeuchi, H. Hori, $TiO_2$-induced heterogeneous photodegradation of a fluorotelomer alcohol in air. Environ. Sci. Technol. **40**, 6824–6829 (2006)
44. H.F. Schroder, R.J.W. Meesters, Stability of fluorinated surfactants in advanced oxidation processes—a follow up of degradation products using flow injection-mass spectrometry, liquid chromatography-mass spectrometry and liquid chromatography-multiple stage mass spectrometry. J. Chromatogr. A **1082**, 110–119 (2005)

45. J. Cheng, C.D. Vecitis, H. Park, B.T. Mader, M.R. Hoffmann, Sonochemical degradation of perfluorooctane sulfonate (PFOS) and perfluorooctanoate (PFOA) in landfill groundwater: environmental matrix effects. Environ. Sci. Technol. **42**, 8057–8063 (2008)
46. J. Cheng, C.D. Vecitis, H. Park, B.T. Mader, M.R. Hoffmann, Sonochemical degradation of perfluorooctane sulfonate (PFOS) and perfluorooctanoate (PFOA) in groundwater: kinetic effects of matrix inorganics. Environ. Sci. Technol. **44**, 445–450 (2010)
47. J. Chen, P. Zhang, L. Zhang, Photocatalytic decomposition of environmentally persistent perfluorooctanoic acid. Chem. Lett. **35**, 230–231 (2006)
48. S.C. Panchangam, A.Y. Lin, J. Tsai, C. Lin, Sonication-assisted photocatalytic decomposition of perfluorooctanoic acid. Chemosphere **75**, 654–660 (2009)
49. S.C. Panchangam, A.Y. Lin, K.L. Shaik, C. Lin, Decomposition of perfluorocarboxylic acids (PFCAs) by heterogeneous photocatalysis in acidic aqueous medium. Chemosphere **77**, 242–248 (2009)
50. C. Song, P. Chen, C. Wang, L. Zhu, Photodegradation of perfluorooctanoic acid by synthesized $TiO_2$-MWCNT composites under 365 nm UV irradiation. Chemosphere **86**, 853–859 (2012)
51. C.R. Estrellan, C. Salim, H. Hinode, Photocatalytic decomposition of perfluorooctanoic acid by iron and niobium co-doped titanium dioxide. J. Hazard. Mater. **179**, 79–83 (2010)
52. F. Quarto, C. Sunseri, S. Piazza, M. Romano, Semiempirical correlation between optical band gap values of oxides and the difference of electronegativity of the elements. Its importance for a quantitative use of photocurrent spectroscopy in corrosion studies. J. Phys. Chem. B **101**, 2519–2525 (1997)
53. W. Wang, D. Wang, W. Qu, L. Lu, A. Xu, Large ultrathin anatase $TiO_2$ nanosheets with exposed {001}facets on graphene for enhanced visible light photocatalytic activity. J. Phys. Chem. C **116**, 19893–19901 (2012)
54. Z. Li, P. Zhang, J. Li, T. Shao, L. Jin, Synthesis of $In_2O_3$-graphene composites and their photocatalytic performance towards perfluorooctanoic acid decomposition. J. Photochem. Photobiol. A **271**, 111–116 (2013)
55. O. Akhavan, M. Abdolahad, A. Esfandiar, M. Mohatashamifar, Photodegradation of graphene oxide sheets by $TiO_2$ nanoparticles after a photocatalytic reduction. J. Phys. Chem. C **114**, 12955–12959 (2010)
56. Q. Li, B. Guo, J. Yu, J. Ran, B. Zhang, H. Yan, J.R. Gong, Highly efficient visible-light-driven photocatalytic hydrogen production of CdS-cluster-decorated graphene nanosheets. J. Am. Chem. Soc. **133**, 10878–10884 (2011)
57. P. Wang, Y. Ao, C. Wang, J. Hou, J. Qian, A one-pot method for the preparation of graphene-$Bi_2MoO_6$ hybrid photocatalysts that are responsive to visible-light and have excellent phtotocatalytic activity in the degradation of organic pollutants. Carbon **50**, 5256–5264 (2012)
58. Z. Li, P. Zhang, T. Shao, X. Li, $In_2O_3$ nanoporous nanosphere: a highly efficient photocatalyst for decomposition of perfluorooctanoic acid. Appl. Catal. B Environ. **125**, 350–357 (2012)
59. Z. Zhang, C.C. Wang, R. Zakaria, J.Y. Ying, Role of particle size in nanocrystalline $TiO_2$-based photocatalysts. J. Phys. Chem. B **102**, 10871–10878 (1998)
60. J. Zhang, K. Sasaki, E. Sutter, R.R. Adzic, Stabilization of platinum oxygen-reduction electrocatalysts using gold clusters. Science **315**, 220–222 (2007)
61. R. Narayanan, M.A. El-Sayed, Shape-dependent catalytic activity of platinum nanoparticles in colloidal solution. Nano Lett. **4**, 1343–1348 (2004)
62. J.C. Yu, J. Yu, J. Zhao, Enhanced photocatalytic activity of mesoporous and ordinary $TiO_2$ thin film by sulfuric acid treatment. Appl. Catal. B **36**, 31–43 (2002)
63. Z. Li, P. Zhang, T. Shao, J. Wang, L. Jin, X. Li, Different nanostructured $In_2O_3$ for photocatalytic decomposition of perfluorooctanoic acid (PFOA). J. Hazard. Mater. **260**, 40–46 (2013)
64. J.C.C. Fan, J.B. Goodenough, X-ray photoemission spectroscopy studies of Sn-doped indium-oxide films. J. Appl. Phys. **48**, 3524–3531 (1977)
65. J. Li, H. Fu, L. Fu, J. Hao, Complete combustion of methane over indium tin oxides catalysts. Environ. Sci. Technol. **40**, 6455–6459 (2006)

66. C. Donley, D. Dunphy, D. Paine, C. Carter, K. Nebesny, P. Lee, D. Alloway, N.R. Armstrong, Characterization of indium-tin oxide interfaces using X-ray photoelectron spectroscopy and redox processes of a chemisorbed probe molecule: effect of surface pretreatment conditions. Langmuir **18**, 450–457 (2002)
67. Y. Zheng, L. Zheng, Y. Zhan, X. Lin, Q. Zheng, K. Wei, Ag/ZnO heterostructure nanocrystals: synthesis, characterization, and photocatalysis. Inorg. Chem. **46**, 6980–6986 (2007)
68. S. Meng, E. Kaxiras, Electron and hole dynamics in dye-sensitized solar cells: influencing factors and systematic trends. Nano Lett. **10**, 1238–1247 (2010)
69. M. Aizawa, Y. Morikawa, Y. Namai, H. Morikawa, Y. Iwasawa, Oxygen vacancy promoting catalytic dehydration of formic acid on $TiO_2(110)$ by in situ scanning tunneling microscopic observation. J. Phys. Chem. B **109**, 18831–18838 (2005)
70. Y.D. Hou, X.C. Wang, L. Wu, Z.X. Ding, X.Z. Fu, Efficient decomposition of benzene over a $\beta$-$Ga_2O_3$ photocatalysts under ambient conditions. Environ. Sci. Technol. **40**, 5799–5803 (2006)
71. T. Shao, P. Zhang, L. Jin, Z. Li, Photocatalytic decomposition of perfluorooctanoic acid in pure water and sewage water by nanostructured gallium oxide. Appl. Catal. B Environ. **142–143**, 654–661 (2013)
72. T. Shao, P. Zhang, Z. Li, L. Jin, Photocatalytic decomposition of perfluorooctanoic acid in pure water and wastewater by needle-like nanostructured gallium oxide. Chin. J. Catal. **34**, 1551–1559 (2013)
73. G. Imoberdorf, M. Mohseni, Degradation of natural organic matter in surface water using vacuum-UV irradiation. J. Hazard. Mater. **186**, 240–246 (2011)
74. W.Y. Han, P.Y. Zhang, W.P. Zhu, J.J. Yin, L.S. Li, Photocatalysis of *p*-chlorobenzoic acid in aqueous solution under irradiation of 254 nm and 185 nm UV light. Water Res. **38**, 4197–4203 (2004)

# Chapter 4
# Electrospun Nanofiber Membranes and Their Applications in Water and Wastewater Treatment

**Shahram Tabe**

**Abstract** Electrospun nanofiber membranes (ENMs) are cutting edge new generation of membranes that offer significantly higher flux at similar rejection rate compared with the conventional membranes. Electrospinning has allowed for fabrication of highly porous ENMs with controllable pore size in the range of microfiltration and ultrafiltration. ENMs could replace conventional water treatment membranes with smaller systems that operate at lower pressures. In this chapter, the fundamentals of ENMs are presented and characterization methods, properties, and applications are discussed.

## 4.1 Introduction

Electrospun nanofiber membranes (ENMs) are a new generation of membranes that could offer a breakthrough in water and wastewater treatment by providing lighter, less expensive, and less energy consuming processes than the existing processes including conventional membranes. The major advantages of ENMs over their conventional counterparts are their higher porosity and surface-to-volume ratio. While these cutting edge membranes have widely been used in commercial air filtration [1–3], their applications in water and wastewater treatment have only been explored recently [4–7].

S. Tabe (✉)
Standards Development Branch, Ministry of the Environment, 40 St. Clair Avenue West, Toronto, ON, Canada M4V 1M2

Department of Chemical Engineering and Applied Chemistry, University of Toronto, 200 College Street, Toronto, ON, Canada M5S 3E5
e-mail: Shahram.Tabe@ontario.ca

A. Hu and A. Apblett (eds.), *Nanotechnology for Water Treatment and Purification*, Lecture Notes in Nanoscale Science and Technology 22,
DOI 10.1007/978-3-319-06578-6_4, 

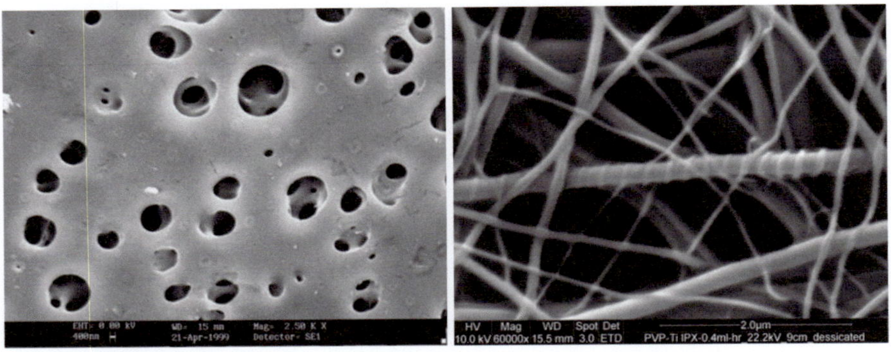


**Fig. 4.1** Comparison of pore structure of a conventional UF membrane (*left*) with that of a nanofiber membrane (*right*)

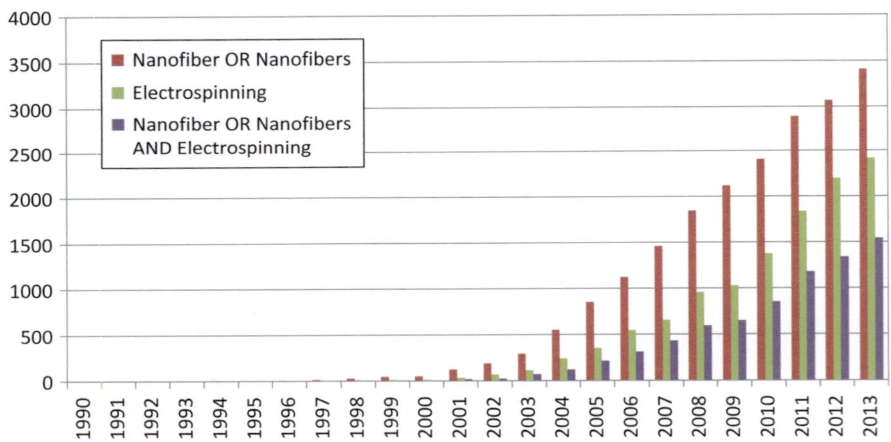


**Fig. 4.2** Number of papers published since 1990 containing keywords "nanofiber OR nanofibers," "electrospinning," and "nanofiber OR nanofibers AND electrospinning." Search made through Scopus

ENMs exhibit porosity in the range of 80 % compared to 5–35 % for conventional membranes. The high porosity and the interconnected pore structure of ENMs allow for permeabilities much higher than conventional membranes. Figure 4.1 compares the pore structure of a conventional UF membrane with that of a nanofiber membrane.

Favorable characteristics of nanofibers have made them suitable for applications as wide as filtration, medicine, bioengineering, electronics, protective clothing, batteries, and catalysis. Figure 4.2 illustrates an exponentially rising number of articles relevant to nanofibers published since 1990 which indicates a booming attention to this technology.

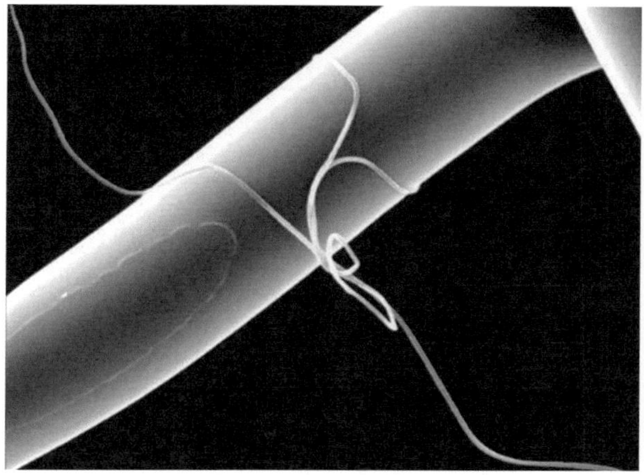

**Fig. 4.3** Size comparison of two nanofibers made by conventional methods (the thicker fiber) and electrospinning method (the thinner fiber). Reproduced with permission of [20]

Conventional spinning techniques, such as melt spinning, wet spinning, dry spinning, and gel spinning, produce fibers with diameters in the range of micrometer. Nanofibers are one or more orders of magnitude smaller in diameter and are in the range of a few ten to a few hundred nanometers. Techniques such as template synthesize [8–11], self-assembly [12, 13], and electrospinning [2, 4, 5, 14–18] are used to produce nanofibers.

Electrospinning is advantageous over conventional nanofiber spinning techniques in being capable of producing fibers that are orders of magnitude thinner. The fiber diameter governs the surface area to volume ratio and impacts membrane porosity. In electrospinning, fiber diameter can be adjusted by varying the process parameters such as solution concentration, applied voltage, surface tension, and spinning distance [19]. Figure 4.3 compares the size of a nanofiber made using conventional techniques and the one made by electrospinning, both spun from identical material [20]. The size of the electrospun fiber in this image is ~300 nm.

How thin is a nanofiber? Approximately one million nanofibers could fit in the cross section of a normal human hair (assuming diameters of 100 nm for a nanofiber and 100 μm for a human hair). Burger et al. [4] also provide a different picture of how thin a nanofiber is: to electrospin a nanofiber that would connect the Earth and its Moon we only need ~3 g of polymer (distance from Earth to the Moon: ~380,000 km, polymer density: 1 g/cm$^3$, fiber diameter: 100 nm).

## 4.2 Electrospinning Process and Preparation of ENMs

Our knowledge of electrospinning roots as far back as the turn of the twentieth century [21]. However, modern use of the technique to fabricate nanofibers for applications such as filtration, tissue engineering, fuel cell, and batteries is only two decades old with the earliest work presented by Doshi and Reneker [14].

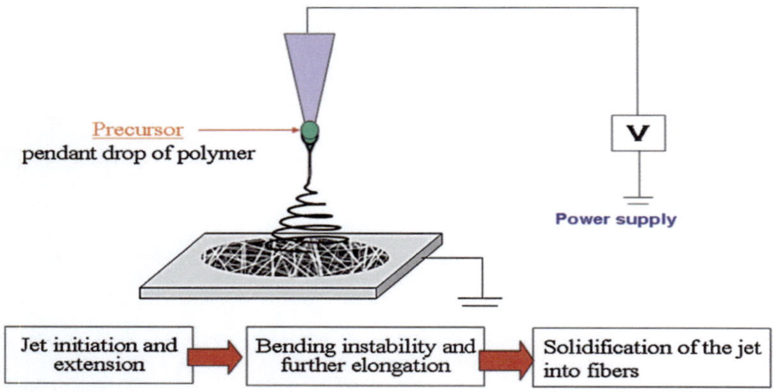


**Fig. 4.4** Schematics of a laboratory-scale electrospinning setup. Reproduced with permission of [18]

In the process of electrospinning, what helps in producing thin fibers is the electrostatic forces applied on the polymer solution (or melt). Figure 4.4 illustrates a laboratory-scale electrospinning apparatus. In a typical process a high voltage is applied between a syringe that holds the polymer solution (spinneret) and a grounded collection plate. At the beginning, the electrostatic force and the surface tension of the polymer solution are in equilibrium. As the external force starts to overcome the surface tension of the fluid a conical droplet—known as Tylor cone—forms at the tip of the nozzle. As the electrostatic force overcomes the surface tension a jet erupts from the tip of the nozzle forming a filament that travels to the opposite electrode. The interaction between the external electric field and the internal repulsive charges stretches the polymer filament thinning the fiber. This causes the diameter of the jet stream to become orders of magnitude smaller than the diameter of the tip of the syringe. To make room for the stretch, the filament undergoes bending instability. The elongated travel path provides more time for the solvent to evaporate and further thinning the fiber. The solvent evaporation also solidifies the filament and forms the nanofiber. The nanofiber is collected on the surface of the collector plate and gradually forms a web much like noodles in a dish. The final shape of the membrane is a thin flat sheet. The thickness of the sheet is controlled by the amount of polymer electrospun.

There are a number of parameters governing the characteristics of electrospun nanofibers. These parameters are commonly divided into three groups of polymer properties, solution properties, and process parameters. The polymer properties include polymer type, molecular weight, and molecular weight distribution. The solution properties include type of solvent, solvent evaporation rate, and solution surface tension, concentration, and viscosity. The process parameters include flow rate, applied voltage, distance from the tip of the nozzle to the collector, ambient temperature, and humidity.

### *4.2.1 Polymer Properties*

Many polymers, both synthetic and natural, have been electrospun. Conventional synthetic polymers such as polyvinylidene fluoride (PVDF) were first used as battery separator as well as in fuel cell and solar cell applications [22, 23] and later in ultrafiltration and membrane distillation applications [18, 24]. Other polymers such as polyacrylonitrile (PAN) [25–30], CA [31–33], and PEO [14, 34] were electrospun alone or together with other copolymers. Natural polymers such as silk [35–37] and chitin [38] have also been electrospun.

### *4.2.2 Solvent Properties*

Volatility of the solvent determines the rate at which the nanofiber dries. A highly volatile solvent might evaporate too quickly plugging the tip of the needle and stopping flow of the polymer solution. A low volatile solvent might not evaporate fast enough for the filament to dry into final nanofiber form [39–40]. An optimized solvent would evaporate before the nanofiber reaches the collector plate.

Solution concentration and viscosity are the two other factors that directly affect the fiber diameter, with lower viscosities producing finer nanofibers [14, 19]. As discussed earlier, the electrostatic repulsive charges in the polymer solution tend to increase the surface area of the filament by stretching it. An opposing force is the surface tension of the solution which tends to reduce the surface area. Because the optimum shape of liquids is a sphere, the opposing force of surface tension tends to break the jet stream into droplets. The viscosity of the solution plays an important role in balancing the two opposing forces. A low viscosity would help the surface tension breaking up the jet stream into droplets. An optimum viscosity would prevent the breakup of the jet stream and would allow for continuous shrinking of the filament until the solvent is fully evaporated, and the fiber is formed before it is collected on the ground plate. In practice, however, an in-between situation is common which is known as "beading." Beads are formed when viscosity is not low enough for the surface tension to fully dominate and breaks up the jet stream, but enough to form spheres along the length of the filament. Raghavan et al. [17] provided a detailed study of beading and factors affecting it. Figure 4.5 illustrates different shapes of beads.

### *4.2.3 Process Parameters*

The flow rate plays a role in the shape and diameter of nanofibers. Higher flow rates would produce thicker nanofibers. If the flow rate exceeds a limit at any given spinning condition the solvent would not have enough time to evaporate and the

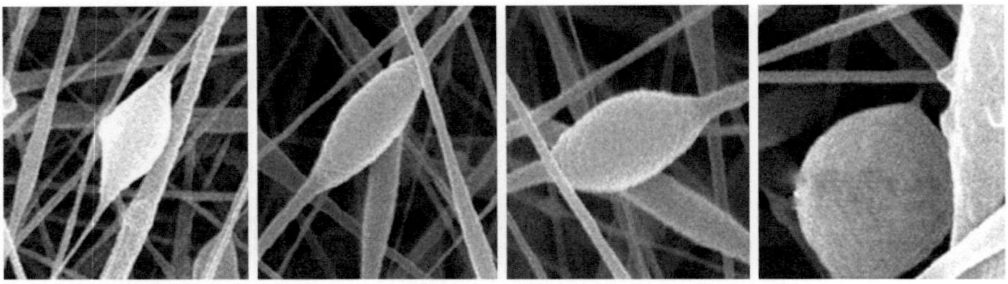

**Fig. 4.5** Formation of beads. Reproduced with permission of [17]

nanofibers would not form [40]. The simplest way to set the flow rate is to let the polymer solution drop from the tip of the needle due to the gravity [41]. In such case, the flow rate could be controlled by placing the syringe in an angle. The highest control of the flow rate is achieved when a pump is used to drive the solution out of the syringe. The latter also allows for horizontal set up of the spinning apparatus [42, 43]. Raghavan et al. [17] reviewed in details the parameters that affect the characteristics of ENMs.

The atmosphere under which electrospinning is performed and the humidity also affect the formation and shape of the nanofibers. Megelski et al. [40] showed that electrospinning cannot take place under helium atmosphere as the electrical breakage of this gas is 2,500 V. They also showed that fibers electrospun under Freon-12 were 1.4–2.6 times thicker than those spun in air. They explained this phenomenon by the higher breakdown voltage of Fereon-12 compared to air.

The distance between the tip of the needle and the collector also affects both the thickness and the shape of the nanofibers [14]. Longer distances would produce thinner fibers. A short distance would not provide enough time for the solvent to evaporate and instead would encourage beads formation.

A typical preparation of ENM has been presented by Feng et al. [18]. In this process, an 18 wt% solution of PVDF in DMF was electrospun at a rate of 2 mL/h. A total of 6 mL of polymer solution produced a 150 μm thick nanofiber membrane. The applied voltage was 18 kV and the distance from the tip of the spinneret to the collection plate was 18 cm. The membrane was kept in a fume hood for 24 h to dry.

## 4.3 Characterization of ENMs

Many techniques are available for characterization of ENMs depending on what we want to learn about them. The characterization techniques could be categorized in groups such as geometry, chemistry, surface properties, and performance.

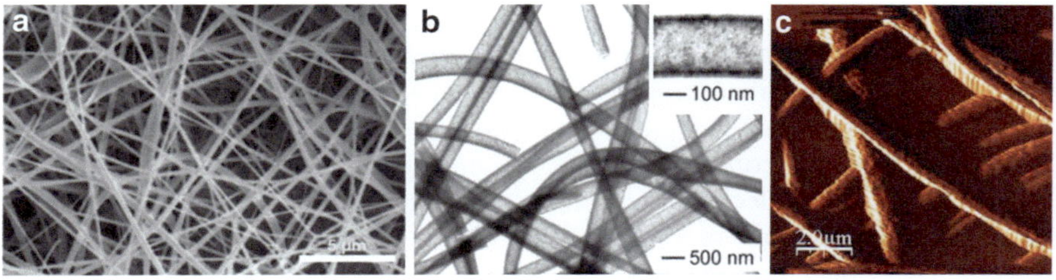


**Fig. 4.6** (**a**) SEM image of a PVDF nanofiber membrane; (**b**) TEM image of nanofibers; and (**c**) AFM image of a PVDF nanofiber membrane. Reproduced with permission of [18, 44]

## *4.3.1 Geometrical Characterization*

The geometry of the ENMs covers characteristics such as fiber size, pore size, and porosity. These parameters control flux and removal efficiency of an ENM. Techniques available for observing and measuring ENMs' fiber size and pore diameter include scanning electron microscopy (SEM), transmission electron microscopy (TEM), and atomic force microscopy (AFM). Figure 4.6 illustrates examples of SEM, AFM, and TEM images of nanofiber membranes. TEM has the advantage of producing images under dry or wet stages. AFM produces 2D images of the fibers. However, the thickness of the fibers might show larger due to the size of the cantilever used in AFM that cannot record smaller objects accurately.

Fiber size and pore diameter could be estimated by direct measurement of a SEM, AFM, or TEM image. Recently, Sambaer et al. [45, 46] have proposed a technique based on digital image analysis applied on nanoscale SEM images. The group suggests an accurate measurement of fiber diameter and pore size by increasing the contrast of the gray areas in a SEM image. They also proposed a model to calculate the pore size and pore size distribution.

The porosity is commonly reported as the ratio of the void space in a membrane to the volume of the membrane. In other words:

$$\text{Porosity} = \frac{\text{Specific volume of membrane} - \text{Specific volume of polymer}}{\text{Specific volume of membrane}}$$

where, specific volumes of membrane and polymer are the inverse of densities of membrane and polymer, respectively. This equation could be expressed in terms of densities as well:

$$\text{Porosity} = \frac{\rho_{\mathrm{p}} - \rho_{\mathrm{m}}}{\rho_{\mathrm{p}}}$$

where $\rho_{\mathrm{p}}$ and $\rho_{\mathrm{m}}$ are densities of polymer and membrane, respectively.

Two common techniques to determine pore characteristics are mercury intrusion porosimetry (MIP) and liquid extrusion porosimetry (LEP) [47]. Mercury is non-wetting to membranes and does not spontaneously penetrate the pores. In mercury intrusion technique, pressure is used to force mercury into the pores. The pore structure is determined through the intrusion pressure and volume. The disadvantages of this technique are high pressure requirement and possibility of distorting the pore structure.

In liquid extrusion technique, the membrane is soaked in a liquid that spontaneously wets the membrane. Such liquid would ideally form zero contact angle with the membrane. Then a relatively small pressure is used to force the liquid out of the membrane pores. The diameter of the pores is then determined through the following equation [48]:

$$d = \frac{4\gamma \cos\theta}{p}$$

where $d$ is the pore diameter, $\gamma$ is the surface tension of the extruding liquid, $\theta$ is the contact angle of the liquid (ideally, zero), and $p$ is the applied pressure.

Both above techniques assume uniform cylindrical pore structure, which is not true in reality. The techniques are also biased in predicting the shape and diameter of the pores. For example, mercury intrusion technique measures both through pores and blind pores (dead-end pores), while liquid extrusion technique can only measure the volume and structure of through pores. However, it is safe to assume that all pores in a nanofiber membrane are through pores and therefore, both techniques should report similarly.

Manickam and McCutcheon [47] used both LEP and MIP to characterize pore size, porosity, and pore size distribution of three different commercial nonwoven polyester (PET) and three different electrospun PAN nanofiber membranes. PET contained pores in the range of a few 10 μm, while pores of the PAN membranes were a few orders of magnitude smaller in the range of a few hundred nanometers. They observed that with one exception, MIP reported smaller porosity than LEP. Also, LEP was more successful in penetrating into smaller pores. The group attributed the difference between LEP and MIP results to the surface tensions of the two liquids. Mercury, due to its higher surface tension, cannot penetrate into fine pores. This is shown in Fig. 4.7.

A simplified version of liquid intrusion technique was used by Savoji et al. [44] in which, a dry membrane was weighed and then immersed in ethanol for many hours to ensure complete wetting. The surface of the wetted membrane was carefully wiped by a cloth to remove excess ethanol and then weighed again. The ratio of the volume of the ethanol entrapped in the membrane to the total volume (membrane + ethanol) was used as the porosity.

Tomadakis and Sotirchosm [49] suggested a simplified model to estimate pore size of fibrous material such as ENMs. According to them, there exist a direct relationship between total porosity ($\varepsilon$), average fiber diameter ($\omega$), and characteristic pore diameter of the structure:

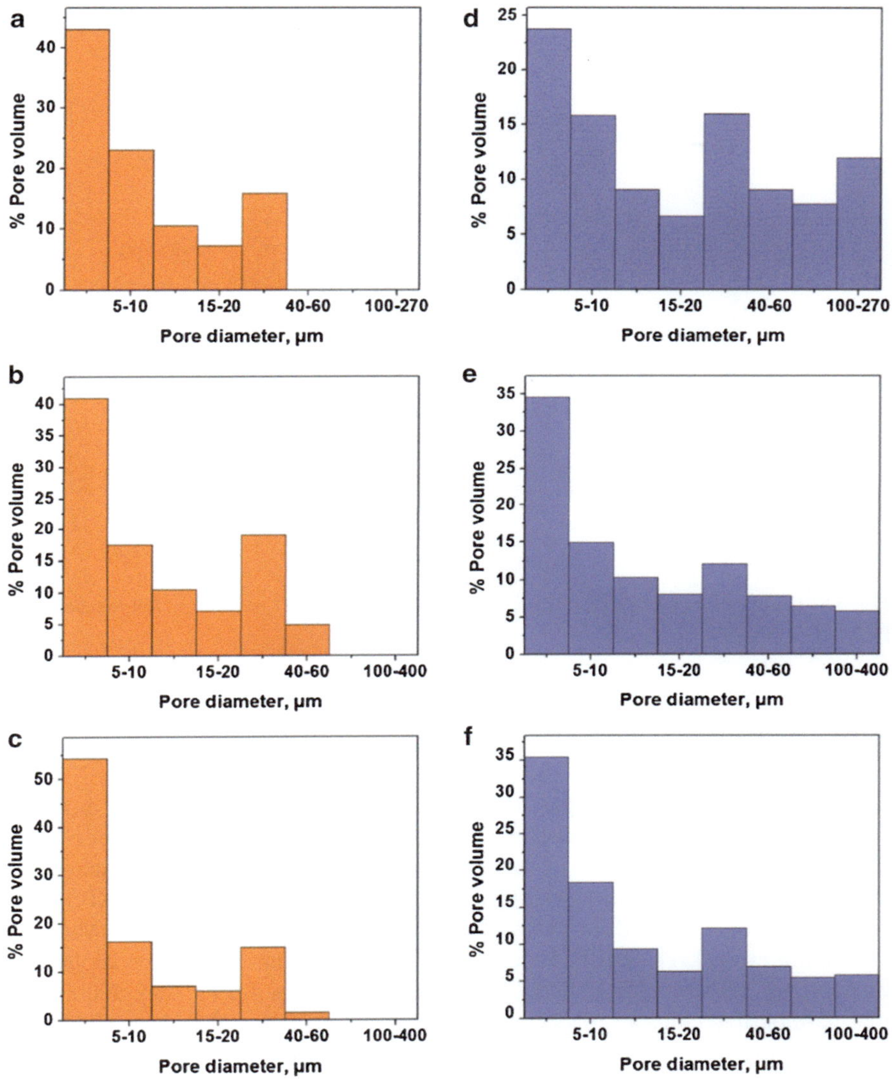


**Fig. 4.7** Pore size distribution of electrospun polyacrylonitrile nanofibers. (**a**) 8 %; (**b**) 10 %; and (**c**) 12 % PAN are from liquid extrusion porosimetry and (**d–f**) are the corresponding distribution from mercury intrusion porosimetry. Reproduced with permission of [47]

$$\text{Pore diameter} = -\omega/\ln(\varepsilon)$$

A capillary flow porometer could also be used to determine the pore throat diameter. The throat diameter of a pore is important because that is where sieving of the passing molecules or particles takes place.

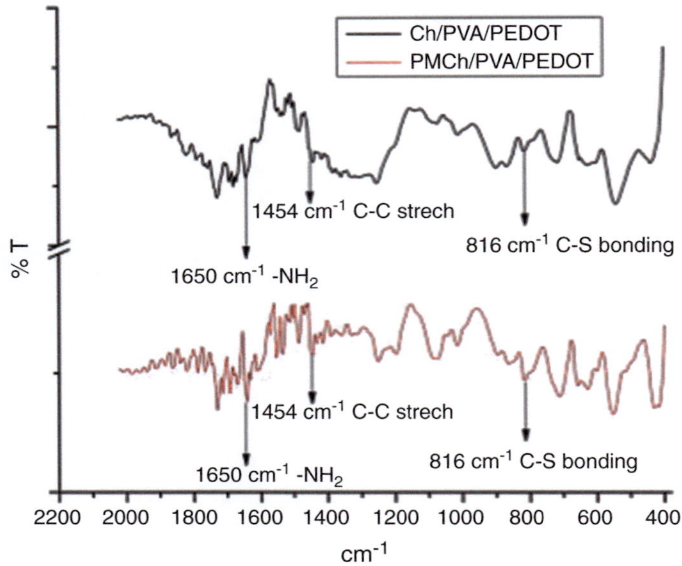


**Fig. 4.8** FTIR spectra of Ch/PVA/PEDOT and PMCh/PVA/PEDOT nanofibers. Reproduced with permission of [51]

### *4.3.2 Chemical Characterization*

Chemical characterization of nanofiber membranes are conducted using a different set of techniques. Fourier transform infrared (FTIR) and nuclear magnetic resonance (NMR) could be used to determine chemical structure and the interaction between polymer molecules [50]. X-ray photoelectron spectroscopy (XPS) and ATR-FTIR could be used for characterization of surface chemistry of ENMs.

Kiristia et al. [51] used FTIR to study chemical compositions and molecular interactions between components of a conducting nanofiber. They blended a plasma-modified chitosan (PMCh) and poly3,4-ethylenedioxythiophene (PEDOT) with polyvinyl alcohol (PVA) as supporting polymer at various volumetric ratios to obtain the conducting nanofibers. The FTIR they obtained is shown in Fig. 4.8 with the major characteristic peaks identified. They concluded that strong interactions among the polymer blends could be attributed to the formation of hydrogen bonds among the constituents.

Singh et al. [27] used ATR-FTIR to study carbonization of a PAN membrane. They compared spectra of PAN membrane samples carbonized at different temperatures and time to conclude that the aromatic rings of original PAN membrane fuse together in carbonized nanofiber membrane. That was a remarkable change in chemical composition that could explain the enhanced adsorptive characteristics of the carbonized nanofiber membranes. The spectra they obtained are

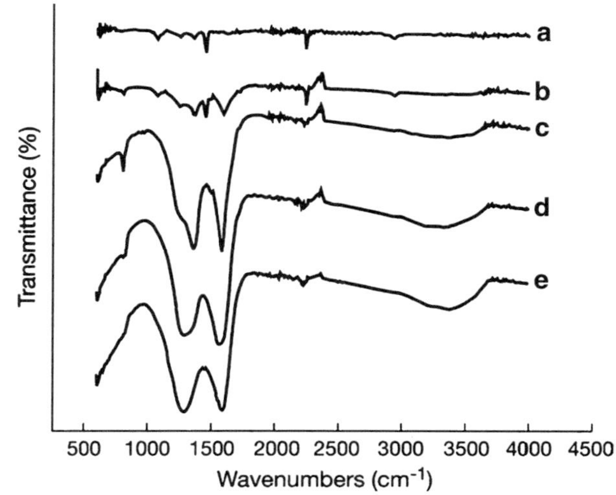


**Fig. 4.9** ATR-FTIR spectra of 10 wt% PAN nanofiber membrane under conditions: (**a**) air dried at 120 °C, (**b**) stabilized at 250 °C for 6 h, (**c**) carbonized at 400 °C for 4 h, (**d**) carbonized at 500 °C for 15 min, and (**e**) carbonized at 600 °C for 15 min. Reproduced with permission of [56]

shown in Fig. 4.9. The same group conducted differential scanning calorimeter (DSC) technique to suggest that carbonization of the membranes only takes place at temperatures in excess of 400 °C.

XPS determines the elemental composition of the surface layer of a membrane. One major application of XPS is to study surface modification of membranes. This is done by comparing the elemental composition of a membrane before and after the modification. Recently, Kayaci et al. [52] used XPS to study surface modification of polyester nanofibers with cyclodextrin. They reported an increase in elemental oxygen on the surface of the membrane which was interpreted as the enhanced presence of cyclodextrin.

### *4.3.3 Characterization of Surface Properties*

Surface properties of nanofiber membranes include roughness, surface charge, and hydrophobicity/hydrophilicity. The latter is determined using contact angle measurement. In this technique, a drop of water is placed on the surface of a membrane and the angle the drop of water makes with the membrane is measured. The water droplet spread easily on a hydrophilic surface making a small angle with the surface. The smaller the contact angle, the more hydrophilic the surface is. In contrast, a hydrophobic surface makes a large contact angle with the water droplet. Similarly, the larger the contact angle, the more hydrophobic the surface is. Commonly, a surface with a contact angle larger than 150° is considered superhydrophobic.

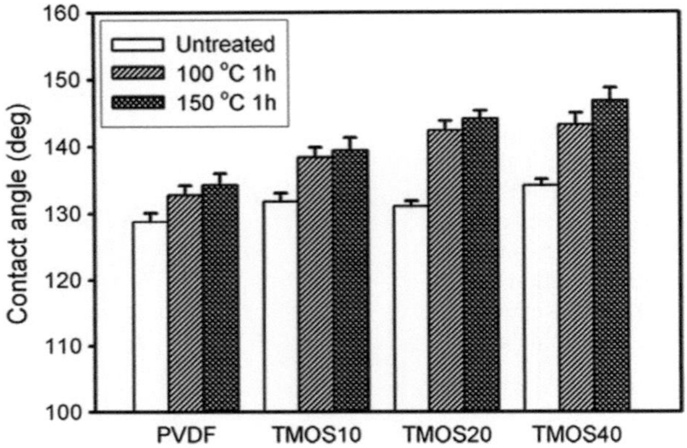


**Fig. 4.10** Use of contact angle to determine hydrophobicity of a nanofiber membrane. As the temperature of heat treatment and the TMOS content of the membrane increased the hydrophobicity enhanced. Reproduced with permission of [53]

An example of using contact angle to determine the hydrophilicity/hydrophobicity of an ENM is provided by Kim et al. [53]. The group used contact angle measurement to understand the change in hydrophobicity of a PVDF-tetramethyl orthosilicate (TMOS) blend as a function of TMOS content of the blend as well as the heat treatment temperature of the membrane. Figure 4.10 shows that both heat treatment and concentration of TMOS enhanced hydrophobicity of the base membrane.

The surface charge of a membrane could be determined by Zeta potential measurement. Dang et al. [54] modified a system to calculate Zeta potential based on the correlation between streaming potential and differential pressure using Helmholtz-Smoluchowski equation [55].

### *4.3.4 Performance Characterization*

The final category of characterization methods to be discussed here is performance of nanofiber membranes. Similar to other categories, there are a number of methods to achieve this, such as pure water permeation (PWP), molecular weight cut-off (MWCO), and liquid entry pressure of water ($LEP_w$) tests.

PWP test involves pushing pure water through a membrane at certain pressure and under steady state. The flux is calculated by dividing the permeation rate by the area of membrane. Also, to obtain a pressure normalized flux, the specific flux is determined by dividing the flux by transmembrane pressure. However, specific flux remains a weak function of pressure.

In MWCO test, solutes with different molecular weights are individually dissolved in feed water and are passed through a membrane. The removal efficiency increases with MW of the solute. At certain MW no more solute passes through the membrane which is known as MWCO for that particular membrane. In practice, the MWCO corresponds to 90 % removal efficiency. Polyethylene glycol

(PEG), with different molecular weights, is commonly used as the standard solute in this test.

$LEP_w$ is known as the pressure at which water penetrates a dry membrane. The $LEP_w$ test is most used in membrane distillation processes. To determine $LEP_w$ a dry membrane is placed in the separation cell. Both feed and permeate sides of the membrane cell are filled with pure water. The transmembrane pressure is increased in small increments (e.g., 0.1 bar) until a steady flow of water starts. This pressure is recorded as $LEP_w$.

## 4.4 Applications of ENMs

Superior characteristics of ENMs have already been realized in a number of applications such as filtration, sensors, energy storage, catalysis, and medicine. Water treatment is the most recent application. Barhate and Ramakrishna [1] listed 20 enterprises around the world that either produce or commercially use nanofiber membranes.

Nanofiber membranes have been found efficient in removing submicron particles, e.g., oil particles, from air in applications such as dust removal [56], and intake air cleaning for engines and turbines [57] and aerosols [58]. The major advantages of ENMs in air cleaning applications are their high filtration efficiency and low air resistance [59].

In the area of fuel cell, nanofibers are used as supporting material for dispersed platinum anode electrode. The high specific surface area of nanofibers provides a suitable support matrix for platinum nanoparticles [60]. Also, carbon nanotubes (CNTs) and carbon nanofibers have been used as the support for the Pt catalyst in a fuel cell [61].

In the tissue engineering ENMs have been used as tissue scaffolds. The material used for this purpose not only should be biocompatible but also should provide mechanical strength, high porosity, and suitable pore diameter and volume [62]. Biocompatibility of nanofibers allows for biomimicking the extracellular matrix (ECM) which is an essential factor for growth and functioning of the surrounding cells. Furthermore, the suitable pore characteristics would provide for nutrient intake and waste disposal. Natural nanofibers such as collagen and chitosan have been successfully used for tissue scaffolding. Mo et al. [63] have recently published a comprehensive review on applications of electrospun nanofibers in tissue engineering.

Enzymatic catalysis is another area where electrospun nanofibers have been applied. Enzymes are recognized as suitable catalysts for certain chemical reactions because of their sensitivity and mild reaction conditions [64]. However, enzymes need to be immobilized to optimize their stability, activity, and availability [65]. Nanofibers offer excellent characteristics as catalyst substrate due to their high catalyst loading and ease of contact because of their high porosity.

## 4.5 Applications of ENMs in Water and Wastewater Treatment

While conventional membranes have been in use in water treatment for a few decades, the potentials of ENMs have been realized just recently. Conventional membranes consist of a selective layer supported by a thick nonselective and highly permeable substrate. The pore size of the selective layer is controlled by different techniques to produce microfiltration (MF), ultrafiltration (UF), nanofiltration (NF), and reverse osmosis (RO) membranes. Conventional membranes suffer from intrinsic limitations such as low flux and susceptibility to fouling. Furthermore, formation of pinholes during preparation steps is common which contribute to separation inefficiency. Generally speaking, depending on the type, membranes have porosities from 5 to 35 %. Nanofiber membranes could easily be made into MF and UF membranes with porosities in the range of 80 % [25]. In addition, the interconnectivity of the pores eliminates the inefficiency of the dead-end pores that are commonly formed in conventional membranes. These characteristics would significantly increase flux and reduce transmembrane pressure drop, hence the overall cost of water production. In the following sections major applications of ENMs in water and wastewater treatment are presented.

### *4.5.1 Particle Removal and Pre-filtration*

The high efficiency of nanofibers in removing microparticles from air opened the door to similar applications in water and wastewater treatment.

In ultrafiltration, nanofiltration, and reverse osmosis applications it is important to remove relatively large particles (submicron to micron size) before the raw water is fed into the membrane system. Particle removal would reduce fouling and deterioration of the membrane system and would increase the life span. ENMs could offer a cost-effective pre-filtration to remove particulates from water.

Gopal et al. [66] electrospun polysulfone into a nanofiber membrane with average pore size in the range of 1.2–1.6 μm and the largest pore at 4.6 μm. The group challenged the membrane samples with polystyrene latex microparticles in 0.1–10 μm size. The removal efficiency was 99+ % for particles larger than 7 μm without any permanent fouling. However, an irreversible fouling was observed with particles at 1–2 μm in size. For smaller particles of 0.1 and 0.5 μm, the membrane performed as a depth filter with removal efficiencies of ~89 %.

Interestingly, the particle removal was achieved at negligible pressure drop of 0.5 psi. The research group concluded that ENMs could be used for particle removal especially for particles larger than the bubble point of the ENMs (corresponding to the largest pore). Figure 4.11 illustrates the removal of microparticles using the polysulfone ENM.

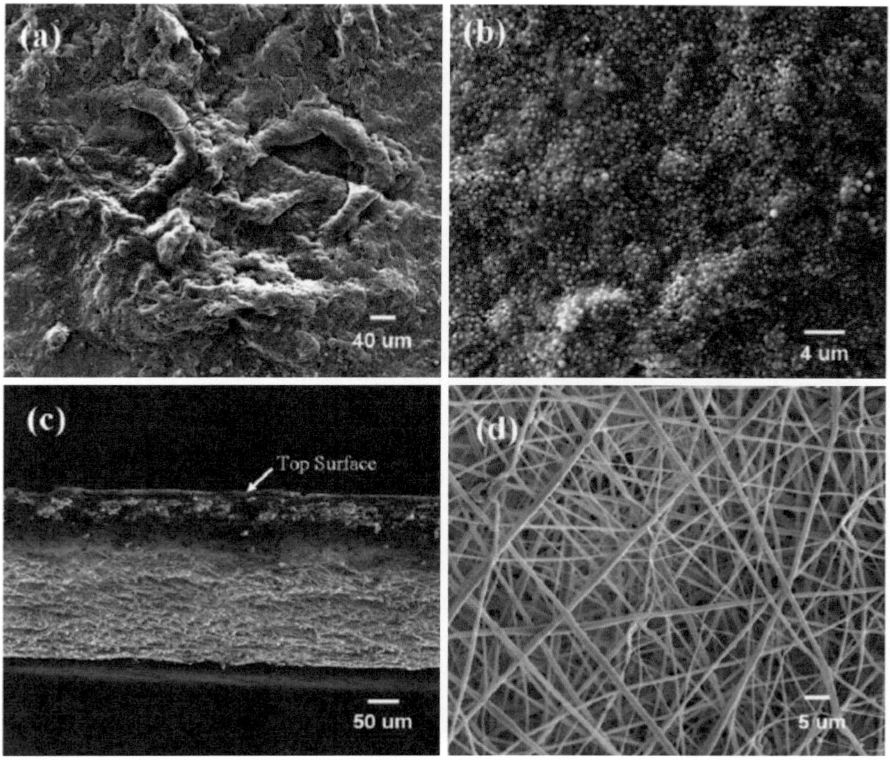


**Fig. 4.11** Electrospun polysulfone nanofiber membrane after particle-challenge test: (**a**) top surface; mag. ×400, (**b**) top surface; mag. ×6,000, (**c**) cross section; mag. ×300, and (**d**) bottom surface; mag ×3,000. Reproduced with permission of [66]

Xu et al. [67] used electrospun polysulfone fiber membrane (EPSFM) to remove particles from secondary bio-treated sewage. The group filtered actual sewage samples using EPSFM with and without precoagulation. Their objective was to reduce chemical oxygen demand (COD), ammonia $NH_3$–N, and suspended solid from a bio-treated wastewater before reusing it for flushing and cleaning purposes.

The group achieved removal efficiencies of 86.7, 71.2, and 91.7 % for suspended solid, CODs, and ammonia, respectively, without precoagulation of the wastewater samples. When polyaluminum chloride was used as a coagulating agent before filtering through the membrane the efficiencies increased to 91.3 %, 85.3 %, and 93.3 %, respectively. Precoagulation-assisted filtration was especially efficient in reducing the COD content of the wastewater as well as the concentrations of a number of VOCs. In addition, the membrane was able to significantly reduce the content of toxic metals such as copper, zinc, and titanium through adsorption process. The group concluded that EPSFM was highly efficient in removing micron and submicron particles from secondary-treated wastewater in addition to removal of several other target contaminants.

Another equally important area is removal of biofoulants that at the same time contributes to fouling by microparticles as well. El-Azizi et al. [68] and Komlenic et al. [69] in two inter-related studies investigated removal of transparent exopolymer particles (TEPs) from fresh water, seawater, and wastewater before feeding the water to UF or RO membranes.

TEPs are sticky organic microgels that accumulate on the surface of the membrane with a double fouling function. They form a cake layer that reduces the water flux and at the same time, they attract and trap suspended solids and microparticles which enhance TEPs' fouling effect. Removing TEPs would increase the life span of the membrane and reduce the energy usage by maintaining the transmembrane pressure low and the flux high.

The two research groups used a deep filter cartridge made of positively charged nanofibers to retain TEPs before they hit the UF and RO membranes. TEPs typically have negative charge which helps attaching to the nanofibers and being removed from water. The long-term tests concluded that the pre-filter would retain the flux of an RO membrane with salt rejection almost intact over 3 months. The same membrane without the pre-filter showed approximately 20 % decline in the flux after 2 months followed by a sharp drop to ~50 % of the initial flux in the third month.

### 4.5.2 Bacteria and Viruses Removal

Removal of bacteria and viruses follow the same principles as particle removal. Bacteria are in micrometer size, while viruses are in the range of tens of nanometers. An ENM such as the one that were discussed above could remove bacteria by size exclusion, but would need much smaller pores for an efficient virus removal. The smaller pore size, however, would significantly reduce the water flux.

Recently, Sato et al. [30] developed a novel nanofiber membrane capable of removing both bacteria and viruses while maintaining the flux. This composite membrane consists of cellulose ultra-fine fibers infused into a PAN ENM on a nonwoven substrate. To remove viruses, the group charged the membrane surface by adding cellulose fibers. The electrostatically positive charged cellulose fibers attracted and trapped the slightly negatively charged viruses. They achieved 6-log (99.9999 %) removal of *E. coli* using this membrane. This performance satisfied the requirement set by the National Sanitation Foundation Standard of max 2 cfu/mL. Figure 4.12 shows the top surface and cross section of the composite nanofiber membrane used in this study. The honeycomb-shaped cake layer shown in Fig. 4.12a, b is dead *E. coli* accumulating on the surface of the membrane. The cake layer would enhance the removal of *E. coli*. At low concentrations of *E. coli*, the bacteria are trapped in the web of the nanofibers as shown in Fig. 4.12c.

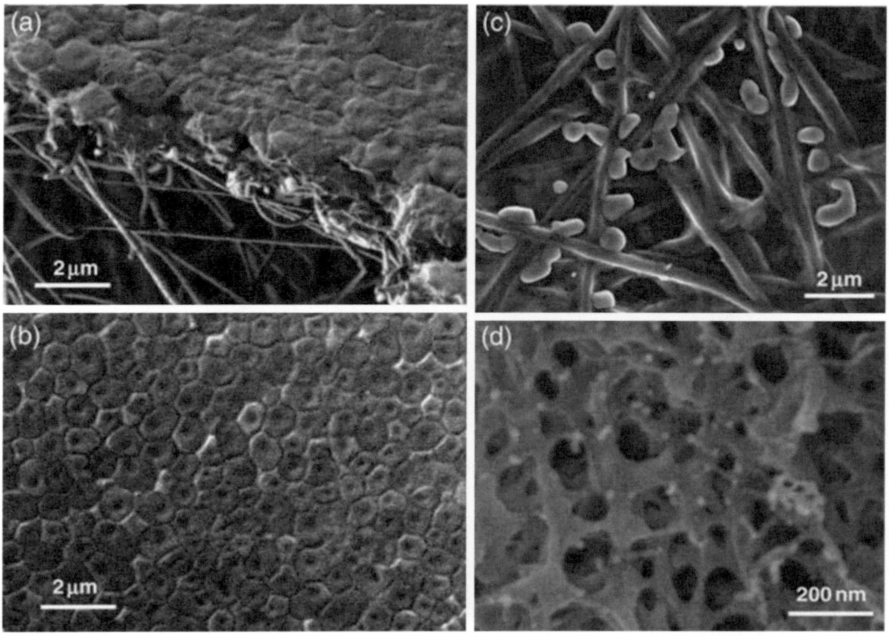


**Fig. 4.12** (**a**–**c**) SEM images of the electrospun PAN nanofibrous scaffold infused with m-UFCNs after the *E. coli* bacteria filtration test. The retained "cake layer" on the surface of the electrospun PAN layer with honeycomb *E. coli* dead cells (at high concentration) is evident in (**a**) cross-sectional view and (**b**) surface view, where (**c**) *E. coli* is trapped in the scaffold (at low concentration). (**d**) SEM image of electrospun PAN scaffold infused with m-UFCNs after MS2 virus adsorption test, showing the MS2 virus (*white dots*) trapped in the m-UFCN network. Nanofibers have also found applications in protective clothing, wound dressing, cosmetics, and prostheses. Reproduced with permission of [30]

## *4.5.3 ENMs as Substrate (Scaffold) in Thin Film Composite Membranes*

Membranes usually benefit from an asymmetric structure in their cross sections. That is, a thin selective layer on top of a porous and nonselective substrate. The flux and rejection of a membrane depends on the characteristics of both layers. A thick top layer would reduce flux as does a nonporous substrate. An optimum configuration would benefit from a thin selective layer which is mechanically supported by a thick, nonselective, and highly permeable substrate. The separation/rejection would take place at the thin selective layer and then the permeant flow with relative ease through the substrate.

The first generation of membranes was integrated asymmetric membranes where the selective layer and the support substrate were prepared from the same material simultaneously. In these membranes, tailoring the thickness of the top layer and the substrate simultaneously was difficult if not impossible. The invention of thin film

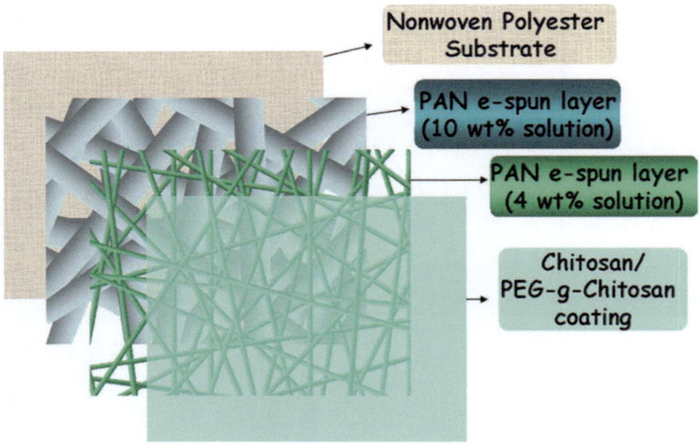


**Fig. 4.13** Schematic diagrams for the assembly of three-tier composite membrane. Reproduced with permission of [25]

composite (TFC) membranes offered the flexibility to tailor both top layer and the substrate independently for optimized flux and rejection [70]. This is done by preparing a thin selective layer and the porous substrate separately and then attaching the selective layer on top of the porous substrate. Finally, the two layers are placed on a mechanically strong non-resistive support layer.

The overall resistance of a TFC towards the permeation of water is the sum of the resistances of the top layer and the porous substrate with a negligible resistance from the mechanical support that could be assumed zero. A thinner selective layer and a more porous substrate would reduce the overall resistance of a TFC. In conventional TFCs, the top selective layer consists of a membrane with desirable pore size, porosity, and thickness, and the substrate is a second membrane with much larger pores that would exhibit smaller resistance to the flow of water. The substrate is usually a MF or UF membrane.

ENMs are excellent choices for the substrate in TFCs. Their large pore size and high porosity would impose negligible resistance to the flow of water while their mechanical strength would provide the required support to the top selective layer. The overall effect would be a membrane with similar rejection rate as the conventional TFCs at higher water flux.

Most of the work in this area has been done by the group at the Stony Brook University. The group [71–73] demonstrated a three-tier composite membrane consisting of a thin hydrophilic water permeable anti-fouling layer coated on top of a nanofiber scaffold and supported by a conventional nonwoven microporous support layer. The separation takes place at the nanofiber layer. Because ENMs are prone to fouling due to their highly porous surfaces, the function of the top coating layer is to protect the ENM from fouling. The third layer imposes negligible resistance to the flow of water and only provides mechanical support to the coating and the ENM layers. A schematic of the membrane is shown in Fig. 4.13.

The thin film nanofiber composite (TFNC) membrane prepared by Wang et al. [71] consisted of a substrate made of crosslinked PVA coated with hydrophilic polyether-b-polyamide copolymer. Surface-oxidized multiwall carbon nanotubes (MWNTs) were also incorporated into the top layer. The group tested the membrane for oil/water emulsion and obtained flux as high as 330 $L/m^2/h$ and a rejection rate of 99.8 % at negligible fouling. It was also demonstrated that the flux could be improved by increasing concentration of MWNT.

Wang et al. [72] also prepared a TFNC in which the coating layer was made from the same material as the scaffold. In this experiment, the group used PVA for both the substrate and the coating layer. The substrate, in this configuration, consisted of an ENM and the coating layer was nonporous and was crosslinked differently. Different molecular weight polymers and different degrees of hydrolysis were examined for preparation of the TFNC. The best results were obtained from a membrane fabricated using 96 % hydrolyzed PVA and relatively high molecular weight (85,000–124,000 g/mol). The membrane exhibited very good mechanical strength and high flux of 130 $L/m^2/h$ and a rejection efficiency of 99.5 %.

The same group [25] improved the performance and mechanical strength of the proposed membrane by using a hydrophobic PAN as the nanofibrous substrate and chitosan as the coating top layer. PAN is resistant to most solvents and therefore a common polymer in preparation of ultrafiltration [74], nanofiltration [75], and reverse osmosis membranes [76]. Chitosan is a hydrophilic biopolymer that is commonly used for its anti-fouling characteristics [77]. A thin layer of chitosan coated on the surface of the PAN ENM would allow water to permeate with minimum resistance and, because of its smooth surface, would minimize fouling.

To prepare a uniform coating layer, the group designed a multilayer substrate. A finer fiber diameter was placed in immediate contact with the chitosan coating, while a layer of larger diameter fibers enhanced the mechanical strength. The performance of two chitosan/PAN TFC membranes, one with 1.37 % and the other with 1.2 % chitosan coatings, was compared with a commercial NF membrane. The test was done with oily water at 130 psi pressure and 30–33 °C temperature. The chitosan/PAN TFNC membranes exhibited fluxes at least one order of magnitude higher than the commercial membrane. Furthermore, the flux decline was slower with the TFNC membranes over a 24 h test period. Figure 4.13 illustrates the three-tire TFNC with multilayer substrate.

Furthermore, the porosity of the PAN ENM was compared with conventional and commercially available MF, UF, and NF membranes. The results, as shown in Fig. 4.14, illustrated that PAN ENM with ~73 % porosity was an order of magnitude more porous than both NF and UF membrane and twice as porous as the MF membrane.

A few years later, Yoon et al. [26] applied the same concept to prepare a NF membrane using ENM. An interfacially polymerized polyamide layer was coated on top of a PAN scaffold similar to the one described above. The polyamide barrier layer contained different ratios of piperazine and bipiperidine. The performance of

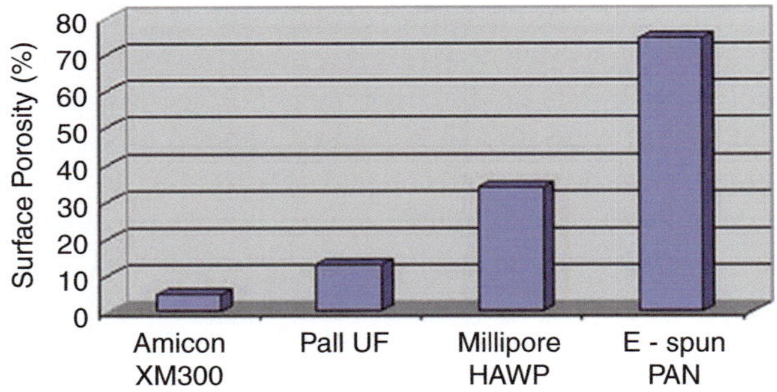


**Fig. 4.14** Surface porosity of conventional membranes and electrospun nanofiber PAN. Reproduced with permission of [25]

the thin film nanofibrous composite (TFNC) was compared to those of laboratory-made conventional TFC as well as a commercial NF TFC membrane. It was demonstrated that the TFNC exhibited over 2.4 times permeate flux compared to the conventional TFC and 38 % higher than that of the commercial NF TFC membrane. The rejection rate of TFNC towards 2,000 ppm of $MgSO_4$ remained the same as that of the tested commercial TFCs at 98 % under similar operating conditions and flow patterns.

Bui et al. [78] prepared a similar membrane with in situ polymerization of polyamide as the top coating layer over a polyethersulfone ENM. The group prepared several TFNC membranes with different polymer/solvent compositions and ratios for engineered osmosis. They observed 2–5 times higher water flux and 100 times salt rejection compared with a standard forward osmosis commercial membrane. Figure 4.15 illustrates the TFNC membrane prepared by this group. The same figure also shows an inadequate adhesion between the supporting and the skin layers in a TFNC membrane.

More recently, You et al. [79] used a novel technique to generate the top barrier layer. They first electrospun both the substrate and the barrier layer using PAN and PVA, respectively. Then, the double layer mats were immersed in an optimized water/acetone solution and were chemically crosslinked by glutaraldehyde. The top layer was swollen and the fibers merged to form an integrated film on the supporting PAN layer. In order to increase the free volume within the top barrier MWNTs were added to the top layer polymer solution at different dosages before electrospinning it. The membranes prepared this way were of UF type and were tested for oil/water emulsion separation. The water flux was markedly improved and was further enhanced with increasing MWNT concentration in the top layer. A flux of 270.1 $L/m^{2/}h$ was obtained at a rejection rate of 99.5 % and at a relatively small applied pressure of 0.1 MPa (~14.5 psi). Both positron annihilation lifetime spectroscopy (PALS) and permeation experiments confirmed increasing free volume and therefore water flux with increasing concentration of MWNTs in the polymer solution.

**Fig. 4.15** SEM images of (**a**) electrospun PES and (**b–d**) PES-based TFNC polyamide membranes at magnifications of (**a**) ×2,200, (**b**) ×460, (**c**) ×8,850, and (**d**) ×5,750. Images (**c**, **d**) show poor adhesion between polyamide and polyethersulfone nanofiber support. Reproduced with permission of [78]

## *4.5.4 ENMs as Adsorption Medium*

Development of nanomaterials was a significant progress in the field of catalysis and adsorption. Nanoparticles could offer advantages to the existing water and wastewater treatment processes. For example, CNTs exhibit adsorption capacities much higher than conventional activated carbon. They are also capable of efficiently removing low molecular weight polar organics that activated carbon cannot remove [80, 81]. In addition, CNTs are excellent adsorbents for heavy metals from water through electrostatic attraction and chemical bonding [82]. However, nanosize materials are most often harmful to health. A major challenge with nanoparticles before they become commercially available to the water treatment processes is to properly anchor them and preventing leakage into the processed water.

Nanofiber membranes could offer a practical solution to this by embedding nanoparticles in the fibers of an ENM. The high porosity and small diameter of nanofibers provide a platform for the nanoparticles—such as CNTs—to exhibit their superior adsorption performance, while bonded to the ENM matrix.

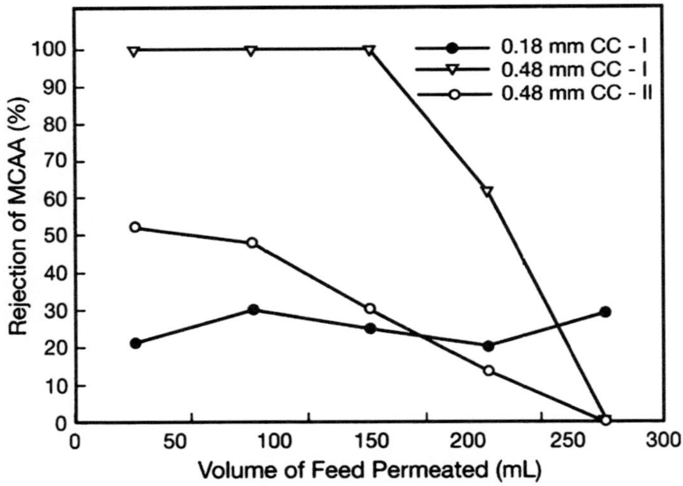


**Fig. 4.16** Removal efficiency of chloroform of concentration 100 mg/L using membranes of thickness and carbonization conditions: (**a**) 0.18 mm CC-I, (**b**) 0.48 mm CC-I, and (**c**) 0.48 mm CC-II. Reproduced with permission of [27]

Singh et al. [27] investigated the adsorption capacity of carbonized ENMs. PAN ENMs were carbonized by heating them under nitrogen atmosphere to 400–500 °C at a rate of 5 °C/min and held at the target temperature between 15 min and 4 h [27]. In addition, separate PAN membranes were prepared by dispersing multi-walled carbon nanotubes (MWCNTs) in the polymer solution before electrospinning it. The MWCNT embedded membranes were carbonized at 400 °C for 4 h. The ENMs prepared this way were used as adsorbents rather than filters. The idea was to remove target contaminants from water by adsorbing them on the carbon nanoparticles.

Water spiked with chloroform and monochloroacetic acid (MCAA) was passed through carbonized ENMs and their adsorption capacities were determined. Concentrations of the target contaminants were between 1 and 100 mg/L. An adsorption capacity of 554 mg/g for chloroform was achieved for a feed solution of 100 mg/L. The adsorption capacity for MCAA with concentrations of 4–18 mg/L was determined to be 287–504 mg/g. The experiments indicated that the adsorption capacity directly increased with the polymer concentration and with the thickness of the membrane. It is interesting to note that unlike conventional membrane separation processes where thinner membranes would perform better, in the case of adsorptive membranes thickness would improve performance. This makes sense because a thicker membrane would provide larger surface area available for adsorption. Figure 4.16 shows the rejection of MCAA by two adsorptive ENMs: CC-I was heated from 250 to 400 °C at 5 °C/min and held at 400 °C for 4 h, and CC-II was heated from 250 to 500 °C at 5 °C/min and held at 500 °C for 15 min. The thickness of the membranes was 0.18 or 0.48 mm.

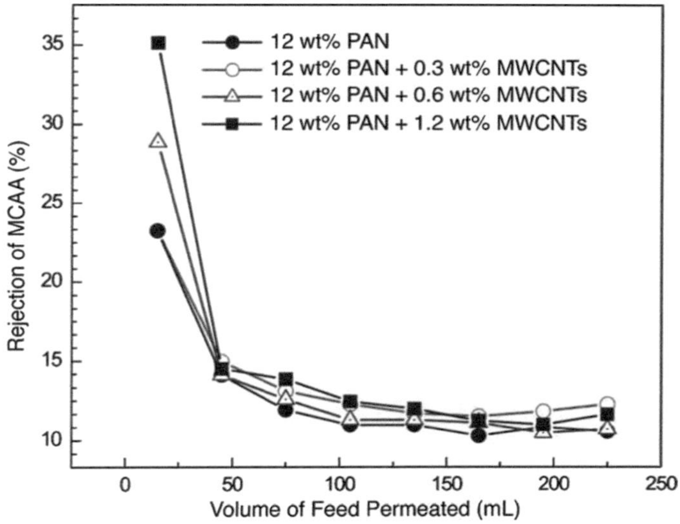


**Fig. 4.17** Rejection of MCAA for carbonized membranes with and without the incorporation of MWCNTs. Reproduced with permission of [27]

The MWCNT blended ENMs were also tested for their adsorption capacity. In this case, MCAA solution at 80 mg/L concentration was passed through MWCNT blended ENMs with 0.20 mm thickness. It was observed that the initial adsorption capacity of membranes enhanced by increasing MWCNT content. However, the capacity dropped sharply to that of a membrane without any MWCNT. This is shown in Fig. 4.17 [27]. The phenomenon was explained by the small portion of the MWCNTs at the surface of the nanofibers, where they were exposed to the contaminants. The exposed portion of the MWCNTs was saturated quickly and therefor the capacity declined sharply.

Regeneration of the carbonized membranes was also studied. It was reported that the carbonized ENMs were fully regenerated by soaking in 0.05 M NaOH.

Liang et al. [83] used a different technique to prepare adsorptive membranes. The group used carbonaceous nanofibers (CNFs) and carbonized them by hydrothermal carbonization (HTC) at 160 °C. The diameter of nanofibers was controlled by regulating the HTC time. The adsorption capacities of the membranes were tested using methylene blue (MB), lead (Pb-II), and Chromium (Cr-VI). The maximum capacity obtained were 818.6 mg/g, 423.7 mg/g, and 221.3 mg/g for MB, Pb-II, and Cr-VI, respectively. These values were 2–3 times larger than the GAC capacity for each of these contaminants. Water flux was 1,580 L/m$^2$ h which was 10–100 times higher than commercial membranes with similar rejection characteristics. The adsorption capacity increased with decreasing fiber diameter. The best results were obtained using a membrane with the fiber diameter of 50 nm.

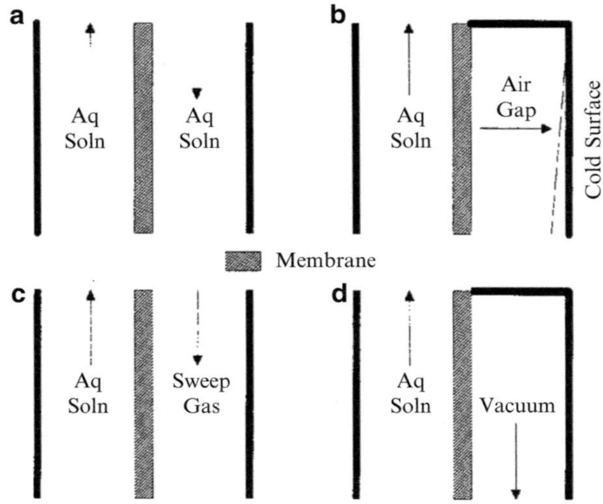


**Fig. 4.18** Simplified illustration of four-membrane distillation configurations: (**a**) direct contact membrane distillation (DCMD), (**b**) air-gap membrane distillation (AGMD), (**c**) sweep gas membrane distillation (SGMD), and (**d**) vacuum membrane distillation (VMD). Reproduced with permission of [84]

### *4.5.5 Applications of ENMs in Membrane Distillation*

Membrane distillation (MD) is a thermally driven process in which a hydrophobic semipermeable porous membrane only allows for passage of water vapor, but not liquid water. The driving force is the difference in water vapor pressure across the membrane. In a typical MD process, water is heated to around 60 °C. The hydrophobic membrane repels liquid water and prevents it from entering the pores. Vapor molecules, on the other hand, cross the membrane and are condensed at the permeate side or are removed by vacuum or a sweep gas in order to maintain the vapor pressure difference across the membrane.

The major advantage of MD is that it works at temperatures well below the boiling point of water, hence saving energy cost. The low operating temperature allows for the use of waste heat or alternative energy sources such as solar and geothermal. Other advantages include smaller footprint and volume compared to a conventional distillation column, and no need for an extensive pretreatment [84]. The major disadvantage of conventional MD processes is low water flux compared with competitive technologies such as reverse osmosis [85]. This has slowed down large-scale industrial applications of MD in areas such as water desalination and water purification. The situation has been reversed in the recent years with the advances made in preparation of highly porous ENMs that allows for high water flux.

There are four different types of membrane distillation processes, including direct contact membrane distillation (DCMD), air-gap membrane distillation (AGMD), sweep gas membrane distillation (SGMD), and vacuum membrane distillation (VMD). Figure 4.18 simplifies the four MD processes.

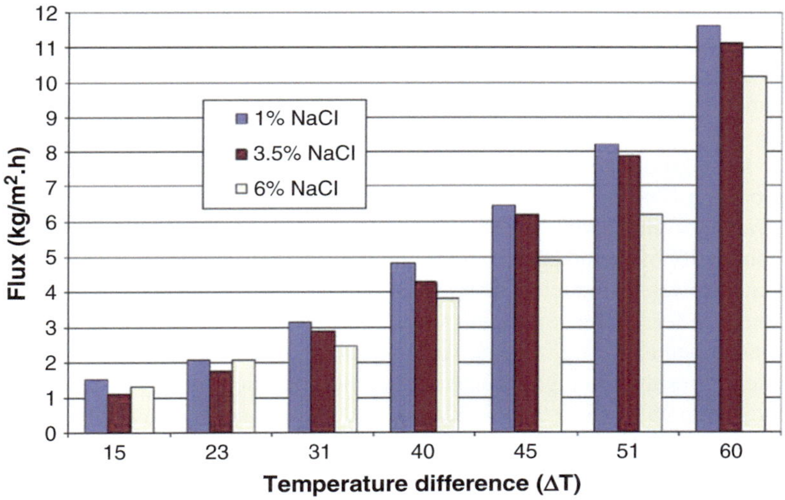


**Fig. 4.19** AGMD flux versus feed NaCl concentration at different temperature difference. Reproduced with Permission of [89]

In DCMD a cold liquid flows at the permeate side of the membrane and in direct contact with it. Water vapor passes through the membrane and condenses and mixes with the cold liquid at the permeate side. In AGMD, a thin layer of air separates the membrane and a cooling surface. The water vapor condenses on the surface of the cooling plate. In SGMD a gas flows at the permeate side and removes the vapor from the surface of the membrane. Finally, in VMD the vaporized liquid is removed by applying a vacuum [86].

Each of the four configurations performs the best for specific applications depending on the characteristics of the feed such as volatility, flux, and composition. For example, DCMD is the simplest of the four and could be applied to desalination where the bulk of water is to permeate. SGMD and VMD could be applied to applications such as VOCs removal from water where the permeate is the volatile contaminant rather than the water. AGMD is the most versatile configuration and could be applied to any of the applications [84].

The membrane used in MD processes should be porous so it imposes a negligible resistance to the flow of water vapor. Usually microfiltration or ultrafiltration membranes are used for this purpose. Also, the membrane should be made of a hydrophobic polymer such as PVDF, polypropylene (PP), and Teflon to repel liquid water.

In the recent years, only a few studies attempted developing new membranes for MD purposes [87, 88]. Recent advances in ENMs with favorable characteristics have renewed the opportunity for developing a cost-effective membrane distillation process. The first attempt was made by Feng et al. [89] who successfully examined PVDF in an AGMD setup for desalination. The group produced drinking water ($<280$ ppm NaCl) from up to 6 % salt solution at fluxes as high as or higher than commercial PVDF membranes. Figure 4.19 illustrates the results. They observed an exponential increase in flux with increasing temperature difference between the feed and the cooling water. Also, flux slightly decreased with increasing salt content of the water, which was attributed to the lower water vapor pressure. Long-term experiments indicated no variation in flux or salt rejection after 25 days of continuous operation.

The same group [85] tried increasing hydrophobicity of the PVDF membrane by incorporating Cloisite® clay in the electrospinning solution at concentrations 2, 4, and 8 %. The contact angle measurements confirmed the direct correlation between hydrophobicity of the prepared ENMs and the clay concentration. The desalination experiments were conducted using a DCMD configuration using a 3.5 % NaCl solution. The feed temperature was varied between 50 and 80 °C and the permeate was maintained at $17 \pm 2$ °C. The results indicated that higher clay concentrations consistently resulted in higher fluxes. Also, incorporation of clay enhanced the salt rejection from 98.27 to 99.95+ %. The DCMD showed consistent performance with time in terms of both flux and salt rejection when clay was incorporated in the membrane.

The number of research papers dealing with the use of ENMs in membrane distillation suddenly jumped in 2013 indicating realization of potential of these membranes for MD applications. Liao et al. [90] built up on the two studies discussed above to improve the PVDF performance by heat-pressing posttreatment of the ENM. The membranes were tested in a DCMD configuration using 3.5 % NaCl solution as feed. They achieved a permeation flux of 21 kg/m$^2$/h which was higher than the ones reported earlier [85, 89]. However, the study does not report salt rejection.

In a follow-up study the same group attempted preparing superhydrophobic PVDF nanofiber membranes for DCMD desalination. The membranes were fabricated by either integral modification (I-PVDF) or surface modification (S-PVDF) methods. In both methods the nanofibers were coated with silver nanoparticles in order to increase roughness which is known to enhance hydrophobicity. The difference between the two methods was that in integral modification the whole surface of the nanofibers was coated by silver nanoparticles, while in surface modification only the top surface was. Poly dopamine (PDA) was used as adhesive in order to enhance the attachment of the nanoparticles. The silver-coated membranes were further coated by 1-dodecanethiol to complete the superhydrophobicity. Figure 4.20 illustrates the modified PVDF ENMs as observed by field emission scanning electron microscope (FSEM) [91].

Characterization of the modified PVDF membranes confirmed achieving enhanced hydrophobicity by both methods. The integrally modified membrane (I-PVDF) exhibited a contact angle of $153 \pm 4$ and the surface modified membrane (S-PVDF) had a contact angle of $158 \pm 3$. Contact angles above 150° are considered superhydrophobic. The permeation tests indicated that the I-PVDF had high and stable flux of 31.6 L/m$^2$/h using a 3.5 % NaCl solution at a salt rejection comparable to commercial PVDF membrane. The performance of S-PVDF was not as impressive and did not show a significant improvement over the commercial PVDF.

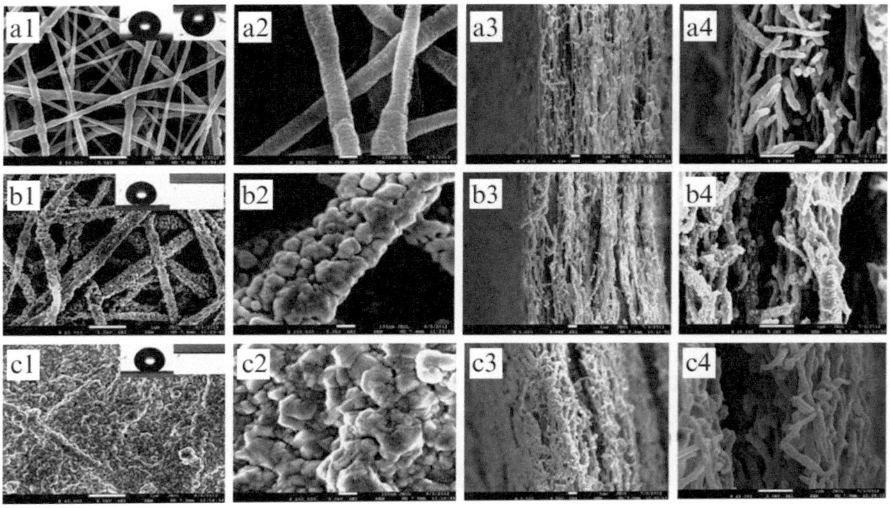


**Fig. 4.20** FSEM images of (**a1**, **a2**): pristine PVDF; (**b1**, **b2**): I-PVDF; (**c1**, **c2**): S-PVDF nanofiber membranes; and (**a3**, **a4**): cross-section images of pristine PVDF; (**b3**, **b4**): I-PVDF; and (**c3**, **c4**): S-PVDF nanofiber membranes. Reproduced with permission of [91]

Removal of VOCs from water and separation of water/ethanol were attempted by Feng et al. [18, 24]. VOCs removal has applications in wastewater treatment while water/ethanol separation by MD could be applied to break azeotropic mixture in conventional distillation processes. In the case of VOCs removal, SGMD was used to remove chloroform (a typical VOC found in wastewater) from water. Nitrogen was used as sweep gas. The feed temperature was varied from room (23 °C) to 60 °C. The feed concentration was between 250 and 2,000 ppm. An impressive 80 % chloroform removal was reported after 1 h of operation by SGMD.

In terms of water/ethanol separation, Feng et al. [18] used a VMD to separate ethanol at concentrations from 20 to 80 %. The feed temperature was 40, 50, and 60 °C. They observed permeate concentrations of ethanol lower than the vapor/liquid equilibrium. The separation factors were between 1.4 and 2.4. They also noticed that feed temperature did not have a significant effect on the separation efficiency.

Essalhi and Khayet [92] examined the effect of thickness of PVDF nanofibers on characteristics of ENMs and performance of a DCMD. The thicknesses of the fibers they produced were between 144.4 and 1,529.3 μm. The saline water concentration was between 12 g/L (1.2 wt%) and 60 g/L (6 wt%). The feed temperature was varied between 40 and 80 °C. It was concluded that nanofiber thickness did not affect the hydrophobicity of the membranes as observed through contact angle measurements. However, higher fluxes were obtained from thinner fiber membranes. They also confirmed the previously known effects such as enhanced flux at higher feed temperature and lower NaCl concentrations. The highest fluxes they achieved were in the range of ~$14 \times 10^{-3}$ to ~$15 \times 10^{-3}$ kg/m$^2$/s (~50–54 L/m$^2$/h).

This level of flux was higher than in any previous study. Salt rejection was 99.39 % as a minimum with 99.9+ % achieved under varying conditions.

Most recently, Prince et al. [93] have proposed a three-layer hydrophobic/hydrophilic membrane configuration where a thin hydrophobic ENM membrane is supported by a conventional microporous layer and a hydrophilic support substrate. The performance of this configuration was tested in an AGMD system for desalination and was compared with the conventional two-layer MD membranes (with no ENM attached) and with a standalone ENM. PVDF was used for both ENM selective layer and the microporous support layer. The substrate was made of poly ethylene terephthalate (PET). The test solution was 3.5 % NaCl in water.

The concept of triple layer configuration is based on the rationale that the top hydrophobic layer would prevent penetration of liquid water into the membrane, hence, high salt rejection. The high porosity of the ENMs would ease transportation of water vapor and would reduce heat loss during the transfer. The middle layer would increase the $LEP_W$ and would delay long-term pore-wetting of the membrane. The hydrophilic substrate would help draw water vapor from the middle layer by absorption.

It was observed that it took longer for the triple layer configuration to be wetted compared with the conventional two-layer membrane. This was confirmed by a 40 h operation under stable conditions for the triple layer versus 10 h for the double layer. Also, the salt rejection was improved with the triple layer (0.02 % salt penetration or 99.8 % salt rejection) compared to the double layer (0.07 % salt penetration or 99.3 % salt rejection).

## 4.6 Summary and Conclusions

Nanofiber membranes have shown promising performances in a number of water-related applications. Electrospinning is the technology that allows for fabrication of nanofibers as thin as a few tens of nanometers. The major characteristics of ENMs are their high porosity, low density, high surface-to-volume ratio, and good mechanical strength. These characteristics make ENMs suitable for many applications such as filtration, adsorption, energy storage, and medicine.

While ENMs have already found commercial applications in a number of areas such as air filtration and protective clothing, they have only recently been considered for water and wastewater treatment applications. However, ENMs have proven superiority over conventional membrane processes in applications such as microfiltration, ultrafiltration, and adsorption, and have revitalized others such as membrane distillation. The number of research papers on ENMs has increased exponentially in the past decade increasing the potential for commercialization of this new generation of membranes in the near future.

## References


1. R.S. Barhate, S. Ramakrishna, Nanofibrous filtering media: Filtration problems and solutions from tiny materials. J. Mem. Sci. **296**, 1–8 (2007)
2. V. Thavasi, G. Singh, S. Ramakrishna, Electrospun nanofibers in energy and environmental applications. Energy Env. Sci. **1**, 205–221 (2008)
3. R. Balamurugan, S. Sundarrajan, S. Ramakrishna, Recent trends in nanofibrous membranes and their suitability for air and water filtrations. Membranes **1**, 232–248 (2011)
4. C. Burger, B.S. Hsiao, B. Chu, Nanofibrous materials and their applications. Annu. Rev. Mater. Res. **36**, 333–368 (2006)
5. T. Matsuura, C. Feng, K.C. Khulbe, D. Rana, G. Singh, R. Gopal, S. Kaur, R.S. Barhate, S. Ramakrishna, S. Tabe, Development of novel membranes based on electrospun nanofibers and their application in liquid filtration, membrane distillation, and membrane adsorption. Maku (Japanese J. Membr.) **35**(3), 119–127 (2010)
6. M. Botes, T.E. Cloete, The potential of nanofibers and nanobiocides in water purification. Crit. Rev. Microbiol. **36**(1), 68–81 (2010)
7. X. Qu, P.J.J. Alvarez, Q. Li, Applications of nanotechnology in water and wastewater treatment. Water Res. **47**(12), 3931–3946 (2013)
8. V.M. Cepak, J.C. Hulteen, G. Che, K.B. Jirage, B.B. Lakshmi, E.R. Fisher, C.R. Martin, Chemical strategies for template syntheses of composite micro– and nanostructures. Chem. Mater. **9**, 1065–1067 (1997)
9. B.B. Lakshmi, P.K. Dorhout, C.R. Martin, Sol-Gel template synthesis of semiconductor nanostructures. Chem. Mater. **9**, 857–862 (1997)
10. M. Ikegame, K. Tajima, T. Aida, Template synthesis of polypyrrole nanofibers insulated within one-dimensional silicate channels: Hexagonal versus lamellar for recombination of polarons into bipolarons. Angew. Che. Int. Ed. **42**(19), 2154–2157 (2003)
11. H. Li, Y. Ke, Y. Hu, Polymer nanofibers prepared by template melt extrusion. J. Appl. Polym. Sci. **99**(3), 1018–1023 (2006)
12. N.I. Kovtyukhova, B.R. Martin, J.K.N. Mbindyo, T.E. Mallouk, M. Cabassi, T.S. Mayer, Layer-by-layer self-assembly strategy for template synthesis of nanoscale devices. Mater. Sci. Eng. C **19**, 255–262 (2002)
13. Z. Yang, B. Xu, Supramolecular hydrogels based on biofunctional nanofibers of self-assembled small molecules. J. Mater. Chem. **17**(23), 2385–2393 (2007)
14. J. Doshi, D.H. Reneker, Electrospinning process and applications of electrospun fibers. J. Electrostat **35**, 151–160 (1995)
15. Z.M. Huang, Y.Z. Zhang, M. Kotaki, S. Ramakrishna, A review of polymer nanofibers by electrospinning and their applications in nanocomposites. Composites Sci. Technol. **63**, 2223–2253 (2003)
16. D. Li, Y.N. Xia, Electrospinning of nanofibers: reinventing the wheel? Adv. Mater. **16**, 1151–1170 (2004)
17. P. Raghavan, D.H. Lim, J.H. Ahn, C. Nah, D.C. Sherrington, H.S. Ryu, H.J. Ahn, Electrospun polymer nanofibers: The booming cutting edge technology. React. Funct. Polym. **72**, 915–930 (2012)
18. C. Feng, K.C. Khulbe, T. Matsuura, S. Tabe, A.F. Ismail, Preparation and characterization of electrospun nanofiber membranes and their possible applications in water treatment. J. Sep. Purif. **102**, 118–135 (2013)
19. S.A. Theron, E. Zussman, A.L. Yarin, Experimental investigation of the governing parameters in the electrospinning of polymer solutions. Polymer **45**(6), 2017–2030 (2004)
20. R. Rangkupan, D.H. Reneker, Electrospinning process of molten polypropylene in vacuum. J. Met. Mater. Miner. Soc **12**, 81–87 (2003)
21. J. Zeleny, The electrical discharge from liquid points, and a hydrostatic method of measuring the electric intensity at their surfaces. Phys. Rev. **3**, 69–91 (1914)

22. J.U. Kim, S.H. Park, H.J. Choi, W.K. Lee, J.K. Lee, M.R. Kim, Effect of electrolyte in electrospun poly(vinylidene fluoride-co-hexafluoropropylene) nanofibers on dye-sensitized solar cells. Sol. Energ. Mater. Sol. Cells. **93**(6–7), 803–807 (2009)
23. S.K. Kim, J.H. Ryu, H.D. Kwen, C.H. Chang, S.H. Cho, convenient preparation of ion-exchange PVDF membranes by a radiation-induced graft polymerization for a battery separator. Polymer (Korea) **34**(2), 126–132 (2010)
24. C. Feng, K.C. Khulbe, S. Tabe, Volatile organic compound removal by membrane gas stripping using electrospun nanofiber membrane. Desalination **287**, 98–102 (2012)
25. K. Yoon, K. Kim, X. Wang, D. Fang, B.S. Hsiao, B. Chu, High flux ultrafiltration membranes based on electrospun nanofibrous PAN scaffolds and chitosan coating. Polymer **47**, 2434–2441 (2006)
26. K. Yoon, B.S. Hsiao, B. Chu, High flux nanofiltration membranes based on internfacially polymerized polyamide barrier layer on polyacrylonitrile nanofibrous scaffolds. J. Membr. Sci **326**, 484–492 (2009)
27. G. Singh, D. Rana, T. Matsuura, S. Ramakrishna, R.M. Narbaitz, S. Tabe, Removal of disinfection byproducts from water by carbonized electrospun nanofibrous membranes. Sep. Purif. Technol. **74**, 202–212 (2010)
28. X. Wang, K. Zhang, Y. Yang, L. Wang, Z. Zhou, M. Zhu, B.S. Hsiao, B. Chu, Development of hydrophilic barrier layer on nanofibrous substrate as composite membrane in a facile route. J. Membr. Sci. **356**, 110–116 (2010)
29. S. Kaur, R. Barhate, S. Sundarrajan, T. Matsuura, S. Ramakrishna, Hot pressing of electrospun membrane and its influence on separation performance on thin film composite nanofiltrtion membrane. Desalination **279**, 201–209 (2011)
30. A. Sato, R. Wang, H. Ma, B.S. Hsiao, B. Chu, Novel nanofibrous scaffolds for water filtration with bacteria and virus removal capability. J. Electron. Microsc. **60**(3), 201–209 (2011)
31. Z. Ma, M. Kotaki, S. Ramakrishna, Electrospun cellulose nanofiber as affinity membrane. J. Membr. Sci. **265**(1–2), 115–123 (2005)
32. T. Christoforou, C. Doumanidis, Biodegradable cellulose acetate nanofiber fabrication via electrospinning. J. Nanosci. Nanotechnol. **10**(9), 6226–6233 (2010)
33. Y. Tian, M. Wu, R. Liu, Y. Li, D. Wang, J. Tan, R. Wu, Y. Huang, Electrospun membrane of cellulose acetate for heavy metal ion adsorption in water treatment. Carbohyd. Polym. **83**, 743–748 (2011)
34. M. Aliabadi, M, M. Irani, J. Ismaeili, H. Piri, M.J. Parnian, Electrospun nanofiber membrane of PEO/Chitosan for the adsorption of nickel, cadmium, lead and copper ions from aqueous solution. J. Chem. Eng. **220**, 237–243 (2013)
35. S. Putthanarat, P. Tapadia, S. Zarkoob, L.D. Miller, R.K. Eby, W.W. Adams, The color of dragline silk produced in captivity by the spider Nephila clavipes. Polymer **45**(6), 1933–1937 (2004)
36. S. Sukigara, M. Gandhi, J. Ayutsede, M. Micklus, F. Ko, Regeneration of Pombyxmori silk by electrospinning-Part 2: Process optimization and empirical modeling using response surface methodology. Polymer **45**(11), 3701–3708 (2004)
37. S. Zarkoob, R.K. Eby, D.H. Reneker, S.D. Hudson, D. Ertleya, W.W. Adams, Structure and morphology of electrospun silk nanofibers. Polymer **45**, 3973–3977 (2004)
38. B.M. Min, S.W. Lee, J.N. Lim, Y. You, T.S. Lee, P.H. Kang, W.H. Park, Chitin and chitosan nanofibers: Electrospinning of chitin and deacetylation of chitin nanofibers. Polymer **45**(21), 7137–7142 (2004)
39. T.J. Sill, H.A. von Recum, Electrospinning: applications in drug delivery and tissue engineering. Biomaterials **29**(13), 1989–2006 (2008)
40. S. Megelski, J.S. Stephens, D. Bruce Chase, J.F. Rabolt, Micro- and nanostructured surface morphology on electrospun polymer fibers. Macromolecules **35**(22), 8456–8466 (2002)
41. S. Koombhongse, W. Liu, D.H. Reneker, Flat polymer ribbons and other shapes by electrospinning. J. Polym. Sci. B Polym. Phys **39**(21), 2598–2606 (2001)

42. A. Buer, S.C. Ugbolue, S.B. Warner, Electrospinning and properties of some nanofibers. J. Textile Res. **71**(4), 323–328 (2001)
43. A.G. MacDiarmid, W.E. Jones Jr., I.D. Norris, J. Gao, A.T. Johnson Jr., N.J. Pinto, J. Hone, B. Han, F.K. Ko, H. Okuzaki, M. Llaguno, Electrostatically-generated nanofibers of electronic polymers. Synth. Met. **119**(1–3), 27–30 (2001)
44. H. Savoji, D. Rana, T. Matsuura, S. Tabe, C. Feng, Development of plasma and/or chemically induced graft co-polymerized electrospun poly(vinylidene fluoride) membranes for Solute Separation. Sep. Purif. Technol. **108**, 196–204 (2013)
45. W. Sambaer, M. Zatloukala, D. Kimmer, The use of novel digital image analysis technique and rheological tools to characterize nanofiber nonwovens. Polym.Test. **29**(1), 82–94 (2010)
46. W. Sambaer, M. Zatloukal, D. Kimmer, Advanced characterization of nanofiber based non-wovens. Proceedings, 68th Annu. Tech. Conf. of the Society of Plastics Engs., ANTEC 2010 Orlando, FL, United States **V1**:779–783 (2010)
47. S. Manickam, J.R. McCutcheon, Characterization of polymeric nonwovens using porosimetry, porometry and X-ray computed tomography. J. Membr. Sci. **407–408**, 108–115 (2012)
48. A. Jena, K. Gupta, Pore structure characterization techniques. Am. Ceram. Soc. Bull. **84**(3), 28–30 (2005)
49. M.M. Tomadakis, S.V. Sotirchosm, Transport properties of random arrays of freely overlapping cylinders with various orientation distributions. J. Chem. Phys. **98**, 616–626 (1993)
50. L. Huang, R.A. McMillan, R.P. Apkarian, B. Pourdeyhimi, V.P. Conticello, E.L. Chaikof, Generation of synthetic elastin-mimetic small diameter fibers and fiber networks. Macromolecules **33**(8), 2989–2997 (2000)
51. M. Kiristia, A.U. Oksuz, L. Oksuz, S. Ulusoy, Electrospun chitosan/PEDOT nanofibers. Mater. Sci. Eng. C Mater. Biol. Appl. **33**(7), 3845–3850 (2013)
52. F. Kayaci, Z. Aytac, T. Uyar, Surface modification of electrospun polyester nanofibers with cyclodextrin polymer for the removal of phenanthrene from aqueous solution. J. Hazard Mater. **261**, 286–294 (2013)
53. Y.J. Kim, C.H. Ahn, M.O. Choi, Effect of thermal treatment on the characteristics of electrospun PVDF-silica composite nanofibrous membrane. J. Eur. Polym. **46**, 1957–1965 (2010)
54. H.T. Dang, R.M. Narbaitz, T. Matsuura, Evaluation of apparatus for membrane cleaning tests. J. Env. Eng. **136**(10), 1161–1170 (2010)
55. M. Nyström, M. Lindstrom, E. Matthiasson, Streaming potential as a tool in the characterization of ultrafiltration membranes. Colloids Surf. **36**(3), 297–312 (1989)
56. K. Kosmider, J. Scott, Polymeric nanofibres exhibit an enhanced air filtration performance. Filt. Sep. **39**(6), 20–22 (2002)
57. T. Grafe, M. Gogins, M. Barris, J. Schafer, R. Canepa, Nanofibers in filtration applications in transportation. Proceedings, Filtration 2001 Int Conf. and Expo of the Assoc. Nonwovens Fabric Industry, Chicago, Illinois, December 3–5 (2001)
58. A. Podgórski, A. Bałazy, L. Gradón, Application of nanofibers to improve the filtration efficiency of the most penetrating aerosol particles in fibrous filters. Chem. Eng. Sci. **61**(20), 6804–6815 (2006)
59. P.P. Tsai, H. Schreuder-Gibson, P. Gibson, Different electrostatic methods for making electret filters. J. Electrostat **54**, 333–341 (2002)
60. F.J. Liu, L.M. Huang, T.C. Wen, A. Gopalan, Large-area network of polyaniline nanowires supported platinum nanocatalysts for methanol oxidation. Synth. Met. **157** (16–17), 651–658 (2007)
61. E.S. Steigerwalt, G.A. Deluga, C.M. Lukehart, Rapid preparation of Pt-Ru/graphitic carbon nanofiber nanocomposites as DMFC anode catalysts using microwave processing. J. Nanosci. Nanotechnol. **3**(3), 247–251 (2003)
62. W. Li, C.T. Laurencin, E.J. Caterson, R.S. Tuan, F.K. Ko, Electrospun nanofibrous structure: A novel scaffold for tissue engineering. J. Biomed. Mater. Res. **60**(4), 613–621 (2002)

63. X. Mo, D. Li, H.A. El-Hamshary, S.S. Al-Deyab, Electrospun nanofibers for tissue engineering. J. Fiber Bioeng. Informatics **6**(3), 225–235 (2013)
64. H. Jia, G. Zhu, B. Vugrinovich, W. Kataphinan, D.H. Reneker, P. Wang, Enzyme-carrying polymeric nanofibers prepared via electrospinning for use as unique biocatalysts. Biotechnol. Prog. **18**(5), 1027–1032 (2002)
65. Y.F. Fang, X.J. Huang, P.C. Chen, Z.K. Xu, Polymer materials for enzyme immobilization and their application in bioreactors. BMB Rep. **44**(2), 87–95 (2011)
66. R. Gopal, S. Kaur, C. Feng, C. Chand, S. Ramakrishna, S. Tabe, T. Matsuura, Electrospun nanofibrous polysulfone membranes as pre-filters: Particulate removal. J. Membr. Sci. **289**, 210–219 (2007)
67. Z. Xu, Q. Gu, H. Hu, F. Li, A novel electrospun polysulfone fiber membrane: Application to advanced treatment of secondary bio-treatment sewage. Environ. Sci. Technol. **29**, 13–21 (2008)
68. I.M. El-Azizi, A. Schmalenberger, R. Komlenic, R.G.J. Edyvean, Study of a depth filter (Disruptor™) for the novel application of reducing SWRO membrane fouling. Desal. Water Treat. **29**(1–3), 20–28 (2011)
69. R. Komlenic, T. Berman, J.A. Brant, B. Dorr, I. El-Azizi, H. Mowers Removal, of polysaccharide foulants from reverse osmosis feedwater using electroadsorptive cartridge filters. Desal. Water Treat. **51**, 1050–1056 (2013)
70. R.J. Petersen, Composite reverse osmosis and nanofiltration membranes. J. Membr. Sci. **83**(1), 81–150 (1993)
71. X. Wang, X. Chen, K. Yoon, D. Fang, B.S. Hsiao, B. Chu, High flux filtration medium based on nanofibrous substrate with hydrophilic nanocomposite coating. J. Env. Sci. Technol. **39**, 7684–7691 (2005)
72. X. Wang, D. Fang, K. Yoon, B.S. Hsiao, B. Chu, High performance ultrafiltration composite membranes based on poly(vinyl alcohol) hydrogel coating on crosslinked nanofibrous poly (vinyl alcohol) scaffold. J. Membr. Sci. **278**, 261–268 (2006)
73. B. Chu, B.S. Hsiao, The role of polymers in breakthrough technologies for water purification. J. Polym. Sci. B Polym. Phys. **47**, 2431–2435 (2009)
74. D.A. Musale, A. Kumar, G. Pleizier, Formation and characterization of poly(acrylonitrile)/ Chitosan composite ultrafiltration membranes. J. Membr. Sci. **154**(2), 163–173 (1999)
75. D.A. Musale, A. Kumar, Solvent and pH resistance of surface crosslinked chitosan/poly (acrylonitrile) composite nanofiltration membranes. J. Appl. Polym. Sci. **77**, 1782–1793 (2000)
76. M.V. Chandorilar, P.C. Bhavsar, Evaluation of polyacrylonitrate and poly (methyl methacrylate) as membrane materials for reverse osmosis. Indian J. Technol. **2**, 124–132 (1983)
77. D.A. Musale, A. Kumar, Effects of surface crosslinking on sieving characteristics of chitosan/poly (acrylonitrile) composite nanofiltration membranes. Sep. Purif. Technol. **21** (1–2), 27–37 (2000)
78. N.-N. Bui, M.L. Lind, E.M.V. Hoek, J.R. McCutcheon, Electrospun nanofiber supported thin film composite membranes for engineered osmosis. J. Membr. Sci. **385–386**, 10–19 (2011)
79. H. You, X. Li, Y. Yang, B. Wang, Z. Li, X. Wang, M. Zhu, B.S. Hsiao, High flux low pressure thin film nanocomposite ultrafiltration membranes based on nanofibrous substrates. Sep. Purif. Technol. **108**, 143–151 (2013)
80. B. Pan, B.S. Xing, Adsorption mechanisms of organic chemicals on carbon nanotubes. Environ. Sci. Technol. **42**(24), 9005–9013 (2008)
81. K. Yang, B.S. Xing, Adsorption of organic compounds by carbon nanomaterials in aqueous phase: Polanyi theory and its application. Chem. Rev. **110**(10), 5989–6008 (2010)
82. G.P. Rao, C. Lu, F. Su, Sorption of divalent metal ions from aqueous solution by carbon nanotubes: A review. Sep. Purif. Technol. **58**(1), 224–231 (2007)
83. H.W. Liang, X. Cao, W.J. Zhang, H.T. Lin, F. Zhou, L.F. Chen, S.H. Yu, Robust and highly efficient free-standing carbonaceous nanofiber membranes for Water purification. Adv. Func. Mat. **21**, 3851–3858 (2011)
84. K.W. Lawson, D.R. Lloyd, Membr. Distil. J. Membr. Sci. **124**, 1–25 (1997)

85. J.A. Prince, G. Singh, D. Rana, T. Matsuura, V. Anbharasi, T.S. Shanmugasundaram, Preparation and characterization of highly hydrophobic poly(vinylidene fluoride) - Clay nanocomposite nanofiber membranes (PVDF-clay NNMs) for desalination using direct contact membrane distillation. J. Membr. Sci. **397–398**, 80–86 (2012)
86. M. Khayet, Membrane and theoretical modeling of membrane distillation: A review. Adv. Coll. Interface Sci. **164**, 56–88 (2011)
87. M. Khayet, T. Matsuura, Application of surface modifying macromolecules for the preparation of membranes for membrane distillation. Desalination **158**, 51–56 (2003)
88. M. Nasir, H. Matsumoto, T. Danno, M. Minagawa, T. Irisawa, M. Shioya, M. Tanioka, Control of diameter, morphology, and structure of PVDF nanofiber fabricated by electrospray deposition. J. Polym. Sci. B Polym. Phys. **44**, 779–786 (2006)
89. C. Feng, K.C. Khulbe, T. Matsuura, R. Gopal, S. Kaur, S. Ramakrishna, M. Khayet, Production of drinking water from saline water by air-gap membrane distillation using polyvinylidene fluoride nanofiber membrane. J. Membr. Sci. **311**, 1–6 (2008)
90. Y. Liao, R. Wang, M. Tian, C. Qiu, A.G. Fane, Fabrication of polyvinylidene fluoride (PVDF) nanofiber membranes by electro-spinning for direct contact membrane distillation. J. Membr. Sci. **425–426**, 30–39 (2013)
91. Y. Liao, R. Wang, A.G. Fane, Engineering superhydrophobic surface on poly (vinylidene fluoride) nanofiber Membranes for direct contact membrane distillation. J. Membr. Sci. **440**, 77–87 (2013)
92. M. Essalhi, M. Khayet, Self-sustained webs of polyvinylidene fluoride electrospun nanofibers at different electrospinning times: 1. Desalination by direct contact membrane distillation. J. Membr. Sci. **433**, 167–179 (2013)
93. J.A. Prince, V. Anbharasi, T.S. Shanmugasundaram, G. Singh, Preparation and characterization of novel triple layer hydrophilic-hydrophobic composite membrane for desalination using air gap membrane distillation. Sep. Purif. Technol. **118**, 598–603 (2013)

# Chapter 5
# Fullerenes and Carbon Nano-onions for Environmental Application

**Yusong Li and Megan Seymour**

**Abstract** Fullerenes and carbon nano-onions both are carbonaceous nanomaterials with different molecular structures. They have emerged as promising candidates for environmental applications due to their unique properties. In this chapter, the potential environmental applications of fullerenes and carbon nano-onions as sorbents, photocatalysists, and membranes are discussed.

## 5.1 Introduction

Research on carbon-based nanomaterials can be traced back to 1950s in Russia, and later surged at the Massachusetts Institute of Technology (MIT) Lincoln Laboratory and also other places in the USA with the unexpected launching of the Sputnik spacecraft by the Russians in 1957 [1]. Carbon element can be transformed into different materials via different degrees of hybridization (sp, sp2, and sp3) [2]. These mutable hybridization states lead to considerable differences among carbon's bulk configurations, from thermodynamically favorable trigonometric sp3 configuration of nanodiamonds to the planar sp2 conformation of graphene [3]. The fullerenes, carbon nano-onions (CNOs), and other carbon-based nanomaterials (e.g., carbon nanotubes) share many properties as graphite, but with distinct and tunable properties due to quantum effects at the nanoscale, and enhanced sp3 character of bonds and closed topology [4].

Y. Li (✉)
Department of Civil Engineering, University of Nebraska-Lincoln,
2200 Vine Street, Lincoln, NE 68583, USA
e-mail: yli7@unl.edu

M. Seymour
HDR Inc., 8404 Indian Hills Drive, Omaha, NE 68114, USA

A. Hu and A. Apblett (eds.), *Nanotechnology for Water Treatment and Purification*, Lecture Notes in Nanoscale Science and Technology 22,
DOI 10.1007/978-3-319-06578-6_5, 

Carbon-based nanomaterials, with their unique physiochemical properties, have emerged in recent years as promising candidates for various environmental applications. The unique properties of carbon-based nanomaterials include their size, shape, surface area, sorption properties, and electric and thermal properties. Based on the traditional application of activated carbon in water and wastewater treatment processes, carbon-based nanomaterials were tested for their capacity as effective nanosorbents [5]. Carbon-based nanomaterials, including carbon nanotube, have been incorporated into traditional reserve osmosis (RO) to produce high flux membrane that can significantly reduce the energy for desalination [6]. Some carbon-based nanomaterials were found to possess antimicrobial properties [7–9], which motivates people to develop antimicrobial agents using these materials [9]. In addition, environmental engineers and scientists are developing sensors based on carbon-based nanomaterials, particularly carbon nanotube [10], to detect and monitor contaminants and microbial ecology in complicated natural environmental settings. Finally, carbon-based nanomaterials also found ways of applications in the area of renewable energy based on the excellent electronic and optical properties of nanotubes and fullerenes [11, 12].

In this chapter, the discussion focuses on the environmental applications of fullerenes and carbon non-onion nanomaterials. For each material, the molecular structure and basic properties of the materials will first be described. The environmental applications based on the unique properties were then discussed.

## 5.2 Fullerenes

### *5.2.1 Basic Properties*

Fullerenes are closed cage carbon molecules, consisting of a number of five-membered rings and six-membered rings. Buckminsterfullerene (simplified as buckballs) are molecules containing 60 carbon atoms ($C_{60}$) in the form of a hollow sphere, arranged as 12 pentagons and 20 hexagons. Figure 5.1 illustrates an example molecular structure of $C_{60}$.

Fullerenes were first discovered in 1985 by Richard Smalley, Robert Curl, James Heath, Sean O'Brien, and Harold Kroto at Rice University [13], which led to the award of the Noble Prize in Chemistry to Curl, Kroto, and Smalley in 1996. Fullerenes have been found to occur in nature by means of high-resolution transmission electron microscopy [14]. More recent research found that fullerenes can and do form efficiently in space when the conditions are right [15].

$C_{60}$ has a molar weight of 720 g/mol and a diameter of approximately 1 nm. The solubility of $C_{60}$ is about 2.8 mg/mL in toluene at room temperature, and the heat of sublimation of $C_{60}$ is 43.3 Kcal/mol [16]. The chemical behavior of $C_{60}$ lies between an aromatic molecule and a straight chained alkene [16, 17]. $C_{60}$ is stabilized by resonance structures supporting an equivalent electronic state and

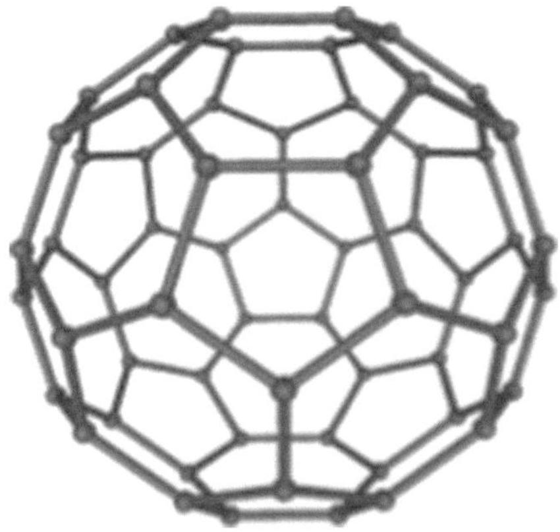

**Fig. 5.1** Buckminsterfullerene (*Source*: http://en.wikipedia.org/wiki/Fullerene)

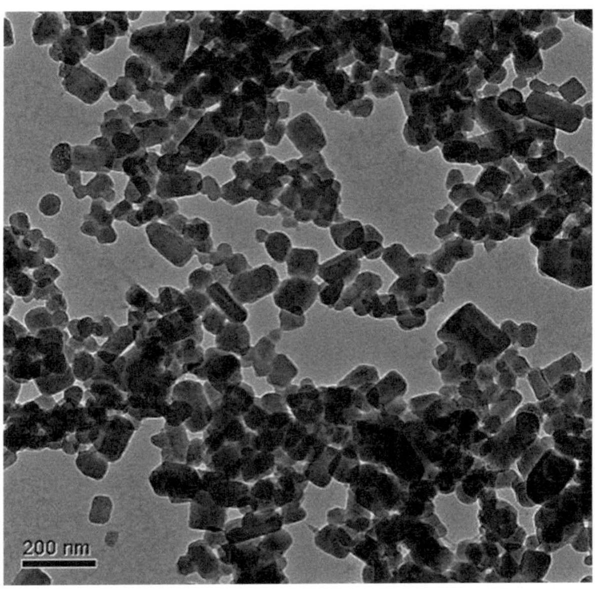


**Fig. 5.2** Transmission electron micrograph of $nC_{60}$ aggregates dried on a 400-mesh carbon-coated copper grid

bonding geometry for each carbon atom [18]. Pristine $C_{60}$ tends to form stable crystalline nanostructures in a variety of solutions, with diameter ranged from 25 to 500 nm [19, 20]. Figure 5.2 shows typical $C_{60}$ aggregates in water.

One attractive aspect of $C_{60}$ is that they may be derivatized into various forms of variations with tailored properties for given applications. In the cage structure of $C_{60}$, up to six electrons can be accommodated in the lowest unoccupied molecular orbital (LUMO), which makes it possible as structural scaffolding for reactive adducts. For example, supramolecular techniques [21] were able to enhance the solubility and reduce the aggregation of hydrophobic fullerene molecules by modifying their surface properties. As another example, $C_{60}$ doped with rare-earth metals was found to be superconductive [22].

## *5.2.2 Fullerenes as Nanosorbents*

Adsorption process is essential for effective water and wastewater treatment. Traditionally, activated carbon, zeolite, or silica have been used in the adsorption process. The sorption capacity of conventional sorbents is limited by the availability of surface active sites and mass transfer rates. Due to the high surface area to volume ratio and manipulatable surface chemistry, fullerenes have been studied for removal of hydrophobic organic contaminants, as well as metals, in aquatic and terrestrial environments [3, 5, 23–32].

### **5.2.2.1 Sorption of Organic Compounds**

#### 5.2.2.1.1 Mechanisms

Although fundamental interactions between the organic sorbates and nanosorbents, including hydrophobic, dispersion, and weak dipolar forces, are similar as in traditional sorption system [33], several aspects of differences between fullerene and activated carbon as sorbents were identified. First, due to the micropores with various sizes on the activated carbons, pore diffusion serves as an intermediate mechanism in sorption processes; on the other hand, sorption sites are located only on the surface of fullerene, promoting faster sorption kinetics [34]. Second, activating processes of activated carbons provided various degrees of saturation and oxidation states, as well as functional groups, leading to a highly heterogeneous sorption energy distribution; while fullerenes only consist of globally conjugated unsaturated carbons in three-dimensional arrays and intentionally uniformly treated surface functional groups. Less heterogeneous sorption energy distribution was considered to promote faster sorption processes [34].

#### 5.2.2.1.2 Sorption of Volatile Organic Compounds

$C_{60}$ fullerenes were tested as sorbent materials for benzene, toluene, ethylbenzene, and xylene isomers in a gas chromatograph–mass spectrometer system [35]. Compared with other sorbents, $C_{60}$ fullerenes were considered as the best choice in terms of sensitivity, accuracy, selectivity, and reusability. In another study [36], dispersed polycrystalline $C_{60}$ fullerites were found to be less efficient for adsorbing benzene, comparing with a mixture of fullerites with the presence of defects in the crystal structure; in another word, the less perfect the lattice, the higher the adsorption capacity. In addition, sorption of benzene on $C_{60}$ was found to be sensitive to the wavelength of light [29] and the preliminary adsorption of water [37].

#### 5.2.2.1.3 Sorption of Polycyclic Aromatic Hydrocarbons

Polycyclic aromatic hydrocarbons (PAHs) are persistent organic contaminants that were introduced into soils due to anthropogenic activities. They typically possess recalcitrant aromatic structure, low polarity, low aqueous phase solubility, and elevated hydrophobicity. After discharged into the environment, PAHs tend to strongly sorb and sequester to soil particulates. The persistence and accumulation of PAHs in soils represent a serious environmental concern due to their toxicity and risk to human and ecological systems.

Research has found that carbonaceous nanomaterials, including $C_{60}$ fullerenes, may efficiently sequester PAHs and reduce their bioaccessibility and extractability [30]. In a study [32] that compared a variety of carbonaceous nanosorbents, Polanyi theory was able to explain the sorption of PAHs on carbonaceous nanomaterials, including fullerenes. Polanyi–Manes model is a nonlinear sorption model, originally developed for gas sorption by activated carbon, which can be written as

$$\log q_e = \log Q^0 + a(\varepsilon_{sw}/V_s)^b$$
$$\varepsilon_{sw} = RT\ln(C_s/C_e)$$

where, $q_e$ [mg/g] is the equilibrium sorbed concentration, $C_e$ [mg/L] is the equilibrium solution phase concentration, $Q^0$ [mg/g] is the sorbed capacity, $C_s$ [mg/L] is the aqueous water solubility, $\varepsilon_{sw}$ [kJ/mol] is the effective adsorption potential, $V_s$ [$cm^3$/mol] is the molar volume of solute, [$(cm^3)^{b+1}/(kg\ J^b)$] and $b$ are fitting parameters, $R$ [$8.314 \times 10^{-3}$ kJ/(mol K)] is the universal gas constant, and $T$ [K] is the absolute temperature.

Another study of pyrene (one type of PAHs) sorption by $nC_{60}$ [38] found that Dubinin–Ashthakov model provided the best fit for the sorption isotherm data. Sorption was strongest for $nC_{60}$ kept under anoxic condition. Both the presence of oxygen and irradiation significantly decrease the sorption capacity of $nC_{60}$.

Desorption behaviors of pyrene, phenanthrene, and naphthalene from fullerenes and other carbonaceous materials (e.g., carbon nanotubes) were studied [31]. Desorption hysteresis was observed for fullerenes, but not for long, cylindrical carbon nanotubes. The hysteresis was attributed to a deformation-rearrangement, or the formation of closed interstitial spaces in spherical fullerene aggregates.

#### 5.2.2.1.4 Sorption of Endocrine Disrupting Chemicals

Endocrine disrupting compounds (EDCs) were used widely as synthetic pharmaceuticals for humans and livestock, which typically have low sorption efficiency. The original chemicals were excreted and discharged into municipal sewage, posing significant responsibility of wastewater treatment plant to remove these compounds. EDCs, however, were detected in the effluents of wastewater treatment plant, indicating the failure of current wastewater treatment techniques to remove these compounds.

Limited work has studied the sorption of EDCs to carbon nanomaterials, including $C_{60}$ fullerenes [27]. Sorption of 17R-ethinyl estradiol and bisphenol A (BPA) on carbon nanomaterials showed significant relationship between sorption coefficient K and specific surface area of nanosorbent. Molecular conformation simulations indicated that BPA was able to sorb on the curvature surface of fullerenes because of its butterfly structure of two benzene rings. $\pi$–$\pi$ electron donor–acceptor system was considered to be a critical mechanism for the sorption of benzene containing chemicals on carbon-based nanomaterials. The high sorption capacity and strong desorption hysteresis indicate the potential of fullerenes as an effective nanosorbent for EDCs.

### 5.2.2.2 Sorption of Metal Ions

Although not as popular as carbon nanotubes, several studies investigated the sorption of metal ions using $C_{60}$ fullerenes [23, 39, 40]. In early 1990s, the potential of $C_{60}$ fullerenes for preconcentration of metal traces were investigated in a minicolumn, where $C_{60}$ fullerene was demonstrated to perform better in terms of both accuracy and selectivity [39]. $C_{60}$ fullerenes were also used to modify the surface of traditional sorbents to improve sorption efficiency. Fullerene/polymer composites were developed by modifying polystyrene film with fullerenes [23], which showed four times higher sorption capacity for copper ions than unmodified samples. The selectivity of sorption on the composite for various metals followed the order: $Cu(II) \geq Cd(II) > Ni(II) > Zn(II)$.

The activated carbon modified with micro amount of fullerenes was found to have improved metal cation sorption capacity [40]. Instead of simple hydrophobic sorption mechanisms for organic sorption on $C_{60}$ surface, mechanisms for metal sorption are more specific. For example, silver ions were reduced on the carbon surface to metal silver; lead cations were sorbed by donor–acceptor interactions; and copper ion sorption may lead to both cation reduction and the formation of complexes with functional groups on the surface of $C_{60}$.

## *5.2.3 Fullerenes for Membranes*

Membranes with high selectivity and permeability have been widely applied for contaminant removal in water and wastewater treatment processes. However, membrane fouling, i.e., deposition and attachment of protein onto membrane surfaces, is still one of the major obstacles for a sustainable treatment. Membrane fouling will result in the clog of membrane pores, and therefore may deteriorate the performance. Recent advances in nanotechnology has rendered additional ways to modify membrane surfaces to mitigate membrane fouling, and increase permeability and selectivity. Fullerenes were incorporated into polymer membrane, which can effectively transform the properties of the polymer while retain the unique properties of fullerenes.

Higuchi, Yoshida et al. [41] prepared homogeneous fullerene (a mixture of C60 and C70)-dispersed poly(1-trimethylsilyl-1-propyne) (PTMSP) membranes using a casting method. The effects of fullerene in PTMSP membrane on the permeation of inorganic, paraffin, and olefin gases were investigated. Pressure dependence of permeability coefficients was reported for the permeation of carbon dioxide, ethylene, ethane, 1-butene, and *n*-butane through the fullerene-modified membranes [41].

Fullerenes were added to polyphenylene oxide (PPO) membranes to improve the efficiency of pervaporation of water mixtures with ethanol and ethyl acetate [28, 42, 43]. Pervaporation is a process of separating binary and multicomponent mixtures by evaporation through a membrane. The structure, chemical nature, and composition of membrane materials determine the selectivity and efficiency of pervaporation. Modification of PPO membranes with fullerenes increases the membrane sorption capacity, their efficiency in pervaporation of water-ethanol mixture [43], and the selectivity with respect to ethyl acetate [28, 42]. The improved performance was attributed to the interactions between $C_{60}$ and PPO, which leads to increased polymer density and decreased free volume of PPO membranes. Another study [44] of using $C_{60}$ fullerenes to modify a PPO film found that gas permeability values are very different between covalently bounded and dispersed PPO-$C_{60}$. The gas permeability of PPO-$C_{60}$ bonded increases up to 80 % with increasing influence concentration, while gas permeability of PPO-$C_{60}$ dispersed decreases in comparison with pure PPO. $C_{60}$ was also used to modify Poly (phenylene-iso-phtalamide) (PA) [45,46], and the PA-$C_{60}$ demonstrated the best values of flux reduced recovery after contact with proteins solution. Very recently [47], ethylcellulose/fullerene$C_{60}$ (EC/C60) hybrid membranes with different $C_{60}$ contents were prepared to examine the sulfur removal efficiency from the model gasoline. It was observed that the hybrid membrane of EC/$C_{60}$ demonstrated improved permeation flux and higher sulfur enrichment when compared with the pure EC membrane.

### 5.2.4 Fullerenes for Photocatalyst

Chemical contaminants, such as textile dyes and fluorescence dyes, are typically very difficult to be naturally degraded. Various researches were conducted to seek stable, low cost, and non-toxicity photocatalysts. Water-soluble $C_{60}$ derivatives displayed a high efficacy for photochemical $^1O_2$ production and associated bacterial/viral inactivation. Based on this property, photoactive $C_{60}$ aminofullerene was immobilized on silica gel (3-(2-succinic anhydride)propyl functionalized silica),which thus enables facile separation of the photocatalyst for recycling and repeated use [48]. With no loss in efficacy of $^1O_2$ production plus insignificant chemical modification of the amino $C_{60}$/silica photocatalyst after multiple cycling, the system offers a promising new visible-light-activated photocatalyst. Under visible-light irradiation, the amino$C_{60}$/silica photocatalyst can effectively and kinetically enhance the oxidation of ranitidine and cimetidine (pharmaceutical pollutants) and inactivate MS-2 bacteriophage, when compared to aqueous solutions of the $C_{60}$ aminofullerene alone. More recently [49], this new photocatalyst was evaluated for oxidative degradation of eleven emerging organic contaminants, including pharmaceuticals, such as acetaminophen, carbamazepine, cimetidine, propranolol, ranitidine, sulfisoxazole, and trimethoprim, and endocrine disruptors such as bisphenol A and pentachlorophenol. Amino $C_{60}$/silica degraded pharmaceuticals faster than common semiconductor photocatalyst, exhibited high target-specificity without significant interference by natural organic matter. In another study [50], $C_{60}$ fullerenes were used to modify pure $Bi_2TiO_4F_2$ photocatalyst, which led to a new photocatalyst with much stronger performance for degrading Rhodamine B (RhB) and Eosin Y (EY) under visible light irradiation.

## 5.3 Carbon Nano-onions

### *5.3.1 Basic Properties*

CNOs, also called onion-like fullerenes and bucky onions, have a cage-in-cage structure, consisting of multilayered concentric graphene spherical shells. Figure 5.3 provides an example TEM image of CNOs. The XRD analysis of CNO powders is shown in Fig. 5.4, which reveals one major peak at 30.7° with counts of 585 and a moderate peak at about 55°. The Raman spectra of CNOs (Fig. 5.5) identified two dominant Raman shifts at 1,330.8 and 1,592.13 $cm^{-1}$ with an excitation of 532 nm [51–53]. The two bands were fitted into five different bands, one of which was ascribed to the amorphous carbon [51].

### *5.3.2 CNO as Sorbent*

Equilibrium sorption capacity of surface modified CNOs for $Zn^{2+}$, $Cu^{2+}$, $Cd^{2+}$, $Ni^{2+}$, and $Pb^{2+}$ was compared with that of $C_{60}$ (oxidized in the same way for comparison). Oxidation of CNO using NaOCl or $HNO_3$ can be described as [55]

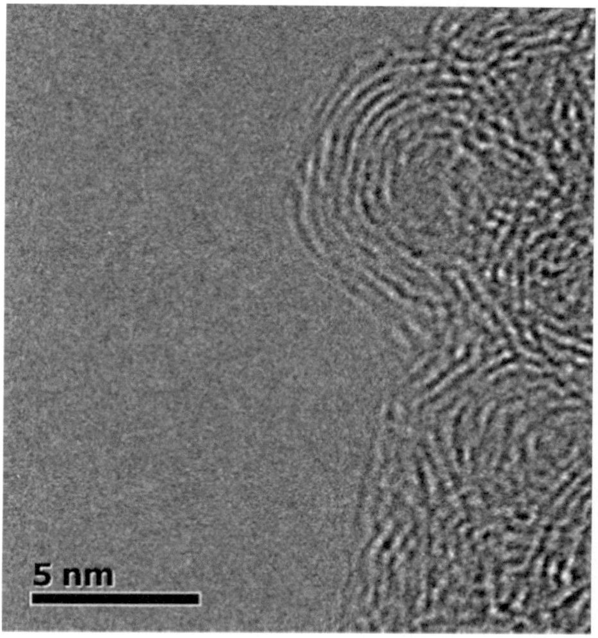


**Fig. 5.3** TEM image of CNOs [54]

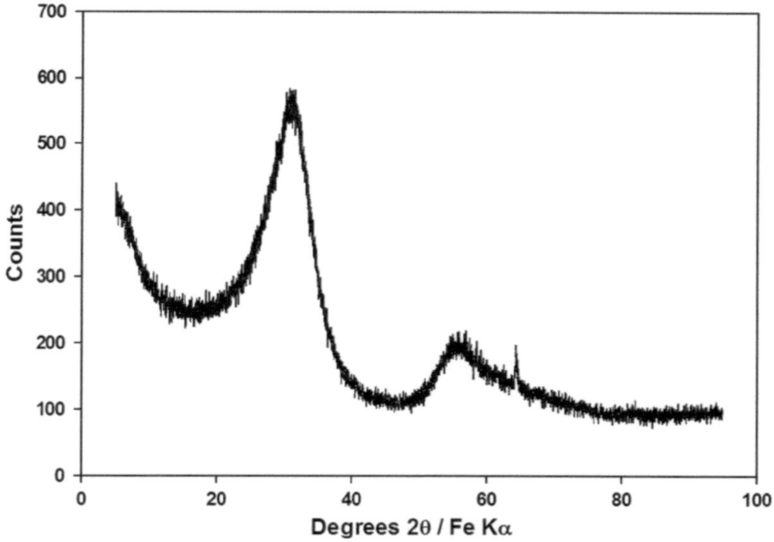


**Fig. 5.4** XRD pattern of CNO [54]

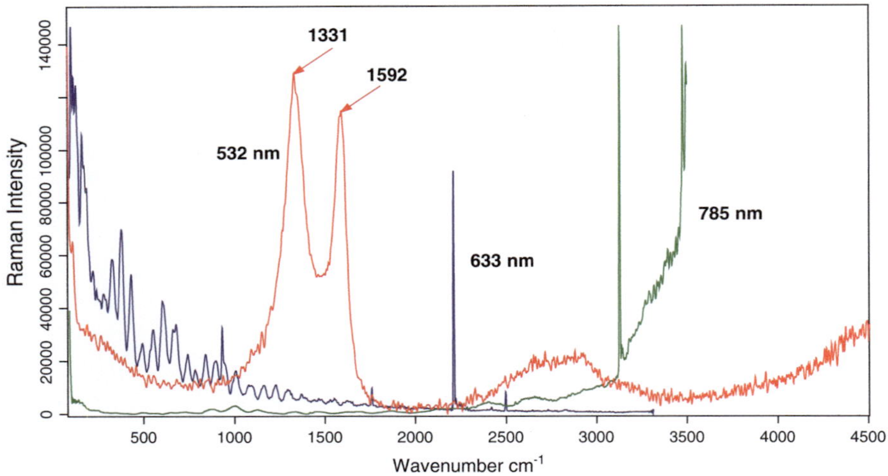


**Fig. 5.5** Raman spectra of CNOs at various excitations of 532, 633, and 785 nm [54]

$$10C_n + 6H^+ + 6NO_3^- + 2H_2O \rightarrow 10C_{n-1} - COOH + 3N_2 \quad (5.1)$$

$$2C_n + 3OCl^- + 2H_2O \rightarrow 2C_{n-1} - COOH + 3Cl^- \quad (5.2)$$

The modification process may introduce various oxygen-containing functional groups onto CNO surfaces, therefore, improving surface hydrophilicity and also increasing available sorption sites [32, 55, 56]. As illustrated in Fig. 5.6, CNO showed stronger sorption capacity comparing with $C_{60}$ for both NaClO and $HNO_3$ modification. NaClO-modified CNOs seemed to have higher sorption capacity for $Zn^{2+}$, $Cd^{2+}$, and $Ni^{2+}$ than $HNO_3$-modified CNOs, which may be due to higher oxidizing ability of NaOCl than that of $HNO_3$. No significant difference was observed for $Cu^{2+}$ and $Pb^{2+}$. NaClO-modified CNOs possessed a relatively constant sorption capacity of ca. 6 mg/g for all of the five metals tested, more than ten times higher than that of the functionalized $C_{60}$ fullerene for $Zn^{2+}$, $Cd^{2+}$, $Ni^{2+}$, and at least equivalent to that of the functionalized $C_{60}$ for $Cu^{2+}$ and $Pb^{2+}$. The significantly higher sorption capacity of CNOs could be attributed to the structural differences between CNO and $C_{60}$. Oxidation process essentially would create some defects on the surfaces, where –COOH groups may locate [55]. XRD results revealed that the structures in the CNOs are not as highly crystallized as in $C_{60}$. Thus, oxidized CNOs might have more –COOH groups on its surface than oxidized $C_{60}$, leading to observed higher sorption capacity.

It was reported [57] that the sorption capacity of functionalized CNTs for $Cd^{2+}$ is about 6.5 mg/g under the similar sorbent mass/liquid ratio (i.e., 10 mg/35 mL). It is anticipated that higher sorption capacity of our CNO will be achieved if we increase the sorbent mass/liquid ratio, as previously reported for CNT sorption of metal contaminants [55]. Oxidation using 6 % NaOCl solution, commonly known as bleach

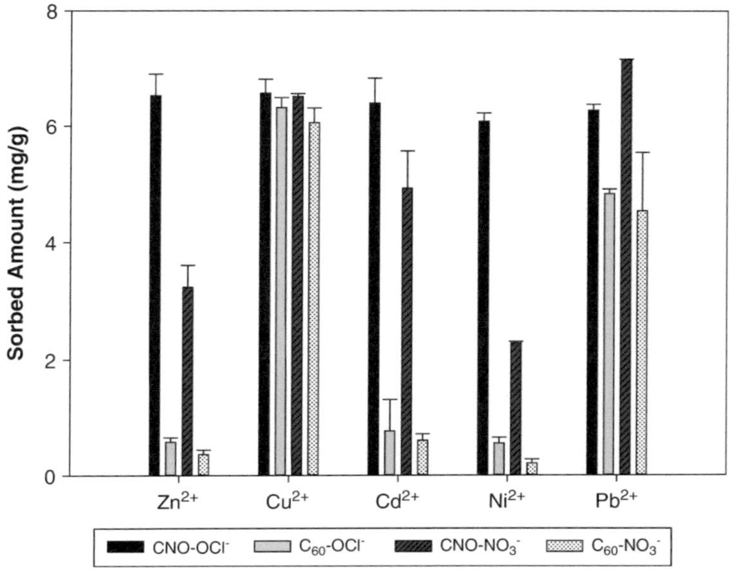


**Fig. 5.6** Comparison of sorption capacity of NaClO-modified and $HNO_3$-modified CNO and $C_{60}$ nanomaterials for $Zn^{2+}$, $Cu^{2+}$, $Cd^{2+}$, $Ni^{2+}$, and $Pb^{2+}$ (sorbent mass/liquid = 10 mg/35 mL) [54]

and frequently used as a disinfectant or a bleaching agent, is a very cost-effective and feasible approach. Functionalized CNOs possess a higher (or at least equivalent) sorption capacity than (as) commercially available carbonaceous nanomaterials, including $C_{60}$ and CNT, for heavy metal contaminant sorption. CNO, however, is 100 times more cost effective than $C_{60}$ and CNT, which makes it feasible for remediation applications.

## References

1. M.S. Dresselhaus, M. Terrones, Carbon-based nanomaterials from a historical perspective. Proc. IEEE **101**, 1522–1535 (2013)
2. Y.H. Hu, O.A. Shenderova, D.W. Brenner, Carbon nanostructures: morphologies and properties. J. Comput. Theoret. Nanosci **4**, 199–221 (2007)
3. E.H.L. Falcao, F. Wudl, Carbon allotropes: beyond graphite and diamond. J. Chem. Technol. Biotechnol. **82**, 524–531 (2007)
4. M.S. Dresselhaus, M. Endo, Relation of carbon nanotubes to other carbon materials. Carbon Nanotubes. **80**, 11–28 (2001)
5. X.K. Cheng, A.T. Kan, M.B. Tomson, Naphthalene adsorption and desorption from aqueous C-60 fullerene. J. Chem. Eng. Data **49**, 675–683 (2004)
6. M.A. Shannon, P.W. Bohn, M. Elimelech, J.G. Georgiadis, B.J. Marinas, A.M. Mayes, Science and technology for water purification in the coming decades. Nature **452**, 301–310 (2008)
7. S. Kang, M. Pinault, L.D. Pfefferle, M. Elimelech, Single-walled carbon nanotubes exhibit strong antimicrobial activity. Langmuir **23**, 8670–8673 (2007)

8. S. Kang, M. Herzberg, D.F. Rodrigues, M. Elimelech, Antibacterial effects of carbon nanotubes: size does matter. Langmuir **24**, 6409–6413 (2008)
9. D.Y. Lyon, D.A. Brown, P.J.J. Alvarez, Implications and potential applications of bactericidal fullerene water suspensions: effect of nC(60) concentration, exposure conditions and shelf life. Water Sci. Technol. **57**, 1533–1538 (2008)
10. S.N. Kim, J.F. Rusling, F. Papadimitrakopoulos, Carbon nanotubes for electronic and electrochemical detection of biomolecules. Adv. Mater. **19**, 3214–3228 (2007)
11. S.S. Mao, X.B. Chen, Selected nanotechnologies for renewable energy applications. Int. J. Energy Res. **31**, 619–636 (2007)
12. P.V. Kamat, Meeting the clean energy demand: Nanostructure architectures for solar energy conversion. J. Phys. Chem. C **111**, 2834–2860 (2007)
13. H.W. Kroto, J.R. Heath, S.C. Obrien, R.F. Curl, R.E. Smalley, C-60—Buckminsterfullerene. Nature **318**, 162–163 (1985)
14. P.R. Buseck, S.J. Tsipursky, R. Hettich, Fullerenes from the geological environment. Science **257**, 215–217 (1992)
15. O. Berne, A.G. Tielens, Formation of buckminsterfullerene (C60) in interstellar space. Proc. Natl. Acad. Sci. U. S. A. **109**, 401–6 (2012)
16. R.C. Haddon, Chemistry of the fullerenes: the manifestation of strain in a class of continuous aromatic molecules. Science **261**, 1545–50 (1993)
17. R. Taylor, D.R.M. Walton, The chemistry of fullerenes. Nature **363**, 685–693 (1993)
18. R.D. Johnson, D.S. Bethune, C.S. Yannoni, Fullerene structure and dynamics—a magnetic-resonance potpourri. Acc. Chem. Res. **25**, 169–175 (1992)
19. S.R. Chae, A.R. Badireddy, J.F. Budarz, S.H. Lin, Y. Xiao, M. Therezien, M.R. Wiesner, Heterogeneities in fullerene nanoparticle aggregates affecting reactivity, bioactivity, and transport. ACS Nano **4**, 5011–5018 (2010)
20. Y.G. Wang, Y.S. Li, K.D. Pennell, Influence of electrolyte species and concentration on the aggregation and transport of fullerene nanoparticles in quartz sands. Environ. Toxicol. Chem. **27**, 1860–1867 (2008)
21. D. Bonifazi, O. Enger, F. Diederich, Supramolecular [60]fullerene chemistry on surfaces. Chem. Soc. Rev. **36**, 390–414 (2007)
22. M. Akada, T. Hirai, J. Takeuchi, T. Yamamoto, R. Kumashiro, K. Tanigaki, Superconductivity appearing from C-60 doped with rare-earth metals. Sci. Tech. Adv. Mater. **7**, S83–S87 (2006)
23. N.A. Bagrovskaya, O.V. Alekseeva, Sorption properties of fullerene-modified polystyrene. Protect. Metals Phys. Chem. Surf. **48**, 217–220 (2012)
24. X.K. Cheng, A.T. Kan, M.B. Tomson, Uptake and sequestration of naphthalene and 1,2-dichlorobenzene by C-60. J. Nanopart. Res. **7**, 555–567 (2005)
25. M.M. Gonzalez, M. Gallego, M. Valcarcel, Effectiveness of fullerene as a sorbent for the determination of trace amounts of cobalt in wheat flour by electrothermal atomic absorption spectrometry. J. Anal. At. Spectrom **14**, 711–716 (1999)
26. D. Li, D.Y. Lyon, Q. Li, P.J.J. Alvarez, Effect of soil sorption and aquatic natural organic matter on the antibacterial activity of a fullerene water suspension. Environ. Toxicol. Chem. **27**, 1888–1894 (2008)
27. B. Pan, D.H. Lin, H. Mashayekhi, B.S. Xing, Adsorption and hysteresis of bisphenol A and 17 alpha-ethinyl estradiol on carbon nanomaterials. Environ. Sci. Technol. **42**, 5480–5485 (2008)
28. G.A. Polotskaya, S.V. Gladchenko, A.V. Pen'kova, V.M. Kuznetsov, A.M. Toikka, Synthesis of fullerene-polyphenylene oxide membranes for separating aqueous-organic mixtures. Russ J Appl Chem. **78**, 1468–1473 (2005)
29. V.V. Samonin, V.Y. Nikonova, E.A. Spiridonova, The influence of optical irradiation on the sorption properties of fullerene materials. Russ. J. Phys. Chem. A. **81**, 1276–1280 (2007)
30. M.G. Towell, L.A. Browne, G.I. Paton, K.T. Semple, Impact of carbon nanomaterials on the behaviour of C-14-phenanthrene and C-14-benzo-[a] pyrene in soil. Environ. Pollut. **159**, 706–715 (2011)

31. K. Yang, B.S. Xing, Desorption of polycyclic aromatic hydrocarbons from carbon nanomaterials in water. Environ. Pollut. **145**, 529–537 (2007)
32. K. Yang, L.Z. Zhu, B.S. Xing, Adsorption of polycyclic aromatic hydrocarbons by carbon nanomaterials. Environ. Sci. Technol. **40**, 1855–1861 (2006)
33. J.J. Pignatello, B.S. Xing, Mechanisms of slow sorption of organic chemicals to natural particles. Environ. Sci. Technol. **30**, 1–11 (1996)
34. V.K. Gupta, T.A. Saleh, Sorption of pollutants by porous carbon, carbon nanotubes and fullerene—an overview. Environ Sci Pollut Res. **20**, 2828–2843 (2013)
35. A. Serrano, M. Gallego, Fullerenes as sorbent materials for benzene, toluene, ethylbenzene, and xylene isomers preconcentration. J. Sep. Sci. **29**, 33–40 (2006)
36. V.I. Berezkin, V.V. Samonin, I.V. Victorovskii, V.Y. Nikonova, M.A. Yagovkina, L.V. Golubev, Adsorption of benzene on dispersed polycrystalline fullerites. Russ. J. Phys. Chem. A. **80**, 1986–1992 (2006)
37. V.V. Samonin, V.Y. Nikonova, E.A. Spiridonova, The influence of the preliminary adsorption of water on the adsorption of organic solvent vapors on fullerene materials. Russ. J. Phys. Chem. A. **81**, 1271–1275 (2007)
38. T. Huffer, M. Kah, T. Hofmann, T.C. Schmidt, How redox conditions and irradiation affect sorption of PAHs by dispersed fullerenes (nC60). Environ. Sci. Technol. **47**, 6935–6942 (2013)
39. M. Gallego, Y. Petit de Pena, M. Valcarcel, Fullerenes as sorbent materials for metal preconcentration. Anal. Chem. **66**, 4074–4078 (1994)
40. V.V. Samonin, V.Y. Nikonova, M.L. Podvyaznikov, Sorption properties of fullerene-modified activated carbon with respect to metal ions. Protect Metals. **44**, 190–192 (2008)
41. A. Higuchi, T. Yoshida, T. Imizu, K. Mizoguchi, Z.J. He, I. Pinnau, K. Nagai, B.D. Freeman, Gas permeation of fullerene-dispersed poly(1-trimethylsilyl-1-propyne) membranes. J. Polym. Sci. B. Polym. Phys. **38**, 1749–1755 (2000)
42. G.A. Polotskaya, A.V. Penkova, Z. Pientka, A.M. Toikka, Polymer membranes modified by fullerene C-60 for pervaporation of organic mixtures. Desalination Water Treatment. **14**, 83–88 (2010)
43. G.A. Polotskaya, A.V. Penkova, A.M. Toikka, Z. Pientka, L. Brozova, M. Bleha, Transport of small molecules through polyphenylene oxide membranes modified by fullerene. Sep. Sci. Technol. **42**, 333–347 (2007)
44. D.M. Sterescu, D.F. Stamatialis, E. Mendes, M. Wubbenhorst, M. Wessling, Fullerene-modified poly(2,6-dimethyl-1,4-phenylene oxide) gas separation membranes: why binding is better than dispersing. Macromolecules **39**, 9234–9242 (2006)
45. A. Penkova, A. Toikka, T. Kostereva, N. Sudareva, G. Polotskaya, Structure and transport properties of fullerene-polyamide membranes. Fullerenes Nanotubes Carbon Nanostruct. **16**, 666–669 (2008)
46. N.N. Sudareva, A.V. Penkova, T.A. Kostereva, A.E. Polotskii, G.A. Polotskaya, Properties of casting solutions and ultrafiltration membranes based on fullerene-polyamide nanocomposites. Expr. Polym. Lett. **6**, 178–188 (2012)
47. S. Sha, Y. Kong, J.R. Yang, Effect of charge-transfer complex between gasoline components/C-60 on desulphurization properties of C-60-filled ethyl cellulose hybrid membranes. J. Membr. Sci. **415**, 835–841 (2012)
48. J. Lee, Y. Mackeyev, M. Cho, L.J. Wilson, J.-H. Kim, P.J.J. Alvarez, C60 Aminofullerene immobilized on silica as a visible-light-activated photocatalyst. Environ. Sci. Technol. **44**, 9488–9495 (2010)
49. J. Lee, S. Hong, Y. Mackeyev, C. Lee, E. Chung, L.J. Wilson, J.-H. Kim, P.J.J. Alvarez, Photosensitized oxidation of emerging organic pollutants by Tetrakis C60 Aminofullerene-derivatized silica under visible light irradiation. Environ. Sci. Technol. **45**, 10598–10604 (2011)
50. G. Li, B. Jiang, X. Li, Z. Lian, S. Xiao, J. Zhu, D. Zhang, H. Li, C60/Bi2TiO4F2 heterojunction photocatalysts with enhanced visible-light activity for environmental remediation. ACS Appl. Mater. Interfaces **5**, 7190–7197 (2013)

51. Y. Gao, Y.S. Zhou, J.B. Park, H. Wang, X.N. He, H.F. Luo, L. Jiang, Y.F. Lu, Resonant excitation of precursor molecules in improving the particle crystallinity, growth rate and optical limiting performance of carbon nano-onions. Nanotechnology **22**(16), 165604 (2011)
52. M.E. Plonska-Brzezinska, A.T. Dubis, A. Lapinski, A. Villalta-Cerdas, L. Echegoyen, Electrochemical properties of oxidized carbon nano-onions: DRIFTS-FTIR and Raman spectroscopic analyses. Chemphyschem **12**, 2659–2668 (2011)
53. D. Roy, M. Chhowalla, H. Wang, N. Sano, I. Alexandrou, T.W. Clyne, G.A.J. Amaratunga, Characterisation of carbon nano-onions using Raman spectroscopy. Chem Phys Lett. **373**, 52–56 (2003)
54. M.B. Seymour, C.M. Su, Y. Gao, Y.F. Lu, Y.S. Li, Characterization of carbon nano-onions for heavy metal ion remediation. J. Nanopart. Res. **14**(9), 1097 (2012)
55. C.H. Wu, Studies of the equilibrium and thermodynamics of the adsorption of $Cu^{2+}$ onto as-produced and modified carbon nanotubes. J. Colloid Interface Sci. **311**, 338–346 (2007)
56. G.C. Chen, X.Q. Shan, Y.S. Wang, B. Wen, Z.G. Pei, Y.N. Xie, T. Liu, J.J. Pignatello, Adsorption of 2,4,6-trichlorophenol by multi-walled carbon nanotubes as affected by Cu(II). Water Res. **43**, 2409–2418 (2009)
57. Y.H. Li, S.G. Wang, Z.K. Luan, J. Ding, C.L. Xu, D.H. Wu, Adsorption of cadmium(II) from aqueous solution by surface oxidized carbon nanotubes. Carbon. **41**, 1057–1062 (2003)

# Chapter 6
# Graphene–Environmental and Sensor Applications

**Sarang V. Muley and Nuggehalli M. Ravindra**

**Abstract** Pollution management has been one of the major global issues being faced by the industrialized economy. In this chapter, applications of graphene-based nanomaterials for water and air remediation via adsorption and the effects of adsorbents on the electronic properties, magnetic properties, and photoconductivity, along with the relevant mechanisms, are discussed. Graphene nanosheets (GNS) and graphene oxides, due to their ease of chemical modification, with the addition of functionalities and stability along with ease of production, have emerged as ideal materials for adsorbents to eliminate heavy metal ions, inorganic ions, as well as organic dyes. Use of graphene as a photocatalyst in water remediation has been discussed. Graphene can help to resolve the challenge of making clean water available for general consumption. These graphene-based novel materials have been found to be suitable for air remediation through the removal of toxic gases, emitted due to the large-scale expansion of industries, including $SO_x$, CO, $NH_3$, and others, which can help to avoid natural calamities on account of global warming. A brief overview of the latest scientific advances in the developments of graphene-based devices to overcome the present shortcomings for fabricating sensors in the field of gas sensing, biosensing, and photonics for image sensing applications, along with their operating principles, is presented.

S.V. Muley • N.M. Ravindra (✉)
Interdisciplinary Program in Materials Science & Engineering, New Jersey Institute of Technology, Newark, NJ 07102, USA
e-mail: n.m.ravindra@njit.edu

A. Hu and A. Apblett (eds.), *Nanotechnology for Water Treatment and Purification*, Lecture Notes in Nanoscale Science and Technology 22,
DOI 10.1007/978-3-319-06578-6_6, 

## 6.1 Introduction

Industrial revolution, urbanization, and increase in population have led to pollution of the natural resources (water, air, and soil). Pollutants released from industrial sources include the release of toxic gases such as $NO_x$, $SO_x$, CO, $NH_3$ (and many others) besides the heavy metals and organic matter (such as dyes) which get mixed with the sources of drinking water and contaminate them. Increased release of greenhouse gases resulting from burning fossil fuels is a global challenge. There have been unpredicted effects of some pollutants on the environment causing natural calamities [1] and disruptions with losses amounting to billions of dollars [2]. Pollution management is essentially at the top of the agenda for sustainability of every nation and this has led to significant research focus on the development of smart ways to tackle it [3]. There have been many international conferences and journal publications highlighting this from time to time [4, 5]. Current studies have presented a constant depletion of natural resources that are suitable for human consumption, which has led to many research initiatives on water and air purification.

Earliest writings about purification of water in India and Greece date back to 2000 BC, in which it was suggested that heating water or filtration by passing the water through sand and gravel could lead to its purification. This was followed by various findings such as coagulation of water, the practice of sieving using bag filters in balance, and subsequently, in about 1500–1700 AD, there was use of water filters containing charcoal, wool, sponge, and sand for domestic applications. With the development of pipes for transportation of water through pipelines in the 1800 s, safe drinking water became accessible to the general population in most of the countries. In 1854, due to cholera epidemic in London, there was increased awareness of purification of water, leading to knowledge of purification of water by chlorination in 1890s. In 1903, water softening process known as ion exchange was implemented for removal of cations for water desalination. In 1970s, due to increased public concern about waterborne diseases due to the presence of microorganisms, pesticide residues, and sludge from industries, various techniques such as aeration and adsorptive removal of pollutants by active carbon were developed, followed by the development of membranes for reverse osmosis for water purification [6–9].

Currently, a number of physical, chemical, and biological techniques have been developed for controlling environmental pollution (of both water and air). With various techniques available in the industry, the adsorption process is broadly implemented and studied due to its ease of operation and simplicity. The advantage of using the adsorption technique is that there is no resultant secondary pollution by production of harmful by-products [5]. However, the use of conventional methods poses a major difficulty in separating the biotic and abiotic pollutants of size in the nanoscale range. Hence, the application of nanotechnological innovations for remediation of water and air is essential. Nanomaterials are excellent pollutant adsorbents due to their size, large surface area, and highly catalytic nature and hence offer multiple sites for adsorption. Latest advancements have suggested significant improvements in the quality of water on application of nano-products such as nano-sorbents, nano-catalysts, bioactive nanoparticles, nanostructured catalytic membranes, and nanoparticle enhanced filtration systems [10]. There have been a number of porous materials that have been developed to serve the application with varying effectiveness in remediation, such as zeolites, polymers and metal-organic framework, activated carbon, and mesoporous oxides. Main carbon-based adsorbents with high adsorption capacity include carbon nanotube (CNT), fullerene, GNS, graphene oxide (GO), and nanocomposites based on GNS [5].

Graphene was isolated from graphite by mechanical exfoliation using scotch tape, by Geim et al. [11, 12] in 2004. Graphene has a thermodynamically stable 2D structure with a network of honeycomb-shaped $sp^2$ hybridized carbon atoms. Graphene oxide (GO) is a new variety of carbonaceous nanomaterial that consists of graphene functionalized with varying oxygen and hydrogen containing groups such as hydroxyl, carboxyl, and epoxy. These functional groups help in forming stable aqueous colloid through sonication and can also interact with positively charged metal ions, dyes, polymers, and biomolecules, which can aid in their removal via adsorption from water [13].

Apart from environmental remediation, there have been numerous areas of applications of graphene and graphene oxide in the field of materials science, physics, chemistry, and biology [5, 13]. These applications are based on its extraordinary electronic and optical properties which are attributed to the conjugation of its long-range $\pi$ electrons. Graphene exhibits ease of functionalization, high carrier mobilities (inferred by the observation of massless Dirac fermions), high thermal conductivity (in the order of few thousands W/mK), and high optical transparency. These characteristics have led to improvement in properties and applications of interest in the areas of electronics and optics, such as high speed FET (field effect transistor) [14–19]. Its lateral quantum confinement, by shaping it to form 1D graphene nanoribbon (GNR), can help to overcome the issue of zero bandgap and lead to improvement in the on/off current ratio of transistors. The bandgap of GNR is dependent on the width of nanoribbon, and hence, it can be controlled [18]. The potential for graphene in photonic applications including transparent conducting electrodes, surface plasmonic devices, polarizers, photodetectors, and others has attracted the attention of researchers and is anticipated to lead to new developments [20]. Chen et al. [21] have developed micrometer size, nanomechanical resonators based on graphene, whose frequencies can be modulated electrostatically up to about 14 %. Voltage-controlled prototype oscillators developed by them exhibit frequency stability and modulation bandwidth which is sufficient for modulation of radio frequency (RF) carrier signals. This research has demonstrated the use of oscillators based on graphene for frequency-modulated signal generation and one can obtain more efficient transmission of audio signals.

With rising needs for industrial health and safety along with process control, one needs to develop better gas detection devices as well as devices for detection of trace levels of ions in water with enhanced performance at optimal cost.

With global market size for detectors exceeding \$2 billion in total and expected to exceed about \$3.3 billion by 2018 [2], research in this direction is taking new shape and magnitude to devise improved sensors through nano-engineering to replace the metal oxide-based sensors that are used traditionally. This enables to obtain better sensitivity, selectivity, and device stability. The use of appropriate sensor can aid in environmental safety; for example, electronic oxygen sensor in automobiles can make it possible to control fuel injection and hence emissions by controlling the air–fuel ratio [22]. Graphene has been reported to be used as an electrochemical sensor to detect gases in concentrations of 1 part per billion (ppb) [23], while exfoliated graphite oxide-modified glassy carbon electrode has been reported to be used for selective determination of picomolar concentration of lead(II) ions in water [24]. Development of electrochemiluminescence (ECL) sensor based on graphene permits supersensitive detection of chlorinated phenols in water [25].

Detection or sensing of light is one of the major challenges in electronic devices based on solid state technology [18]. Image sensors have seen vast improvements since the 1990s from simple single array CCD (charge-coupled device) to complex 3D (three-dimensional) digital SLR (single lens reflector) sensors. Besides cameras, image sensors have seen large-scale applications in medical devices, security, surveillance, and machine vision. These applications are in line with the demand for system and process automation in the developing industrial scenario today. Low-cost CMOS (complementary metal oxide semiconductor) technology has resulted in the development of high speed, high-resolution cameras. The superior properties of graphene over conventional silicon and the possibility for easy replacement within the current setup of fabrication can see an emergence of new generation of materials for the photonics industry. Graphene-based photodetectors, invented recently [26], have been found to be 1,000 times more photosensitive than the presently used IR detectors (such as HgCdTe). Bilayer graphene-based absorption photodetector is reported to be used as an infrared bolometer [18, 27]. Touch screen displays of graphene and integrated circuits based on graphene have been developed recently. The photoconductivity of graphene is found to vary significantly by adsorption of gases such as oxygen and nitrogen at terahertz frequencies and there exists a signature associated with the stimulated emission from gas-adsorbed graphene, which can be used to design and fabricate sensors [20].

Application of graphene in the field of energy storage has been presented by researchers at Rice University [28], who have developed a hybrid cathode material by combining GNRs and vanadium oxide ($VO_2$), which is expected to revitalize lithium ion batteries, by fulfilling the lack of electrode that displays ultrafast cycles of charging and discharging and hence results in high power batteries. The electrode is reported to need only 20 s for fully charging and discharging. The fabrication involves a challenging process of baking GNR into $VO_2$. Graphene is shown to contribute its high conductivity properties to the hybrid material, even though $VO_2$ accounts for 84 % of the electrode weight. Even the resulting electrode is shown to retain more than 90 % of its initial capacity after 1,000 cycles at an ultrahigh rate of 190 °C (~37.2 A/g).

Another upcoming field of interest in materials science focuses on the applications of nanotechnology for medical purposes termed as "Biosensing" [29]. Graphene, reduced graphene oxide (RGO), and various functionalized groups attached to graphene have been shown to be extremely useful in detecting various biomolecules, especially DNA and analyzing them. Although the literature shows some controversial reports about biocompatibility of graphene and its derivatives, more focused research on different species of graphene should result in improved understanding of its applications [30].

The chapter is organized in the following manner:

Section 6.2 of this chapter focuses on a discussion of the properties of carbon-based adsorbents (including CNT, GNS, GNR, and RGO) and the relevant mechanisms for water purification. Sections 6.3–6.5 present adsorption studies of various pollutant elements and compounds (organic as well as inorganic) in graphene and their effects on its electronic and magnetic properties as well as its photoconductivity. These effects have been successfully utilized to develop graphene-based devices, such as gas sensors, photodetectors, chemical sensors, and biosensors (including DNA sensors, FET sensors), by proper calibration, as discussed in Sects. 6.6 and 6.7. A detailed overview on the development of graphene-based photocatalysts for water remediation is reported in Sect. 6.8. A comparative study of commonly used photocatalysts for purification of water such as $TiO_2$, CdS, and their doped forms with graphene-based nanocomposite photocatalysts has also been presented in Sect. 6.8. Conclusions and outlook, based on these studies, are presented in Sects. 6.9 and 6.10.

## 6.2 Carbon-Based Adsorbents

Adsorption is a surface phenomenon which involves accumulation of gases, liquids, or solutes on the surface of another solid or liquid; the former is termed as adsorbate and the latter as adsorbent. Forces governing the adsorption are either physical in nature (governed by universal Van der Waals interactions) termed as physisorption or chemical in nature (governed by nature of chemical process) termed as chemisorption [31]. Graphene is a highly efficient adsorbent, has a large surface area, has high porosity, is physically stable, and works effectively. The adsorption process in graphene does not result in secondary pollution, i.e., there are no other harmful by-products during adsorption [32]. Adsorption is affected by various parameters such as temperature of pollutant, pH, concentration, contact time, and the size of pollutant particles [31].

When a solid carbon material is immersed in aqueous solution, a surface charge is developed due to the dissociation of surface groups or due to adsorption of ions from the solution [33]. To date, the most studied adsorption processes on carbon-based materials are the adsorption of phenolic compounds. Complex interplay between the electrostatic and non-electrostatic interactions, as well as $\pi$–$\pi$ dispersion interaction, contributes to adsorption of organic molecules on carbon materials from dilute aqueous solutions [34]. It is reported that the increase in the surface acidity of activated carbon, due to oxidation, results in decrease of extent of adsorption of phenolic compounds from dilute aqueous solutions [35]. The pH of the solution and the surface chemistry of carbon determine the surface charge. Mechanism of adsorption of natural organic matter (NOM) on CNT is based on accessible surface area, and the high molecular weight component of NOM is adsorbed relatively strongly. GNS and graphene oxide (GO) are found to exhibit comparable or better adsorption capacities for phenols (phenathrene and biphenyl) in the presence of NOM. The presence of NOM also reduces the adsorption capacity of the synthetic organic compounds (SOC), where the reduction in adsorption capacity on graphenes is less than that in CNT and activated carbon [34]. It should be noted that graphenes may be considered as alternative adsorbents for removal of organic molecules from water.

The adsorption of gases on carbon-based nanomaterials is affected by the site of adsorption relative to the structure. It has been noted that oxygen is chemisorbed on armchair sites of both CNT and graphene even at room temperature. The adsorption step is predicted by density functional theory (DFT) calculations to be more exothermic and spontaneous on CNT than on graphene. For zigzag configuration, the activation energy for adsorption of oxygen is reported to be higher on surface of CNT than that on graphene to the extent of about 2 kcal/mol, while the activation energy for adsorption of oxygen on graphene is found to be about 6 kcal/mol higher than CNT for armchair configuration. This implies easier adsorption of oxygen on armchair edge CNT than in graphene but not on the zigzag edge, which is in agreement with experimental observations [36]. It is reported that increasing diameter of nanotube decreases adsorption enthalpy of CNT [33]. Thus, one can understand the importance of adsorption site in relation to the physical structure and therefore the adsorption behavior.

Carbon-based materials such as activated carbon, CNTs, and fullerene have been shown to have excellent transport properties along with high adsorption capability as well as thermal stability [37, 38]. The cavity inside a CNT forms a natural pore which, in some cases, is of diameter smaller than 1 nm. The smoothness and hydrophobicity of the inner surfaces of the nanotubes have been found to provide almost frictionless flow of water through them, resulting in rates that are orders of magnitude higher than in case of conventional pores. The peculiar characteristic of the structure of CNTs is that they permit specific modifications of pores without affecting the properties of the inner surface of the nanotube. All of the above-stated characteristics enable CNTs to become the next generation of membranes with higher transport efficiency, rejection properties, and lifetimes that are drastically higher than those of the currently used membranes. However, the CNT technology is still in its infancy because of the fact that there are some major characteristics of membranes (e.g., fouling properties) that need to be addressed and this will help to establish CNT as a membrane on a large scale. It is possible to implement single-stage treatment process for water purification as CNT has uniformity of pore sizes and one can modify them as per requirements. Accordingly, it can be inferred that, for CNT technology to have a higher share in markets of water

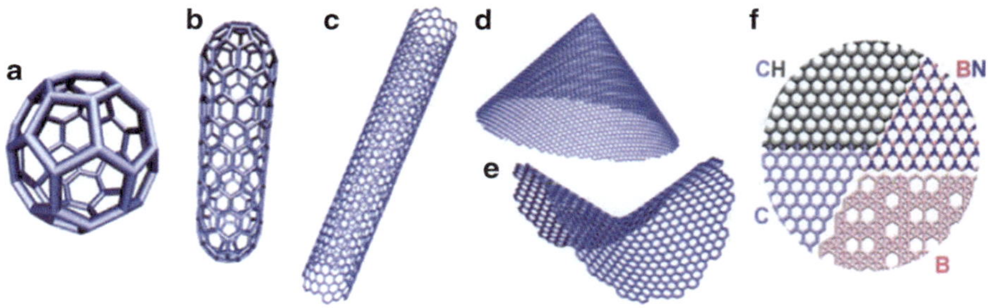


**Fig. 6.1** Various carbon forms: from (**a**) the C 60 Buckyball to (**b**) a "stretched" version of a fullerene: (10,0) nanotube cage, and to (**c**) long chiral (13,7) carbon nanotube; (**d**) Carbon cone, with a lone pentagon in fully hexagonal lattice (one of five possible cones, discovered experimentally) [40] and (**e**) a Pringle shape, with a lone heptagon [41] (**f**) A patch-quilt of close 2D family: graphEne (C), graphAne (CH), hexagonal boron nitride (h-BN), and pure boron layer (B) [42]

purification, a large-scale production of CNT is required which can make the process economical [39].

Graphene is highly stable and can be large in size (mm to inch). In order to avoid aggregate formation of bare nanoparticles, graphene has been used as a supporting material in multiple studies. Figure 6.1 shows various forms of carbon in nanostructured form. Graphene can be a 0D material termed as buckyball, rolled into 1D nanotube—termed as carbon nanotube (CNT) or stacked one over the other to form 3D graphite. The layered structure of graphene, its hydrophobic surface along with its excellent high strength considering its size, makes it suitable as an excellent adsorbent. Graphene has a theoretical value of specific surface area of 2,630 $m^2/g$ and its surface is shown to have high chemical adsorption because of strong $\pi$–$\pi$ interaction. It has large pore volume and can have functional groups attached to it. This indicates its enormous potential for being used as adsorbents.

Functionalized graphene can be produced economically and in large scale. Thus, it finds application as adsorbent for purification of water [43]. GO can be formed by oxidizing crystalline graphite using sodium nitrate, sulfuric acid, and $KMnO_4$ by a technique known as Hummers method [5]. Due to the highly acidic nature of GO, it is found to exhibit high adsorption of basic compounds and cations [5]. Since oxygen containing groups have high affinity to water, GO is hydrophilic and hence it can be easily dispersed in water. Hence, GO can readily form macroscopic flakes, most likely one atomic layer thick. It is convenient for processing and has high ionic adsorption capacity. One can even modify graphene oxide or graphene with organics or metal oxides and can form varieties of nanocomposites, leading to enhancement of its adsorption capacity and separating efficiency. Graphene-based nanocomposites have high porosity and are found to exhibit less aggregation tendency. For effective separation of dyes and organic compounds, 3D RGO composites can be produced for use as adsorbents, which are hydrophobic in nature [44]. Several papers have been published in the literature on applications of graphene or its composites being used for removal of pollutants from air and water.

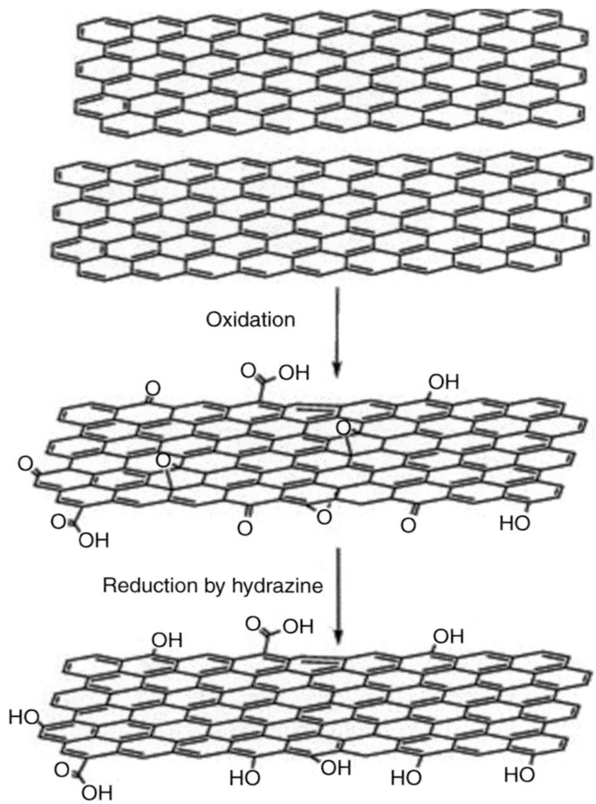


**Fig. 6.2** Oxidation of graphite to graphene oxide and reduction by hydrazine to reduced graphene oxide. "Reprinted with permission from Progress in Materials Science, Vol 56, Virendra Singha, Daeha Jounga, Lei Zhaia, Soumen Dasa, Saiful I. Khondakera, Sudipta Seal, Graphene-based materials: Past, present, and future, Pages No. 1178–1271, Copyright (2011), with permission from Elsevier"

All fields of science have seen various reports that describe applications of GO and GNS. A brief process showing the synthesis of graphene oxide and its further reduction by hydrazine is as shown in Fig. 6.2 [45].

## 6.3 Fundamentals of Graphene-Based Adsorption

### *6.3.1 Effects of Adsorption on Electronic Properties*

Despite a plethora of available studies, there are not enough theoretical studies on the physics and chemistry of the adsorption of small molecules onto the surface of pristine graphene. The electronic band structure of pristine graphene is as shown in Fig. 6.3.

It is extremely important to know the role of metal graphene interface [47–49] when studying electron transport through a graphene sheet, especially when the field of interest is focused on the influence of the adsorption of metal or impurities

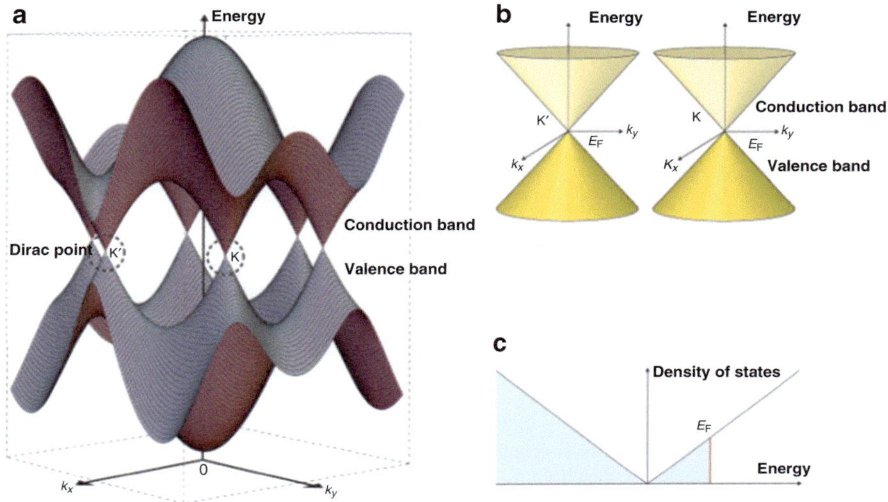


**Fig. 6.3** (**a**) Energy bands near Fermi level in graphene. Overlap of conduction and valence bands at Fermi level is evident. Points of intersection are K and K′ termed as Dirac points. (**b**) Cone-shaped energy bands at K and K′, (**c**) Density of states near Fermi level having energy $E_F$. "Reprinted by permission from Macmillan Publishers Ltd. [NPG Asia Materials] [46], copyright (2009)"

on the surface of graphene [50–53]. This can help in modifying graphene substrates for a specific application. The fundamental science associated with the interaction of adsorbed atoms with graphene is developing at a rapid rate [54–59]. Chemisorption of metal atoms changes the electronic properties and the magnitude of the change depends on the nature of the bonds formed with graphene as well as on the doping of graphene with alkali/transition metals.

As shown in Fig. 6.3, intrinsic graphene has a characteristic zero bandgap with its bonding $\pi$ and antibonding $\pi^*$ bands crossing each other at a single point termed as Dirac point at the Fermi level ($E_F$). This point is at the corner of the Brillouin zone [60], and a linear dispersion at this point can be observed. This is the basic reason for the large charge carrier mobility of about 15,000 $cm^2/V.s$ [11]. This can be seen as an extraordinary value addition as compared with silicon, a semiconductor with charge carrier mobility of about 1,400 $cm^2/V.s$. Graphene has invariably higher strength and rigidity. It is harder than diamond and about 300 times harder than steel [61]. This is another property that makes graphene more suitable than conventional silicon-based materials for a variety of applications. An increase in the charge carrier concentration of graphene, due to the adsorption of gas molecules, has been reported in the literature. This significant property of graphene, along with its high conductivity even for low charge density, makes it suitable for the fabrication of highly sensitive gas sensors.

Using DFT calculations, Ali Fathalian et al. [62] studied the effects of adsorption of oxygen on the semiconducting armchair GNRs and noted that the adsorption

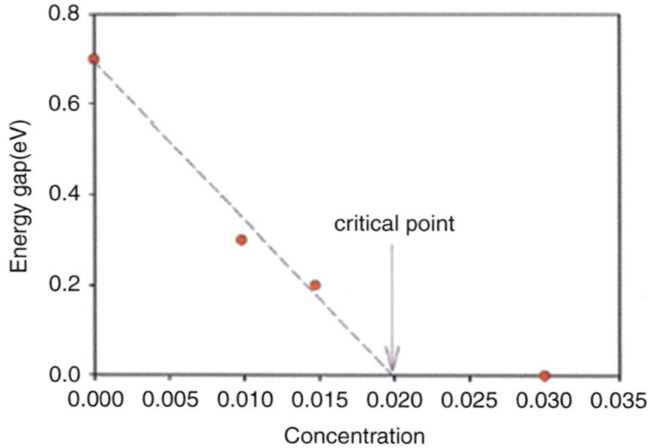


**Fig. 6.4** Reduction in semiconducting energy gap of hydrogenated edges in graphene nanoribbon versus adsorbed oxygen concentration. "Reprinted from Physica B: Condensed Matter, Vol 417, Ali Fathalian, Jaafar Jalilian, Sahar Shahidi, Effects of gas adsorption on the electronic properties of graphene nanoribbons, Pages No. 75-78, Copyright (2013), with permission from Elsevier"

energy in the edges is lower than in other positions. Moreover, on increasing the adsorption of gas concentration, its energy gap eventually decreases, and, at a critical concentration of adsorbate as shown in Fig. 6.4, semiconductor-to-metal transitions are observed, which can, in turn, be useful for designing sensors for oxygen detection [62].

The structural and electronic properties of oxygen-adsorbed graphene have been studied theoretically by Nakamura et al. [63] and Ito et al. [64], whose results indicate that the adsorption of oxygen molecules onto graphene produces epoxy and ether group phases which are almost bistable. Moreover, they have concluded that the ether structure is the most energetically preferable for adsorption involving both sides of the sheet, while on one side, the adsorption structure appears only as a metastable phase, with a finite energy gap at the K point and its value increases as the number of oxygen molecules increases with respect to the number of carbon atoms.

Uchoa et al. [65] studied the effects of metal doping (K and Pd) on the electronic properties of graphene using DFT. The charge transfer from a single layer of potassium on top of graphene was estimated, and it was found that when graphene is coated with a nonmagnetic layer of palladium, it leads to a magnetic instability in the coated graphene due to the hybridization between the transition metal and the carbon orbitals. Bermudez et al. [66] have studied the interactions of $CCl_4$, $CS_2$, water, and acetone with single-layer graphene, under steady state conditions, with IR spectroscopy and noted that there is a broad and intense absorption band spanning from 200 to 500 meV range, attributed to the electronic excitations.

Further, Sevinçli et al. have theoretically investigated the electronic and magnetic properties of graphene and GNRs by functionalizing with 3*d* transition metal (TM) atoms [67]. It was found that, except Cr, the adsorption on the hollow sites gave the minimum energy for all TM atoms and the properties of graphene and GNRs could be strongly modified through the adsorption of 3*d* TM atoms. Graphene is found to become a magnetic metal after the adsorption of TM atoms. Nonmagnetic armchair GNR (AGNR) is found to behave metallic or like semiconductor depending on its ribbon width. Fe or Ti adsorption converts some AGNRs into partially metallic with a 100 % spin polarization at the Fermi level. Aktürk et al. studied the adsorption of Si and Ge on graphene [68]. The authors found that the atoms are bonded to graphene at bridge site with high binding energy. Semimetallic graphene has been found to become magnetic due to the adsorption of these atoms, the phenomena for which have been attributed to the combination of $\pi$ and $\pi^*$ orbitals.

Till now, we have seen various investigations involving changes in electronic band structure of graphene upon adsorption. Zahrani [69] performed ab initio simulations taking benzene and naphthalene as adsorbates on the graphene surface and noticed that there was no significant change in its electronic structure, though the adsorbed configuration is more stable than the pristine hollow structure. This shows that the Dirac point of benzene and naphthalene on graphene coincides with the Fermi level of graphene, and hence there is no charge transfer between the molecules. The same is evident in Fig. 6.5 [69].

### 6.3.2 Effect of Adsorption on the Magnetic Properties

There has been a considerable interest in the magnetism of nano-graphite particles for quite some time for spintronic applications. Controlling magnetism by applied voltage is the most important criteria for these applications [70]. Graphene is a suitable material for spintronics, because it does not have very strong spin–orbit interaction [71] and long spin relaxation lengths [70, 72, 73] and because of the fact that it is feasible to form paramagnetic centers by using controlled defect introduction. It has been demonstrated in recent experiments [74, 75] that magnetic moments can be carried by both vacancies and adsorbed atoms in graphene. In graphene, the prospect of defect ferromagnetism is more prominent on account of two reasons: Firstly, the conduction electrons in graphene provide a medium for coupling between localized spins and, second, the defect-induced moments are believed to be present because of the same $\pi$ electrons which cause electronic transport [70, 76–83]. Hence, the electric field in graphene can be used for controlling the coupling of localized magnetic moments and even to control the magnetic moments [70].

Recently, Nair et al. [70] showed that the magnetism of adsorbed atoms in graphene is itinerant and it is possible to be controlled by doping, so that the magnetic moments can be tailored. The vacancy magnetism is reported to have dual origin, with two approximately equal contributions: one from itinerant magnetism and the other from dangling bonds. Their work reports that the spin transport in graphene is possible to be controlled using the field effect. This is similar to the

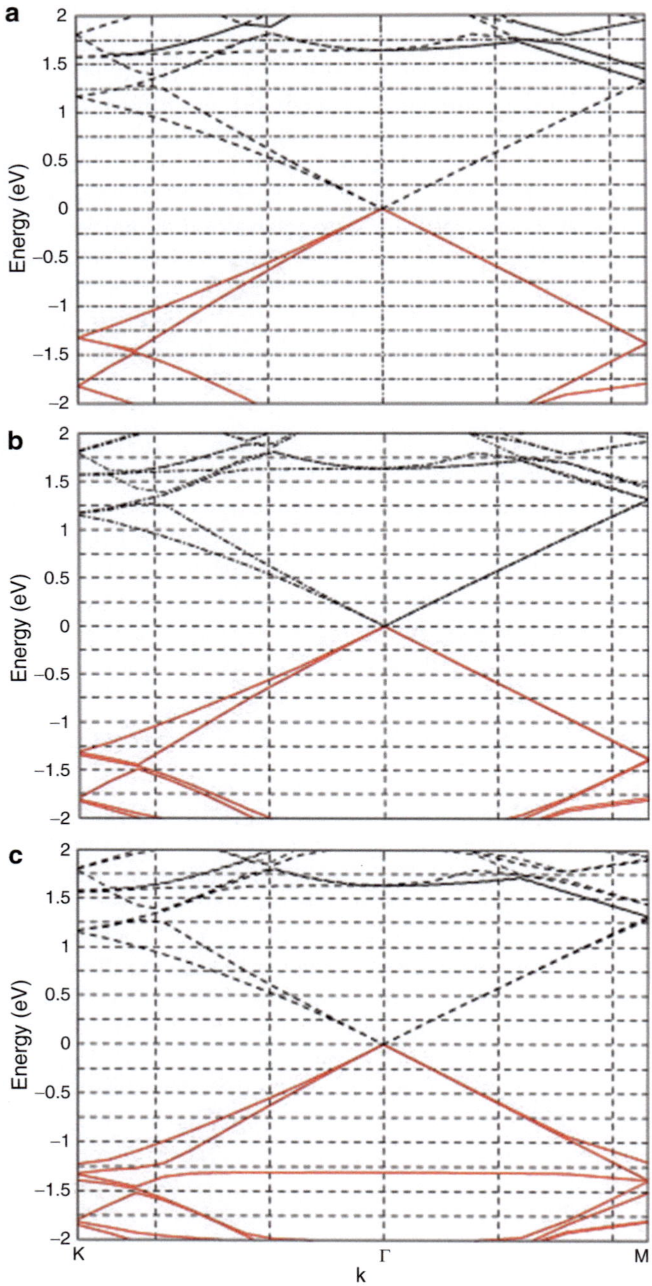


**Fig. 6.5** Electronic band structure of (**a**) pristine graphene, (**b**) benzene/graphene, and (**c**) naphthalene/graphene along the high symmetry directions of the hexagonal Brillouin zone of graphene. The Fermi level is set to 0. "Reprinted from Journal of Applied Surface Science, Vol 257, A.Z. AlZahrani, First-principles study on the structural and electronic properties of graphene upon benzene and naphthalene adsorption, Pages No. 807-810, Copyright (2010), with permission from Elsevier"

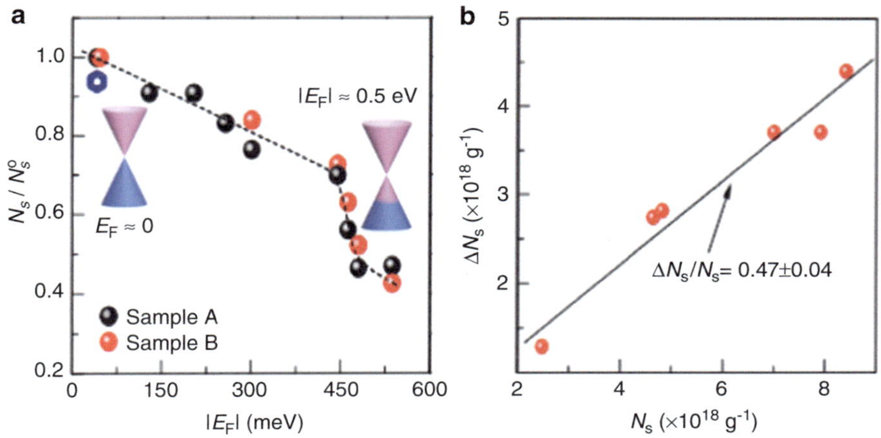


**Fig. 6.6** (**a**) Reversible control of vacancy magnetism in graphene, (**b**) Universal halving of the vacancy magnetic moment by doping graphene. "Reprinted by permission from Macmillan Publishers Ltd. Nature Communications [70], copyright (2013)"

concept of controlling the electronic and optical properties by doping graphene. Figure 6.6a shows the reversible control of vacancy magnetism in graphene laminate samples that were irradiated with 350 keV protons (to produce the required density of vacancies) at temperature between 300 and 1.8 K in a magnetic field up to 7 T. The normalized density of spins ($s = 1/2$) in two irradiated samples is evident in this figure. $N^0_S$ is the spin concentration before doping and $N_S$ is extracted from the Brillouin function fit for M(H) dependence for each doping level and H is the applied magnetic field. Dashed line is just connecting the various points of reference. The open symbol (in blue) is for sample B after its maximum doping and annealing at 100 °C for 2 h. Insets are schematic diagrams of graphene in undoped and heavily doped states. $E_F$ is the Fermi level. Figure 6.6b shows the universal halving of the vacancy magnetic moment by doping in graphene. $\Delta N_S = N^0_S - N^F_S$ shows change in number of spins after maximum doping in samples with different vacancy concentrations. $N^F_S$ is the number of spins measured at $E_F \geq 0.5$ eV. Solid line is the best fit showing the trend.

Enoki et al. [84] pointed out that edge states as well as adsorbed or intercalated species play an important role in determining the magnetic properties of these nanographite particles. In zigzag edges, the presence of nonbonding π-electron state (termed as "edgestate") is observed while, in armchair edged graphene, there are no such edge states. There is a strong spin polarization of the edge states that induces nonconventional magnetism in nanographene. Recently, it was shown that semimetallic bare graphene could be changed into either ferromagnetic or antiferromagnetic state upon transition due to metal adsorption [67]. The electronic properties of graphene have been shown to depend on the nature of the edges [85].

In another similar study, Zhou et al. [86] assessed the changes in magnetic and electronic structures of graphene with adsorbing elements such as Cl, S, and P,

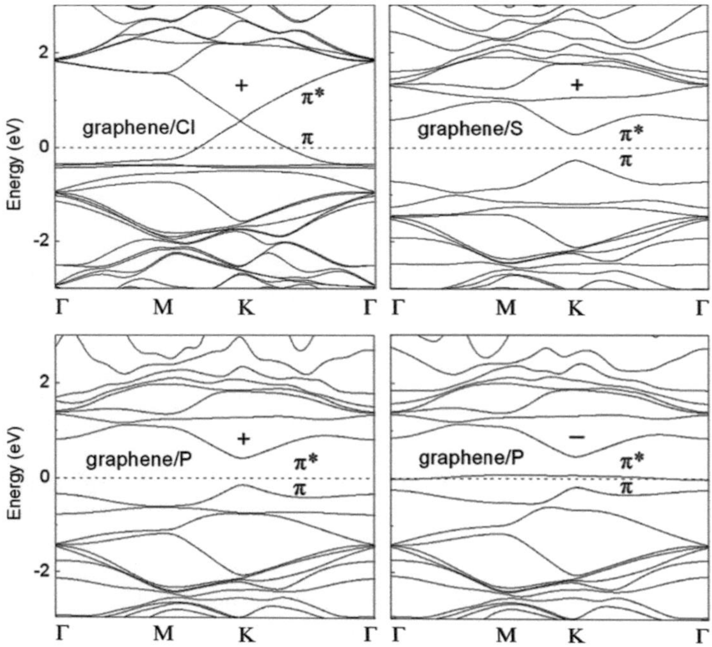


**Fig. 6.7** Bands of graphene with Cl, S, and P atoms absorbed at most stable configurations. "Reprinted with permission from [X. T. Zu, Y. G. Zhou, F. Gao, H. F. Lv, H. Y. Xiao, Appl. Phys. Lett. 95, 3 (2009)]. Copyright [2009], AIP Publishing LLC"

using ab initio and first principles calculations. Cl adsorption endows graphene with metallicity. S is found to induce a bandgap of about 0.6 eV in graphene, hence resulting in the formation of a semiconductor. Unpaired electrons in P-adsorbed graphene are magnetized and this ultimately leads to a magnetic moment of 0.86 $\mu_B$, while there is no magnetic moment observed after Cl and S adsorption. In particular, P-absorbed graphene may be useful for spintronic applications, such as tunneling magnetoresistance. Their effect on the bandstructures of graphene is as shown in Fig. 6.7 [86].

Li et al. [87] studied the adsorption of monovalent and bivalent adsorbates and proposed a mechanism for their magnetic coupling with graphene. They used spin-polarized DFT for the calculations to explain the effects of concentration and electronegativity of adsorbents for chemisorption of the species. For monovalent chemisorptions, the origin of the magnetic order was attributed to the instability of π electrons that are produced on account of the adsorption, hence opening up a narrow energy gap and ultimately resulting in antiparallel spin directions on any of the adjoining carbon atoms on the graphene sheet. Furthermore, it was found that the magnetic order is only observed for the separation between the adsorbing sites $<10$ Å which is in contrast with the long-range magnetic interaction and coupling for divalent chemisorptions. The origin for divalent chemisorptions is localized nonbonding (i.e., spin polarized) and is attributed to $\pi$ electrons mediated by the conduction $\pi$ electrons around the Fermi energy, which can be understood to be similar to the *s–d* interactions in transition metals. Knowing the mechanism permits to calibrate a graphene-based sensor for the separation of adsorbates such as H, OH, CN, CH, F, or other groups present in water by analyzing their magnetic coupling with graphene.

Knowing the fact that graphene can only physisorb various gas molecules [88–91] and that it has no spin polarization (being its drawback), Dai and Yuan [91] studied the adsorption of oxygen on B-, N-, Al-, Si-, P-, Cr-, and Mn-doped graphene theoretically. Using DFT, they noticed that oxygen is physisorbed on B- and N-doped graphene with negligible adsorption energy, while it is chemisorbed on Al-, Si-, P-, Cr-, or Mn-doped graphene. Chemisorption, which results from local curvature arising from large bond length of C–X (X = Al, Si, P, Cr, or Mn), is found to induce dramatic changes in the electronic structures as well as localized spin polarization of doped graphene. Chemisorption of $O_2$ on Cr-doped graphene is antiferromagnetic. Thus, changes in the electronic and magnetic properties of graphene can be used for calibrating a device based on doped graphene substrates (used for sensing) and determining the concentrations of the adsorbents.

### 6.3.3 *Effect of Adsorption on Photoconductivity of Graphene*

Photoconductivity is defined as the increase in electrical conductivity caused by incident light, due to high intrinsic carrier mobility. Graphene has high mobility of charge carriers and this has made it suitable for significant applications in high speed devices [92, 93].

Among the characterization techniques, terahertz (THz) spectroscopy has been found to be ideal for surface-sensitive materials such as graphene. Response of graphene junctions to incident photons has been attributed to having an origin to its thermoelectric and photovoltaic effects (in which electric field accelerates photo-excited carriers, i.e., holes and electrons in opposite directions) or bolometric effect (incident photons raise graphene's local temperature, which in turn changes the device resistance and hence its current conduction) [94], along with hot carrier transport and carrier multiplication. In contrast, photo-response of intrinsic biased graphene is reported to have no influence on its thermoelectric effects. THz spectroscopy permits detection of material properties with electromagnetic radiation in the frequency range (about few hundred GHz to 43 THz) and operates in a noncontact mode [95]. Because of the rapid development of graphene in the form of large sheets, using chemical vapor deposition (CVD) [96], research began in understanding the nonequilibrium carrier dynamics of devices based on graphene [97]. Docherty et al. [20] reported on the sensitivity of graphene to environmental gases (mainly nitrogen and oxygen) upon their adsorption. It was noted that there is a signature THz emission on adsorption of gases, which can be used to design and fabricate high speed devices such as gas sensors. Earlier, Geim [14] had predicted

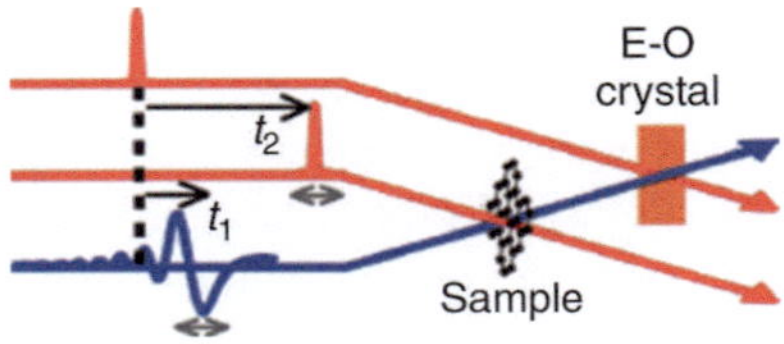


**Fig. 6.8** Schematic of measurement for THz Spectroscopy of graphene. "Reprinted by permission from Macmillan Publishers Ltd. Nature Communications [20], copyright (2012)"

graphene's susceptibility to environmental conditions. Figure 6.8 shows a schematic representation of measurements for THz spectroscopy of graphene [20].

Acoustic phonons are quanta arising out of coherent movements of atoms and are responsible for intrinsic energy loss of electrons [94]. Optical phonons can detect interband charge dynamics, while THz frequency photons are excellent probes for intraband dynamics. CVD-grown graphene can either display transparency at post-photo-excitation optical frequencies (photo-induced bleaching) or can have increase in THz conductivity (photo-induced absorption) [20].

The type and density of gas adsorbed on the surface of graphene can lead to increased density of charge carriers [20]. George et al. [98] have shown that after interband excitation, there is an increase in absorption of terahertz photons, whose occurrence, after photoexcitation, is attributed to increase of mobile charge carrier density. Studies have shown that heating of charge carriers, due to photo-excitation, occurs in tens of femtoseconds, while longer timescale of order of about 2.5 ps is known to be associated with decay of hot phonons [99]. Particularly, orders of magnitudes in photoconductivity lead to an important technology in producing graphene-based gas detectors. Theoretically, it was demonstrated by Ryzhii et al. that, in a nonequilibrium 2D in optically pumped graphene, at THz frequencies, population inversion can be observed leading to negative conductivity [100]. Figure 6.9 shows the photoconductivity spectra of graphene with varying environment before and after photoexcitation [20].

Moreover, it is noted that charge dynamics in graphene is highly affected by adsorbed oxygen, while it is found to have negative photoconductivity of higher magnitude than in case of vacuum. The spectral photoconductivity of graphene in the presence of $N_2$, air, and $O_2$ can be described in a Lorentzian form [20]:

$$\sigma_{\mathrm{Lorentz}}\alpha(\mathrm{i}\omega)/\left(\omega^2-{\omega_0}^2+\mathrm{i}\omega\Gamma\right),$$

where $\Gamma$ is the linewidth and $\omega$ is the resonant frequency.

The proposed mechanism for charge dynamics is that post photo-excitation, the charge carriers relax and then populate and combine around Dirac point and emit excess radiation. Thus, one can expect THz emission over a wide spectral range [100].

It has been noted that physisorption of $O_2$ and $N_2$ on graphene makes it act as a doping element [101], which leads to opening of bandgap at the Dirac point, which is zero for intrinsic graphene, where bonding $\pi$ orbitals overlap with antibonding $\pi^*$ orbitals [14, 64, 102]. This may even modify the density of states [46, 91] and

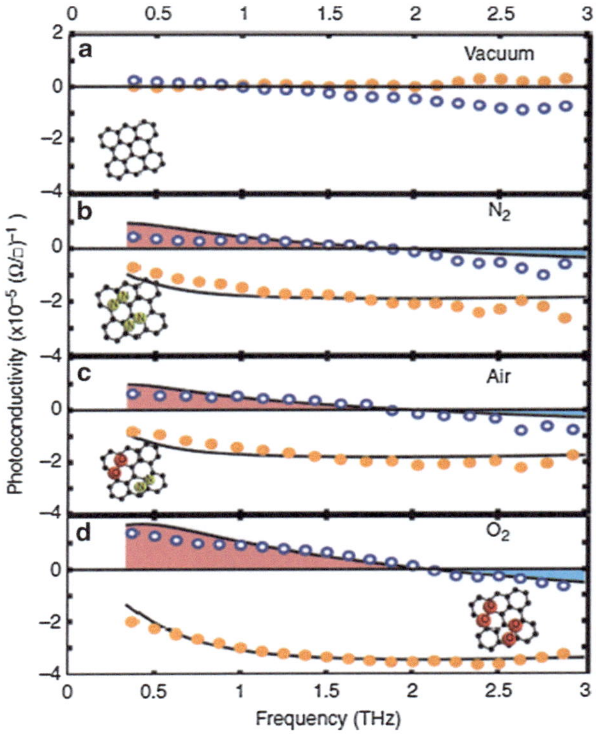


**Fig. 6.9** Photoconductivity spectra of graphene in different environments: (a) Vacuum, (b) $N_2$, (c) Air, (d) Oxygen at 2 ps after photo-excitation [20]. "Reprinted with permission from Macmillan Publishers Ltd. Nature Communications [20], copyright (2012)"

hence its electronic properties. The fact that chemisorption directly affects the electronic properties of graphene makes it an ideal material for sensing applications. It is interesting to note that the spectral width of about 10 THz can be described using high interband transition rate, fast energy loss of electrons via interaction with other electrons, and cooling limits of energies of optical phonons. Here, it should be noted that CVD-grown graphene is found to have characteristics of heavily p-doped material.

Sato et al. [103] studied tuning of Fermi level in bilayer graphene (BLG) due to adsorption of oxygen gas molecules in FET configuration. Here, the point to be noted is that there are three processes that can cause change in the Fermi level: one is charge transfer between graphene and the adsorbed gas molecules; second is back gating of graphene with the $SiO_2$/Si substrate, which can lead to change in polarization of gas molecules [104, 105] due to additional charge across the interface; and third is electric field gradient generated on account of application of external voltage across the gate, which leads to modification of charge distribution on surface as well as on adsorbed molecules [106–109]. This can even lead to corresponding changes in the adsorption process at the surface [110]. It is reported that when $SiO_2$/Si substrate is used as back-gate insulator of FET, tuning of the Fermi level of graphene by applying gate voltage is about $\pm 0.1$ eV [15, 103], which would be enough to change the chemical reactivity on the surface. Figure 6.10

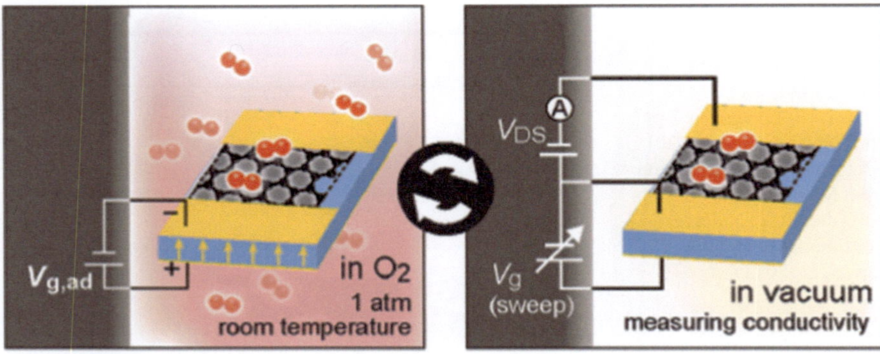


**Fig. 6.10** Schematic diagram of measurement cycles for conductivity of graphene. "Reprinted (adapted) with permission from (Y. Sato, K. Takai, T. Enoki, Nano Letters 11, 3468-3475 (2011)). Copyright (2011) American Chemical Society"

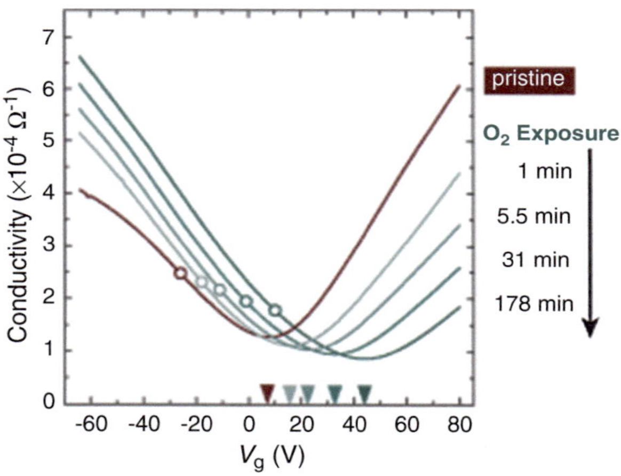


**Fig. 6.11** Change in field effect behavior in BLG FET by exposure to oxygen for various times, at different gate voltages. "Reprinted (adapted) with permission from (K. T. Yoshiaki Sato, and Toshiaki Enoki, Nano Letters 11, 3468 (2011)). Copyright (2011) American Chemical Society"

shows a schematic representation of the conductivity measurement cycles for graphene and Fig. 6.11 shows the changes in conductivity of BLG FET by exposure to oxygen for different times, at varying gate voltages.

Another study of carrier transport on molecular gas adsorption on epitaxial graphene with IR spectroscopy is discussed by Daas et al. [111]. They observed changes in conductance upon adsorption of $NH_3$ and $NO_2$, on graphene epitaxially deposited on SiC, using amperometric measurements. Conductance is directly related to the carrier concentration and scattering; hence they used IR reflection spectra to develop relationships between the two frequency domains resulting from each.

The sensitivity of exfoliated graphene to $NH_3$ [112] and $NO_2$ [113] and that of epitaxial graphene (EG) have been established by molecular adsorption doping. Adsorption leads to increase or decrease of electron concentrations, depending on the electronic structure of the adsorbent. Moreover, adsorbents have an effect on carrier scattering. Both the factors cumulatively change the conductivity of graphene surface. $NO_2$, being a strong oxidizer, is expected to reduce the number of free electrons at EG surface, while the converse is true for $NH_3$ as it donates electrons. Polar molecules [113] change EG conductivity by (a) inducing carriers in the EG and (b) increasing scattering, i.e., decreasing mobility. Surface carrier concentration in EG is given by:

$$n_s = \int_0^{\infty} D(E)f(E - E_F)dE,$$

where $D(E)$ is the density of states and Fermi energy function is $f(E{-}E_F)$, and $E_F$ is the Fermi level.

Scattering ($\tau$) can be of types—intraband (within the same band) and interband (within two different bands) scattering [114]. Hence, one needs to study the role of Fermi level ($E_F$) and $\tau$ as a result of gas adsorption. In the case of ammonia and $NO_2$, interband and intraband scattering time shortens due to the fact that the impurity concentration at the surface is higher than in the case of $N_2$ atmosphere. Hence, the overall conductivity reduces in both cases of $NH_3$ and $NO_2$ compared to $N_2$ environment. Thus, for describing IR carrier transport in EG, one needs to include the influence of interband scattering.

## 6.4 Water Treatment by Using Graphene-Based Adsorbents

Fluorides, phosphates, sulfides, nitrates, heavy metal ions, and organic compounds are the most important ones amongst the polluting groups in water. GO and GNS have been shown to present a very strong adsorption tendency toward these groups, which is higher in magnitude than activated carbon or CNT.

### *6.4.1 Inorganic Metal/Metalloid Cations*

Polluting cationic groups, normally released by industries, include Hg, Pb, Ag, Cu, Cd, Cr, Zn, Ni, Co, and Mn. They are found to affect the ecosystem (that includes humans, aquatic life, plants, and animals). The anionic groups polluting ecosystem include As, Hg, Cd, Cr, Co, Cu, Se, Pb. A few of them that are significant are discussed below.

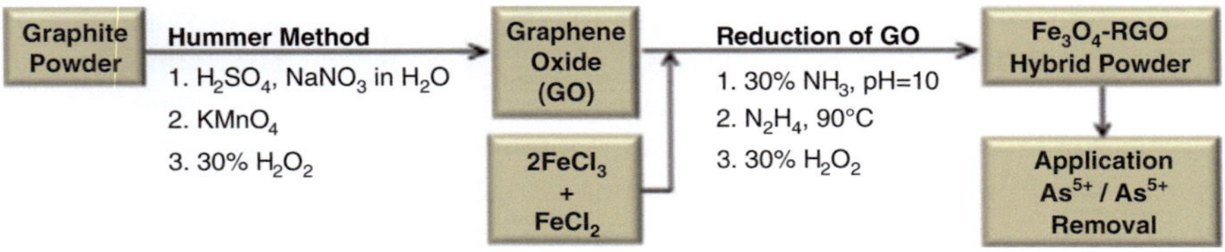


**Fig. 6.12** Schematic diagram of synthesis and application of magnetite-reduced graphene oxide composites. "Reprinted (adapted) with permission from [115]. Copyright (2010) American Chemical Society"

One of the initial studies of functionalized graphene as adsorbent of heavy ions from water was by Chandra et al. who showed that it is easy to remove arsenic from contaminated water [115]. Arsenic is identified by the US Environment Protection Agency (EPA) [29] as a carcinogenic element and is dangerous even if consumed in levels higher than 10 ppb. In this study, a composite of $Fe_3O_4$-RGO was made in accordance with steps shown in Fig. 6.12. The composite was found to form 10 nm magnetite nanoparticles on RGO sheets even at room temperature and displayed a high capacity of binding to As(III) and As(V), which lead to almost 99.99 % removal. High binding capacity is attributed to increased sites of adsorption in magnetite-RGO composite, which takes place by reduction of aggregation of bare magnetite.

In 2012, Zhu et al. [116] developed magnetic graphene nanoplatelet composites (MGNCs), which have core-shell Fe–$Fe_2O_3$ nanoparticles and demonstrated higher adsorption capacity instead of conventional iron oxide, $Fe_2O_3$, based adsorbents. The material shows strong magnetization enabling separation of Fe core. Furthermore, it is found to have higher binding capacity of arsenic(III), on account of higher number of adsorption sites.

Simultaneous removal of As(III) and As(V) from water was demonstrated by Luo et al. [117]. They used nanocomposites of graphene oxide-hydrated zirconium oxide, which were made by hydrothermal co-precipitation method. Excellent dispersion of ZrO $(OH)_2$ particles on graphene oxide (GO) substrate led to high capacity to adsorb. The substrate was found to show good anti-interference from other ions in the solution, along with the capability to recycle it, making it an ideal method for purifying water.

Reduced GO-metal/metal oxide composites have been synthesized by means of redox reaction, by Sreeprasad et al. [43]. Metal ion scavenging application of reduced GO (RGO)-$MnO_2$ and (RGO)-Ag has been demonstrated for Hg(II) as model pollutant. Figure 6.13 shows a TEM image of RGO-Ag with dispersion of nanoparticles of Ag [43]. Main reason for composite formation between RGO and metal precursor is due to redox-like reaction. RGO, eventually, undergoes oxidation to form GO and hence forms metal nanoparticles that are closely attached to the sheets. Size of nanoparticles that are being formed can be tuned by controlling the added precursor concentration.

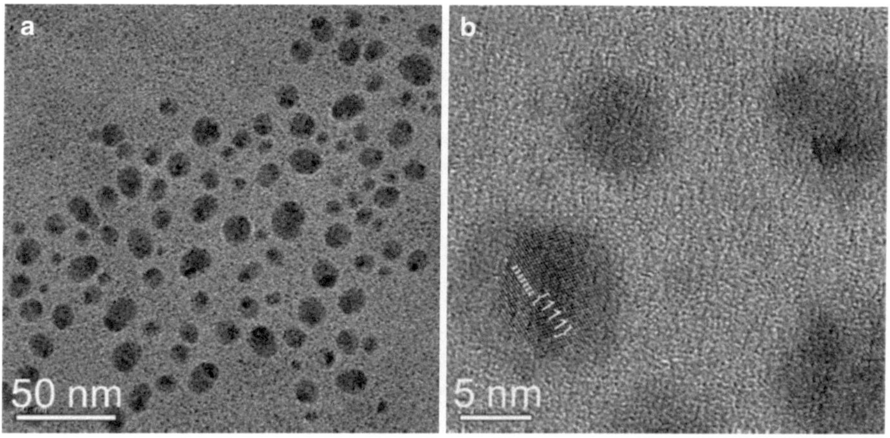


**Fig. 6.13** TEM image of reduced graphene oxide—Ag (0.05 mM) showing well-dispersed nanoparticles over RGO sheet. (**a**) The particles are well separated and do not exhibit any aggregation in the size range of 10–15 nm. (**b**) A lattice resolved image with [111] plane of Ag nanoparticle, which is characteristic of cubic Ag. Some polydispersity can also be observed in the figure. "Reprinted from Journal of Hazardous Materials, Vol 186, T.S. Sreeprasad, Shihabudheen M. Maliyekkal, K.P. Lisha, T. Pradeep, Reduced graphene oxide–metal/metal oxide composites: Facile synthesis and application in water purification, Page No. 921–931, Copyright (2011), with permission from Elsevier"

Musico et al. [61] have shown improved removal of lead(II) from water using PVK (Poly *N*-vinylcarbazole)–GO nanocomposite by adsorption process. Their results have shown increase in adsorption capacity of $Pb^{2+}$ by nanocomposite with increasing amount of GO. This phenomenon has been attributed to increased concentration of availability of functional groups containing oxygen in the nanocomposite. Also, pH is one of the factors affecting adsorption, with higher capacity of adsorption with increased value of pH, as it is linked with changes in surface properties of nanocomposites.

Fan et al. [118] have shown selective adsorption of lead(II) ions by water dispersible magnetic chitosan–GO (MCGO) nanocomposites. MCGO is shown to be used as sorbents for large volumes of aqueous solutions. It is stable and environmentally friendly with high adsorption capacity of about 76.94 mg/g and about 90 % desorption capability. $Pb^{2+}$ ions adsorption on MCGO is reported to be strongly dependent on pH.

Machida et al. [119] have highlighted the study of lead(II) adsorption on graphene layer from aqueous solutions. It has been proposed that not only acidic oxygen sites such as carboxylic and lactonic groups but also basic sites are important in Pb(II) adsorption process. Basic site for adsorption is on graphene layer, where C π–cation interactions will be operative for the Pb(II) adsorption, with no ion exchange. Moreover, they have proposed that effective and simultaneous removal of both organic substances and heavy metals can be possible if acidic oxygen functional groups are totally eliminated.

Wu et al. [120] have shown efficient removal of $Cu^{2+}$ ions from aqueous solutions with an upper edge of GO over activated carbons or CNTs. They have noted that GO was responsible for the sorption of the metal ions, due to oxygenous adsorption sites of GO and its large surface area. $Cu^{2+}$ ions could be desorbed to the extent of over 90 % using HCl solution. Adsorption capability of GO can reach about 117.5 mg/g, which is found to be much higher than CNT and activated carbon.

Hu et al. [121] have synthesized sulfonated magnetic GO (SMGO) for the removal of $Cu^{2+}$ ions from wastewater. It is to be noted that introduction of sulfo groups to GO surface can improve efficiency of Cu(II) adsorption. Study of rate controlling mechanism brings out the fact that film diffusion and intraparticle diffusion are occurring simultaneously during the process of adsorption.

Deng et al. [122] have shown simultaneous removal of $Cd^{2+,}$ methylene blue (MB) and Orange G by synthesizing magnetic GO (MGO) nanocomposite. In a binary system of Cd(II) and MB, the capacity to adsorb Cd(II) gets suppressed with increase in MB concentration, while the sorption capacity for MB is not affected by increasing Cd(II) concentration. On the other hand, in binary system of Orange G and Cd(II), the capacity to adsorb Orange G (OG) increases by increase in the concentration of Cd(II), while sorption of Cd(II) has been reported to be independent of the concentration of OG with increasing Cd(II). Tap water samples have been reported to have very small interference toward performance of MGO due to MB and OG. Deng et al. [123] have studied adsorption properties of $Pb^{2+}$ and $Cd^{2+}$ on functionalized graphene and have reported adsorption capacities of 406.6 mg/g ($pH = 5.1$) for Pb(II) and 73.42 mg/g ($pH = 6.2$) for Cd(II), which is much better than that observed for CNT.

Sitko et al. [124] have studied adsorption properties of GO for divalent metal ions (copper, zinc, lead, and cadmium) and reported the trend showing the affinity toward GO in the order of Pb(II) > Cu(II) ≫ Cd(II) > Zn(II). Sorption of metal ions on nanosheets of GO is monolayer coverage and it is controlled by chemical adsorption, which involves strong surface complexation of metal ions on GO surface with oxygen-containing groups. Greater dispersibility of GO and higher tendency of GO-divalent ions to precipitate are the major factors leading to the technique for removal of heavy metals from water.

Wang et al. [125] have developed a technique for synergistic removal of Pb(II), Cd(II), and humic acid, using magnetic mesoporous silica–GO nanocomposites. The development of bifunctionalized adsorbents for simultaneous removal of two different types of environmental pollutants is an upcoming field in the treatment of water [126]. Humic acid is found to enhance adsorption of heavy metals by these nanocomposites due to their interactions with the aqueous solutions. Figure 6.14 shows a TEM micrograph of polyethyleneimine-modified magnetic mesoporous silica and graphene oxide composites. Inset is a magnified TEM image of individual magnetite-mesoporous silica microsphere on GO nanosheet.

Zhang et al. [127] have worked on developing magnetic GO/$Fe_3O_4$ composites, using copper-catalyzed azide–alkyne cycloaddition reaction. Subsequent modification with polyacrylic acid (PAA) led to formation of water-soluble PAA/GO/$Fe_3O_4$ nanocomposites, which were used for recyclable removal of $Cu^{2+}$, $Cd^{2+}$, and $Pb^{2+}$ ions from aqueous solution. Because of high specific surface area,

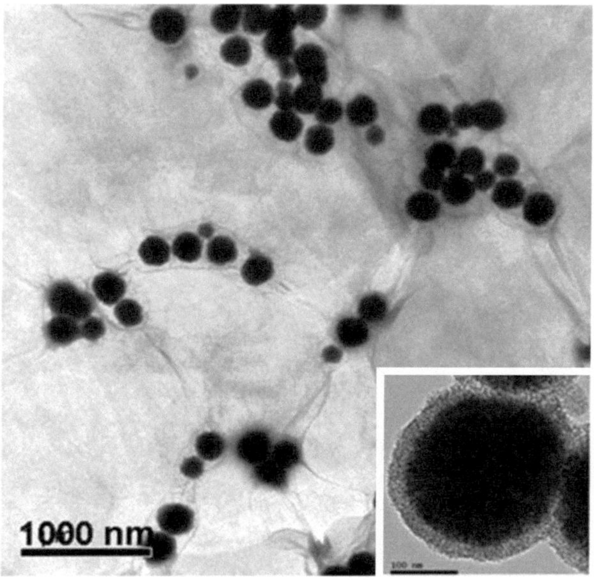


**Fig. 6.14** TEM image of composites [125]

superparamagnetism, and even excellent complex ability, PAA/GO/$Fe_3O_4$ nanocomposites show good removal efficiency of over 85 % even after five cycles. In addition, on account of superparamagnetic nature, these nanocomposites are easy to separate and get recycled, proving it to be a safe platform for purifying water of heavy metal ions.

Li et al. [128] have reported removal of $Cu^{2+}$ and fulvic acid by synthesizing GO nanosheets and decorating it with $Fe_3O_4$ nanoparticles. It is reported that sorption of $Cu^{2+}$ on GO/$Fe_3O_4$ is highly dependent on pH while it is independent of the ionic strength which indicates that sorption is mainly dominated by the inner-sphere surface complexation or ion exchange. The presence of fulvic acid is shown to lead to immense increase in Cu(II) sorption at low pH and a reduction at high pH, while the presence of Cu(II) is found to lead to increase in fulvic acid sorption. It is observed that adsorption of fulvic acid leads to modification of surface properties of the sorbent and there is partial complexation of Cu(II) with adsorbed fulvic acid. The results are helpful in estimating and optimizing the removal of metal ions and organic substances from water.

Zhang et al. [129] have reported selective removal of Cr (VI) from aqueous solution in the presence of other ions such as $Na^+$, $K^+$, $Ca^{2+}$, $Cl^-$, and $Br^-$ using GO which is chemically reduced and functionalized by refluxing GO with ethylenediamine (ED) using dimethyl formamide (DMF) as a solvent. The adsorbent ED-DMF-RGO is highly efficient for removal of Cr(VI) and even has an ease of separation.

Wu et al. [130] have studied adsorption of Cr(VI) from aqueous solutions by using graphene modified with cetyltrimethylammonium bromide. Results have shown that the optimal pH for adsorption is about 2 and the suitable temperature is about 293 K. In this study, it is reported that the processes for Cr(VI) adsorption are exothermic and spontaneous and they reach equilibrium state faster than other carbon-based adsorbents.

Ren et al. [131] have developed GNS/δ-$MnO_2$ (GNS-$MnO_2$) composite by microwave-assisted method for removal of nickel ions from wastewater. This composite can be used for five times with high recovery of ~91 %. Capacity of adsorption of Ni(II) for GNS/$MnO_2$ is found to be about 1.5 and 5 times higher than δ-$MnO_2$ and GNS, respectively, which is mainly attributed to the growth of $MnO_2$ nanoparticles (5–10 nm) on the surface of GNS.

Chandra and Kim [132] have synthesized polypyrrole–RGO composite for highly selective removal of $Hg^{2+}$. It is possible to have gram scale production, high capacity of adsorption, and recycling with this technique for the treatment of wastewater. Zhao et al. [133] have reported selective adsorption of toxic metal ions (specifically $Hg^{2+}$) by studying electrochemical reactions on polypyrrole/RGO nanocomposite. This research is of practical significance as compared to functionalization methods on the surface and biotechniques.

Song et al. [134] have synthesized β-cyclodextrin-modified graphene oxide (β-CD-GO) in order to make aqueous solution devoid of Co(II) ions. Their results have shown that sorption of Co(II) on the surface of β-CD-GO is strongly dependent on pH and it is independent of ionic strength; thermodynamic studies have shown that sorption of β-CD-GO is spontaneous and exothermic. This indicates that β-CD-GO is shown to be an ideal material for the removal of Co(II) from wastewater.

Liu et al. [135] have synthesized magnetite/GO (M/GO) nanocomposite with size of magnetite being 10–15 nm for removal of heavy ions such as Co(II) from aqueous solutions. Thermodynamically, it is shown that the process of adsorption of Co(II) on the M/GO composite is endothermic and spontaneous process. Sorption of Co(II) on M/GO is ionic strength independent and pH dependent, which indicates that, at low values of pH, inner-sphere surface complexation is the mechanism of sorption, while at high values of pH, Co(II) removal was achieved by combination of simultaneous precipitation and inner-sphere surface complexation is attributed to be the mechanism of sorption. Thus, M/GO-bound Co(II) can be separated and recovered by easy magnetic separation from the solution.

Leng et al. [136] have reported excellent removal of $Sb^{3+}$ from water using graphene as the adsorbent. In this work, highest adsorption (99.5 %) is achieved at a pH of 11.0, which is attributed to Van der Waals interactions that are possibly dominating the electrostatic interactions.

### 6.4.2 *Inorganic Anions*

Unlike heavy metal ions which are highly hazardous, some inorganic anions are present in water, such as $F^-$, $Cl^-$, $Br^-$, and phosphate ions. It has been known that anion–π interactions are generally not very strong; for example, halide anions such as $F^-$ present on benzene are found to have negligible anion–π interaction, unless there is substitution of one of H of benzene by highly negatively charged F. Using DFT, it has been found by Shi et al. that there are strong anion–π interactions between halide ions and hydrogenated graphene flakes. This phenomenon has been attributed to the effective electron-accepting power of the inherent π orbitals. This has paved a way to the design of a new family of neutral anion receptors and detectors [137].

Being nonbiodegradable, fluorides get accumulated in natural resources, if not treated sufficiently before release from industrial wastes [138]. Conventional methods for treating fluorides include liming and consequent precipitation of fluoride, use of ion-exchange resins, activated alumina, and many more. But, due to the high operational and maintenance costs, not all the techniques are adopted in the industry. Tomar and Kumar [138] have performed critical studies on techniques for de-fluoridizing water and have reported that nano-adsorbents are found to have highest adsorption capacities with respect to other forms such as natural or biomaterials. Graphene, in particular, has been shown to have the overall best performance with respect to other carbon-based adsorbents. Process of adsorption is found to be endothermic and proceeds by ion-exchange process.

Vasudevan and Lakshmi have studied systematically that graphene can be used in the effective removal of phosphate ions from aqueous solutions with adsorption capacity of up to 89.4 mg/g [139]. In this study, graphene was made by using liquid phase exfoliation of graphite, which helped to introduce functionalization of graphene surfaces, and was beneficial during adsorption of phosphate.

### *6.4.3 Organic Dyes*

Among the water pollutants, dyes are significantly toxic for living organisms, which are discharged from many sources such as printing, textiles, dyeing, paper and pulp, tannery, and paint industries. Dyes are either in cationic or anionic forms. Due to ionic forms of dye molecules in water, major interaction can be expected during dye adsorption on GO and GNS due to the cationic–anionic forces. GO exhibits high adsorption of cationic dyes, but low affinity for anionic dyes, due to the strong electrostatic repulsion between GO and dyes. GNS and composites, due to π–π stacking, can be good adsorbents for both types of ions [140].

Graphene-based materials have been successfully applied in the adsorption of organic pollutants in the form of dyes, polycyclic aromatic hydrocarbons, and gasoline. Zhu et al. [141] have shown that graphene can be efficient in the adsorption of various dyes from polluted water. In the study, Bisphenol A (BPA) has been removed from water by using graphene, whose adsorption capacity has been reported to be higher than that of any of the carbon-based adsorbents. Fourier transform infrared spectra (FTIR) of GO, graphene, BPA, and graphene, before and after BPA adsorption, is as shown in Fig. 6.15. It is helpful in the identification of certain functional groups on the solid surfaces. Features in the FTIR spectra show that most of the oxygen-containing groups get removed after reduction of GO by

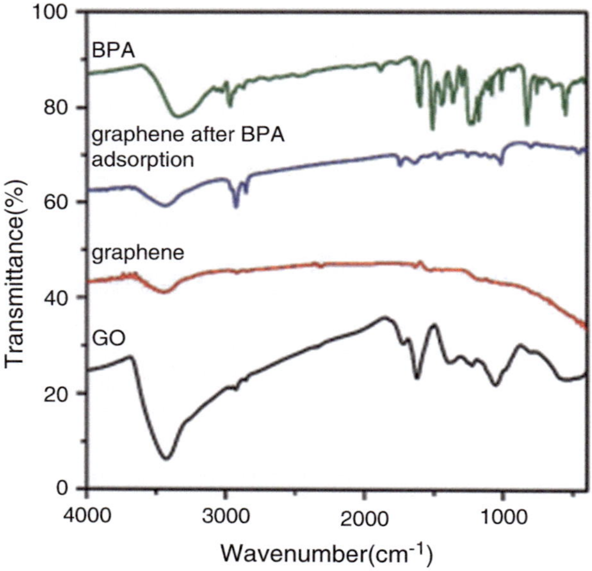


**Fig. 6.15** FTIR spectra of GO, Graphene, BPA, and graphene after BPA adsorption "Reprinted with permission from [141]. Copyright (2013) American Chemical Society"

hydrazine hydrate. Compared to graphene, many new peaks get introduced in the FTIR spectra of graphene after BPA adsorption. FTIR spectra are the proof of the presence of π–π bonding and hydrogen bonding between graphene and BPA. Besides, in comparison to other carbonaceous materials, graphene is found to have larger and smooth surfaces for formation of π–π bonding, which in turn can easily adsorb the organic contaminants. Adsorption process could be facilitated by addition of NaCl, while it is found that high temperature is not favorable.

Li et al. have synthesized hematite nanocubes anchored on polymer PVP-modified RGO nanosheets (termed HPG) nanocomposites which have been shown to have excellent adsorption performance of Rhodamine B (RhB) dye molecules from water [142]. Shi et al. [143] have investigated the removal of azo dye orange II from water using supported cobalt oxide on GO as a catalyst system. Sharma and Das [144] have studied the removal of cationic species (specifically methyl green) from aqueous solutions using GO nanosheets. GO with functionalities such as carboxyl, epoxy, ketone, and hydroxyl groups can acquire negative charge in aqueous solutions, thus enabling the removal of cationic species.

Junyong et al. [145] have developed magnetic graphene oxides, which are functional hybrids of $Fe_3O_4$–GO, for the effective removal of cationic dye (RhB) from aqueous solutions. Wang et al. [146] have developed graphene-based magnetic nanocomposite, as adsorbent, for the removal of organic dye, fuchsine, from wastewater. Fuchsine is a weakly basic compound which contains amido groups. It is observed that the adsorption of dye increases from pH values of 3–5.5, with less significant increase beyond this value. This is attributed to the fact that at lower values of pH, the protonation of amido groups increases its solubility in water, thus leading to decreased uptake on magnetic graphene nanocomposite. The adsorption process has been found to be fast and achieves equilibrium within 30 min. Table 6.1 shows the metal adsorption capacity of various graphene-based materials.

**Table 6.1** Metal adsorption capacity of graphene-based materials [5]

| Adsorbent | GO reduction | Metal | Adsorption (mg/g) |
|---|---|---|---|
| GO | – | Cu(II) | 46.6(RT) |
| GO–gelatin/chitosan | – | Cu(II) | 120 (RT) |
| GO | – | Cd(II) | 106.3 (30 °C) |
| GO | – | Cd(II) | 14.9 (RT) |
| $TiO_2$/GO | – | Cd(II) | 72.8 |
| $GNs^{C8P}$ | – | Cd(II) | 30.0 (RT) |
| $_{GNs}$PF6 | – | Cd(II) | 73.4 (RT) |
| GO | – | Co(II) | 68.2 (30 °C) |
| $Fe_3O_4$/GO | – | Co(II) | 13.0–22.7 (30–70 °C) |
| GO | – | Zn(II) | 30.1 (RT) |
| $TiO_2$/GO | – | Zn(II) | 88.9 |
| GO | – | Pb(II) | 35.6 (RT) |
| GO | – | Pb(II) | 842–1850 (20–60 °C) |
| GO | – | Pb(II) | 367 (25 °C) |
| GO–EDTA | – | Pb(II) | 479 (25 °C) |
| GO/chitosan | – | Pb(II) | 99 (RT) |
| GO–gelatin/chitosan | – | Pb(II) | 100 (RT) |
| $TiO_2$/GO | – | Pb(II) | 65.6 |
| GNs | Vacuum, 200 °C | Pb(II) | 22.4 (30 °C) |
| GNs-500 | Vacuum, 200 °C | Pb(II) | 35.2 (30 °C) |
| GNs-700 | Vacuum, 200 °C | Pb(II) | 35.5 (30 °C) |
| GNs–EDTA | Water | Pb(II) | 228 (25 °C) |
| $GNs^{C8P}$ | – | Pb(II) | 74.2 (RT) |
| $GNs^{PF6}$ | – | Pb(II) | 405.9 (RT) |
| $SiO_2$–GNs | Hydrazine | Pb(II) | 113.6 (25 °C) |
| GO | – | Hg(II) | 35 (RT) |
| GO–SH | – | Hg(II) | 190 (RT) |
| GNs–SH | Sodium dodecylsulfate | Hg(II) | 22 (RT) |
| GNs–polypyrrole | Hydrazine | Hg(II) | 980 (20 °C) |
| $MnO_2$/GNs | Hydrazine | Hg(II) | 10.8 (30 °C) |
| Ag/GNs | Hydrazine | Hg(II) | 10.9 (30 °C) |
| delta-$MnO_2$/GNs | Hydrazine | Ni(II) | 46.5–66.0 (25–45 °C) |
| $Fe_3O_4$/GNs/LDH | – | As(V) | 73.1 |
| GNs | Hydrogen | As(III) | 138 (RT) |
| $Fe_3O_4$/GNs | Hydrazine | As(III) | 13.1 |
| GNs | Sodium borohydride | Cr(VI) | 43 (20 °C) |
| $Fe^0$–GNs | Sodium borohydride | Cr(VI) | 162.6 (20 °C) |
| ED–RGO | – | Cr(VI) | 5.0 (33 °C) |
| $Fe^0$/$Fe_2O_3$/Si–S–O–GNs | – | Cr(VI) | 1.0 (RT) |



*GO-SH* Surface modified GO, *GNs-SH* Surface modified GNs, $GNs^{C8P}$ GNs functionalized with 1-octyl-3-methyl-imidazolium hexafluorophosphate, $GNs^{PF6}$ GNs functionalized with potassium hexafluorophosphate solution

Yang et al. [147] have shown that GO is effective in the removal of methylene blue (MB) from aqueous solutions, with a process removal efficiency of about 99 % and can help in decolorizing the solution. The process of removing MB is faster and more efficient at lower temperature and high values of pH. The increase in ionic strength and dissolved organic matter is found to further improvise when concentration of MB is high. The fast absorption of MB on GO is its merit over other carbon-based absorbents. The main factor that is responsible for the absorption of MB on GO is reported to be electrostatic interaction as well as a minor contribution of π–π-stacking interaction.

Li et al. [148] have studied the adsorption properties of methylene blue onto calcium alginate immobilized graphene oxide composites. These composites have been shown to have high efficiency of adsorption. The process is spontaneous and exothermic.

Ai et al. [149] have synthesized graphene/magnetite composite using solvothermal method for removal of methylene blue from aqueous solutions with high efficiency. The composite is found to show characteristics of both graphene and magnetite and hence its capacity to remove MB along with fast rates of adsorption. This is attributed to electrostatic interactions between cationic MB and anionic surface oxygen-containing groups, as well as π–π interactions between MB molecules and graphene.

Wu et al. [150] have produced superparamagnetic graphene–$Fe_3O_4$ ($G/Fe_3O_4$) nanocomposite by facile solvothermal process and have examined the removal of organic dye, pararosaniline, from aqueous solution. The hybrid composite can be easily manipulated in magnetic field and, hence, can be removed from the dye from polluted water. The mechanism of adsorption of organic dye from water using $G/Fe_3O_4$ composites is proposed to be due to two reasons: the first one is Van der Waals interaction between aromatic backbones of the dye and hexagonal arrays of carbon atoms in graphite sheet of the composite, and the second one is strong π-stacking interaction between the delocalized π electrons of graphene and benzene ring of the dye.

From the above-mentioned literature, it can be seen that the effective removal of arsenic using graphene-based magnetic nanocomposites is possible to the extent of 99.99 %. Simultaneous removal of different valencies of arsenic (III and V) is possible with high recycling ability due to good dispersion of hydrated zirconium hydroxide on GO, which can be attributed to high specific surface area of graphene and ease of functionalization.

In case of lead(II) removal, the conditions that are suitable for its removal are higher pH and increased availability of concentration of functional groups containing oxygen in the nanocomposite. Moreover, increasing the amount of GO with Poly *N*-vinylcarbazole in the nanocomposite leads to increase in adsorption capacity of lead ions. The adsorption of lead(II) and Cd(II) using GO on functionalized graphene is found to be better than CNT and activated carbon, the reason being attributed to the large exposed surface area and oxygen adsorption sites of GO. In case of $Cu^{2+}$ and fulvic acid in aqueous solution, the sorption is found to be mainly dominated by the inner-sphere surface complexation or ion exchange as well as its strong dependence on pH of aqueous solution. $MnO_2$ is found to improve adsorption characteristics of GNS for nickel(II) removal.

Removal of inorganic ions from water by adsorption is found to be the best for nano-adsorbents, especially graphene, as compared to other carbon-based adsorbents. Here, principal mechanism of adsorption is ion exchange.

On the other hand, for the removal of dyes from water, GO is found to have high adsorption capacity for cationic dyes in an aqueous solution due to its negatively charged surface on account of various attached functionalities including carboxylic, epoxy, ketone, and others. Graphene is found to be the best adsorbent for removal of BPA from water as compared to other carbon-based adsorbents, where the mechanism could be mainly $\pi$-stacking interaction as well as hydrogen bond formation. Here, the smooth surface of graphene is an additional advantage with respect to other forms of carbon. Interaction of delocalized $\pi$ electrons of graphene and benzene ring of the dye, along with weak Van der Waals force between hexagonal arrays of carbon atoms and aromatic backbones of composite, is mainly responsible for the removal of organic dye especially pararosaniline from water. Even in some cases, the electrostatic interactions, between anionic oxygen containing surface and the dyes, are found to be responsible for the adsorption and its subsequent remediation from water. Table 6.2 shows dye adsorption capacity of various graphene-based materials.

## 6.5 Adsorptive Removal of Gas Pollutants

Leenaerts [58], using DFT, has reported that the charge transfer between the considered adsorbates and graphene is found to be almost independent of the adsorption site, but it does depend strongly on the orientation of the adsorbate with respect to the graphene surface. The removal of some of the toxic gases which are important in industrial applications is discussed in this section.

Graphene oxide (GO) has a layered structure with strong acidity and can be used for reactive adsorption of basic gases, as it contains a range of reactive oxygen functional groups, making it suitable for this application. The transport properties of graphene are reported to be sensitive to some gases ($NH_3$, $NO_2$, etc.) that are adsorbed on the materials' surface [15, 151, 152].

Seredych and Bandosz [140, 153] observed a very high adsorption of ammonia on GO (61 mg/g); this has been attributed to the surface chemistry and geometry. Besides, as ammonia is basic, the acid–base interactions with surface functional groups, located at the edges of the carbon layers, play important role in its adsorption.

Romero et al. [154] analyzed the adsorption/desorption of $NH_3$ on graphene experimentally, and proposed a model describing a small charge transfer due to the adsorption of $NH_3$ at the pristine surface of graphene. Peng et al. [155] studied adsorption of $NH_3$ on GO using ab initio simulations and found that the binding energy of graphene is independent of the site and orientation of adsorbate atoms.

**Table 6.2** Dye adsorption capacity on graphene-based materials [5]

| Adsorbent | GO reduction | Dye | Adsorption (mg/g) |
|---|---|---|---|
| GO | – | MB | 351 |
| GO | – | MB | 714 |
| GO | – | MB | 1939 |
| GO sponge | – | MB | 397 |
| $Fe_3O_4$–GO | – | MB | 167.2 (25 °C) |
| $Fe_3O_4/SiO_2$–GO | $NH_3$ | MB | 97.0–111.1(25–60 °C) |
| GNs | Hydrazine | MB | 154–204 (20–60 °C) |
| G-aerogel | Vitamin C | MB | 152.8 (RT) |
| $Fe_3O_4$–GNs | Hydrothermal | MB | 43.8 (25 °C) |
| $Fe_3O_4$–GNs | $NH_3$ | MB | 45.3 |
| $TiO_2$–GNs | NaOH | MB | 83.3 |
| $MnFe_2O_4$–GNs | Polyethylene glycol | MB | 34.7 (25 °C) |
| MCGO | – | MB | 180.83 (30 °C) |
| CNT–GNs | | MB | 81.97 |
| GNs | Hydrazine | X-GRL | 217–238 (15–60 °C) |
| GO | – | AO | 1428 |
| GNs | Sodium hydrosulfite | AO | 3333 |
| $Fe_3O_4$–GO | – | NR | 171.3 (25 °C) |
| G-aerogel | Vitamin C | RB | 133.6 (RT) |
| $Fe_3O_4$–GNs | Hydrothermal | RB | 13.2 (RT) |
| $MnFe_2O_4$–GNs | Polyethylene glycol | RB | 22.5 (25 °C) |
| Sand–GNs | – | R6G | 55(30 °C) |
| $Fe_3O_4$–GNs | Hydrothermal | MG | 22 (RT) |
| GO | – | MG | 248 |
| GO sponge | – | MV | 467 |
| $CoFe_2O_4$–GNs | NaOH | MO | 71.5 |
| G-aerogel | Vitamin C | Fuchsine | 142.7 (RT) |
| $Fe_3O_4$–GNs | Hydrazine/$NH_3$ | Fuchsine | 89.4 |
| G-aerogel | Vitamin C | Acid Fuchsine | 43.2 (RT) |
| $Fe_3O_4$–GNs | $NH_3$ | CR | 33.7 |
| Fe–GNs | $NaBH_4$ | Methyl blue | 75–375 (RT) |
| GO | – | Methyl blue | 43.5 (30 °C) |
| MCGO | – | Methyl blue | 95.2 (30 °C) |


*CR* Congo Red, *MV* Methyl Violet

Further, they reported that the process can be promoted by having additional surface epoxy and hydroxyl groups; this enhancement is greater in the presence of hydroxyl groups than epoxy species on the surface.

Using DFT calculations, Tang and Cao [156] made a similar observation about enhanced adsorption, during their study of adsorption and dissociation of ammonia on GO. They found that the adsorption of $NH_3$ on GO is generally stronger than that on graphene, due to the presence of various active defect sites such as the hydroxyl and epoxy functional groups and their neighboring carbon atoms. These active oxygen sites can form bonds with $NH_3$ and hence enhance charge transfer from $NH_3$ to GO. This shows good agreement with the above experimental observations.

Petit et al. [157] compared two GO samples that were synthesized using the Hummers–Offeman and Brodie methods for ammonia adsorption and concluded that ammonia adsorption is strongly related to the type of GO. Furthermore, they observed intercalation of ammonia in the interlayer space of GO and reactive adsorption. For GO prepared by the Brodie method, ammonia was retained by reaction with the carboxylic groups at the edges of the graphene layers while, for the GO from the Hummers method, the reaction with the epoxy and carboxylic groups was responsible for the adsorption.

Seredych et al. [158] made two types of graphite oxide by oxidation of commercial graphite and reduced them using hydrazine hydrate. It is reported that the difference in texture and chemistry of the materials significantly affects their capacity for the adsorption of ammonia. This adsorption capacity is linked with the presence of surface epoxides, which can react with ammonia leading to the formation of amines. Porosity creation after reduction did not enhance ammonia adsorption, because of the absence of reactive oxygen (epoxides) sites.

Chen et al. [159] have studied adsorption/desorption and flipping behavior of ammonia molecules on a graphene surface by observing the shift in the Fermi level, which has been inferred to be originating from partial charge transfer from the $NH_3$ molecules to graphene. They have devised a model to evaluate the rates of adsorption and desorption of $NH_3$ molecules on graphene from the measured shift in the Fermi level with respect to exposure time. Further, it is shown that the electric field can flip molecular dipoles of $NH_3$ molecules by measuring the shift of the back-gate voltage at Dirac point, after application of the electric field.

Petit and Bandosz [160] made GO/polyoxometalate nanocomposites and reported an improvement in ammonia retention compared to the parent GO. This is as shown in Fig. 6.16. The materials studied are GO samples and subject to breakthrough dynamic tests in which GO is packed in glass column and exposed to ammonia diluted either in stream of moist air (EM) or dry air (ED) or pre-humidification of GO samples followed by ammonia adsorption from dry air stream (EPD). Here, steepness of the desorption curves suggests fast kinetics of interaction of ammonia with the adsorbent. The area under the curves in the desorption regime is indicative of the fact that less ammonia is removed by air purging, hence showing strong interactions between the material studied and ammonia gas molecules.

Seredych and Bandosz [161] made GO/aluminum–zirconium polycation composites for adsorption of $NH_3$ and found that pre-humidification could play an important role in the adsorption. Even the surfactant used for dispersion of the composite affected the performance. More surfactant leads to a higher dispersion of graphene-like layers and intercalation of ammonia can occur. GO/$MnO_2$ composites have been used for ammonia removal and exhibit an enhanced adsorption capacity.

Ren et al. [162] have fabricated graphene FET and assessed its sensing properties for $SO_2$ gas. It was found that $SO_2$ strongly p-dopes the graphene, as noted from

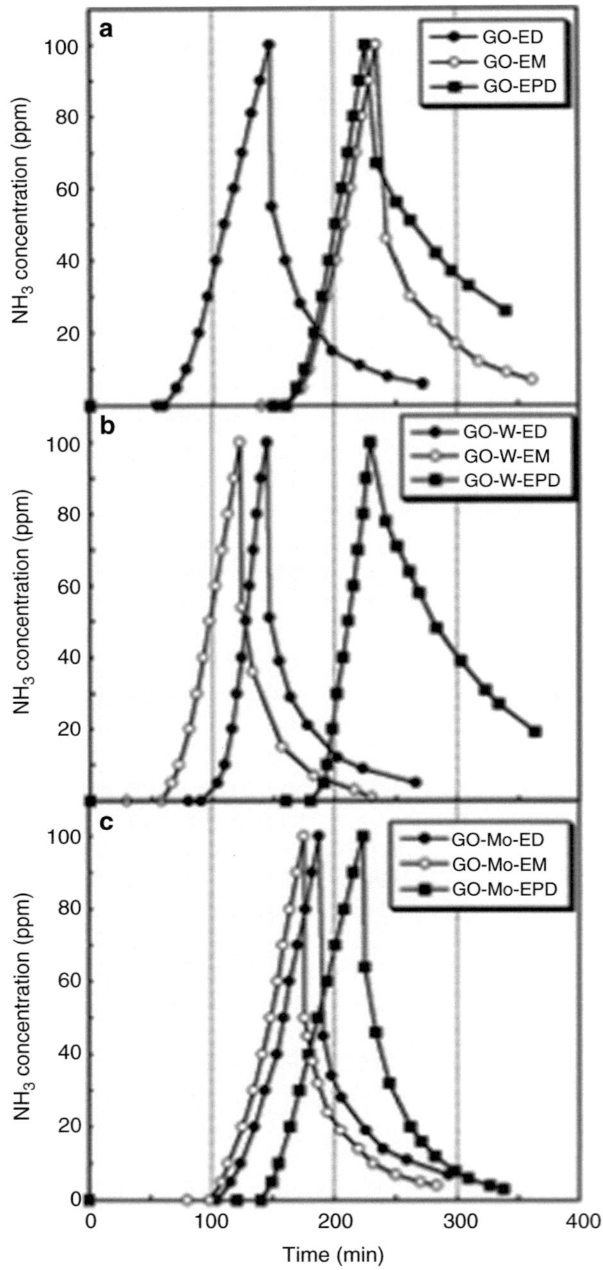


**Fig. 6.16** Breakthrough curves and their desorption parts for (**a**) GO, (**b**) GO-W, (**c**) GO-Mo samples. "Reprinted with permission from [160]. Copyright (2009) American Chemical Society"

the shift in its Dirac point. This effect can be used to monitor $SO_2$ gas. It is reported that increasing the temperature of the detector to about 100 °C in high vacuum can even reset it, with its sensitivity for detection increasing with temperature.

Seredych and Bandosz [140, 163] reported zirconium–hydroxide/graphene composites for sulfur dioxide ($SO_2$) removal at ambient conditions. It is noted that formation of new basic sites and porosity, developed due to interactions between zirconium-hydroxide units and oxygen groups attached to graphene, are responsible for the enhanced adsorption of $SO_2$ on composites. Both physical as well as reactive adsorption of $SO_2$ were responsible and graphene is shown to catalyze the oxidation of $SO_2$ to form sulfates.

Zhang et al. [164] assessed the structural and electronic properties of $H_2S$–graphene systems by tuning the geometries of $H_2S$ molecule toward 2D nanosheets of pristine, defective, and doped graphene. They found that Ca-, Co-, and Fe-doped and defective GNS show much higher affinities to $H_2S$ molecules in comparison to pristine graphene. The electronic transport behavior of Fe-doped GNS has been shown to exhibit higher sensitivity for detecting $H_2S$ gas, as compared with that of devices made with pristine graphene. It is possible to design $H_2S$ chemical sensors with highly improved performance, using GNS as sensing materials with appropriate metal dopants or defects. Moreover, graphene doped with Si is found to absorb $H_2S$ molecule by forming Si–S bond, compared to the weak physisorption of $H_2S$ molecule in case of pristine and the B- and N-doped graphene.

Mishra et al. [165] have demonstrated that $CO_2$ molecules are physically adsorbed on graphene. Hydrogen exfoliated graphene (HEG) is shown to have higher adsorption for $CO_2$ than other carbon nanostructures and zeolites, attributing it to the higher surface area in graphene. Cost-effectiveness and regeneration capability along with high adsorption capacity make HEG a suitable technology for $CO_2$ capture and storage.

From the above-mentioned studies, one can note that the functional groups on layer-structured GO and GNS are responsible for the adsorption of gases ($NH_3$, $SO_2$, $H_2S$). Introduction of metal centers into the surface of GO or GNS can promote adsorption of gases like $H_2S$ in comparison to pristine graphene. The adsorption behavior of GO and GNS is different from other carbons in which porosity plays an important role in adsorption. The higher adsorption surface area of GNS is also found to play significant role in adsorption of greenhouse gases like $CO_2$ and it is possible to recycle them. This leads to an economical process. Ability to functionalize GNS leads to better adsorption of various gases such as $SO_{2,}$ by forming a nanocomposite that has hydroxyl oxide groups attached to GNS. Physical as well as chemical adsorption is found to be an active mechanism for adsorption of $SO_{2.}$ Even pre-humidification and the surfactant used for dispersion of composite play a role in improving efficiency of adsorption of gases like ammonia. Moreover, for ammonia adsorption, in spite of the presence of porosity in GO, proper effectiveness of ammonia adsorption can be achieved only in the presence of surface epoxides or carboxylic groups at edges of GNS. This shows significance of surface texture and chemistry at the edges of graphene in adsorption.

The understanding of the effects of adsorption on the surface of graphene and on its electronic, magnetic, and photo-conducting properties can enable an in-depth analysis of the efficiencies of the processes and hence lead to their specificity to applications. Many of the hazardous elements and compounds enter the environment from various sources and there is an urgent need to design processes for removal of these species. In order to be commercially viable, the processes must be efficient and cost-effective.

## 6.6 Graphene-Based Sensors for Water Monitoring

Use of graphene as a metal electrode in Schottky diodes has been demonstrated to display change in electrical properties upon chemical doping [23]. This has led to the development of novel graphene-based diode sensors (GDSs) by researchers in Ireland [166]. In GDS, graphene is laterally in contact with n- and p-type silicon substrates over its active area, which allows exposure to liquids and gases at the top. Changes in work function and doping levels of graphene, on account of exposure to different molecules, can be monitored using this diode-type configuration, which makes it suitable to operate as a sensor for these gases and liquids.

Graphene–silicon diode junctions have been reported by Chen et al. [167] and with improved characteristics by Tongay et al. [168]. Yang et al. [169] have reported graphene variable-barrier "barrister" (GB) in which the barrier height is tuned by applying gate voltage to graphene. Kim et al. [166] have used GDS and shown that droplets of aromatic molecules such as nitrobenzene, anisole, benzene, and chlorobenzene lead to formation of n- or p-type doping of graphene, depending on the nature of the dopant. This is analyzed by observing the changes in Schottky barrier height at the interface. Figure 6.17 shows the process of fabrication of GDS, and Fig. 6.18 shows the schematic band diagram of graphene/n-Si interface doped with electron donor and acceptor molecules.

Because of the highly toxic nature of phenols to living organisms, it is critical to determine the concentrations of phenols in water before they are consumed. The other methods used for the determination of concentrations of phenols include chromatography, capillary electrophoresis, and spectrophotometry. The disadvantages of these methods such as pretreating samples and longer times for analyses have been overcome to a large extent using electrochemical methods. However, even in this case, using conventional electrodes in electrochemical methods leads to poor detection sensitivity as well as high values of over-potential. Hence, Chen et al. [170] have designed glassy carbon electrodes (GCE) coated with a film of graphene/polymer that overcome these problems. These functional materials, with electrocatalytic activity, are found to have selective detection of phenols in aqueous solutions. High conductivity, along with the strong adsorption surface area of graphene, is found to be responsible for the electrocatalytic activity of the substrate. It has been shown that the designed sensor is economical, is easy to prepare, and has operational simplicity, as compared to HPLC (high-performance liquid

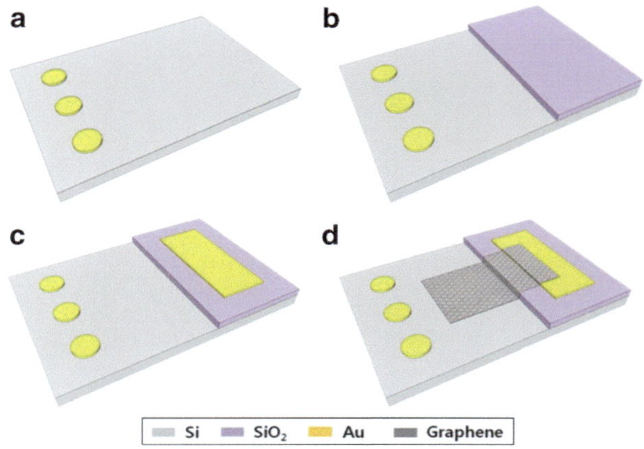


**Fig. 6.17** Process of fabricating GDS—(**a**) Depositing source electrodes immediately after native silicon oxide removal to form Ohmic contacts, (**b**) 150 nm $SiO_2$ insulation layer is deposited by sputtering, (**c**) The drain electrode is deposited on $SiO_2$, (**d**) transfer of graphene after removal of the native oxide from Si. "Reprinted (adapted) with permission from [166]. Copyright (2013) American Chemical Society"

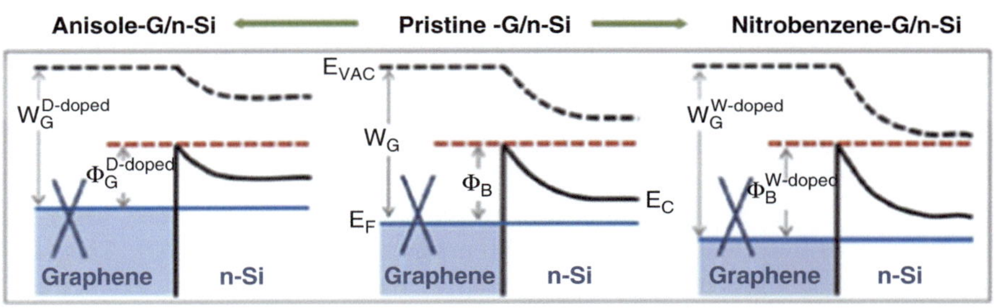


**Fig. 6.18** Schematic band diagram of graphene/*n*–Si interface with anisole (electron donor) on left, pristine graphene/*n*–Si (*middle*), and graphene/*n*–Si interface with nitrobenzene (electron acceptor) on the right. $E_{vac}$ indicates vacuum energy, $E_c$ indicates conduction band, $W_G$ is graphene work function, and $\Phi_B$ is the Schottky barrier height. "Reprinted (adapted) with permission from [166]. Copyright (2013) American Chemical Society"

chromatography). In addition, it is found to display about 97–100 % recovery of hydroquinone, phenol, and p-chlorophenol.

Yang et al. [25] have devised an ECL sensor for supersensitive detection of chlorinated phenols in water with multistage amplification effect. The device is based on carbon quantum dots that were immobilized on the surface of graphene. ECL is emission of light generated in electrochemical reactions, in which intermediates produced go through a reaction to produce electronically excited state, which emits light on relaxation to lower energy state (corresponding to the difference of energies between those states) [171]. Chlorinated phenols are organic pollutants

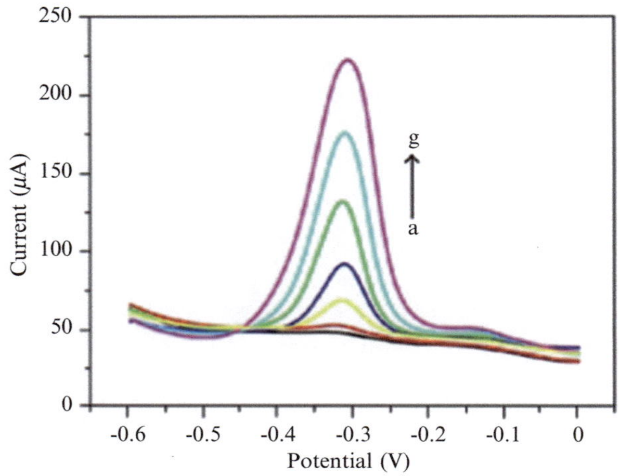


**Fig. 6.19** DPV response for different concentrations of o-NP (a–g: 5, 10, 50, 100, 200, 300, and 400 μM) [172]

classified to be carcinogenic by the International Agency for Research on Cancer. Pentachlorophenol (PCP) has been studied as it is considered to be an important indicator of chlorinated phenols in environmental samples. The designed ECL sensor has been found to have a sensitivity of $1.0 \times 10^{-12}$ M concentration in water, which is near ideality. Graphene is shown to have a major role in multistage amplification effect by facilitating some specific ion production.

Nitrophenols which are extensively used in the manufacture of dyes, pesticides, plastics, and explosives are significantly toxic for human, animal, and aquatic life. Liu et al. [172] devised an electrochemical sensor to quantify o-nitrophenol (o-NP) in water. This technique is based on glassy carbon electrode which has been modified with β-cyclodextrin and has been functionalized with GNS. This substrate, with excellent electrochemical properties, has a detection limit of o-NP up to 0.3 μM in water, based on the signal-to-noise ratio. Recoveries of o-NP from lake water have been found to be varying from 97.4 % to over 100 %. Differential pulse voltammetric (DPV) method is used to estimate o-NP under optimum conditions. Figure 6.19 shows DPV response for varying concentrations of o-NP. Overall, the sensor is found to have good selectivity along with stable nature and repeatability.

## 6.7 Graphene-Based Sensors

### *6.7.1 Graphene-Based Photodetectors*

Graphene, with its unique properties, offers a huge potential as a new photonic material as a transparent electrode for application in displays. Due to its peculiarity of band structure with zero bandgap, the possibility to modify it with

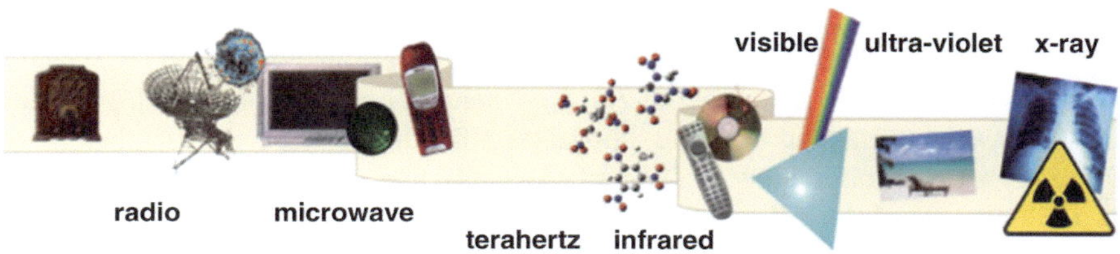


**Fig. 6.20** Schematic representation of the electromagnetic spectrum illustrating the position of the THz frequency range. "Reprinted from Materials Today, Vol 11, A. Giles Davies, Andrew D. Burnett, Wenhui Fan, Edmund H. Linfield, John E. Cunningham, Terahertz spectroscopy of explosives and drugs, Pages No. 18-26, Copyright (2008), with permission from Elsevier"

voltage application or doping, and its high charge carrier mobility, graphene leads to a new generation of devices that can operate within THz domain. THz frequency domain, i.e., long wavelengths (invisible to eyes), are useful for a variety of applications such as security screening, medical imaging, high altitude communications, process monitoring in manufacturing industries, and many others. Figure 6.20 shows a schematic representation of electromagnetic spectrum, with applications [173].

Researchers from IBM [174, 175] have shown that graphene can be positively charged because of photovoltaic effect or negatively charged due to bolometric effect. Bolometric effect can be inferred as heating of lattice due to photo generation of carriers, which while traveling though the graphene lattice, emit quanta of lattice vibration (phonons) which in turn get scattered due to their interaction with electrons in the lattice. This leads to a reduction in the charge mobilities and it is reported that this effect dominates the photo-response of graphene and leads to a photocurrent in the opposite direction. This property has the potential for application in the development of wavelength-sensitive and frequency-sensitive graphene plasmonics-based photodetectors.

Metamaterials are artificially engineered materials, which are assemblies of multiple individual elements, in which microscopic materials (metals or plastics) are arranged in periodic patterns usually, and their properties are derived from their designed structures and not compositions. These materials are much smaller in size than the corresponding wavelength and so they exhibit optical properties that vary from their bulk counterpart. Yan et al. [175, 176] studied the fundamental characteristics of plasmons (collective oscillation of electrons) in graphene-based metamaterials using optical methods. This has led to a detailed understanding of the mechanisms of dispersion and damping in graphene and thus to the development of graphene IR photodetectors driven by intrinsic plasmons.

Ju et al. [177] have reported tuning of the plasmon resonances over terahertz frequency range by controlling GNR width and electrostatic doping. Their results have implied that it is possible to control plasmon resonances in graphene-based terahertz materials (simplest of which is micro-ribbon of graphene which has engineered structure). The width of the ribbon and the dependence on carrier doping follow power law behavior, peculiar to 2D massless Dirac Fermions. This light plasmon coupling can lead to devices based on graphene terahertz materials.

Due to its ability to absorb about 2.3 % of incident light photons [178], graphene can be used to make broadband photodetectors that exceed the bandwidth of currently used IR photodetectors (mainly HgCdTe and quantum well IR detectors) in a big way. These detectors are reported to be 1,000 times more sensitive than the image sensors (CCD and CMOS) which are being used presently. It is to be noted that they require about ten times less energy and are about five times cheaper than the presently available detectors [179].

Electronic properties of graphene have been widely controlled using band structure engineering. Creation of bandgap in graphene is possible with various techniques such as by patterning GNR by electron-beam lithography [180], etching of nanowire shadow mask [181], unzipping of CNTs [182], and/or block copolymer lithography [26, 183]. Besides, it is also possible to create defect states in graphene by defect engineering for controlling electrical, magnetic, and chemical properties [26]. Zhang et al. [26] have invented monolayer graphene photodetector which has high photoresponsivity (i.e., photo-generated current per incident optical power) of 8.6 A/W by introduction of electron trapping centers by engineering defect types and defect densities, and leading to creation of bandgap through band structure engineering. Strategies such as surface plasmons have led to photoresponsivity of only few tens of mA/W. Along with covering a broad wavelength range, including mid-infrared, graphene has the potential for applications in medical, remote sensing, monitoring environment, and telecommunications. Moreover, it is reported that the proposed graphene photodetector is compatible with the current manufacturing setup for CMOS and has the capability to work with lower operational voltage. FET-based graphene photodetectors have two drawbacks: the first one is the low photoresponsivity due to low optical absorption and the second one is extremely small recombination time of photo-excited charge carriers (about few picoseconds), resulting in quantum efficiency of about 6–16 %. With a view to increase its photosensitivity, attempts have been made to increase the lifetime of charge carriers without affecting the electron mobilities. The energy band diagram and concepts for graphene-based photodetector are as shown in Fig. 6.21.

Device processing involves making graphene quantum dot (GQD) like array structures followed by introducing defects—midgap states band (MGB) [184] through titanium sacrificial layer fabrication method at the boundary and surface of GQD, which act as electron trapping canters, leading to the formation of bandgap on account of quantum confinement. However, there are some issues with the proposed graphene photodetector. These include the following: low operation speed (associated with long lifetime of trapped electrons due to the present fabrication technique) and the lifetime of trapped electrons in MGB depends on types of defect, density of trap state, and defect depths.

Compared with other bandstructure engineering methods, this method is quite easier and it has the potential to fabricate larger scale devices based on graphene. Further, with immense performance management of the device through defect engineering, one can have a wide variety of optoelectronic applications.

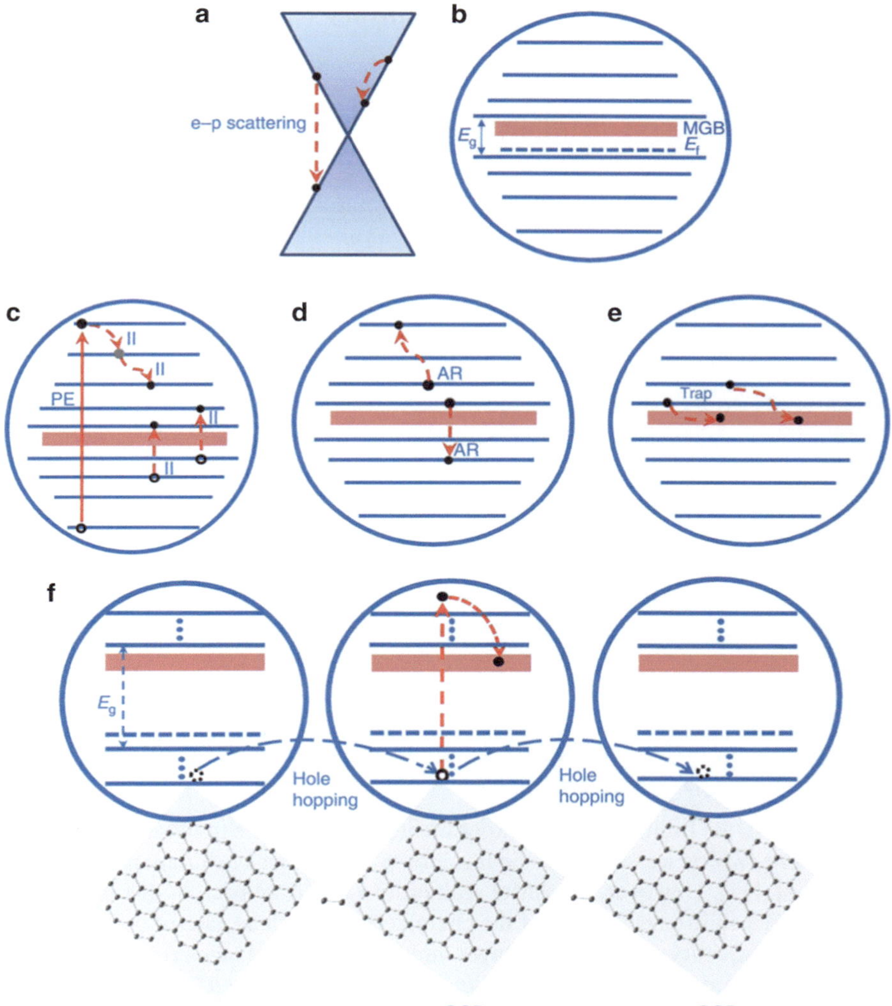


**Fig. 6.21** Energy band diagram and concept of a graphene photodetector [26]. (**a**) Energy band diagram of pure graphene sheet. *Red arrows* show intraband and interband electron phonon scattering. (**b**) Energy band diagram of GQD (quantum dot) including MGB. $E_g$ is quantum confined bandgap. Shadow region indicates MGB and $E_f$ is the Fermi level. (**c**) Photon-excitation (PE) leads to generation of electron–hole pair, followed by impact ionization process. When electron is excited into much higher energy in the conduction band, cascade of carriers occurs that involves electron–hole pair generation. Owing to high efficiency of carrier–carrier scattering in graphene, MEG (Multi-excitation generation, called as inverse auger recombination (AR) as well) effect exists. (**d**) Excitation of electron to higher energy level in conduction band via AR process, followed by recombination of another electron with hole in the valence band. (**e**) Trapping/relaxation process of excited electrons to MGB. (**f**) Graphene photoconductive channel consists of GQD arrays. Photo excited and secondary electrons are captured by MGB having long lifetime. Before the trapped electrons are recombined, corresponding holes can circulate many times in photoconductive channel, hence achieving high photoconductive gain. "Reprinted by permission from Macmillan Publishers Ltd. [Nature Communications] [26], copyright (2013)"

### 6.7.2 Graphene-Based Chemical Sensors

Solid state sensors have been popular and unique due to their high sensitivity to toxic gases which is essential for environmental safety [185]. Ideal characteristics of gas sensors include but are not limited to the following: robustness, must be reusable, have high detection sensitivity, must be economical, and must have minimum intrinsic noise and ease to miniaturize. Carbon is one of the most commonly used materials for electrocatalysis [186, 187], gas sensors, as well as biosensors [187, 188]. In recent past, CNTs have been known for their use as biosensors and polymer electrolyte membrane (PEM) fuel cells [185, 189, 190]. However, unlike the other forms of carbon such as fullerene, CNT, or graphite, graphene [185] has been found to be suitable for being used as a sensor at room temperature, for detecting very small changes in local concentration (1 ppb resolution limit) [23] due to the following reasons:

1. The 2D structure of graphene with high surface area to volume ratio (theoretically 2,630 $m^2/g$ [185] for single-layer graphene).
2. High metallic conductivity and, hence, low Johnson (intrinsic) noise in the limit of no charge carriers [17, 19, 191]. This is a regime for noise, where even a few extra electrons lead to relatively noteworthy changes in carrier concentration, while in conventional materials, bulk properties like resistivity are not changed considerably.
3. Having very few defects in crystal structure [17, 19, 191], thus ensuring low noise levels on account of thermal switching.
4. Possibility of taking four-probe measurements with low resistance Ohmic contacts on a single crystal device [23].

In the past, the limiting step for experiments was to be able to obtain single layer of graphene, which was shown to be feasible by mechanical exfoliation in 2004 by Geim et al. [11]. However, industrial scalability is the drawback of this technique. Industrial feasibility needs each of the characteristics such as ease of processing to achieve high-quality single-crystal layer, uniform crystallite sizes for homogenous properties of product/s, and compatibility with the currently used CMOS technology [185], hence avoiding the need for specific cost-intensive apparatus for production. Due to ease of mass production, graphene oxide and its reduced form have been found to be suitable for promising scientific applications such as electrochemical capacitors [185]. Graphene oxide is amenable to making composites as well that show good electrocatalytic properties [192] for making electrochemical sensors. The ease of processing graphene and functionalizing with other groups has made it attractive for use in photovoltaic applications [193].

The principle behind sensing of various gases by graphene is that gas molecules that are adsorbed or desorbed are acting as electron donors or acceptors, by creating change in local carrier concentration in graphene, one electron each time, ultimately leading to changes in conductance of graphene [23]. $H_2$, CO, $NH_3$, Cl, $NO_2$, and $O_2$ are gas molecules that have been detected with high sensitivity making it

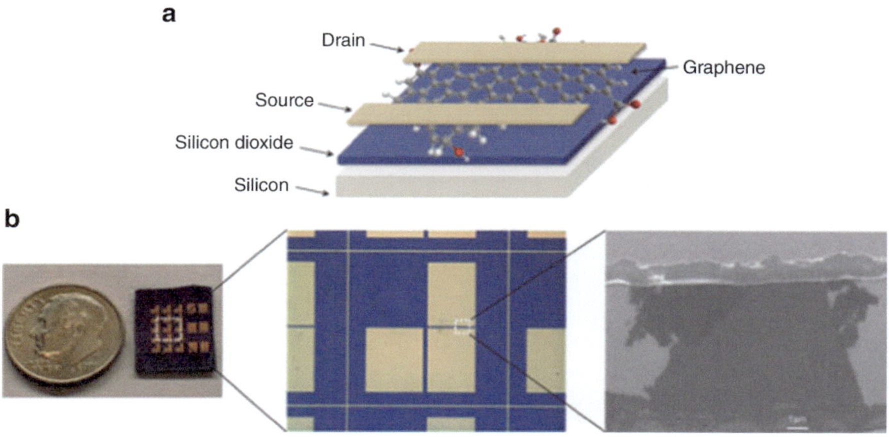


**Fig. 6.22** (**a**) Spin-coated well-dispersed chemically converted graphene-based FET onto $SiO_2$ substrate (**b**) Photograph (*left*), Optical image (*middle*), SEM Image (*right*) of working device with 7 μm channel. "Reprinted by permission from Macmillan Publishers Ltd. [Nature Nanotechnology] [196], copyright (2008)"

possible for creating gas sensors. Many of the gases and some other volatile organic compounds are known to have toxicity. In recent years, GO-, GNS-, and MO-modified composites have been mainly investigated for removal of ammonia but not limited to it [23, 38, 185].

Due to exceptional mobilities of charge carriers along with single atomic thickness and ballistic transport mechanism even at a micron level, graphene FETs have been proposed for holding intense potentials for sensitive detection of chemical or biological species [194]. The only issue is that, due to zero bandgaps, it is not possible to turn off such a FET, restricting its applications in nano electronics. However, for gas sensing applications, this issue is not of great significance. Various studies have been performed on the interaction of gas molecules with graphene and these studies have indicated the applicability of graphene as a gas sensor [11, 15, 195]. Ruoff et al. produced large graphene flakes, as shown in Fig. 6.22, and demonstrated scalable arrays of FETs using conventional photolithography for patterning the parts of graphene onto the substrate [196].

Graphene is expected to lead to a new breed of chips replacing silicon, as silicon is predicted to reach its limit in the next 20 years [197]. The first demonstration of graphene-based sensors was experimentally shown by Schedin et al. [23]. In this work, graphene was assessed and it was found to have excellent sensitivity toward adsorption of $NO_2$, $NH_3$, $H_2O$, and CO. As shown in Fig. 6.23, the sensor responds to the gases in concentrations of 1 ppb, which puts graphene as best amongst the gas sensors.

But, in order to demonstrate the fundamental limit of the sensitivity of graphene-based sensors, Schedin et al. optimized their devices by using high driving currents to suppress the Johnson noise and annealing the samples close to

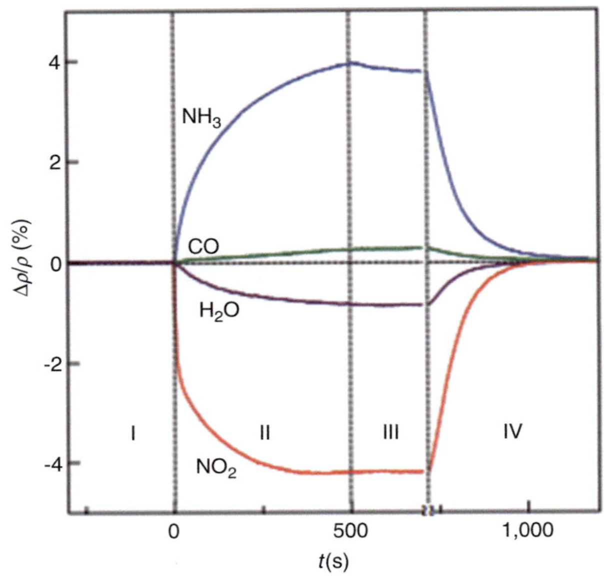


**Fig. 6.23** Lack of surface states in graphene makes it possible for detecting single molecule of adsorbate. Direction of change in the picture shows sign of induced carriers (holes for $H_2O$ and $NO_2$; electrons for CO and $NH_3$). "Reprinted by permission from Macmillan Publishers Ltd. [Nature materials] [23], copyright (2007)"

the neutrality point. In all the cases of adsorption, free carriers were induced which led to decrease in the resistivity of single-layer flake. Hall resistivity plots confirmed that opposite sign charge carriers were generated by two gases; $NO_2$ induced conduction by generating holes, while $NH_3$ induced current by generating electrons. Hence, Hall resistivity changes can be used to study the fundamental sensitivity of graphene-based sensors [23, 37].

It is reported that the increase in graphene charge carrier concentration, induced by adsorbed gas molecules, could be used to make highly sensitive sensors [152]. Furthermore, chemical doping of graphene by both holes and electrons, in high concentration, without affecting the mobility of graphene and hence allowing for micro-fabrication of *p–n* junctions, has been demonstrated [23, 198]. Ghosh et al. [194] have studied characteristics of a few layers of graphenes for sensing $NO_2$ and humidity. Graphene has been made by three methods including thermal exfoliation of graphitic oxide (EG), conversion of nanodiamond (DG), and by arcing graphite rods in hydrogen atmosphere (HG). The sensitivity for $NO_2$ is reported to be highest with DG. Relative to pure HG, n-type nitrogen-doped HG shows increased sensitivity for $NO_2$, while the highest sensitivity for humidity has been observed with HG. The sensing characteristics of graphene for number of aliphatic alcohols have been studied and they showed sensitivity dependence with chain length and branching.

Using DFT, Ao et al. have shown that Al-doped graphene can be used for detection of carbon monoxide [199]. It was reported that CO adsorption introduces large number of shallow acceptor states, leading to increased electrical conductivity. Chi and Zhao noted that Al-doped graphene is an excellent candidate for sensing formaldehyde ($H_2CO$) [200]. Leenaerts et al. [88] studied adsorption of $NH_3$, $NO_2$,

NO, CO, and $H_2O$ on graphene sheet. Functionalization of graphite nanostructures enables it to sense traces of pollutant gases such as $NO_2$. Qazi et al. [113] studied water vapor sensing characteristics of RGOs. RGO has been shown to form good graphene electronic sensor for a wide variety of applications. RGO has been found to be a novel sensor for toxic gases [152, 201]. RGO is electronically hybrid, whose properties are changed from insulator to semiconductor, using reduction chemistry [202].

Sensor materials are reported to be reactivated on application of external electric field, as demonstrated in the case of fully hydrogenated graphene sheet, where hydrogen atoms are unloaded on one side [152]. GNRs-based sensors, with edges decorated metal clusters, have been shown to exhibit enhanced sensitivity and selectivity. It is known that GNRs can be either metallic or semiconducting, depending on the zigzag or armchair edge configurations [203]. The edge disorder gives rise to Coulomb blockade and a mobility gap in the transport properties of GNRs [204]. Drastic changes have been observed in the adsorption energies and equilibrium molecule/graphene distance along with charge transfer values [203]. Zhang et al. [16], using DFT techniques, studied the binding of organic electron donor and acceptor molecules on sheets of graphene. They found that the molecule–graphene interaction could be dramatically enhanced by introducing metal atoms to construct molecule/metal/graphene sandwich structures. A chemical sensor based on iron modified graphene has been shown to have a sensitivity two orders of magnitude higher than that of pristine graphene [152]. Some et al. [205] have demonstrated the use of new hydrophilic 2D GO nanosheets with oxygen functional groups, with high sensitivity in highly unfavorable environments (high humidity, strong acidic or basic) for detection of volatile organic gases. The sensor has two physically different surfaces GO and RGO in order to differentiate simultaneously tetrahydrofuran (THF) and dichloromethane (MC), respectively, which is the most challenging issue in the area of gas sensors. Figure 6.24 shows a schematic representation of polymer optical fiber (POF) sensors based on graphene. It shows a schematic of the transition of GO into RGO with sunlight for sensor applications [205].

### *6.7.3 Graphene-Based Biosensors*

The development of enzyme electrodes, in 1962, led to enormous research in various fields of physics, materials science, and biotechnology resulting in highly advanced and accurate biosensors. Applications of biosensors in medical, agricultural, and military and in detection and prevention of bioterrorism have been of special interest [29]. Biosensors have a selective interface located closely with transducer that relays information about analyte–electrode surface interaction directly or with the aid of a mediator [206]. Properties of graphene which make it suitable for use as biosensor are [207–209]:

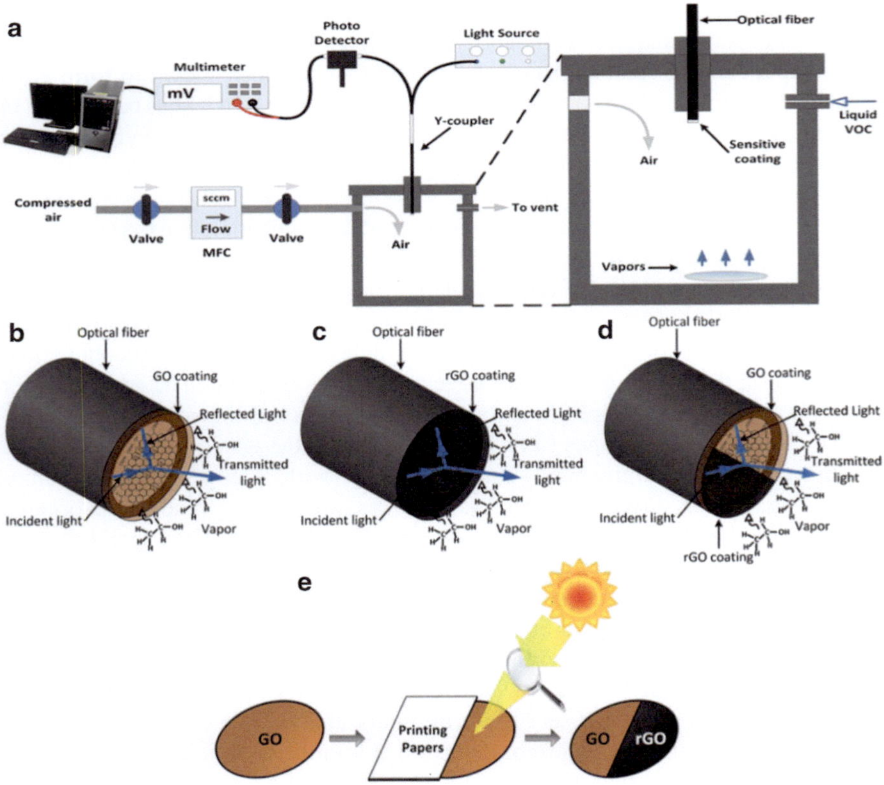


**Fig. 6.24** (**a**) Schematic representation for fabrication of one-headed polymer optic fiber (POF) sensor with (**b**) GO, (**c**) rGO, and (**d**) GO-rGO. (**e**) Schematic representation for fabrication processes of GO-rGO POF sensor by converting GO into rGO with sunlight. "Reprinted by permission from Macmillan Publishers Ltd. [Scientific Reports] [205], copyright (2013)"

1. High electron mobilities
2. High thermal conductivity
3. Flexibility to be shaped
4. Good biocompatibility

The use of graphene can avoid the problems associated with metal alloy nanoparticles and CNT [29]. Sreekanth et al. [210] proposed graphene-based 1D photonic crystal to be used as a highly sensitive biosensor, with alternate layers of high (graphene) and low (PMMA) refractive index materials. The schematic is as shown in Fig. 6.25 [210]. This can be used as a replacement of metallic thin film biosensors.

Earlier research in the field of biosensing applications used graphene FET as a transducer under vacuum or atmospheric conditions, while, recently, there has been some work using solution-gated FETs (SGFETs) [211, 212]. 2D nature of graphene

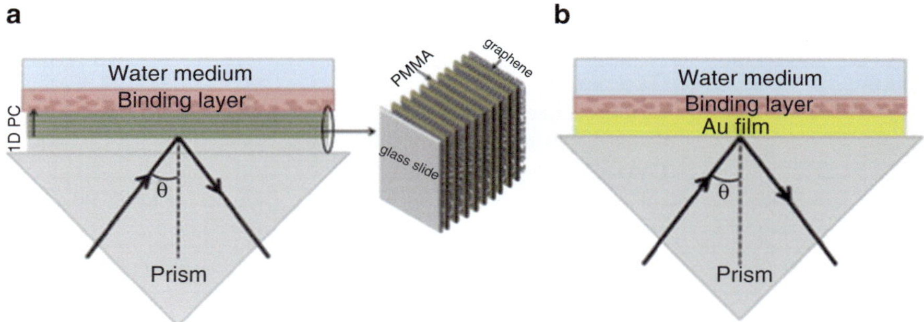


**Fig. 6.25** Schematic of (**a**) proposed biosensor configuration and (**b**) surface plasmon resonance biosensor. Graphene-based 1D photonic crystal consists of alternating layers of graphene and PMMA layers with an additional graphene termination layer. "Reprinted from Sensors and Actuators B: Chemical, Vol 182, Kandammathe Valiyaveedu Sreekanth, Shuwen Zeng, Ken-Tye Yong, Ting Yu, Sensitivity enhanced biosensor using graphene-based one-dimensional photonic crystal, Pages No. 424-428, Copyright (2013), with permission from Elsevier"

**Fig. 6.26** Graphene-based solution gated FET (SGFET) [211]. "From Reference [211], M. Dankerl, M. V. Hauf, A. Lippert, L. H. Hess, S. Birner, I. D. Sharp, A. Mahmood, P. Mallet, J.-Y. Veuillen, M. Stutzmann, and J. A. Garrido, Advanced Functional Materials 20, 3117-3124, (2010). Reprinted with permission of John Wiley & Sons, Inc."

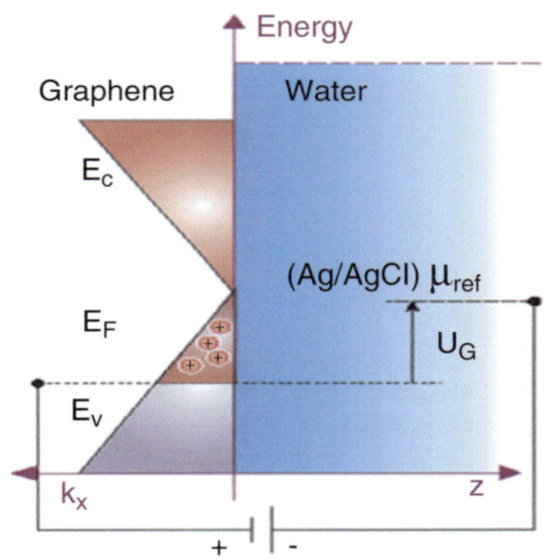


with $sp^2$-bonded carbon atoms is found to be ideal for the development of high sensitivity SGFETs, which depend on the mobility of charge carriers and the distance between conductive channel and surface of graphene. A schematic representation of the modulation of charge carriers in graphene film, induced by application of external voltage, which shifts the Fermi level either above or below the Dirac point, is shown in Fig. 6.26. With extremely low intrinsic noise and high transconductance of the system, it is proposed as a suitable material for biosensor applications [211].

Graphene-based biosensors can be classified based on the electrode types, i.e., enzymatic or nonenzymatic. Enzymatic electrode biosensors include glucose oxidase, cytochrome C, NADH, hemoglobin, HRP, cholesterol, or catechol biosensor. Nonenzymatic electrodes based biosensors can be using hydrogen peroxide,

**Table 6.3** Various types of enzymatic electrodes made from graphene and GO

| Type of biosensor | Sensor materials | RSD (%) | Detection limit |
|---|---|---|---|
| Glucose | Graphene-PPy | – | 3 μM ± 0.5 μM |
| | Metal decorated graphene | – | 1 μM |
| | Graphene/Nafion | 4.21 | $1 \times 10^{-6}$ M |
| | CVD-grown graphene | – | 0.1 mM |
| | CS–GR | 5.3 | 0.02 nM |
| | GO | 5.8 | – |
| | PVDF-protected graphene | 3.2 | 2–14 mM |
| | CS–GR-AuNP | 4.7 | 180 μM |
| | GO | – | – |
| | Graphene-CdS | 5.3 | 0.7 mM |
| | CS–GR-AuNP | 6 | 0.6 M |
| | Graphene | 2.5 | 10 ± 2 μM |
| | CMG | – | 0.376 mM |
| | Graphene | – | 0.168 mM |
| | CR-GO | 4.3 | 2 μM |
| | Graphene | 3.2 | 5 μM |
| NADH | IL functionalized graphene | 4.2 | 5 μM |
| | Graphene | 3.5 | – |
| Hb | $Fe_3O_4$–graphene | 1.6 | 0.5 μM |
| | Graphene | – | $5.1 \times 10^{-7}$ M |
| HRP | Graphene | 4.48 | $1.05 \times 10^{-7}$ M |
| | SDBS–graphene | – | $1 \times 10^{-7}$ M |
| | AuNP–graphene | 3.6 | $1 \times 10^{-6}$ M |
| | CS–GR/$Fe_3O_4$ /HRP | – | $6 \times 10^{-7}$ M |
| Cholesterol | PtNP–graphene | – | 0.5 nM |

Reprinted from Biosensors and Bioelectronics, Vol 26, Tapas Kuila, Saswata Bose, Partha Khanra, Ananta Kumar Mishra, Nam Hoon Kim, Joong Hee Lee, Recent advances in graphene-based biosensors, Pages No. 4637–4648, Copyright (2011), with permission from Elsevier

In all the cases, the mode of detection is change in electrical current [29]

*PPy* polypyrrole, *CS* chitosan, *GR* graphene, *AuNP* gold nanoparticle, *CdS* cadmium sulfide, *CMG* chemically modified graphene, *CR-GO* chemically reduced graphene oxide, *IL* ionic liquid, *SDBS* sodium dodecyl benzene sulfonate, *NADH* Nicotinamide Adenine Dinucleotide, *Hb* Hemoglobin

ascorbic acid, uric acid, or dopamine with graphene-based nanocomposites. Proper conjugation needs to be developed between graphene-based sensing electrodes and biomolecules, which necessitates appropriate functionalization of graphene and immobilization of any other biomaterials on electrodes before use [29]. Tables 6.3 and 6.4 represent the recent research done on enzymatic and nonenzymatic electrodes, respectively, along with corresponding detection limits.

#### 6.7.3.1 Addressing Biocompatibility of Graphene and GO

Liu et al. reported that GO sheets are biocompatible and can be used in regular drug delivery [212]. GO is found to decrease cell adhesion when it enters nucleus and

**Table 6.4** Various types of nonenzymatic electrodes made from graphene and GO

| Sensor materials | Detection mode | Detected element | Detection limit |
|---|---|---|---|
| CMG | Change in electrical current | DA | 0.01 μM |
| PtNP/IL/GR | Change in electrical current | AA and DA | – |
| $MnO_2$ /GO | Change in electrical current | $H_2O_2$ | 0.8 μM |
| MGNFs | Change in electrical current | AA, DA and UA | 0.17 μM DA |
| β-CD/GR | Change in electrical current | DA | 0.2 μM |
| Graphene | Change in electrical current | DA | – |
| Nafion/Hb-IL-GR/CCE | Change in electrical current | $H_2O_2$ | 30 μM |
| PSSA-g-PPY/GR | Change in electrical current | $H_2O_2$ | 10 nM |
| CMG | Change in electrical current | AA, DA and UA | – |

Reprinted from Biosensors and Bioelectronics, Vol 26, Tapas Kuila, Saswata Bose, Partha Khanra, Ananta Kumar Mishra, Nam Hoon Kim, Joong Hee Lee, Recent advances in graphene-based biosensors, Page No. 4637–4648, Copyright (2011), with permission from Elsevier
*CMG* chemically modified graphene, *PtNP* platinum nanoparticles, *IL* ionic liquid, *GR* graphene, *MGNF* multilayer graphene nanoflake, β-*CD* β-cyclodextrin, *PSSA-g-PPY* poly (styrenesulfonic acid-g-pyrrole), *PPy* polypyrrole, *AA* ascorbic acid, *DA* dopamine, *UA* uric acid, *Hb* hemoglobin

cytoplasm [29]. Magnetic resonance imaging (MRI) can use magnetic nanoparticles with graphene composites, as seen from the magnetization curve of GO-$Fe_3O_4$ composites at room temperature, which displays that the system has a superparamagnetic behavior [115]. Reduced graphene nanowalls are sharp and hence reduce growth of Gram-positive, Gram-negative, and *Escherichia coli* bacteria [29]. As noted by Yang et al. [213], female mice were found to be biologically safe at a dose of 20 mg/kg. Because of controversies in the reported literature on cytotoxicity of graphene and its various forms (bare or functionalized), it can be concluded that research needs to be done with different animal models.

Furthermore, the high chemical and mechanical stability of graphene suggest its use in aqueous environments, for example, in biological systems. Due to the exceptional band structure of graphene, both electrons and holes can be observed and tuned by electrolytic gating. As micrometer-sized transistors are beneficial for many biophysical experiments, large-scale graphene films are grown by CVD on copper foil. The graphene film is transferred to various substrates and microscopic graphene SGFET arrays are fabricated. The devices are characterized in aqueous environments using standard transistor characterization as well as advanced techniques such as in-liquid Hall-effect and low-frequency noise measurements.

#### 6.7.3.2 DNA Sensors

Electrochemical DNA sensors are known to be highly sensitive, economical, faster, and miniaturizable method to analyze biomolecules, typically to detect sequences or mutated genes in humans [29, 30]. Noncovalent interactions have been found to play an important role in studying a range of biological issues such as ligand binding to DNA and RNA. CNT, carbon nanodot, and carbon nanofiber have been studied for quite some time for this purpose [29]. Lu et al. (2009a) [29, 214] worked on fluorescence quenching of GO which is applied in DNA biosensing. DNA sensor based on oxidation process is known to be the simplest [30]. Zhou et al. [215] studied DNA sensors based on chemically rGO and have shown to detect all four bases of DNA (i.e., guanine (G), adenine (A), thymine (T), and cytosine (C), which is neither feasible with glassy carbon or graphite. The reason for this has been attributed to the high electron mobilities and antifouling properties for oxidation of the bases of DNA on chemically reduced graphene sheets. This may have a potential application in forensic serology. Figure 6.27 shows the differential pulse voltammograms (DPVs) at various electrodes based on carbon-based materials [215].

#### 6.7.3.3 Field Effect Transistor Sensor

This type of sensor uses electric fields to control the shape and conductivity of the channel of one type of charge carrier in the semiconducting material, which, in this case, is usually reduced graphene. This is used for the detection of DNA hybridization. Aptamers are functional nucleic acids with specific and high affinity to their targets and electrochemical aptamer-modified graphene FET has been shown to be used for sensing immunoglobulin E (IgE) due to its stability, low production cost, and ease of chemical modifications [216]. Dong et al. [217] have synthesized rGO in order to make graphene–Au nanocomposites in aqueous solutions. These nanocomposites are found to have high hole mobilities and conductance, as compared to pristine rGO sheets. They have demonstrated detection of label-free DNA hybridization using rGO–Au nanocomposite, with a high efficiency. Table 6.5 describes various types of sensors based on graphene and GO, along with the respective detection modes based on the method of detection of the elements.

Table 6.6 discusses and compares the efficiency of detection of CNTs and graphene-based biosensors along with their respective detection limits.

## 6.8 Graphene as a Photocatalyst for Water Remediation

Photocatalysis is the acceleration of a photoreaction in the presence of a catalyst. Materials in which irradiation of UV and/or visible light on the surface leads to excitation of their valence electrons to the conduction band, leaving holes in the valence band (leading to creation of negative-electron ($e^-$) and positive-hole ($h^+$) pairs), are termed as photocatalysts. Photocatalysis is of two types: the first is homogenous photocatalysis in which the reactants and photocatalysts are in the same phase, and the second is heterogeneous photocatalysis in which the reactants and photocatalysts are in different phase. Commonly observed homogeneous photocatalysts are ozone, transition metal oxides, etc. while heterogeneous photocatalysis is observed in semiconductors.

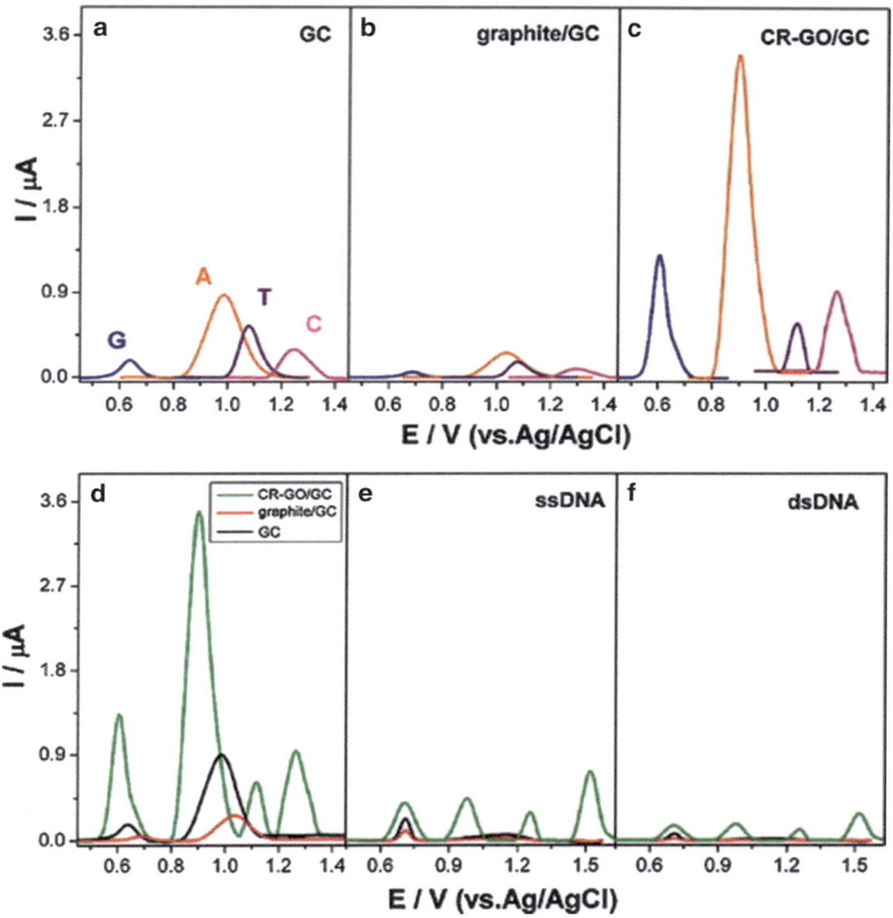


**Fig. 6.27** (**a**) DPVs at the glassy carbon (GC) electrode for G (*blue*), A (*orange*), T (*violet*), and C (*magenta*), respectively. (**b**) DPVs at the graphite/GC electrode for G (*blue*), A (*orange*), T (*violet*), and C (*magenta*), respectively. (**c**) DPVs at the chemically reduced (CR)-GO/GC electrode for G (*blue*), A (*orange*), T (*violet*), and C (*magenta*), respectively. (**d**) DPVs for a mixture of G, A, T, and C at CR-GO/GC (*green*), graphite/GC (*red*), and GC electrodes (*black*). (**e**) DPVs for single-stranded (ss) DNA at CR-GO/GC (*green*), graphite/GC (*red*), and GC electrodes (*black*). (**f**) DPVs for dsDNA at CR-GO/GC (*green*), graphite/GC (*red*), and GC electrodes (*black*). Concentrations for different species (A–F): G, A, T, C, ssDNA or dsDNA: 10 μg/mL. Electrolyte: 0.1 M pH 7.0 phosphate buffer solution (PBS). "Reprinted (adapted) with permission from (M. Zhou, Y. Zhai, and S. Dong, Analytical Chemistry 81, 5603 (2009)). Copyright (2009) American Chemical Society"

The stage in which electron-hole pairs are created is termed as "photo-excited" state of a semiconductor. The difference of energies between the valence band and conduction band is called as "bandgap," and it represents the state of a semiconductor in which no energy levels are available for promoting electron–hole

**Table 6.5** Various types of sensors made from graphene and GO

| Sensor materials | Detection mode | Detected element | Detection limit |
|---|---|---|---|
| Hydrazine-reduced GO | Change in frequency of surface acoustic waves | $H_2$, CO | ~125 ppm CO; <600 ppm $H_2$ |
| GR/PANIw | Changes in electrical current | DNA | $3.25 \times 10^{-13}$ M |
| Graphene | Changes in electrical current | $Cd^{2+}$ | 200 μg\L |
| Hydrazine-reduced GO | Changes in electrical current | $NO_2$, $NH_3$, DNT | <5 ppm $NO_2$ and $NH_3$; 28 ppb DNT |
| Graphene nanoribbons | Changes in electrical current | CO, NO, $NO_2$, $O_2$, $N_2$, $CO_2$, and $NH_3$ | – |
| COOH functionalized graphene | Changes in electrical current | Adenine and guanine | $2.5 \times 10^{-8}$ M and $5.0 \times 10^{-8}$ M |
| Chemically and thermally reduced GO | Changes in electrical current | $H_2O$ | Less than 0.1 Torr of $H_2O$ vapor |
| GO | Change in fluorescence intensity | DNA | 103 pfum |
| Graphene | Changes in electrical current | $Pb^{2+}$ and $Cd^{2+}$ | 0.02 μg |
| EG | Changes in electrical current | DNA | 1 mL |
| GO | Change in fluorescence intensity | DNA | 200 nM |
| GO | Change in fluorescence intensity | DNA | – |
| Thermally reduced GO | Changes in electrical current | $NO_2$ | 2 ppm |
| GO | Change in fluorescence intensity | DNA | 2.0 nM |
| Hydrazine-reduced GO | Changes in electrical conductance | HCN, CEES, DMMP, DNT | 70 ppb HCN; 0.5 ppb CEES 5 ppb DMMP; 0.1 ppb DNT |
| Mechanically exfoliated graphene | Changes in electrical resistivity | $NO_2$, $H_2O$, $I_2$, $NH_3$, CO, ethanol | 1 ppb |
| GO | Change in fluorescence intensity | $Ag^+$ | 5 nM |
| CVD-grown graphene | Changes in electrical resistivity | $H_2$ | – |
| GO | Change in fluorescence intensity | $Ag^+$ and $Hg^{2+}$ | 20 and 5.7 nM |
| CR-GO | Changes in electrical current | DNA | – |

Reprinted from Biosensors and Bioelectronics, Vol 26, Tapas Kuila, Saswata Bose, Partha Khanra, Ananta Kumar Mishra, Nam Hoon Kim, Joong Hee Lee, Recent advances in graphene-based biosensors, Pages No. 4637–4648, Copyright (2011), with permission from Elsevier

*DMMP* dimethylmethylphosphonate, *HCN* hydrogen cyanide, *CEES* chloroethylethyl sulfide, *DNT* dinitrotoluene, *PANIw* polyaniline nanowire, *GR* graphene, *CRG* chemically reduced graphene

**Table 6.6** Comparative study of sensing efficiency between CNT and graphene-based biosensors

| Sensor type | Detected element | Sensing materials | Detection limit |
|---|---|---|---|
| Glucose biosensor | Glucose | Graphene-PPy | 3 μM ± 0.5 μM |
| | Glucose | CNT-PPy | 0.2 mM |
| | Glucose | Graphene/Nafion | $1.0 \times 10^{-6}$ M |
| | Glucose | CNT/Nafion | $3.0 \times 10^{-9}$ M |
| | Glucose | CS–GR | 0.02 nM |
| | Glucose | CS-CNT | 0.05 mM |
| | Glucose | CS–GR-AuNP | 0.6 M |
| | Glucose | CS-CNT-AuNP | $3.0 \times 10^{-6}$ M |
| Cholesterol oxidase | Cholesterol | PtNP-Graphene | 0.5 nM |
| | Cholesterol | PtNP-CNT | – |
| NADH | | Graphene | – |
| | NADH | GO | 0.1 μm |
| | NADH | CNT | $5 \times 10^{-3}$ M |
| | NADH | CNT | 2 μm |
| HRP | $H_2O_2$ | Graphene | $1.05 \times 10^{-7}$ M |
| | $H_2O_2$ | AuNP-Graphene | $1.0 \times 10^{-6}$ M |
| | $H_2O_2$ | CNT | $1.0 \times 10^{-7}$ M |
| | $H_2O_2$ | CNT-Mb | 1.0 μm |
| Hb | $H_2O_2$ | Graphene | $5.1 \times 10^{-7}$ M |
| | $H_2O_2$ | CNT | 2.41 μm |
| | $H_2O_2$ | CNT-sodium alginate | 16.41 μm |
| | Nitrite | Hb-CS-DMF/GR | $1.8 \times 10^{-7}$ M |
| Electrochemical sensor | AA, UA, DA | MGNF | – |
| | DA | β-CD/GR | 0.2 μM |
| | DA | CMG | 0.01 μM |
| | AA, UA, DA | PSS-CNT | $5.0 \times 10^{-7}$ M, $1.5 \times 10^{-7}$ M, $5.0 \times 10^{-7}$ M |
| | AA, UA, DA | PtNP-CNT | $1.9 \times 10^{-5}$ M, $3.2 \times 10^{-8}$ M, $2.7 \times 10^{-8}$ M, |
| DNA sensor | DNA | EG | 1 μM |
| | DNA | GO | 200 nM |
| | DNA | GR/PANIw | $3.25 \times 10^{-13}$ M |
| | DNA | CNT-Mb | 0.252 nM |
| | Phenolic pollutants | CNT-DNA | $1.6 \times 10^{-5}$ M |
| | DNA | CNT | $1.0 \times 10^{-10}$ M |
| Gas sensors | $NO_2$ | Thermally reduced GO | 2 ppm |
| | $NO_2$ and $NH_3$, DNT | Hydrazine reduced GO | <5 ppm ($NO_2$ and $NH_3$) 28 ppb DNT |
| | $NO_2$ and $NH_3$ | CNT | 2 ppm |
| | $NO_2$ | CNT | 44 ppb |
| | $NH_3$ | CNT | 10 ppm |
| | $NO_2$ | CNT | 5–100 ppb |

(continued)

**Table 6.6** (continued)

| Sensor type | Detected element | Sensing materials | Detection limit |
|---|---|---|---|
| | $NH_3$, CO, ethanol | Mechanically exfoliated graphene | 1 ppb |
| | HCN | CNT | 4000 |
| | CEES | | 0.5 |
| | DMMP | | 0.1 |
| | DNT | | 0.1 |
| | HCN | Reduced GO | 70 |
| | CEES | | 0.5 |
| | DMMP | | 5 |
| | DNT | | 0.1 |



*PPy* polypyrrole, *CS* chitosan, *GR* graphene, *PtNP* platinum nanoparticles, *NADH* nicotinamide adenine dinucleotide, *Hb* hemoglobin, *MGNF* multilayer graphene nanoflake, β-*CD* β-cyclodextrin, *CMG* chemically modified graphene, *PSS* polystyrene sulfonate, *CNT* carbon nanotube, *EG* epitaxial graphene, *PANIw* polyaniline nanowire, *Mb* methylene blue, *AA* ascorbic acid, *UA* uric acid, *DA* dopamine, *DNT* dinitrotoluene, *HCN* hydrogen cyanide

combination produced in the process of photoactivation. Light of a given wavelength would be effectively absorbed only if the bandgap of photocatalyst is less than the energy of incident light and both the energy of bandgap as well as incident light are equal to each other. Photocatalytic activity (PCA) depends on the ability of the photocatalyst to create electron-hole pairs. Once the excitation occurs across the bandgap, electron–hole pairs are transferred from the semiconductor to the adsorbed species. This process is continuous and exothermic. Photo-induced electron transfer to adsorbed organic molecules/inorganic species or to the solvent is a result of migration of electrons and holes to the surface of semiconductors.

On being excited, electrons and holes can follow several paths, as shown in Fig. 6.28 [218]. Semiconductors can donate electrons to reduce an electron acceptor, usually oxygen in aerated aqueous solutions (path C), and holes can travel to the surface where an electron from donor species can combine with the surface hole, hence oxidizing the donor species (Path D). In competition with the charge transfer to the adsorbed species, electrons and holes can either combine in the volume of the semiconductor (path B) or at the surface (path A), resulting in the generation of heat. This process is known as recombination [218]. Figure 6.28 shows the pathways of deactivation followed by electrons and holes. Rate and probability of transfer of electrons and holes depend on the respective positions of band edges for valence and conduction bands and levels of redox potential of the adsorbate species.

Holes (positive charges) are responsible to cause oxidation and reactions with electron transfer are linked with reduction. Reaction with toxic pollutants on the surface, leading to their degradation, causes the formation of hydroxyl ions ($OH^-$) which further drive pollutant's oxidation till completion of mineralization.

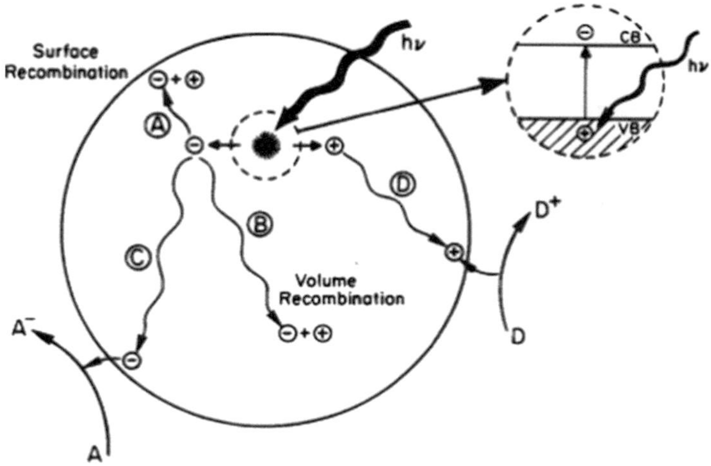


**Fig. 6.28** Principle of photo-excitation in solids and subsequent de-excitation [218]. "Reprinted (adapted) with permission from (A. L. Linsebigler, G. Lu, and J. T. Yates, Chemical Reviews 95, 735 (1995)). Copyright (1995) American Chemical Society"

Heterogeneous photocatalysis with semiconductors is found to have immense potential in remediation of organic pollutants in water and air. This process is named as advance oxidation process (AOP) and is known to be highly efficient to degrade recalcitrant organic compounds [219].

The discovery of photocatalytic splitting of water, in photoelectrochemical cell, using $TiO_2$ electrodes as anodes and inert cathode, by Fujishima and Honda in late 1960s, led to deluge of research in this field by chemists, physicists, as well as material scientists and engineers [220]. Most active areas of application, using this technology, have been for the removal of organic compounds from water and air, resulting in their improved quality.

Besides $TiO_2$, a number of semiconductors such as ZnO, $Fe_2O_3$, CdS, ZnS can act as photocatalysts but $TiO_2$ has been actively studied due to its hydrophilic nature, high reactivity, less toxicity, chemical stability, and costs [219]. Anatase, $TiO_2$ (rutile) semiconductor electrode, is found to have sufficiently positive valence band edge in order to produce oxygen from water, and it is an extremely stable material in the presence of aqueous solutions, as compared to other semiconductors [221]. $TiO_2$ has a large bandgap of 3.2 eV [222]; it can respond only to ultraviolet light which consists of about 3–5 % of the solar spectrum [223, 224]. CdS is a widely researched photocatalyst with a relatively narrow bandgap of 2.42 eV [225] and hence has better ability to absorb incident light. Besides, redox potentials of water lie within the bandgap of CdS; hence CdS can be used for effective hydrogen evolution. However, the rapid recombination rates of electron–hole pairs along with easy aggregation of nanoparticles of CdS (resulting in less specific surface area and photo-corrosion during light irradiation) cause deterioration of its performance as a photocatalyst [226].

After intense research, it has been established that the use of nano-hybrids of graphene with semiconductors can help to resolve the issues encountered with $TiO_2$ and CdS [226, 227]. Enhancement of photocatalysis in graphene-based nanocomposites is due to the reduction of the recombination of the electron–hole pairs. It is even attributed to extended range of light absorption and increased intensity of light absorption, along with greater number of active sites at surface and higher chemical stability of photocatalysts [228]. These photocatalysts are shown to have excellent performance in the field of photo-driven chemical conversion with the aid of solar energy, on account of commendable physical and chemical properties of graphene. Understanding the mechanism of electron transport in graphene nanocomposite-based photocatalysts has led to the development of technologies for toxic elimination of heavy metal ions, photo-reduction of $CO_2$ into renewable fuels, and degradation of organic compounds in water as well as antibacterial applications [228].

Shet et al. [229] have synthesized ZnO and mixed ZnO:GaN thin films on fluorine-doped tin oxide substrates by reactive radio frequency (RF) magnetron sputtering in mixed $N_2$ and $O_2$ atmosphere. Their results have shown an enhancement in the crystallinity and narrowing of the bandgap in the case of mixed ZnO: GaN thin films as compared to the ZnO thin films. The enhanced properties of mixed ZnO:GaN films are expected to be of help in improving their photoelectrochemical performance. Shet et al. [230] have synthesized p-type ZnO films with varying carrier concentrations through codoping of Cu and Ga. The bandgap reduction is caused by Cu and the carrier concentrations can be tuned by varying Ga concentration. The optimization of carrier concentration can enhance the photoelectrochemical response for p-type ZnO thin films. Shet et al. [231] synthesized ZnO thin films, on F-doped tin-oxide-coated glass, with significantly reduced bandgaps by doping N and codoping Al and N at 100 °C using RF magnetron sputtering. They have found that codoped ZnO:(Al, N) thin films show significantly enhanced crystallinity compared to ZnO doped solely with N, ZnO:N, under the same growth conditions. ZnO:(Al, N) films have been shown to exhibit better photocurrents than ZnO:N films as well, suggesting charge-compensated donor–acceptor codoping to be a potential method for bandgap reduction of wide-bandgap oxide materials to improve their photoelectrochemical performance.

Yang et al. [232] fabricated nanocomposites of functionalized graphene sheets (FGS)/ZnO via thermal treatment. They used GO as a precursor for graphene, Zn $(NH_3)_4CO_3$ as a precursor of ZnO, and poly(vinyl pyrrolidone) as an intermediate in order to combine Zn with carbon-based materials. The nanocomposites showed highly efficient PCA to degrade RhB under low power UV light and the process has been reported to be highly economical.

Xu et al. [233] reported hybridization of ZnO with graphene by coating ZnO nanoparticles with appropriate amount of GO, followed by in situ reduction of GO to form ZnO/graphene composite. This composite is found to have high PCA for the decomposition of organic dye, with the degree dependent on the coverage of graphene on nanoparticles of ZnO. Samples with 2 wt% graphene-hybridized ZnO showed highest PCA, about four times that of pristine graphene, which is attributed to higher efficiency of migration of photo-induced electrons and suppressed electron–hole recombination due to interaction of ZnO with graphene.

Štengl et al. [234] have used graphene oxide (GO) sheet as substrate for $TiO_2$ deposition using liquid phase deposition method. GO works as electron acceptor as well as photosensitizer for efficient photodecomposition of butane. Moreover, photocatalytic property of $TiO_2$-GO does not depend on light sources. This is attributed to the following facts: (a) GO has high specific surface area and (b) the formation of n–n conjugations between butane molecules and aromatic rings as well as ionic interactions between butane- and carbon-based nanosheets.

Liang et al. [235] have shown that $TiO_2$ nanocrystals, grown on graphene, have superior PCA in the degradation of RhB, as compared with $TiO_2$. The enhancement in PCA has been attributed to strong coupling between $TiO_2$ and GO, facilitating in the interfacial charge transfer. GO acts as electron acceptor and hence stops the process of electron–hole recombination. The hybrid material has high specific surface area. The binding of conjugated dye molecules to GO sheets via $\pi$–$\pi$ bonds leads to higher reactivity.

Rao et al. [236] have studied photocatalytic decomposition of MB and RhB adsorbed over composites of $TiO_2$ nanoparticles with pure graphene and even boron- and nitrogen-doped graphenes. Being a good donor of electrons and having lower ionization energy, MB is found to have strong interaction with electron-deficient boron-doped graphene, ultimately leading to decomposition of the dye. RhB which is not a good electron donor is found to strongly interact with electron-rich nitrogen-doped graphene, with its resultant quicker degradation.

Gao et al. [237] synthesized graphene oxide-$Bi_2WO_6$ composite using in situ hydrothermal reaction in the presence of GO, followed by reduction of GO by ethylene glycol, leading to the formation of graphene-$Bi_2WO_6$ (G-BWO). This composite is found to have high PCA for degrading RhB under visible light with wavelength more than 420 nm. It is to be noted that the electronic interaction and equilibration of charge between graphene and $Bi_2WO_6$ causes negative shift of the Fermi level, i.e., decrease of the conduction band potential. This is reported to cause higher migration efficiency of photo-induced electrons and hence reduces the electron–hole recombination, ultimately leading to enhanced PCA.

Seema et al. [32] have demonstrated higher efficiency of graphene–$SnO_2$ composites for photocatalytic degradation of organic dye, methylene blue, under sunlight, as compared to bare $SnO_2$. They have synthesized RGO modified with $SnO_2$ via redox reaction between GO and $SnCl_2$ at 90 °C. Gawande and Thakare [238] have synthesized graphene wrapped $BiVO_4$ photocatalyst using sol-gel method. Enhanced performance of PCA for degradation of methylene blue (MB) under visible light for this composite has been attributed to effective quenching of photo-generated carriers (electron–hole pairs). Zhou and Zhu [239] have synthesized hybrid of graphene with $Bi_2WO_6$ photocatalyst and have shown enhancement of about four times over PCA of pristine $Bi_2WO_6$ for removal of MB from water. Optimal amount for hybridization to have maximum enhancement in PCA is about 1.5 wt% graphene, which is nearly equivalent to depositing about single layer of graphene over $Bi_2WO_6$. This enhanced activity is due to the rapid photo-induced charge separation and subsequent inhibition of recombination of charge carriers.

Some spinel materials including $ZnFe_2O_4$ have narrow bandgaps. They have been found to show excellent response to visible light and photochemical stability along with favorable magnetism [240]. Fu and Wang [241] have made magnetically separable $ZnFe_2O_4$-graphene nanocomposite photocatalyst with varying graphene content via facile hydrothermal method. As compared to pure $ZnFe_2O_4$ catalyst, it was found that this hybrid nanocomposite serves two roles—one for photo-electrochemical degradation of MB and the other is to help the generation of highly oxidant hydroxyl ions ($OH^-$) by photo-electrochemical decomposition of $H_2O_2$ upon visible light irradiation. $ZnFe_2O_4$ has inherent magnetic properties. It can be used to make nanocomposites that are magnetically separable from water solution. Similar studies by Fu et al. [242] show that the combination of cobalt ferrite ($CoFe_2O_4$) and graphene can be used as a catalyst for photo-degradation of methylene blue (MB), RhB, methyl orange(MO), active black under visible light. Increased activity has been attributed to the formation of RGO by transfer of photo-generated electrons of $CoFe_2O_4$ to GO and hence effectively preventing recombination of charge carriers. Here, $CoFe_2O_4$ has strong magnetic property and can enable magnetic separation.

Zhang et al. [243] have synthesized a series of CdS–graphene nanocomposites and shown that they can serve as selective visible light photocatalysts toward oxidation of alcohols and reduction of heavy ions Cr(IV) in water. Besides, there is enhanced PCA of GR (either solvent-exfoliated graphene (SEG) or GO), with decrease in defect density of this hybrid substrate by the use of SEG as compared to GO. This is because of the improved electron conductivity as compared to that of RGO. The interfacial interaction, due to hybridization of CdS with GR, helps in stopping photocorrosion of CdS due to redox reaction occurring during photocatalysis.

Wang et al. [244] prepared three different nanocarbon-$TiO_2$ nanocomposites by self-assembly of colloidal $TiO_2$ nanoparticles on nanocarbon surfaces (respectively SWCNT, C60, graphene) by using thermal reaction. It is reported that RGO-$TiO_2$ sample exhibits highest PCA, the reason being attributed to increased conductivity of graphene substrate and more uniform distribution of nanoparticles on the surface of graphene as compared to other forms of nanocarbon. Here, the photocatalytic performance is enhanced by about 3.46 times in the visible region in comparison with P25 (available commercially).

Liu et al. [245] prepared $TiO_2$-RGO composites using microwave-assisted reduction of GO in suspension of $TiO_2$. Reduction of Cr(VI) with removal rate of 91 %, under irradiation of UV light with $TiO_2$-RGO composites, is in comparison to pure TiO2 (83 %) and commercial P25(70 %). With introduction of RGO in $TiO_2$, there is profound reduction in electron–hole pair recombination.

Zhang et al. have reported an additional covalent linkage of GO with $TiO_2$ through esterification reaction between carboxyl group attached to GO sheets with hydroxyl group of $TiO_2$, which ultimately leads to enhancement in the removal of Cr(VI), even in the presence of only about 5 wt% GO in the composite.

From the above-mentioned works in the literature, it is seen that the reduced recombination of electron–hole pairs and increased conductivity lead to an increase in the PCA of graphene. In the case of removal of organic pollutants in water, ZnO should be doped with either metals (reportedly Cu and Ga) or semiconductors such as GaN to form composites which show improved PCA due to enhanced crystallinity and narrowing of bandgap or with graphene to form nanocomposites which have improved rate of migration of photo-induced electrons and suppressed electron–hole recombination even in low intensity UV light. By using spinel in combination with graphene as photocatalyst for water remediation, there is the possibility of magnetic separation of spinel and increased PCA by transfer of photo-generated electrons and effective prevention of recombination of charge carriers. RGO-$TiO_2$ is found to show best PCA as compared to other forms of nanocarbon due to increased conductivity of graphene and uniform distribution of nanoparticles in graphene. GO with $TiO_2$ has covalent linkage and is found to show good PCA under UV light, while CdS-graphene nanocomposites are found to show selective photocatalysis under visible light for Cr(VI) removal from water.

## 6.9 Conclusions

Graphene-based compounds are a set of emerging carbonaceous materials. With their high specific area, they can be used for the efficient removal of heavy metal ions, hazardous compounds, and dyes from aqueous solutions and gaseous pollutants in the environment. The highly acidic and anionic behavior of graphene oxide (GO) makes it suitable as an effective adsorbent for cationic heavy ions in water as well as chemicals such as ammonia. While getting devoid of oxygen species, GNS follows $\pi$–$\pi$ interaction as the mechanism for adsorption of various pollutants. Improvement in efficiency of adsorption in graphene-based compounds is attributed to the ease of chemical functional modification leading to formation of nanocomposites. High stability, high specific surface area, and ease of separation from the aqueous medium make it ideal for applications of graphene in remediation of water. Layered structures of GO and GNS are useful for their adsorptive behavior. Toxicity of graphene is a matter of concern, which can be addressed by investigating its biocompatibility and safety for the ecosystem [246].

Mechanical exfoliation route to extract graphene seems to be the cost-effective way of extracting single layer. However, it is found that mass production of graphene is not feasible by this way [11]. Hence, CVD route of making graphene could be used for large-scale pattern production for applications such as stretchable transparent electrodes [247]. The chemical route can be followed for production of GO and rGO [248], both of which have been shown to have applications in selective adsorption of toxic metal ions. Sensors based on graphene have been developed. These include diodes for detection of aromatic molecules such as phenols, which are based on specific changes in the work function and hence the Schottky barrier height at the interface [23]. Functional materials such as GCE coated with film of graphene/polymer are found to have high sensitivity toward phenols in aqueous solutions [166]. Electrochemiluminescence sensors have been devised for supersensitivity toward chlorinated phenols in water [170]. HEG has been shown to have high adsorption capacity of $CO_2$, relative to other carbon nanostructures [164]. Shifts in Fermi level with respect to exposure times have been used as mechanisms for evaluating adsorption rates of gases such as ammonia on GO [158]. Observation of changes in electronic band structures, changes in magnetic moments, and photoconductivity of graphene, upon doping (with adsorption), have been the basis for specific detection of gases as well as pollutants in water.

Use of graphene-based nanocomposites as photocatalysts in purifying water that initially contains organic compounds and bacteria has been an area of emerging research. GO deposited over $TiO_2$ has been used for efficient photodecomposition of butane [228]. RhB has been decomposed by $TiO_2$ nanocrystals grown on graphene, which are shown to have superior PCA [234]. Spinel materials form composites with graphene for efficient photocatalytic removal of methylene blue (MB) from water [241]. Composites of cobalt ferrite with graphene have been shown to be efficient for photodegradation of various contaminants such as MB, RhB, and methyl orange [242]. In spite of all the advantages of graphene-based nanomaterials, low-cost production along with biocompatibility can be challenging for its commercial use.

## 6.10 Outlook

Use of nanotechnology for water purification is a promising method that offers significant potential to obtain clean potable water using environmentally friendly techniques. Nanomaterials, based on graphene, exhibit important physicochemical properties, such as higher surface area than their bulk counterparts, ease of functionalization, as well as hydrophobicity. This helps in making graphene- and graphene oxide-based nano-sorbents with high capacity, affinity, and selectivity for common radicals such as cations, anions, and organic compounds along with recyclability. Due to the ability to vary electronic, optical, magnetic, and photoconductivity properties of graphene-based nanocomposites up on adsorption, there are number of opportunities for developing highly efficient catalysts for water remediation. Engineered graphene has been shown to be an excellent material for applications in the fabrication of "smart" nano-sensing devices such as chemical sensors, photodetectors, and biosensors with improved precision and sensitivity. These devices have smaller sizes and capability for detecting multiple analytes simultaneously, which makes them more economical. These devices can be made even in the current manufacturing setup for silicon-based devices, which reduces additional equipment cost. It is known that the currently used disinfectants such as chlorine and ozone can potentially form toxic by-products such as trihalomethanes, which limit their applications in the elimination of emerging viruses and microbial organisms in water. A recently proposed promising application of CVD-grown graphene membranes, with varying pore sizes, is to separate microscopic contaminants from water as well as to separate specific types of molecules from biological samples [249].

Despite a deluge of research in graphene technologies, real-world applications are in their infancy and need to overcome challenges of mass production of graphene, large and pure enough, to be useful in applications [250]. It is anticipated that advances and effective integration of nanotechnology, biotechnology and information technology can lead to the development and implementation of cost-effective nano-devices for water purification. The driving force to make these developments possible is the sponsorship and investments towards this innovative research. Large funds have been deployed by various sources including governmental organizations across the globe as well as investors in the private sector, which is expected to narrow the commercialization timelines of graphene-based nanotechnology.

## References


1. E. S. Blake, T. B. Kimberlain, R. J. Berg, J. P. Cangialosi, J. L. B. II, Tropical Cyclone Report Report No. AL182012, 2013
2. Global Gas Sensors, Detectors, and Analyzers Markets 2013, 2013
3. M. L. Clark, J. L. Peel, K. Balakrishnan, P. N. Breysse, S. Chillrud, L. P. Naeher, C. E. Rodes, A. F. Vette, J. M. Balbus, in *Environmental Health Perspectives* (National Institute of Environmental Health Sciences, 2013), p. 38
4. J. Tollefson, N. Gilbert, Nature **486** (2012)
5. S. Wang, H. Sun, H.M. Ang, M.O. Tadé, Chem. Eng. J. **226**, 336–347 (2013)
6. J.C. Crittenden, R.R. Trussell, D.W. Hand, K.J. Howe, G. Tchobanoglous, *Water Treatment—Principles and Design*, 2nd edn. (Wiley, New York, 2005)
7. EPA, The History of Drinking Water Treatment, 2000
8. A.B. Outwater, *Water: A Natural History* (NY, New York, 1996) (BasicBooks, c1996)
9. D. Siculus, *Library of History* (Loeb Classical Library, Harvard University Press, Cambridge, 1939), Vol. III
10. N. Savage, M. Diallo, J. Duncan, A. Street, R. Sustich, in *Nanotechnology Applications for Clean Water*, ed. by N. Savage (William Andrew, New York, 2009).
11. K.S. Novoselov, A.K. Geim, S.V. Morozov, D. Jiang, Y. Zhang, S.V. Dubonos, I.V. Grigorieva, A.A. Firsov, Science **306**, 666–669 (2004)
12. N.M. Ravindra, Emerg. Mater. Res. **1**(1-3), 1 (2012) (Editorial)
13. G.K. Ramesha, A. Vijaya Kumara, H.B. Muralidhara, S. Sampath, J. Colloid Interface Sci. **361**, 270–277 (2011)
14. A.K. Geim, Science **324**, 1530–1534 (2009)
15. K.S. Novoselov, A.K. Geim, S.V. Morozov, D. Jiang, M.I. Katsnelson, I.V. Grigorieva, S.V. Dubonos, A.A. Firsov, Nature **438**, 197–200 (2005)
16. Y.-H. Zhang, K.-G. Zhou, K.-F. Xie, J. Zeng, H.-L. Zhang, Y. Peng, Nanotechnology **21**, 065201 (2010) (6 pages)
17. A.K. Geim, K.S. Novoselov, Nat. Mater. **6**, 183–191 (2007)
18. A.V. Klekachev, A. Nourbakhsh, I. Asselberghs, A.L. Stesmans, M.M. Heyns, S.D. Gendt, Interface **22**, 63–68 (2013)
19. K.S. Novoselov, D. Jiang, F. Schedin, T.J. Booth, V.V. Khotkevich, S.V. Morozov, A.K. Geim, Proc. Natl. Acad. Sci. U. S. A. **102**, 10451 (2005) (3 pages)

20. C.J. Docherty, C.-T. Lin, H.J. Joyce, R.J. Nicholas, L.M. Herz, L.-J. Li, M.B. Johnston, Nat. Commun. 1-6 (2012)
21. C. Chen, S. Lee, V.V. Deshpande, G.-H. Lee, M. Lekas, K. Shepard, J. Hone, Nat Nano (2013) (5 pages)
22. J.W. Grizzle, K.L. Dobbins, J.A. Cook, IEEE Trans. Veh. Technol **40**, 280–286 (1991)
23. F. Schedin, A.K. Geim, S.V. Morozov, E.W. Hill, P. Blake, M.I. Katsnelson, K.S. Novoselov, Nat. Mater. **6**, 652–655 (2007)
24. G.K. Ramesha, S. Sampath, Electroanalysis **19**, 2472–2478 (2007)
25. S. Yang, J. Liang, S. Luo, C. Liu, Y. Tang, Anal. Chem. **85**, 7720–7725 (2013)
26. Y. Zhang, T. Liu, B. Meng, X. Li, G. Liang, X. Hu, Q.J. Wang, Nat. Commun. **4**, 1811–1815 (2013)
27. J. Yan, M.-H. Kim, J.A. Elle, A.B. Sushkov, G.S. Jenkins, H.M. Milchberg, M.S. Fuhrer, H.D. Drew, Nat. Nanotechnol. **7**, 472–478 (2012)
28. S. Yang, Y. Gong, Z. Liu, L. Zhan, D.P. Hashim, L. Ma, R. Vajtai, P.M. Ajayan, Nano Lett. **13**, 1596–1601 (2013)
29. T. Kuila, S. Bose, P. Khanra, A.K. Mishra, N.H. Kim, J.H. Lee, Biosens. Bioelectron. **26**, 4637–4648 (2011)
30. Y. Shao, J. Wang, H. Wu, J. Liu, I.A. Aksay, Y. Lin, Electroanalysis **22**, 1027–1036 (2010)
31. A. Dąbrowski, Adv. Colloid Interface Sci. **93**, 135–224 (2001)
32. S. Humaira, K.C. Kemp, C. Vimlesh, S.K. Kwang, Nanotechnology **355705** (2012) (8 pages)
33. A.B. Silva-Tapia, X. García-Carmona, L.R. Radovic, Carbon **50**, 1152–1162 (2012)
34. O.G. Apul, Q. Wang, Y. Zhou, T. Karanfil, Water Res. **47**, 1648–1654 (2013)
35. C. Moreno-Castilla, Carbon **42**, 83–94 (2004)
36. A. Kutana, K.P. Giapis, J. Phys. Chem. C **113**, 14721–14726 (2009)
37. M.J. Allen, V.C. Tung, R.B. Kaner, Chem. Rev. **110**, 132–145 (2010)
38. C.N.R. Rao, A.K. Sood, K.S. Subrahmanyam, A. Govindaraj, Angew. Chem. Int. Ed. **48**, 7752–7777 (2009)
39. O. Bakajin, A. Noy, F. Fornasiero, C.P. Grigoropoulos, J.K. Holt, J.B. In, S. Kim, H.G. Park, in *Nanotechnology Applications for Clean Water*, ed. by M.D. Nora Savage, Jeremiah Duncan, Anita Street, Richard Sustich (William Andrew, Norwich, NY, 2009)
40. A. Krishnan, E. Dujardin, M.M.J. Treacy, J. Hugdahl, S. Lynum, T.W. Ebbesen, Nature **388**, 451–545 (1997)
41. Y. Liu, B.I. Yakobson, Nano Lett. **10**, 2178–2183 (2010)
42. E.S. Penev, V.I. Artyukhov, F. Ding, B.I. Yakobson, Adv. Mater. **24**, 4956–4976 (2012)
43. T.S. Sreeprasad, S.M. Maliyekkal, K.P. Lisha, T. Pradeep, J. Hazard. Mater. **186**, 921–931 (2011)
44. Z. Niu, J. Chen, H.H. Hng, J. Ma, X. Chen, Adv. Mater. **24**, 4143–4143 (2012)
45. V. Singh, D. Joung, L. Zhai, S. Das, S.I. Khondaker, S. Seal, Progr. Mater. Sci. **56**, 1178–1271 (2011)
46. T. Ando, NPG Asia Mater. **1**, 17–21 (2009)
47. Y.M. Blanter, I. Martin, Phys. Rev. **B76**, 155433 (2007) (6 pages)
48. G. Giovannetti, P.A. Khomyakov, G. Brocks, V.M. Karpan, J. van den Brink, P.J. Kelly, Phys. Rev. Lett. **101**, 026803 (2008) (4 pages)
49. V.M. Karpan, G. Giovannetti, P.A. Khomyakov, M. Talanana, A.A. Starikov, M. Zwierzycki, J. van den Brink, G. Brocks, P.J. Kelly, Phys. Rev. Lett. **99**, 176602 (2007) (4 pages)
50. Y.S. Dedkov, A.M. Shikin, V.K. Adamchuk, S.L. Molodtsov, C. Laubschat, A. Bauer, G. Kaindl, Phys. Rev. B **64**, 035405 (2001) (6 pages)
51. G. Giovannetti, P.A. Khomyakov, G. Brocks, P.J. Kelly, J. van den Brink, Phys. Rev. B **76**, 073103 (2007) (4 pages)
52. A.V. Krasheninnikov, P.O. Lehtinen, A.S. Foster, P. Pyykkö, R.M. Nieminen, Phys. Rev. Lett. **102**, 126807 (2009) (4 pages)
53. M. Wu, E.-Z. Liu, M.Y. Ge, J.Z. Jiang, Appl. Phys. Lett. **94**, 102505 (2009) (3 pages)
54. K.T. Chan, J.B. Neaton, M.L. Cohen, Phys. Rev. B **77**, 235430 (2008) (12 pages)

55. I. Gierz, C. Riedl, U. Starke, C.R. Ast, K. Kern, Nano Lett. **8**, 4603–4607 (2008)
56. A. Ishii, M. Yamamoto, H. Asano, K. Fujiwara, J. Phys. Conf. Ser. **100**, 052087 (2008) (4 pages)
57. P.O. Lehtinen, A.S. Foster, A. Ayuela, A. Krasheninnikov, K. Nordlund, R.M. Nieminen, Phys. Rev. Lett. **91**, 017202 (2003) (4 pages)
58. O. Leenaerts, B. Partoens, F.M. Peeters, Microelectron. J. **40**, 860–862 (2009)
59. H. Şahin, R.T. Senger, Phy. Rev. B **78**, 205423 (2008) (8 pages)
60. A.Z. Alzahrani, Nanotechnology and nanomaterials, in *Graphene Simulation*, ed. by J.R. J. R. (InTech, Rijeka, 2011)
61. Y.L.F. Musico, C.M. Santos, M.L.P. Dalida, D.F. Rodrigues, J. Mater. Chem. A **1**, 3789–3796 (2013)
62. A. Fathalian, J. Jalilian, S. Shahidi, Phys. B Condens. Matter **417**, 75–78 (2013)
63. J. Nakamura, J. Ito, A. Natori, J. Phys. Conf. Ser. **100**, 052019 (2008) (4 pages)
64. J. Ito, J. Nakamura, A. Natori, J. Appl. Phys. **103**, 113712 (2008) (5 pages)
65. B. Uchoa, C.Y. Lin, A.H. Castro Neto, Phys. Rev. B **77**, 035420 (2008) (5 pages)
66. V.M. Bermudez, J.T. Robinson, Langmuir **27**, 11026 (2011) (11 pages)
67. H. Sevinçli, M. Topsakal, E. Durgun, S. Ciraci, Phys. Rev. B **77**, 195434 (2008) (7 pages)
68. E. Aktürk, C. Ataca, S. Ciraci, Appl. Phys. Lett. **96**, 123112 (2010) (3 pages)
69. A.Z. AlZahrani, Appl. Surf. Sci. **257**, 807–810 (2010)
70. R.R. Nair, I.-L. Tsai, M. Sepioni, O. Lehtinen, J. Keinonen, A.V. Krasheninnikov, A.H.C. Neto, M.I. Katsnelson, A.K. Geim, I.V. Grigoriev, *Nature Communications, 1-15* (2013)
71. D. Pesin, Nat. Mater. **11**, 409–416 (2012)
72. N. Tombros, C. Jozsa, M. Popinciuc, H. T. Jonkman, B. J. v. Wees, Nature **448**, 571–574, (2007)
73. C. Józsa, M. Popinciuc, N. Tombros, H.T. Jonkman, B.J. van Wees, Phys. Rev. B **79**, 081402 (2009) (4 pages)
74. R.R. Nair, M. Sepioni, I.-L. Tsai, O. Lehtinen, J. Keinonen, A.V. Krasheninnikov, T. Thomson, A.K. Geim, I.V. Grigorieva, Nat. Phys. **8**, 199–202 (2011)
75. K.M. McCreary, A.G. Swartz, W. Han, J. Fabian, R.K. Kawakami, Phys. Rev. Lett. **109**, 186604 (2012) (5 pages)
76. P.O. Lehtinen, A.S. Foster, Y. Ma, A.V. Krasheninnikov, R.M. Nieminen, Phys. Rev. Lett. **93**, 187202 (2004) (4 pages)
77. V.M. Pereira, F. Guinea, J.M.B. Lopes dos Santos, N.M.R. Peres, A.H. Castro Neto, Phys. Rev. Lett. **96**, 036801 (2006) (4 pages)
78. O.V. Yazyev, L. Helm, Phys. Rev. B **75**, 125408 (2007) (5 pages)
79. B. Uchoa, V.N. Kotov, N.M.R. Peres, A.H. Castro Neto, Phys. Rev. Lett. **101**, 026805 (2008) (4 pages)
80. J.J. Palacios, J. Fernández-Rossier, L. Brey, Phys. Rev. B **77**, 195428 (2008) (14 pages)
81. P. Haase, S. Fuchs, T. Pruschke, H. Ochoa, F. Guinea, Phys. Rev. B **83**, 241408 (2011) (4 pages)
82. J.J. Palacios, F. Ynduráin, Phys. Rev. B **85**, 245443 (2012) (8 pages)
83. E.J.G. Santos, S. Riikonen, D. Sánchez-Portal, A. Ayuela, J. Phys. Chem. C **116**, 7602–7606 (2012)
84. C.N.R. Rao, K.S. Subrahmanyam, H.S.S.R. Matte, A. Govindaraj, in *Graphene and Its Fascinating Attributes*, ed. by S.K. Pati, T. Enoki, C.N.R. Rao (World Scientific, Singapore, 2011), p. 287
85. T. Enoki, V.L. J. Joly, K. Takai, in *Graphene and its Fascinating Attributes*, ed. by S.K. Pati, T. Enoki, C.N.R. Rao (World Scientific, Singapore, 2011)
86. Y.G. Zhou, X.T. Zu, F. Gao, H.F. Lv, H.Y. Xiao, Appl. Phys. Lett. **95**, 123119 (2009) (3 pages)
87. W. Li, M. Zhao, Y. Xia, R. Zhang, Y. Mu, J. Mater. Chem. **19**, 9274–9282 (2009)
88. O. Leenaerts, B. Partoens, F.M. Peeters, Phys. Rev. B **77**, 125416 (2008)

89. P. Giannozzi, R. Car, G. Scoles, J. Chem. Phys. **118**, 1003–1006 (2003)
90. J. Dai, P. Giannozzi, J. Yuan, Surf. Sci. **603**, 3234–3238 (2009)
91. J. Dai, J. Yuan, Phys. Rev. B **81**, 165414 (2010) (7 pages)
92. Y.-M. Lin, A.J. Keith, V.-G. Alberto, P.S. Joshua, B.F. Damon, A. Phaedon, Nano Lett. **9**, 422–426 (2008)
93. Y. Wu, Y.-m. Lin, A. A. Bol, K. A. Jenkins, F. Xia, D. B. Farmer, Y. Zhu, and P. Avouris. Nature **472**, 74–78 (2011)
94. M. Freitag, T. Low, F. Xi, P. Avouris, Nat. Photon. **7**, 53–59 (2013)
95. R. Ulbricht, E. Hendry, J. Shan, T.F. Heinz, M. Bonn, Rev. Mod. Phys. **83**, 543–586 (2011)
96. X. Li, W. Cai, J. An, S. Kim, J. Nah, D. Yang, R. Piner, A. Velamakanni, I. Jung, E. Tutuc, S.K. Banerjee, L. Colombo, R.S. Ruoff, Science **324**, 1312–1314 (2009)
97. H. Wang, J.H. Strait, P.A. George, S. Shivaraman, V.B. Shields, M. Chandrashekhar, J. Hwang, F. Rana, M.G. Spencer, C.S. Ruiz-Vargas, J. Park, Appl. Phys. Lett. **96**, 081917 (2010) (3 pages)
98. P.A. George, J. Strait, J. Dawlaty, S. Shivaraman, M. Chandrashekhar, F. Rana, M.G. Spencer, Nano Lett. **8**, 4248–4251 (2008)
99. P.J. Hale, S.M. Hornett, J. Moger, D.W. Horsell, E. Hendry, Phys. Rev. B **83**, 121404 (2011) (4 pages)
100. V. Ryzhii, M. Ryzhii, T. Otsuji, J. Appl. Phys. **101**, 083114 (2007) (4 pages)
101. Y. Shi, W. Fang, K. Zhang, W. Zhang, L.-J. Li, Small **5**, 2005–2011 (2009)
102. J. Berashevich, T. Chakraborty, Phys. Rev. B **80**, 033404 (2009) (4 pages)
103. Y. Sato, K. Takai, T. Enoki, Nano Lett. **11**, 3468–3475 (2011)
104. J. Zhou, Q. Wang, Q. Sun, P. Jena, X.S. Chen, Proc. Natl. Acad. Sci. **107**, 2801–2806 (2010)
105. P. Saalfrank, J. Chem. Phys. **113**, 3780–3791 (2000)
106. W. Liu, Y.H. Zhao, J. Nguyen, Y. Li, Q. Jiang, E.J. Lavernia, Carbon **47**, 3452–3460 (2009)
107. Y.-H. Lu, L. Shi, C. Zhang, Y.-P. Feng, Phys. Rev. B **80**, 081601 (2009) (4 pages)
108. X. Zhao, X. Zhai, A. Zhao, B. Wang, J.G. Hou, Appl. Phys. Lett. **102**, 201602 (2013) (4 pages)
109. R. Ifuku, K. Nagashio, T. Nishimura, A. Toriumi, J. Appl. Phys. **103**, 1–5 (2013)
110. A.M. Suarez, L.R. Radovic, E. Bar-Ziv, J.O. Sofo, Phys. Rev. Lett. **106**, 146802 (2011) (4 pages)
111. B.K. Daas, W.K. Nomani, K.M. Daniels, T.S. Sudarshan, G. Koley, M.V.S. Chandrashekhar, *Journal of Materials Science, 1-4*, 2012
112. M. Gautam, A.H. Jayatissa, J. Appl. Phys. **111**, 9094317 (2012) (8 pages)
113. M. Qazi, T. Vogt, G. Koley, Appl. Phys. Lett. **91**, 233101 (2007) (3 pages)
114. J.M. Dawlaty, S. Shivaraman, J. Strait, P. George, M. Chandrashekhar, F. Rana, M.G. Spencer, D. Veksler, Y. Chen, Appl. Phys. Lett. **93**, 131905 (2008) (3 pages)
115. V. Chandra, J. Park, Y. Chun, J.W. Lee, I.-C. Hwang, K.S. Kim, ACS Nano **4**, 3979–3986 (2010)
116. J. Zhu, R. Sadu, S. Wei, D.H. Chen, N. Haldolaarachchige, Z. Luo, J.A. Gomes, D.P. Young, Z. Guo, ECS J. Solid State Sci. Technol. **1**, M1–M5 (2012)
117. X. Luo, C. Wang, L. Wang, F. Deng, S. Luo, X. Tu, C. Au, Chem. Eng. J. **220**, 98–106 (2013)
118. L. Fan, C. Luo, M. Sun, X. Li, H. Qiu, Colloids Surf. B Biointerfaces **103**, 523–529 (2013)
119. M. Machida, T. Mochimaru, H. Tatsumoto, Carbon **44**, 2681–2688 (2006)
120. W. Wu, Y. Yang, H. Zhou, T. Ye, Z. Huang, R. Liu, Y. Kuang, Water Air Soil Pollut. **224**, 1 (2012) (9 pages)
121. X.-J. Hu, Y.-G. Liu, H. Wang, A.-w. Chen, G.-M. Zeng, S.-M. Liu, Y.-M. Guo, X. Hu, T.-T. Li, Y.-Q. Wang, L. Zhou, S.-H. Liu, Separ. Purif. Technol. **108**, 189–195, (2013).
122. J.-H. Deng, X.-R. Zhang, G.-M. Zeng, J.-L. Gong, Q.-Y. Niu, J. Liang, Chem. Eng. J. **226**, 189–200 (2013)
123. X. Deng, L. Lü, H. Li, F. Luo, J. Hazard. Mater. **183**, 923–930 (2010)
124. R. Sitko, E. Turek, B. Zawisza, E. Malicka, E. Talik, J. Heimann, A. Gagor, B. Feist, R. Wrzalik, Dalton Trans. **42**, 5682–5689 (2013)

125. Y. Wang, S. Liang, B. Chen, F. Guo, S. Yu, Y. Tang, PLoS ONE **8**, e65634 (2013) (18 pages)
126. H. Tokuyama, J. Hisaeda, S. Nii, S. Sakohara, Sep. Purif. Technol. **71**, 83–88 (2010)
127. W. Zhang, X. Shi, Y. Zhang, W. Gu, B. Li, Y. Xian, J. Mater. Chem. A **1**, 1745–1753 (2013)
128. J. Li, S. Zhang, C. Chen, G. Zhao, X. Yang, J. Li, X. Wang, ACS Appl. Mater. Interfaces **4**, 4991–5000 (2012)
129. Y. Zhang, H.-L. Ma, J. Peng, M. Zhai, Z.-Z. Yu, J. Mater. Sci. **48**, 1883–1889 (2013)
130. Y. Wu, H. Luo, H. Wang, C. Wang, J. Zhang, Z. Zhang, J. Colloid Interface Sci. **394**, 183–191 (2013)
131. Y. Ren, N. Yan, Q. Wen, Z. Fan, T. Wei, M. Zhang, J. Ma, Chem. Eng. J. **175**, 1–7 (2011)
132. V. Chandra, K.S. Kim, Chem. Commun. **47**, 3942–3944 (2011)
133. Z.-Q. Zhao, X. Chen, Q. Yang, J.-H. Liu, X.-J. Huang, Chem. Commun. **48**, 2180–2182 (2012)
134. W. Song, J. Hu, Y. Zhao, D. Shao, J. Li, RSC Adv. **3**, 9514–9521 (2013)
135. M. Liu, C. Chen, J. Hu, X. Wu, X. Wang, J. Phys. Chem. C **115**, 25234 (2011) (7 pages)
136. Y. Leng, W. Guo, S. Su, C. Yi, L. Xing, Chem. Eng. J. **211–212**, 406–411 (2012)
137. Y.D. Guosheng Shi, Haiping Fang. J. Comput. Chem. **33**, 1328–1337 (2012)
138. V. Tomar, D. Kumar, Chem. Cent. J **7**, 51 (2013) (15 pages)
139. S. Vasudevan, J. Lakshmi, RSC Adv. **2**, 5234–5242 (2012)
140. H.S. Shaobin Wang, H.M. Ang, M.O. Tadé, Chem. Eng. J. **226**, 336–347 (2013)
141. J. Xu, L. Wang, Y. Zhu, Langmuir **28**, 8418–8425 (2012)
142. J. Liu, H. Cao, J. Xiong, Z. Cheng, CrystEngComm **14**, 5140–5144 (2012)
143. P. Shi, R. Su, S. Zhu, M. Zhu, D. Li, S. Xu, J. Hazard. Mater. **229–230**, 331–339 (2012)
144. P. Sharma, M.R. Das, J. Chem. Eng. Data **58**, 151–158 (2012)
145. C. Junyong, H. Yongmei, L. Yan, G. Jiajia, RSC Adv. **3**, 7254–7258 (2013)
146. C. Wang, C. Feng, Y. Gao, X. Ma, Q. Wu, Z. Wang, Chem. Eng. J. **173**, 92–97 (2011)
147. S.-T. Yang, S. Chen, Y. Chang, A. Cao, Y. Liu, H. Wang, J. Colloid Interface Sci. **359**, 24–29 (2011)
148. Y. Li, Q. Du, T. Liu, J. Sun, Y. Wang, S. Wu, Z. Wang, Y. Xia, L. Xia, Carbohydr. Polym. **95**, 501–507 (2013)
149. L. Ai, C. Zhang, Z. Chen, J. Hazard. Mater. **192**, 1515–1524 (2011)
150. Q. Wu, C. Feng, C. Wang, Z. Wang, Colloids Surf. B Biointerfaces **101**, 210–214 (2013)
151. P.R. Wallace, Phys. Rev. **71**, 622–634 (1947)
152. J. Tyagi, R. Kakkar, in *Advanced Materials*, ed. by A. Tiwari (University of Delhi, Delhi, India, 2013)
153. C. Petit, T.J. Bandosz, Micropor. Mesopor. Mater. **118**, 61–67 (2009)
154. H.E. Romero, P. Joshi, A.K. Gupta, H.R. Gutierrez, M.W. Cole, S.A. Tadigadapa, P.C. Eklund, Nanotechnology **20**, 245501 (2009) (8 pages)
155. Y. Peng, J. Li, Front. Environ. Sci. Eng. **7**, 403–411 (2013)
156. S. Tang, Z. Cao, J. Phys. Chem. C **116**, 8778–8791 (2012)
157. C. Petit, M. Seredych, T.J. Bandosz, J. Mater. Chem. **19**, 9176–9185 (2009)
158. M. Seredych, J.A. Rossin, T.J. Bandosz, Carbon **49**, 4392–4402 (2011)
159. S. Chen, W. Cai, D. Chen, Y. Ren, X. Li, Y. Zhu, J. Kang, R.S. Ruof, *New Journal of Physics 12* (2010)
160. C. Petit, T.J. Bandosz, J. Phys. Chem. C **113**, 3800–3809 (2009)
161. M. Seredych, S. Bandosz, J. Teresa, Mater. Chem. Phys. **117**, 99–106 (2009)
162. Y. Ren, C. Zhu, W. Cai, H. Li, H. Ji, I. Kholmanov, Y. Wu, R.D. Piner, R.S. Ruoff, Appl. Phys. Lett. **100**, 4 (2012)
163. M. Seredych, T.J. Bandosz, J. Phys. Chem. C **114**, 14552 (2010) (9 pages)
164. Y.-H. Zhang, L.-F. Han, Y.-H. Xiao, D.-Z. Jia, Z.-H. Guo, F. Li, Comput. Mater. Sci. **69**, 222–228 (2013)
165. A.K. Mishra, S. Ramaprabhu, AIP Adv. **1**, 032152 (2011) (6 pages)
166. H.-Y. Kim, K. Lee, N. McEvoy, C. Yim, G.S. Duesberg, Nano Lett. **13**, 2182–2188 (2013)
167. C.-C. Chen, M. Aykol, C.-C. Chang, A.F.J. Levi, S.B. Cronin, Nano Lett. **11**, 1863–1867 (2011)

168. S. Tongay, T. Schumann, X. Miao, B.R. Appleton, A.F. Hebard, Carbon **49**, 2033–2038 (2011)
169. H. Yang, J. Heo, S. Park, H.J. Song, D.H. Seo, K.-E. Byun, P. Kim, I. Yoo, H.-J. Chung, K. Kim, Science **336**, 1140–1143 (2012)
170. K. Chen, Z.-L. Zhang, Y.-M. Liang, W. Liu, Sensors **13**, 6204–6216 (2013)
171. R.J. Forster, P. Bertoncello, T.E. Keyes, Annu. Rev. Anal. Chem. **2**, 359–385 (2009)
172. J. Liu, Y. Chen, Y. Guo, F. Yang, F. Cheng, J. Nanomater. **2013**, 632809 (2013) (6 pages)
173. A.G. Davies, A.D. Burnett, W. Fan, E.H. Linfield, J.E. Cunningham, Mater. Today **11**, 18–26 (2008)
174. D. Johnson, *in IEEE Spectrum* (Spain, IEEE Spectrum, Madrid, 2013)
175. M. Freitag, T. Low, W. Zhu, H. Yan, F. Xia, P. Avouris, *Nature Communications* ***4***, 2013
176. http://newscenter.lbl.gov/news-releases/2011/09/04/graphene-thz/. Accessed 31 Aug 2013
177. L. Ju, B. Geng, J. Horng, C. Girit, M. Martin, Z. Hao, H.A. Bechtel, X. Liang, A. Zettl, Y.R. Shen, F. Wang, Nat. Nanotechnol. **6**, 630–634 (2011)
178. R.R. Nair, P. Blake, A.N. Grigorenko, K.S. Novoselov, T.J. Booth, T. Stauber, N.M.R. Peres, A.K. Geim, Science **320**, 1308–1308 (2008)
179. http://www.gadgetreview.com/2013/05/new-image-sensor-made-from-graphene-could-revolutionize-cameras.html. Accessed 31 Aug 2013
180. M.Y. Han, B. Özyilmaz, Y. Zhang, P. Kim, Phys. Rev. Lett. **98**, 206805 (2007) (4 pages)
181. J. Bai, X. Duan, Y. Huang, Nano Lett. **9**, 2083–2087 (2009)
182. J. Liying, Z. Li, W. Xinran, D. Georgi, D. Hongjie, Nature **458**, 877–880 (2009)
183. J. Bai, X. Zhong, S. Jiang, Y. Huang, X. Duan, Nat. Nanotechnol. **5**, 190–194 (2010)
184. J.-H. Chen, W.G. Cullen, C. Jang, M.S. Fuhrer, E.D. Williams, Phys. Rev. Lett. **102**, 236805 (2009) (4 pages)
185. M.J. Allen, V.C. Tung, R.B. Kaner, Chem. Rev. **110**, 132–145 (2009)
186. B.R. Stoner, A.S. Raut, B. Brown, C.B. Parker, J.T. Glass, Appl. Phys. Lett. **99**, 183104 (2011) (3 pages)
187. J. Wang, Electroanalysis **17**, 7–14 (2005)
188. K. Balasubramanian, M. Burghard, Anal. Bioanal. Chem. **385**, 452–468 (2006)
189. A.F.S. Capone, L. Francioso, R. Rella, P. Siciliano, J. Spadavecchia, D.S. Presicce, A.M. Taurino, J. Optoelectron. Adv. Mater **5**, 1335–1348 (2003)
190. P.T. Moseley, Meas. Sci. Technol **8**, 223–237 (1997)
191. Z. Yuanbo, T. Yan-Wen, L.S. Horst, K. Philip, Nature **438**, 201–204 (2005)
192. S. Alwarappan, R.K. Joshi, M.K. Ram, A. Kumar, Appl. Phys. Lett. **96**, 263702 (2010) (3 pages)
193. A. N. Pal, A. Ghosh, in *Graphene and Its Fascinating Attributes*, ed. by T. E. Swapan K. Pati, C.N.R. Rao (World Scientific, Singapore, 2011)
194. A. Ghosh, D.J. Late, L.S. Panchakarla, A. Govindaraj, C.N.R. Rao, J. Exp. Nanosci. **4**, 313–322 (2009)
195. C. Berger, Z. Song, X. Li, X. Wu, N. Brown, C. Naud, D. Mayou, T. Li, J. Hass, A.N. Marchenkov, E.H. Conrad, P.N. First, W.A. de Heer, Science **312**, 1191–1196 (2006)
196. V.C. Tung, M.J. Allen, Y. Yang, R.B. Kaner, Nat. Nanotechnol. **4**, 25–29 (2009)
197. R.V. Noorden, *The graphene challenge*, 2008
198. H. Liu, Y. Liu, D. Zhu, J. Mater. Chem. **21**, 3335–3345 (2011)
199. Z.M. Ao, J. Yang, S. Li, Q. Jiang, Chem. Phys. Lett **461**, 276–279 (2008)
200. M. Chi, Y.-P. Zhao, Comput. Mater. Sci **46**, 1085–1090 (2009)
201. J.T. Robinson, F.K. Perkins, E.S. Snow, Z. Wei, P.E. Sheehan, Nano Lett. **8**, 3137–3140 (2008)
202. C.E. Kehayias, S. MacNaughton, S. Sonkusale, C. Staii, Mesosc. Nanosc. Phys **24**, 245502 (2013) (13 pages)
203. H. Wang, K. Li, Y. Cheng, Q. Wang, Y. Yao, U. Schwingenschlogl, X. Zhang, W. Yang, Nanoscale **4**, 2920–2925 (2012)
204. A.D. Alhaidari, H. Bahlouli, A. Mouhafid, A. Jellal, Eur. Phys. J. B Condens. Matter **86**, 1–12 (2013)

205. S. Some, Y. Xu, Y. Kim, Y. Yoon, H. Qin, A. Kulkarni, T. Kim, H. Lee, Nature **3**, 01868 (2013) (8 pages)
206. U. Yogeswaran, S. Thiagarajan, S.-M. Chen, Sensors **8**, 7191–7212 (2008)
207. M. Pumera, Chem. Rec. **9**, 211–223 (2009)
208. M. Pumera, Chem. Soc. Rev. **39**, 4146–4157 (2010)
209. D.A.C. Brownson, C.E. Banks, Analyst **135**, 2768–2778 (2010)
210. K.V. Sreekanth, S. Zeng, K.-T. Yong, T. Yu, Sens. Actuators B **182**, 424–428 (2013)
211. M. Dankerl, M.V. Hauf, A. Lippert, L.H. Hess, S. Birner, I.D. Sharp, A. Mahmood, P. Mallet, J.-Y. Veuillen, M. Stutzmann, J.A. Garrido, Adv. Funct. Mater. **20**, 3117–3124 (2010)
212. N. Mohanty, V. Berry, Nano Lett. **8**, 4469–4476 (2008)
213. K. Yang, J. Wan, S. Zhang, Y. Zhang, S.-T. Lee, Z. Liu, ACS Nano **5**, 516–522 (2010)
214. X. Li, G. Wang, X. Ding, Y. Chen, Y. Gou, Y. Lu, Phys. Chem. Chem. Phys. **15**, 12800–12804 (2013)
215. M. Zhou, Y. Zhai, S. Dong, Anal. Chem. **81**, 5603–5613 (2009)
216. F.J. Hernandez, V.C. Ozalp, Biosensors **2**, 1–14 (2012)
217. X. Dong, W. Huang, P. Chen, Nanoscale Res. Lett. **6**, 60 (2011) (6 pages)
218. A.L. Linsebigler, G. Lu, J.T. Yates, Chem. Rev. **95**, 735–758 (1995)
219. M. Umar, H.A. Aziz, Photocatalytic degradation of organic pollutants in water, in *Organic Pollutants—Monitoring, Risk and Treatment*, ed. by M. Nageeb Rashed (InTech, Rijeka, 2013)
220. F. Akira, H. Kenichi, Nature **238**, 37–38 (1972)
221. K. Hashimoto, H. Irie, F. Akira, Jpn. J. Appl. Phys **44**, 8269–8285 (2005)
222. W.-J. Yin, H. Tang, S.-H. Wei, M.M. Al-Jassim, J. Turner, Y. Yan, Phys. Rev. B **82**, 045106 (2010) (6 pages)
223. Y. Lin, Z. Jiang, C. Zhu, X. Hu, X. Zhang, H. Zhu, J. Fan, Appl. Phys. Lett. **101**, 062106 (2012). 5 pages)
224. Q. Li, B. Guo, J. Yu, J. Ran, B. Zhang, H. Yan, J.R. Gong, J. Am. Chem. Soc. **133**, 10878–10884 (2011)
225. J.C. Tristão, F. Magalhães, P. Corio, M.T.C. Sansiviero, J. Photochem. Photobiol. A Chem. **181**, 152–157 (2006)
226. G. Peng, L. Jincheng, L. Siewsiang, Z. Tong, S.D. Delai, J. Mater. Chem. Phys **22**, 2292–2298 (2011)
227. X. Lan, W. Yilong, Y. Shuili, T. Yulin, Prog. Chem. **25**, 419–430 (2013)
228. W. Tu, Y. Zhou, Z. Zou, Adv. Funct. Mater. **23**, 4996–5008 (2013)
229. S. Shet, Y. Yan, H. Wang, N. Ravindra, J. Turner, M. Al-Jassim, Emerg. Mater. Res **1**, 201–204 (2012)
230. S. Shet, K.-S. Ahn, Y. Yan, T. Deutsch, K.M. Chrustowski, J. Turner, M. AlJassim, N. Ravindra, J. Appl. Phys. **103**, 073504–073505 (2008)
231. S. Shet, K.-S. Ahn, T. Deutsch, H. Wang, N. Ravindra, Y. Yan, J. Turner, M. Al-Jassim, J. Mater. Res. **25**, 69–75 (2010)
232. Y. Yang, L. Ren, C. Zhang, S. Huang, T. Liu, ACS Appl. Mater. Interfaces **3**, 2779–2785 (2011)
233. T. Xu, L. Zhang, H. Cheng, Y. Zhu, Appl. Catal. Environ. **101**, 382–387 (2011)
234. V. Štengl, S. Bakardjieva, T.M. Grygar, J. Bludská, M. Kormunda, *Chemistry Central Journal* *7*, 2013
235. Y. Liang, H. Wang, H. Sanchez Casalongue, Z. Chen, H. Dai, Nano Res. **3**, 701–705 (2010)
236. K. Gopalakrishnan, H.M. Joshi, P. Kumar, L.S. Panchakarla, C.N.R. Rao, Chem. Phys. Lett **511**, 304–308 (2011)
237. E. Gao, W. Wang, M. Shang, J. Xu, Phys. Chem. Chem. Phys. **13**, 2887–2893 (2011)
238. S. Gawande, S. Thakare, Int. Nano Lett **2**, 11–18 (2012)
239. F. Zhou, Y. Zhu, J. Adv. Ceram. **1**, 72–77 (2012)
240. L. Sun, R. Shao, L. Tang, Z. Chen, J. Alloys Compd. **564**, 55–62 (2013)
241. Y. Fu, X. Wang, Ind. Eng. Chem. Res. **50**, 7210–7218 (2011)

242. Y. Fu, H. Chen, X. Sun, X. Wang, Appl. Catal. Environ. **111–112**, 280–287 (2012)
243. N. Zhang, M.-Q. Yang, Z.-R. Tang, Y.-J. Xu, J. Catal. **303**, 60–69 (2013)
244. F. Wang, K. Zhang, Curr. Appl. Phys **12**, 346–352 (2012)
245. X. Liu, L. Pan, T. Lv, G. Zhu, T. Lu, Z. Sun, C. Sun, RSC Adv. **1**, 1245–1249 (2011)
246. K.C. Kemp, H. Seema, M. Saleh, N.H. Le, K. Mahesh, V. Chandra, K.S. Kim, Nanoscale **5**, 3149–3171 (2013)
247. K.S. Kim, Y. Zhao, H. Jang, S.Y. Lee, K.S. Kim, J.-H. Ahn, P. Kim, J.-Y. Choi, B.H. Hong, Nature **457**, 706–710 (2009)
248. S. Stankovich, D.A. Dikin, R.D. Piner, K.A. Kohlhaas, A. Kleinhammes, Y. Jia, Y. Wu, S.T. Nguyen, R.S. Ruoff, Carbon **45**, 1558–1565 (2007)
249. http://web.mit.edu/newsoffice/2012/graphene-pores-and-membranes-1023.html. Accessed 12 Jan 2014
250. N. Savage, Nature **483**, S30-S31 (10.1038/483S30a), (2012)

# Chapter 7
# Magnetically Recyclable Nanomaterials for Water Treatment

**Tim Leshuk and Frank Gu**

**Abstract** Nanoparticle colloidal dispersions are highly promising for use in water purification, but practical and cost-effective options to separate the dispersed nanoparticles from the treated water remain a critical roadblock to industrial adoption. Magnetic separation of superparamagnetic nanoparticles from water for recycling and reuse has the potential to be an efficient, practical, and low-cost slurry-type water treatment method. In this chapter we review the fundamental theory and concepts of magnetic nanoparticle separations, and present both a synthesis process for size-tunable superparamagnetic iron oxide nanospheres, as well as the application of these nanospheres as a core material for the immobilization of $TiO_2$, to be used in photocatalytic water treatment as magnetically recyclable composite particles.

## 7.1 Introduction

Nanotechnologies based on magnetic separations show huge promise in the field of water treatment and purification, and magnetic nanomaterials are being heavily researched for scrubbing water of a wide variety of contaminants. This chapter is primarily focused on the magnetic separation or recovery of nanoscale particles from a colloidal dispersion or slurry, and their use in water purification, with the potential for recyclability. Slurry-type water treatment technologies involving

T. Leshuk • F. Gu (✉)
Department of Chemical Engineering, University of Waterloo, 200 University Avenue West, Waterloo, ON, Canada N2L 3G1

Waterloo Institute for Nanotechnology, University of Waterloo, 200 University Avenue West, Waterloo, ON, Canada N2L 3G1
e-mail: frank.gu@uwaterloo.ca

A. Hu and A. Apblett (eds.), *Nanotechnology for Water Treatment and Purification*, Lecture Notes in Nanoscale Science and Technology 22,
DOI 10.1007/978-3-319-06578-6_7, 

colloidal particles show distinct advantages compared with other treatment methodologies, such as the use of membranes, flocculation, or chemical methods such as chlorination. The concept of a slurry is the dispersion of (nano)particles throughout the volume of contaminated water to be treated, purifying the water using these dispersed particles, and then recovering them afterwards from the treated water (potentially for reuse/recycling of the particles in subsequent rounds of treatment). Such a nano-colloid allows a huge specific surface area to be mobilized and dispersed throughout the water volume in an energy-efficient manner, allowing for easy mixing of the particles with the polluted water (due to the enhanced diffusion coefficients of fine particles) and efficient mass transfer with the target contaminants, which has the potential to exhibit superior treatment kinetics at lower applied energy and cost compared to membrane technology for example. If the nanoparticles can be reused multiple times, this only adds to the cost savings. Nanoparticles in a slurry formulation can treat water through adsorption or chelation of contaminants, catalytic degradation, flocculation, or other means dependent on accessible surface area of the colloidal particles. Furthermore, magnetic nanoparticles can be used as the support or substrate for any conventional water treatment materials; similarly, composite materials containing a magnetic component can be used as magnetically active particles.

An obvious and critical concern when using nanoparticle slurries however is the difficulty in fully recovering the nanoparticles from the water in a cost-efficient manner. Indeed, due to concerns of nanotoxicology or environmental release of "nanopollutants," it is imperative to avoid leaving any residual nanoparticles in the final treated water, which is why having a fast and cost-effective method to recover the nanoparticles from dispersion is so important. The essential concept of magnetic separation is that when a magnet or magnetic field gradient is applied to a slurry of magnetic nanoparticles in the treated water, the particles will move towards the magnet for collection, resulting in nanoparticle-free treated water for discharge. Using permanent magnets for the separation step is also low energy and cost efficient. Thus the magnetic separation paradigm has the potential to enable efficient and cheap water purification, with the ability to address many of the same water contaminants treated by conventional means, but without requiring the use of chemical additives (which can be relatively expensive or leave degradation by-products in the treated water). Magnetic separations for water treatment purposes could be envisioned to be applied at an industrial or civic scale, for the purification of industrial effluent or drinking water; for environmental cleanup; in niche applications such as aerospace; or in point-of-use processes where more advanced water treatment infrastructure is nonexistent. Indeed, it should be emphasized that magnetic separation technology has already been extensively employed in industry (e.g., mining, food quality control) [1, 2]. From a research perspective, the benefits of magnetic separation technology simply need to be compelling enough to justify the costs of implementing such infrastructure in an industrial water treatment context.

In this chapter, we begin our discussion with the physical concepts of superparamagnetism and magnetic particle separation. A synthesis method for size-controlled, magnetically separable superparamagnetic nanospheres is then presented, followed by an application of these nanospheres as a substrate or core material for the immobilization of $TiO_2$, to be used in photocatalytic water treatment as magnetically recyclable composite particles.

## 7.2 Magnetophoresis: The Physics of Magnetic Particle Separation

Conceptually, the process of magnetic separation is easily envisaged—many of us have played with magnets and iron filings as children, and the principle seems clear and concrete: magnetic particles simply move towards a magnet. Indeed, several recent reviews on magnetically recyclable particles make a similar assumption that simply by virtue of their magnetic nature, magnetic particles will be automatically separated from dispersion upon application of a magnetic field [3, 4]. In actuality however, magnetic separation of particles from a suspension involves complex physics, and magnetic separability is far from guaranteed, but rather depends on the nature of the magnetic materials used, the concentration of particles and their colloidal stability, the magnitude of magnetic field gradients, among many other variables. The purpose of this section is to introduce some of the physics governing magnetophoretic phenomena, and considerations to be taken into account when designing magnetic separations.

### *7.2.1 Magnetism and Magnetization*

In the context of nanotechnology for water treatment, we are primarily interested in nanoparticles of magnetic materials, or nanomaterials and composites comprised of such particles, due to their small size and high specific surface area. Therefore the following discussion will relate primarily to ultrafine particles of magnetic materials, and is drawn from [5–8]. Due to considerations of brevity, only the fundamentals are presented herein, and readers are referred to specialized magnetic materials textbooks for a more complete description of the involved physics.

For magnetic separations in a water treatment, oxide nanoparticles are typically used, as fine metallic particles are highly reactive in aqueous environments and oxidize readily under environmental conditions, which would lead to variable or unpredictable magnetic properties as the particles oxidize over time. Indeed, this high reactivity is the basis of nano zero-valent iron (nZVI) in water purification, as reviewed elsewhere in this book. Iron oxides have been especially studied and implemented as magnetic nanomaterials due to economic considerations (iron is cheap and abundant, compared to cobalt for example, or rare earth elements), their

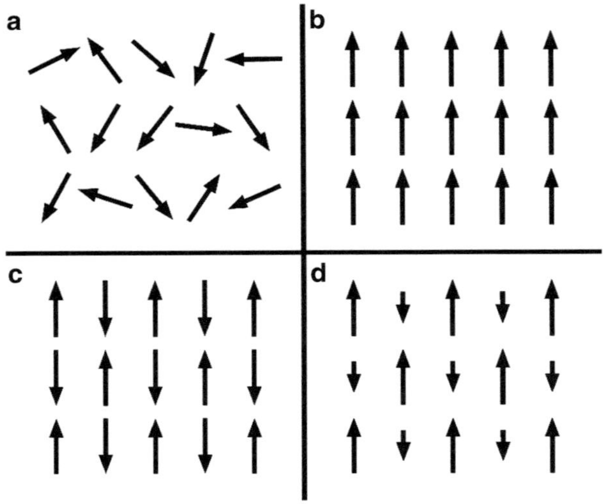


**Fig. 7.1** Simple schematic of the alignment of atomic magnetic domains in different magnetic materials: (**a**) paramagnetic, (**b**) ferromagnetic, (**c**) antiferromagnetic, and (**d**) ferrimagnetic. Figure adapted from [8]

relatively low toxicity and low environmental threat, along with their ease of synthesis and well-understood chemistry.

The iron ion possesses a strong magnetic moment due to the unpaired spin magnetic moments of its valence electrons. In crystals containing iron, these permanent atomic magnetic moments can align or orient themselves with respect to each other in various arrangements, giving rise to paramagnetic, ferromagnetic, antiferromagnetic, or ferrimagnetic materials (Fig. 7.1). In a paramagnetic material, the individual atomic magnetic moments are randomly oriented in the absence of a magnetic field, irrespective of temperature, giving the bulk material a net magnetic moment of zero. On application of an external magnetic field however, some of the atomic moments will align in the direction of the field, imparting a small net magnetic moment to the material, which disappears on removal of the external field. In a ferromagnetic crystal, the atomic magnetic moments are aligned parallel with each other even in the absence of an applied field (Fig. 7.1b), while in an antiferromagnetic crystal the moments are aligned but in opposite directions (Fig. 7.1c), canceling each other out and rendering a net zero magnetic moment to the material as a whole. A ferrimagnetic material is similar to an antiferromagnetic material, but where the antiparallel magnetic moments are of different magnitudes (Fig. 7.1d), and thus the material retains a net magnetic moment, behaving like ferromagnets.

Thermal energy typically disrupts alignment of these atomic magnetic moments, and the ordered alignment breaks down above a certain temperature, namely the Curie temperature $T_C$ for ferromagnetic and ferrimagnetic materials, or the Néel temperature $T_N$ for antiferromagnetic materials, above which temperatures the magnetic moments of the materials are randomly oriented like a paramagnetic material, retaining no net magnetic moment.

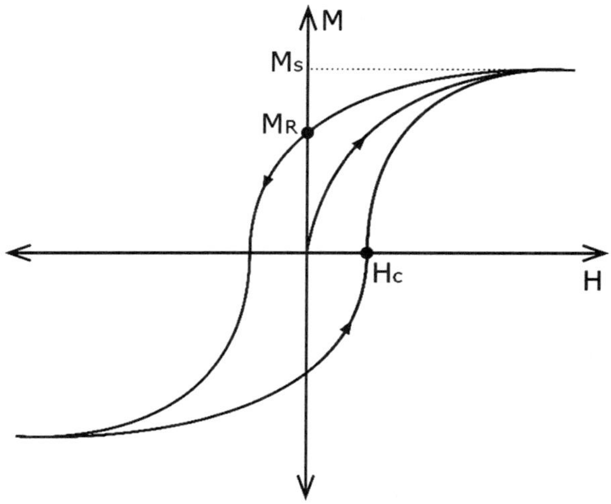


**Fig. 7.2** Magnetization curve of a typical ferromagnetic material, adapted from [5]

A bulk ferromagnetic material is comprised of multiple microscopic magnetic domains, where the atomic magnetic moments within each domain are aligned with each other (as in Fig. 7.1b), but not necessarily with adjacent domains. Crystal grain boundaries delineate magnetic domains in a polycrystalline material, although a single crystal grain can be composed of multiple magnetic domains.

The magnetization of a material $M$ is defined as its net magnetic moment per unit volume, and is the sum of the individual magnetic moments composing the material, per unit volume. On a bulk scale, the magnetization of a material is thus the sum of the magnetization of the constituent magnetic domains, while on the microscopic level of the domains, the magnetization is the sum of the atomic magnetic moments in that domain. Within a single domain, due to the alignment of all the atomic magnetic moments (in a ferromagnet), the magnetization of the material is saturated at a maximum value $M_s$ (in the direction of the alignment of the atomic moments), which is limited by the magnitude of the atomic moments and their arrangement and packing within the crystal lattice. Thus $M_s$ is a material-specific property, dependent on the atomic composition and crystal structure of the particular material. However, on the bulk scale again, the net magnetization of a ferromagnetic material can be zero, as all the domain magnetic moments within the material, despite being each magnetized at $M_s$, can be oriented in random directions, canceling each other out.

On application of an external magnetic field $H$, the magnetization of the material can be altered, producing a magnetization curve (Fig. 7.2). Starting from zero, the magnetization of a bulk ferromagnetic material increases with $H$ until a saturation value of $M_s$ is reached. This process corresponds to the reorientation and alignment of the individual domain moments composing the material in the direction of the externally applied field, converting a multi-domain material to effectively a single domain (alignment of all microscopic domains). Thus $M_s$ is naturally reached when all the domains, each magnetized at $M_s$, are oriented in the same direction. Thus magnetization of a material does not alter the magnitude of magnetization in any domain, but rather the direction of their magnetization.

However, when $H$ is removed, the magnetization of the material does not return to net zero, since the magnetic domains do not all return to their original random orientations, but rather remain aligned with each other in the previous direction of $H$. This residual magnetization, $M_R$, is called the remanence. To reduce the net magnetization of the material $M_R$ back to zero, an externally applied field of magnitude $H_C$ must be applied in the opposite direction of $M_R$, where $H_C$ is called the coercivity. This effect leads to hysteresis in the magnetization curve. The susceptibility of the material is the relation between $H$ and $M$, where the differential susceptibility $\chi_d = dM/dH$, with an initial susceptibility $\chi_0$ given by the value of $\chi_d$ at $M = 0$, i.e., the slope of the magnetization curve at the origin.

Magnetite ($Fe_3O_4$) and maghemite ($\gamma$-$Fe_2O_3$) are common magnetic iron oxides, and are ferrimagnetic at room temperature, with $T_C$ of 850 K (although $T_C$ is difficult to measure for maghemite, as it is unstable at high temperatures, irreversibly transforming to hematite at ~670 K). Magnetite contains a 2:1 ratio of $Fe^{3+}$:$Fe^{2+}$ ions, where maghemite is the typical oxidation product of magnetite, produced through removal of the $Fe^{2+}$ ions from the magnetite crystal lattice. Both materials have an average susceptibility of ~3 (SI units), although magnetite has a higher saturation magnetization of 92 A m$^2$ kg$^{-1}$, compared to maghemite's 76 A m$^2$ kg$^{-1}$. Despite the higher $M_s$ of magnetite, maghemite is also frequently employed in nanotechnology applications, since the spontaneous oxidation of magnetite under standard conditions is very difficult to prevent, and from an engineering perspective it can be useful to pre-oxidize magnetite nanomaterials to maghemite in order to avoid oxidation-induced iron leaching over time, and the inevitable change in magnetic properties that this oxidation process implies.

Superparamagnetism is a nano-specific phenomenon that arises when ferromagnetic (or ferrimagnetic) crystals are reduced in size to the nanoscale. Below a certain size, such nanocrystals can only support a single magnetic domain, and thus some of the above properties of bulk magnetic materials emerging from the collective ensemble of many domains disappear in magnetic nanocrystals (e.g., superparamagnetic nanocrystals do not possess hysteresis in their magnetization curve). Simply put, thermal energy is sufficient to demagnetize small crystals. As stated above, single magnetic domains can be thought of as possessing a single large magnetic moment (the sum of the atomic moments aligned with each other), however after application of an external magnetic field $H$, for the single domain particle (of volume $V$) to maintain its magnetic polarization alone against the randomizing thermal background energy requires that its magnetic anisotropy energy, $KV$, be much greater than thermal energy $k_BT$, or $KV \gg k_BT$. Since $KV$ is a function of volume, below a certain size and above a certain temperature, magnetic anisotropy energy is too weak to prevent the thermal fluctuation of the particle's magnetic moment, and thus on removal of $H$, the magnetization of the nanoparticle rapidly flips directions, rendering a net zero magnetization for the particle at zero applied field (i.e., no hysteresis). However, upon application of an external field, the particle's magnetic moment will align in the direction of $H$ and will approach $M_s$ like a bulk ferromagnetic material. Hence this effect was termed superparamagnetism, as the thermal spin randomization is conceptually similar to paramagnetic materials, although the attainable magnetization values in an externally applied field are much higher than for paramagnets. The implication is that superparamagnetic nanocrystals possess no permanent magnetic dipole moment at zero applied field, but when placed in an externally applied field can be temporarily magnetized with strength equivalent to a bulk ferromagnetic material.

It should also be noted that possessing a single magnetic domain is not sufficient to render a particle superparamagnetic; superparamagnetism is an independent function of size. Conveniently, magnetite and maghemite are superparamagnetic at room temperature below crystal sizes of ~30 nm, above which threshold size they regain ferrimagnetic properties, although they may still be single domain. Interfacial effects are also important in nanocrystals, and the measured saturation magnetization of nanocrystals is typically smaller than that of their corresponding bulk materials, due to surface spin canting, undercoordination, or crystal defects.

Although useful in magnetic data recording media, ferromagnetic materials are typically impractical for magnetic separation-based water treatment. Due to their permanent magnetic moments at zero applied field, ferromagnetic particles tend to strongly aggregate in water due to dipolar attraction, reducing the accessible surface area and diffusion coefficient of the particles, and even leading to gravity settling of the aggregates (i.e., the particles removing themselves from the water treatment volume). These problems have been observed for ferromagnetic nanoscale zero-valent iron particles used in water treatment [9]. Thus the primary advantages of using nanoparticles and nanostructured materials for water treatment (viz. high specific surface area, excellent dispersibility throughout the water matrix) are counteracted by magnetic aggregation.

From an engineering perspective, the ideal particles for magnetic separation-based water treatment processes would behave as "on-off" magnetic switches: unmagnetized when dispersed in throughout the water matrix for treatment (in order to avoid the aforementioned deleterious magnetic dipolar aggregation), but which can be easily and strongly magnetized after the treatment process in order to be magnetically separated, and then demagnetized again prior being employed in subsequent treatment cycles, etc. This is of course a description of superparamagnetism, where the magnetism of the particles is "switched on" by application of an external magnetic field. Thus the significance of superparamagnetism for magnetic water treatment is primarily to minimize uncontrolled nanoparticle aggregation during the treatment process, yet allow for easy magnetic recovery afterwards. Iron oxide nanoparticles of magnetite or maghemite below 30 nm in diameter are superparamagnetic at room temperature (termed SPIONs, superparamagnetic iron oxide nanoparticles), and are thus heavily employed in water treatment nanotechnologies, although other ferrite materials, or even non-superparamagnetic nanoparticles, can be used if properly stabilized against aggregation.

### 7.2.2 *Forces in a Magnetic Colloid*

As referred to above, typically the ideal magnetic separation-based water treatment nanotechnology can be described as a colloidal dispersion (slurry) of non-interacting nanoparticles or nanomaterials, diffusing freely throughout the volume of water to be treated, which can be rapidly separated from the water once the treatment is complete through application of an externally applied magnetic field. However, achieving such a system is nontrivial, due to the complexity of designing for the number of forces acting on the particles at the different stages of the separation process. Particles may be either too magnetic (leading to extensive interparticle magnetic aggregation) or not magnetic enough (requiring impractically long times for nanoparticle magnetic separation).

In order to achieve directional magnetophoresis, the magnetic force must be greater than the forces opposing motion, namely Brownian motion in the case of nanoparticles. The magnetic force acting on a spherical magnetic particle of volume $V$ (radius $r$) in a magnetic field gradient is

$$\vec{F_m} = V\left(\vec{M}\cdot\vec{\nabla}\right)\vec{B} = \frac{4}{3}\pi r^3\left(\vec{M}\cdot\vec{\nabla}\right)\vec{B} \tag{7.1}$$

where $\vec{M}$ is the magnetization of the particle per unit volume and $\vec{B}$ is the externally applied magnetic field.

For nanoparticles, gravitational forces are typically insignificant, but may be more important for dense iron-based particles, and should thus be considered for completeness:

$$\vec{F_g} = m\vec{g} = \frac{4}{3}\pi r^3 \Delta\rho\vec{g} \tag{7.2}$$

where $\Delta\rho$ is difference of the particle's density from that of the surrounding medium (e.g., water), and $\vec{g}$ is the acceleration due to gravity.

The motion of the nanoparticles is opposed by viscous drag forces, given by Stokes' law as

$$\vec{F_d} = -3\pi\eta D_{\mathrm{H}}\vec{v} \simeq -6\pi\eta r\vec{v} \tag{7.3}$$

where $\eta$ is the dynamic viscosity of the medium ($\eta = 10^{-3}$ Pa s for water at room temperature), $D_{\mathrm{H}}$ is the hydrodynamic diameter of the particle (which can be approximated by $D_{\mathrm{H}} = 2r$), and $\vec{v}$ is the velocity of the particle. The average magnetophoretic velocity of a single particle can thus be estimated by balancing Eqs. (7.1) and (7.3):

$$\vec{F_m} + \vec{F_d} = 0$$

$$\frac{4}{3}\pi r^3\left(\vec{M}\cdot\vec{\nabla}\right)\vec{B} - 6\pi\eta r\vec{v} = 0$$

$$\vec{v} = \frac{2r^2\left(\vec{M}\cdot\vec{\nabla}\right)\vec{B}}{9\eta} \tag{7.4}$$

More generally however, the motion of the particle in any single direction is opposed by Brownian motion, where a simplified representation of the Brownian force at a particular moment in time can be given by [10]

$$\vec{F_B} = \vec{\xi}\left(12\pi\eta r k_B T/dt\right)^{1/2} \tag{7.5}$$

where $\vec{\xi}$ is a fluctuating random vector with a Gaussian distribution, $k_B$ is the Boltzmann constant, $T$ is the temperature, and $dt$ is the time increment. An approximation of the time average Brownian force can be simply taken as the thermal energy in opposition to motion of the particle by a significant amount (e.g., distance $r$) in any specific direction:

$$k_B T = \left(\vec{F_m} + \vec{F_g} + \vec{F_d}\right)\cdot\vec{r} \tag{7.6}$$

When the right-hand side of Eq. (7.6) is significantly greater than the thermal energy of order $k_B T$, the particle can have directional motion, and if significantly less than $k_B T$, the net directional forces acting on the particle are insufficient to overcome the randomizing Brownian force, and the particle remains dispersed in suspension. Thus as a rough approximation [11], the threshold Brownian force for the particle to overcome is

$$F_B \approx \frac{k_B T}{r} \tag{7.7}$$

This accounts for the size dependence of the Brownian force, where large microparticles are less susceptible to such randomizing molecular motions and can settle under the force of gravity, while for small nanocrystals Brownian forces dominate [12]. Alternatively, comparing Eqs. (7.1) and (7.2) with Eq. (7.5), it can be seen that the directional forces scale with $r^3$, while the Brownian force only scales as $r^{1/2}$, indicating the diminished significance of Brownian forces for larger particles.

In summary, to achieve directional magnetophoresis, $\vec{F_m}$ should be $\gg \left(\vec{F_B} + \vec{F_g} + \vec{F_d}\right)$. From Eq. (7.1) it can be seen that for a constant applied magnetic field gradient there are only two ways to increase $\left|\vec{F_m}\right|$ through engineering of the nanoparticles: either by selection of a material with higher magnetization, or for a fixed material, by increasing the particle size. However, as stated above, superparamagnetism is a size-dependant nanoscale phenomenon, and upon increasing the size of magnetic nanocrystals above a particular threshold diameter they transition to ferromagnetic behavior, resulting in deleterious uncontrolled magnetic dipolar aggregation in suspension. Since superparamagnetism is an essential property for many magnetic separation-based water treatment processes, there would appear to be an upper limit on the per-particle $\left|\vec{F}_m\right|$, since superparamagnetism is limited to nanocrystals $\lesssim$ 30 nm in size (for magnetite/maghemite). However, as discussed in Sect. 7.2.3 below, such small nanocrystals are often found to be magnetically inseparable from suspension (i.e., $\left|\vec{F}_m\right| \ll k_B T/r$), thus requiring a larger $\left|\vec{F}_m\right|$ to enable magnetophoresis, yet apparently requiring the sacrifice of superparamagnetism in order to achieve it. However, this apparent paradox can be resolved through the use of polycrystalline particles, where every constituent crystal in the polycrystalline particle is below the superparamagnetic threshold in size (i.e., nanocrystals), yet where the $\vec{F}_m$ of all these crystals sum together to render a larger net $\vec{F}_m$ for the particle as a whole, enabling magnetophoresis while preserving superparamagnetism. Such polycrystalline or supraparticles can be composed exclusively of magnetic nanocrystals (as described in Sect. 7.3), or multiple SPIONs dispersed in a matrix or composite material (e.g., Dynabeads® as used in biological magnetic separations).

### *7.2.3 Interparticle Interactions and Cooperative Magnetophoresis*

Magnetic nanocrystals < 20 nm in diameter (SPIONs) should not be magnetically separable at low field gradients (<100 T m$^{-1}$) according to the above considerations. For example, for a 10 nm magnetite nanocrystal ($M_s \simeq 4.8 \times 10^5$ A m$^{-1}$) in a 100 T m$^{-1}$ field gradient at 300 K, $F_m = 2.5 \times 10^{-17}$ N, while $F_B \approx k_B T/r = 8.3 \times 10^{-13}$ N; that is, $F_B \gg F_m$. Indeed, by this logic only magnetite crystals larger than ~135 nm should be magnetically separable. Even without considering Brownian motion, from Eq. (7.4) the average magnetophoretic velocity of the above 10 nm SPIONs should be only ~0.3 μm s$^{-1}$ in water. However, this is several orders of magnitude slower than experimentally observed SPION magnetophoretic velocities under similar conditions [11]. This experimental observation that SPIONs can in fact be readily magnetically separable even under low field gradients has thus prompted the expansion of magnetophoretic theory to include interparticle interactions.

To explain the low field gradient magnetic separability of superparamagnetic nanoparticles, it was theorized that when magnetized, nanocrystals would be expected to exhibit very large field gradients near their surface, which could lead to the magnetic aggregation of the particles into larger "effective particles" or aggregate structures with sufficient net magnetic force to overcome Brownian motion [11]. Thus nanoparticles could be expected to accelerate or catalyze their own magnetic separation in a cooperative manner. For superparamagnetic particles, removal of the externally applied magnetic field should dissipate magnetic interactions, resulting in aggregate dissociation. Such transient magnetic aggregation behavior (chain formation) has been experimentally observed [13].

The group of Faraudo and Camacho has perhaps done the most work in describing the physical concepts of cooperative magnetophoresis, and the following discussion is derived primarily from [13–15].

The dipole–dipole interaction energy between two superparamagnetic colloidal particles in an externally applied magnetic field is given by

$$U_{\mathrm{dd}} = \frac{\mu_0 {m_{\mathrm{d}}}^2}{4\pi x^3}\left(1 - 3\cos^2\theta\right) \tag{7.8}$$

where $\mu_0$ is the magnetic permeability of free space, $x$ is the distance separating the particles (center-to-center), $\theta$ is the angle between the direction of the externally applied field and the line joining the centers of the two particles, and $m_{\mathrm{d}}$ is the dipole moment of the particles, given by

$$m_{\mathrm{d}} = \frac{4}{3}\pi r^3 M_{\mathrm{s}} \tag{7.9}$$

where the assumption is made that under typical fields applied during magnetic separation, the magnetization of the sample reaches saturation. The minimum energy configuration is obtained when $x = d$ (the diameter of the particles) and $\theta = 0$, or

$$U_{\mathrm{dd}}^{\max} = -\frac{\mu_0 {m_{\mathrm{d}}}^2}{2\pi d^3} \tag{7.10}$$

A useful means by which to characterize the relative strength of this magnetic interaction is to take its ratio with the thermal energy, defining a magnetic coupling parameter $\Gamma$ as

$$\Gamma = \frac{\left|U_{\mathrm{dd}}^{\max}\right|}{k_{\mathrm{B}}T} = \frac{\mu_0 {m_{\mathrm{d}}}^2}{2\pi d^3 k_{\mathrm{B}}T} = \left(\frac{\lambda_{\mathrm{B}}^m}{d}\right)^3 \tag{7.11}$$

where $\lambda_{\mathrm{B}}^m$ is called the magnetic Bjerrum length, the characteristic interparticle distance at which magnetic interactions are significant compared to thermal energy, given by

$$\lambda_B^m = \left( \frac{\mu_0 m_d{}^2}{2\pi k_B T} \right)^{1/3} \tag{7.12}$$

Physically, interparticle magnetic interactions are significant (i.e., magnetic aggregation and cooperative magnetophoresis are possible) for $\Gamma \gg 1$, while for $\Gamma \ll 1$ thermal energy dominates. For the above example of 10 nm SPIONs, $\Gamma \simeq 3.05$, indicating the possibility of cooperative magnetic separation where the non-interacting model predicted magnetophoresis to be impossible, thus demonstrating the importance of considering interparticle interactions.

With this understanding of the effect of interparticle distance on magnetic interactions, the kinetics of cooperative magnetophoretic separation can be elucidated in terms of particle concentration in suspension (i.e., the higher the particle concentration, the closer their spacing and thus higher their magnetic interactions). The mass concentration of particles in a suspension is given as

$$C = \frac{Nm}{V_t} = \frac{4\pi r^3 \rho N}{3V_t} = \frac{4\pi r^3 \rho}{3x^3} \tag{7.13}$$

where $N$ is the total number of particles in a suspension of volume $V_t$, $m$ is the mass of a single particle of density $\rho$, and $x = \sqrt[3]{V_t/N}$ is the average interparticle distance. A characteristic concentration $C^*$ can be defined when $x = \lambda_B^m$,

$$C^* = \frac{4\pi r^3 \rho}{3\lambda_B^{m3}} = \frac{\pi \rho}{6\Gamma} \tag{7.14}$$

Experimentally [13], the cooperative magnetophoretic separation time $t_s$ of superparamagnetic particles has been found to have an exponential dependence on particle concentration of the form:

$$t_s = t_0 \left( \frac{C^*}{C} \right)^{\alpha} \tag{7.15}$$

where $t_0$ (in units of time) and $\alpha$ are experimentally determined parameters dependant on the dimensions of the magnetic separator, field gradient applied, etc., although an ab initio argument for $\alpha \simeq \frac{1}{4}$ (which was the experimentally fitted value) is given in [15], which relates to the growth kinetics and magnetophoretic velocity of the particle aggregates/chains.

The obvious assumption in the above discussion is that, apart from their magnetic interactions, the particles are otherwise non-interacting. However, for colloidal nanoparticles, this assumption is generally untrue, and other colloidal interactions can be significant in comparison to the magnetic dipolar energy. In magnetic separation-based water treatment applications, the ideally desirable situation is for the transient magnetic aggregation to behave as a reversible "on-off switch," i.e., for the particles to be relatively non-interacting and highly colloidally stable in the absence of an externally applied magnetic field (in order for the particles to diffuse freely, expose a high surface area, and exhibit good mass transfer kinetics in the volume of water to be treated), to rapidly form transient aggregates when an external field is applied (for efficient magnetic separation), and then for the aggregates to reversibly dissociate and return the particles to a highly dispersed colloidal state when the external field is removed again. Actually achieving such behavior in realistic matrices is non-trivial however, since frequently (due to nonmagnetic interactions discussed below) the particles will either be too colloidally stable to form the transient aggregates in the first place (thus remaining magnetically inseparable), or they will form irreversible aggregates which cannot be re-dispersed back to a stable colloid when demagnetized. Avoiding such undesirable outcomes requires understanding and balancing all the colloidal forces present in the system.

Apart from the magnetic interaction energy given above (Eq. (7.8)), other interactions to consider are electrostatics (DLVO theory) and dispersion forces. The van der Waals potential between two spherical particles with radii $r_1$ and $r_2$ is given by [16]

$$U_{\mathrm{vdW}} = -\frac{A_{\mathrm{H}}}{6}\left[\frac{2r_1r_2}{x^2-(r_1+r_2)^2}+\frac{2r_1r_2}{x^2-(r_1-r_2)^2}+\ln\left(\frac{x^2-(r_1+r_2)^2}{x^2-(r_1-r_2)^2}\right)\right] \tag{7.16}$$

where $A_{\mathrm{H}}$ is the Hamaker constant between the particles through the surrounding medium ($A_{\mathrm{H}} = 33 \times 10^{-21}$ J for magnetite nanoparticles interacting across water [17]) and $x$ is the distance between the centers of the spheres (where the distance between the surfaces is $D = x - r_1 - r_2$). For equally sized particles ($r_1 = r_2 = r$), the above equation can be simplified to

$$U_{\mathrm{vdW}} = -\frac{A_{\mathrm{H}}}{6}\left[\frac{4x^2r^2-8r^4}{x^4-4x^2r^2}+\ln\left(\frac{x^2-4r^2}{x^2}\right)\right] \tag{7.17}$$

Calculation of the electric double layer interaction between two particles is often a complex consideration, typically involving many approximations. For more rigorous equations (beyond Derjaguin's approximation) describing the double layer interaction between nanoparticles at low ionic strengths, refer to [18]. More accessible (and hence more approximate) equations for the interaction potential between two equally sized spheres with equal surface potentials are given below [18, 19]. For large interparticle separations ($\kappa D > 4$, where $\kappa$ is the inverse Debye-Huckel length), assuming the surface potential of one particle does not affect the other, thin double layers and symmetrical electrolytes, the linear superposition approximation can be used:

$$U_{\mathrm{DL}} = \frac{64\pi\varepsilon_0\varepsilon_r r^2}{x}\left(\frac{\gamma k_{\mathrm{B}}T}{ze}\right)^2 \exp(-\kappa D) \tag{7.18}$$

$$\gamma = 4\tanh\left(\frac{\Phi}{4}\right), \text{ for } \kappa x \geq 10 \text{ and } \Phi < 8 \tag{7.19}$$

$$\Phi = \frac{ze\psi}{k_{\mathrm{B}}T} \tag{7.20}$$

$$\kappa = \sqrt{\frac{2c_0 z^2 e^2}{\varepsilon_0\varepsilon_r k_{\mathrm{B}}T}} \tag{7.21}$$

where $\varepsilon_0$ is the vacuum permittivity, $\varepsilon_r$ is the relative permittivity of the solvent, $z$ is the charge number (electrolyte valence), $c_0$ is the concentration of the symmetrical electrolyte (in units of $m^{-3}$), $e$ is the elementary charge, and $\psi$ is the surface potential of the particles (which can be estimated from the Grahame equation if only the charge of the colloid is known [14]). For small interparticle separation distances, the Derjaguin approximation can be used:

$$U_{\mathrm{DL}} = \frac{\pi\varepsilon_0\varepsilon_r r\psi^2}{2}\ln[1 + \exp(-\kappa D)] \tag{7.22}$$

If the particles are surfactant-stabilized (the above assumes simple electrostatic stabilization), additional repulsion forces must be taken into account. However these above equations are sufficient for a basic energy balance on the system during magnetic separation, where the total interaction energy between two nanoparticles in suspension is the sum of Eqs. (7.8), (7.17), and (7.18), given by $U = U_{\mathrm{dd}} + U_{\mathrm{vdW}} + U_{\mathrm{DL}}$. This results in a potential energy curve as a function of interparticle separation distance, examples of which are given in Fig. 7.3.

The desirable situation in most magnetic separations (the "on-off switch" as described above) is for the potential energy to be everywhere positive in the absence of an applied magnetic field (Fig. 7.3a, repulsive force to stabilize the colloidal dispersion and prevent aggregation), and exhibit a secondary minimum in the potential energy curve when the particles are magnetized (Fig. 7.3b, to allow for transient magnetic aggregation), with an energy barrier (provided by electrostatic repulsion) to prevent the particles from approaching each other too closely (resulting in irreversible aggregation at the primary energy minimum). This way, when the magnetic field is removed, the secondary minimum disappears, and the particles repel each other once again, returning to a dispersed colloid. The undesirable situation occurs when the magnetic dipolar interaction is too strong, or electrostatic repulsion insufficient, in which case there is no energy barrier to prevent irreversible aggregation of the particles at the primary minimum when magnetized (Fig. 7.3c). As can be seen, this delicate energy balance is a function of particle size, surface charge, applied field gradient, and ionic strength of the

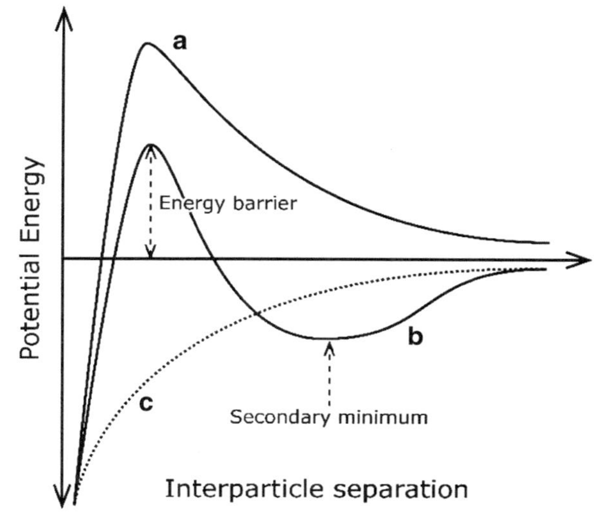


**Fig. 7.3** Potential energy profiles of magnetic particles in colloidal dispersion: (**a**) desirable curve for non-magnetized particles, (**b**) desirable curve for magnetized particles to enable reversible magnetic aggregation, (**c**) undesirable curve (*dotted line*) for magnetized particles, resulting in irreversible magnetic aggregation

water to be treated, and thus many factors must be taken into account to ensure timely and practical reversible magnetic separation is achievable in the system of interest.

## 7.2.4 Challenges for Magnetic Water Treatment

Due to the cooperative magnetophoretic mechanism and balance of forces required to prevent irreversible aggregation, designing a reusable magnetic separation-based water treatment system can be challenging from an engineering perspective, and it is usually not obvious to predict whether the magnetic particles of interest will even separate when a magnetic field is applied to the suspension. Several challenges when designing water treatment systems based on magnetic separation nanotechnology are discussed below.

Firstly, as discussed above, the energy balance of the colloidal dispersion of nanoparticles should be considered under both magnetized and demagnetized conditions (assuming demagnetization is possible, given superparamagnetic particles). If the repulsive interparticle interactions are too strong, the particles may not be able to transiently aggregate even when magnetized, inhibiting cooperative magnetophoresis and resulting in magnetic inseparability of the particles from suspension. Conversely, if repulsive interactions are low, irreversible aggregation of the particles could result during magnetic separation, rendering them essentially single use. Although only electrostatic stabilization was considered above, steric stabilization of the colloidal particles with polymers can either help or hinder magnetic separability according to the same principles (i.e., if particles cannot approach each other closely enough, cooperative magnetophoresis will be impossible). When not magnetized, high colloidal stability of the nanoparticle dispersion is typically desired, in order to expose a high surface area throughout the volume of water to be treated. Ensuring colloidal stability in realistic matrices (i.e., water with ionic strength, dissolved contaminants, humic acids) is challenging for even nonmagnetic particles, as adsorption of contaminants or flocculation with dissolved substances can change the surface energy balance of the particles, or result in premature flocculation and sedimentation. Such destabilizing of otherwise too-stable nanoparticle dispersions can be desirable in some cases (enabling magnetic separation of otherwise inseparable particles through the formation of larger aggregates/supraparticles), such uncontrolled aggregation is difficult to account for or predict through modeling.

Another engineering challenge is the different length scales at work in typical magnetic separations, from within angstroms from the surface of single nanoparticles to the scale of meters in a magnetic separator. Although this brief review has focused primarily on the nanoscale, predicting and designing a system for a particular required magnetic separation duration or magnetophoretic velocity will also incorporate macroscopic considerations, such as the geometry of the magnetic separator and magnetic fields. For example, macroscopic time scales for separating magnetic nanoparticles are influenced by the cooperative magnetic separation mechanisms discussed above, while these mechanisms are simultaneously influenced by field gradients which are a function of separator design; the micro- and macroscopic are convoluted. Furthermore, many of the equations provided herein assume static fluid, although for magnetic separations involving flowing water (e.g., a flow-through magnetic separator), additional fluid dynamics must be brought to bear on the system.

It should also be emphasized that the cooperative magnetophoresis mechanism discussed herein, while enabling the separation of nanoparticles that would be individually nonmagnetically separable, still has a strong built-in size dependence, and does not guarantee the separation of magnetic nanoparticles. For example, it is still controversial whether SPIONs are magnetically separable under low field conditions [20]. Indeed, the main industrial application of SPIONs is as ferrofluids, which are designed to be magnetically inseparable. Such dependence of separation kinetics on particle size is well illustrated in Fig. 7.6 below. Thus although use of nanoparticles (SPIONs) may be especially desirable in water treatment applications due to their extraordinary specific surface area, care should be taken to ensure that they will still be magnetically separable from the target matrix.

Finally, an additional concern of the cooperative magnetic separation paradigm is the potential for trace residual nanoparticles remaining in suspension after the magnetic separation process. That is, if nanoparticles require cooperative assembly into transient magnetic aggregates to enable magnetic separability, there is the possibility that particles which are not incorporated into the transient aggregates in time would be "stranded" in suspension, unable to magnetically separate on their own. Given the concerns about environmental contamination by nanoparticles or nanoparticle toxicity, the potential for adding low concentrations of nanoparticles to treated water in the process of magnetic separation is clearly of concern. Thus for practical applications of magnetic separation nanotechnology, at least until this potential problem of nanoparticles left behind is investigated further, magnetic separators should probably be coupled with a filtration process, to separate any residual trace nanoparticles which were not recovered in the "first pass" magnetic separation step.

## 7.3 Synthesis of Size-Tunable Superparamagnetic Spheres

### *7.3.1 Introduction*

Recent years have seen an explosion in the application of magnetic nanomaterials in water treatment [21–23]. Iron oxide magnetic nanomaterials, specifically of the magnetite ($Fe_3O_4$) and maghemite ($\gamma$-$Fe_2O_3$) phases, have been the most extensively developed and implemented, due primarily to their useful property of room-temperature superparamagnetism, that is, the ability of nanocrystals smaller than ~30 nm in diameter to lose a net magnetization except on application of an external magnetic field [5, 6], as well as for their biocompatibility, low cost, and facile chemistry. Superparamagnetism allows for the formation of stable colloidal dispersions of iron oxide particles, which can be quickly magnetically separated from solution on demand. Indeed, many high profile and desirable applications of magnetic particles are magnetophoretic in nature, dependent on the magnetic separation principle [2]. While individually dispersed superparamagnetic iron oxide nanoparticles (SPIONs) have been demonstrated to be magnetically separable in principle [11], the low per-particle magnetic moment and strong Brownian forces often make quick and efficient separations impossible, as discussed above [13]. Simply increasing the size of the nanocrystals is impractical, as the essential feature of superparamagnetism is lost above ~30 nm, whereupon the particles become ferromagnetic, rendering the formation of stable colloidal dispersions impossible.

Researchers have addressed this challenge by designing larger hierarchical particles composed of multiple SPIONs as controlled assemblies or clusters, thereby maintaining superparamagnetism while increasing per-particle magnetic moment, diminishing the relative effect of Brownian forces (inversely proportional to particle size), and thus allowing for much more efficient and reliable magnetic separation. Several synthesis strategies for such hierarchical magnetic particles have been described, including emulsion approaches [24–26], microgels [27], and polyol solvothermal techniques [28, 29], although these methods often employ excessive use of surfactants (lowering the saturation magnetization of the particles), toxic solvents and reagents, or complex and tedious protocols.

Hydrothermal synthesis offers a cost-effective, nontoxic, and environmentally preferable solution to many of these issues. Hydrothermal chemistry has recently been used to prepare hierarchical spheres from a wide variety of oxide materials.

Cheng et al. reported the hydrothermal synthesis of monodisperse submicron $Fe_3O_4$ superparamagnetic spheres with excellent crystallinity and high magnetization [30], although a means by which to vary or control the size of the particles in a hydrothermal process has not yet been described. Particle size control is essential for magnetophoretic applications to achieve uniform and specific magnetic properties, and to balance surface area and particle diffusivity with efficient magnetic separability.

Herein we describe a straightforward, one-pot hydrothermal technique for tuning the size of hierarchical $Fe_3O_4$ submicron spherical nanoclusters over an order of magnitude in scale, wherein particle size is simply controlled by the concentration of the precipitating agent through a novel process of controlled nucleation and growth kinetics. This technique represents a significant improvement to the hydrothermal synthesis of $Fe_3O_4$ particles for a variety of applications. This section is reproduced from [31].

### *7.3.2 Experimental*

$Fe_3O_4$ submicron spheres were synthesized according to a similar hydrothermal process as previously [30, 32]. $FeCl_3{\cdot}6H_2O$ (99 %, Sigma), sodium citrate dihydrate (99 %, Fisher), polyacrylamide (PAM, $M_W$ ~5–6 MDa, Polysciences), ammonium hydroxide ($NH_4OH$, 28 % solution $NH_3$ basis, Sigma), and urea (99 %, Fisher) were purchased and used without further purification.

$FeCl_3{\cdot}6H_2O$ (in a pre-prepared 1 M aqueous solution), sodium citrate dihydrate, and PAM were mixed and dissolved in Millipore deionized water through stirring for 24 h at room temperature. A small amount of precipitating agent, either $NH_4OH$ or urea, was then added to the solution under vigorous stirring, which was then allowed to continue to stir for 30 min. After addition of the precipitating agent, the final concentrations of reagents in solution were 50 mM $FeCl_3{\cdot}6H_2O$, 100 mM sodium citrate, 7.5 g $L^{-1}$ PAM, and a known concentration of $NH_3$ or urea. This mixture was then poured into a 125 mL PTFE-lined stainless steel pressure vessel (Parr) at a 66 % fill fraction, which was then heated in an oven (Binder) at 2 °C $min^{-1}$ to 200 °C, and held at this temperature for 12 h, before removal from the oven to cool naturally in room temperature air. Once cool, the product was recovered magnetically and washed thoroughly with deionized water and ethanol by magnetic decantation, and then dried under $N_2$ and in a vacuum desiccator at room temperature.

Magnetic separation experiments were performed with a rare earth homogenous magnetic separator (Sepmag Lab 2142, inner bore diameter 31 mm, radial magnetic field gradient 45 T $m^{-1}$). A 20 mL vial containing 10 mL of 0.13 mg $mL^{-1}$ aqueous particle dispersion was placed in the central cylindrical bore of the magnetic separator, and aliquots were drawn from the center of the vial at regular time intervals and their absorbance measured at 368 nm with a UV/Vis spectrophotometer, which was compared against a prepared standard curve to assess the concentration of particles remaining in suspension over time.

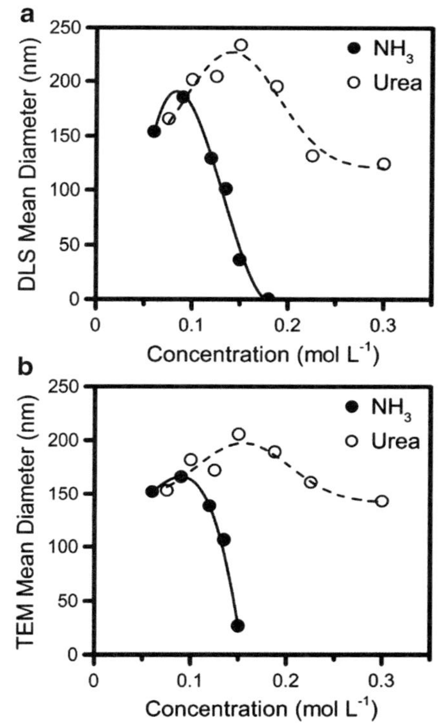


**Fig. 7.4** Mean diameters measured from (**a**) DLS and (**b**) TEM images for different particle preparations as a function of the concentration of ammonia or urea used in the synthesis reaction

The size and surface morphology of the particles were analyzed with transmission electron microscopy (TEM, Philips CM-10, 60 keV). The hydrodynamic diameter of the particles suspended in deionized water was measured with dynamic light scattering (DLS, Brookhaven 90Plus Particle Size Analyzer, MSD number-weighted mean diameter). The crystallinity of the powdered samples was assessed through X-ray diffraction analysis (XRD, Bruker D8 Focus, 1.542 Å Cu Kα radiation). The surface area of the materials (pre-dried at 300 °C in $N_2$) was calculated from the Brunauer-Emmett-Teller (BET) equation with data from $N_2$ adsorption isotherms obtained at 77 K (Micrometrics Gemini VII 2390 Surface Area Analyzer). Fourier-transform infrared spectroscopy (FTIR, Bruker Tensor 27) was measured for powdered samples in KBr pellets.

### *7.3.3 Results and Discussion*

The measured sizes of particles prepared at different concentrations of precipitating agent, either urea or ammonia, are presented in Fig. 7.4 (unless otherwise specified, references herein to particle size or diameter refer to the hierarchical nanoclusters as a whole, rather than the subunit nanocrystals composing the structure). It can be observed that the size of the particles is strongly dependent on concentration of the

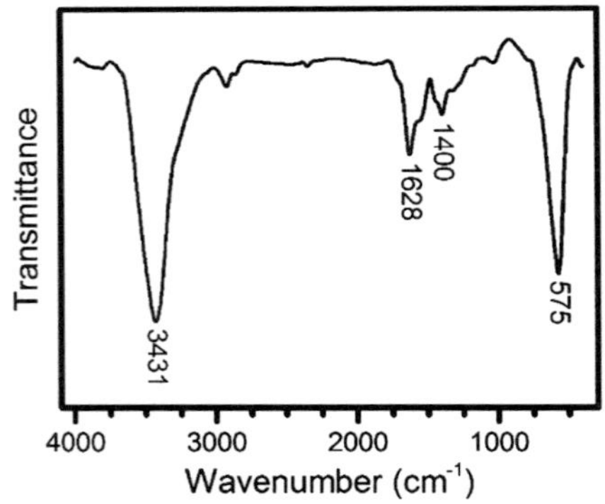


**Fig. 7.5** FTIR spectrum of particles synthesized with 0.09 M $NH_3$

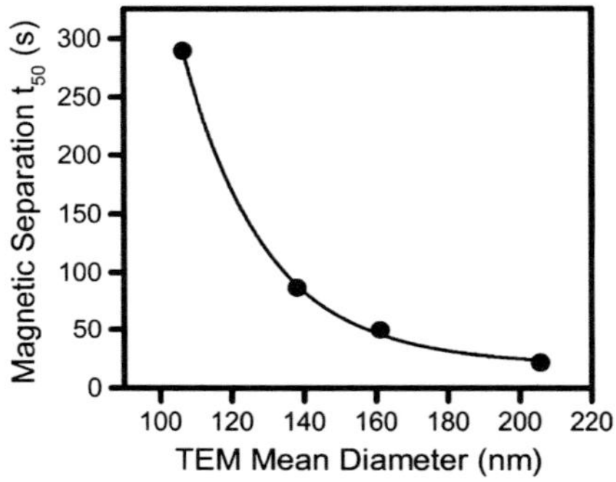


**Fig. 7.6** Time for the concentration of particles to be reduced by half ($t_{50}$) during magnetic separation from an aqueous dispersion, as a function of particle size. The fitted curve (an exponential function) is simply a guide for the eye and is not intended to represent any physical model

precipitating agent, and passes through a maximum for each trend. Notably, using ammonia in the synthesis reaction allows broad tuning of the particle size to below 100 nm, while use of urea generally yielded particles $\geq$ 150 nm in size, in accordance with previous observations [30]. All particle preparations formed stable aqueous colloidal dispersions.

While the TEM diameters matched closely to the DLS hydrodynamic diameters, indicating minimal aggregation of the particles in an aqueous dispersion, the hydrodynamic particle sizes were sometimes larger (Fig. 7.4). This is possibly due to residual PAM chains on the surface of the particles extending into solution. Bound PAM was detected with FTIR, with characteristic peaks at 1,400 $cm^{-1}$ ($CH_2$ deformation and wagging) and 1,628 $cm^{-1}$ (C–O stretching and $NH_2$ deformation) (Fig. 7.5). The residual PAM likely serves as a steric stabilizer for the particles, responsible for the excellent dispersibility and colloidal stability of the particles in water [33]. Indeed, it should be noted that the smallest particle samples prepared could not be readily magnetically separated from deionized water, although the larger nanoclusters exhibited rapid separation kinetics, thus reinforcing the importance of particle size for magnetic separations (Fig. 7.6).

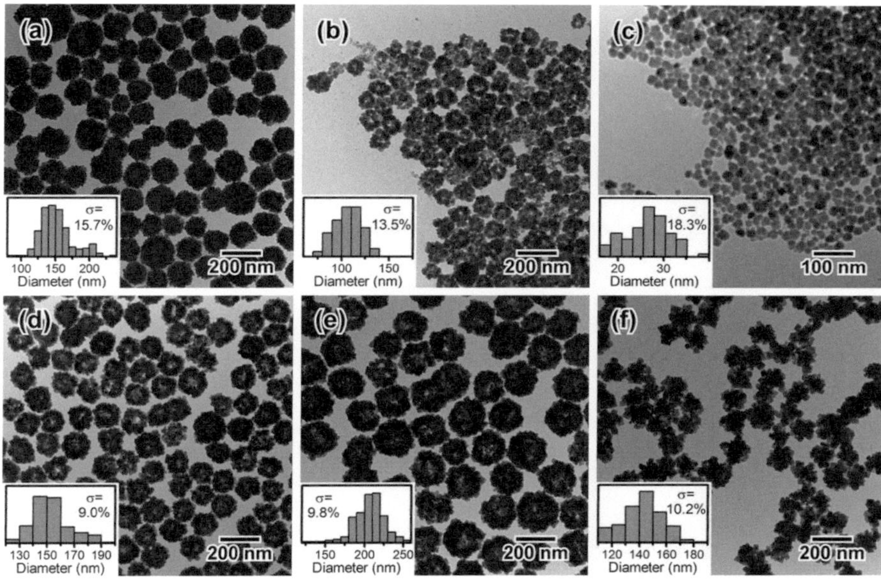


**Fig. 7.7** TEM images of particles synthesized with (**a**) 0.06 M $NH_3$, (**b**) 0.135 M $NH_3$, (**c**) 0.15 M $NH_3$, (**d**) 0.075 M urea, (**e**) 0.15 M urea, and (**f**) 0.3 M urea. *Insets* show normalized frequency histograms of the particle size distribution of each sample as measured from TEM, where $\sigma$ is the standard deviation

The prepared particles were observed to be spherical in morphology with good monodispersity in their size distributions (Fig. 7.7). The particles synthesized with urea possessed somewhat narrower size distributions than those prepared with ammonia. The spherical nanoclusters are clearly hierarchical and polycrystalline, comprised of aggregated smaller nanocrystals (Fig. 7.8). They can also be observed to be highly porous, with some spheres even assuming a hollow structure (e.g., Fig. 7.7d, e), as has been observed previously [30].

While the particles all possess a similar morphology at lower concentrations of urea and ammonia, at high concentrations notable changes can be seen. The size of the particles prepared at 0.15 M $NH_3$ is greatly reduced, although they are still generally spherical, and also polycrystalline (Figs. 7.7c and 7.8c). The particles prepared at 0.3 M urea contain a core of fine, aggregated nanocrystals, and an outer shell of larger, well-faceted nanocrystals (Fig. 7.8f).

The crystallinity of the particles was characterized with XRD, and all samples were indexed to the magnetite phase ($Fe_3O_4$) of iron oxide (Fig. 7.9), in accordance with their black coloration. The largest particles (0.09 M $NH_3$ and 0.15 M urea) appeared to be the most crystalline. The mean sizes of the nanocrystals composing the hierarchical spheres were calculated according to the Scherrer formula ($d = 0.9\lambda/\beta \cos \theta$, where $d$ is the crystal size, $\lambda$ is the wavelength of the X-rays, $\beta$ is the full width at half maximum of the peak, and $\theta$ is the diffraction

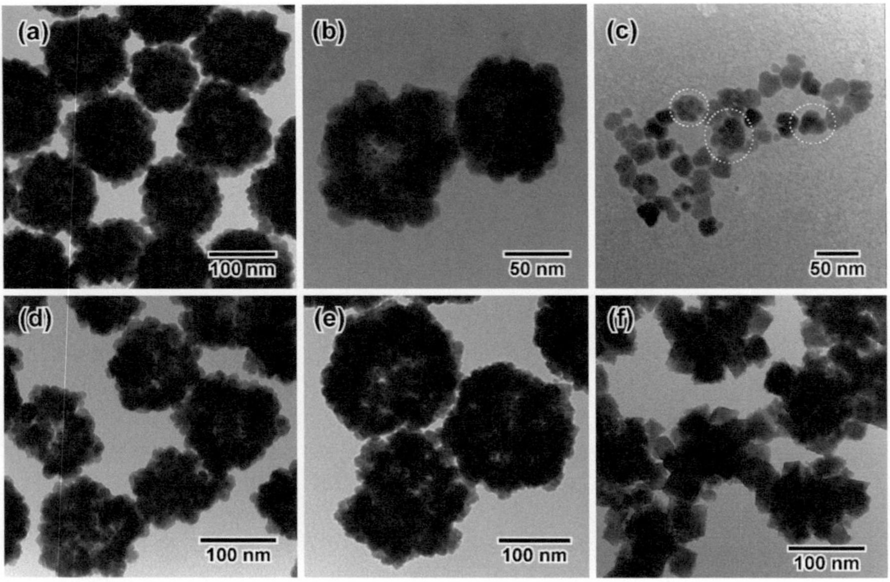


**Fig. 7.8** Higher magnification TEM images of particles synthesized with (**a**) 0.06 M $NH_3$, (**b**) 0.135 M $NH_3$, (**c**) 0.15 M $NH_3$, (**d**) 0.075 M urea, (**e**) 0.15 M urea, and (**f**) 0.3 M urea. *Dotted lines* added to (**c**) indicate polycrystalline particles

angle), and are given in Table 7.1. As can be seen, the largest particles in each series also possessed the largest crystal sizes, and the urea samples generally contained larger nanocrystals than the ammonia samples. The crystallite size for all samples was within the range for room temperature superparamagnetism (<30 nm), as expected from the particles' colloidal stability and lack of magnetic aggregation.

We studied the surface area of representative particle samples with $N_2$ adsorption. The BET surface areas of the largest hierarchical spheres were 63.9 $m^2\ g^{-1}$ for the 0.09 M $NH_3$ particles and 59.0 $m^2\ g^{-1}$ for the 0.15 M urea particles. These results are comparable to the theoretical specific surface area of 64.7 $m^2\ g^{-1}$ for 18 nm magnetite nanoparticles, indicating that surface area is not significantly sacrificed upon hierarchical nanoparticle assembly, despite dramatically improved magnetophoretic velocity for these larger spheres (Fig. 7.6). The relatively high surface area of the nanoclusters is understandable given their porous and hollow structures. Improving magnetic separability while preserving high surface area is especially important for applications involving adsorption or binding, such as in water treatment or biosensing.

The particle size trends and above observations can be understood in terms of a kinetics controlled LaMer model [34, 35]. In the synthesis reaction, iron oxide precipitation follows the generation of activated precursors, or the formation of hydroxide to deprotonate aquo-hydroxo iron complexes and initiate condensation.

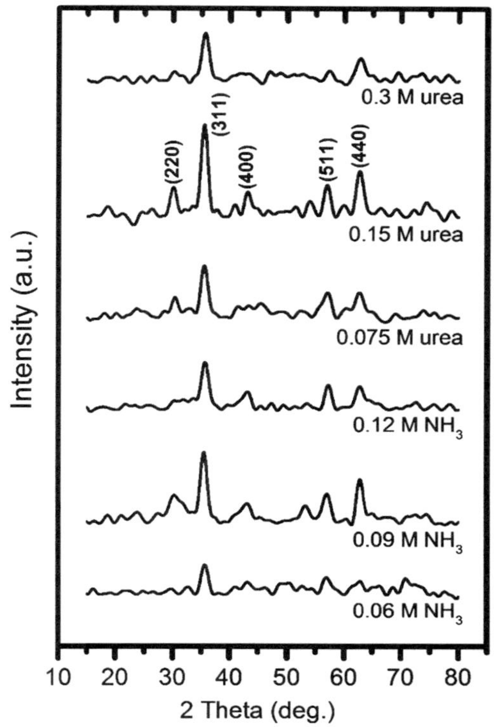


**Fig. 7.9** Measured XRD patterns of the prepared powders. Labeled peaks correspond to $Fe_3O_4$ (JCPDS No. 71-6336)

**Table 7.1** Nanocrystal size as calculated from the magnetite (311) XRD peak using the Scherrer formula

| | $NH_3$ | | | Urea | | |
|---|---|---|---|---|---|---|
| Concentration (M) | 0.06 | 0.09 | 0.12 | 0.075 | 0.15 | 0.3 |
| Crystal size (nm) | 5.0 | 15.0 | 11.9 | 15.7 | 24.1 | 13.1 |

Hydroxide anions are provided by $NH_3$ or urea, where urea undergoes thermal decomposition in the hydrothermal system to release $NH_3$. Similarly, at high temperature citrate can complex and reduce dissolved ferric ions to $Fe^{2+}$ (both $Fe^{2+}$ and $Fe^{3+}$ are necessary for the formation of magnetite) [30]. Particle nucleation and growth thus follows the generation of hydroxide-activated precursors according to classic nucleation and growth theory. The nanocrystals formed by this process are highly unstable due to the high ionic strength of the medium and thus aggregate into hierarchical spheres to minimize surface energy. This aggregation process can also be understood to follow hierarchical nucleation and growth mechanisms. Over the course of the reaction further crystallization and Ostwald ripening of the particles occurs according to a dissolution-recrystallization process, due to the high reactivity of the hydrothermal medium. This ripening process can also account for the hollow structure of some of the spheres (Fig. 7.7b, d, e), as the poorly crystalline cores of the particles are dissolved and reprecipitated on their outer surfaces as a well-crystallized magnetite corona throughout the course of the reaction, upon exposure to hydroxide. For high concentrations of precipitating agent, equilibrium size and crystallinity of the particles is quickly achieved early in the reaction process, with low driving force for ripening towards a hollow structure.

The main differences between the $NH_3$ and urea samples are explained by the different generation rate of hydroxide anions in the two systems. In the samples with added $NH_4OH$, hydroxide is immediately available to the reaction, and thus the generation rate of activated precursors and the particle nucleation rate are both high (indeed, in the $NH_3$ samples a color change could already be observed in the 30 min before heating, due to the precipitation of iron oxides). Increasing $NH_3$ concentration increases the rate of nucleation, increasing the number of particles for additional $Fe_3O_4$ to grow onto, and thus decreasing particle size. In the urea reactions the generation rate of hydroxide is much slower, as this depends on the thermal decomposition of urea, thus resulting in slower nucleation and growth, and larger particles. This is also why the urea synthesized particles cannot achieve as broad variability in size as the $NH_3$ reaction, since even at high urea concentration the initial nucleation rate is tied to the heating rate of the system. The slower particle growth process also explains the better monodispersity of the particles synthesized with urea, as well as the generally larger crystal sizes.

We therefore postulate that particle dimension can be precisely tuned as a function of hydroxide concentration, up to a peak size (Fig. 7.4), the existence of which is understandable because at lower concentrations of precipitating agent the reactions are hydroxide limited, with insufficient precursors for the particles to grow any larger. This is also attested to by the fact that in the samples with low concentration of precipitating agent (e.g., 0.06 M $NH_3$) some excess dissolved iron could be seen in solution upon magnetic separation of the particles. At higher concentrations of precipitating agent, nucleation is accelerated, thus providing an upper limit to particle size.

### 7.3.4 Conclusions

We have presented a simple, one-pot hydrothermal synthesis method for preparing size-controlled $Fe_3O_4$ submicron spheres over a broad size range. The diameter of the particles can be tuned by simply varying the concentration of precipitating agent added to the reaction, where a greater range of particle sizes can be realized when using ammonia rather than urea as the precipitating agent. The particles are highly dispersible in water, maintain a large surface area, and are readily magnetically separable. This new protocol offers a relatively nontoxic and environmentally friendly method to produce size-controlled superparamagnetic $Fe_3O_4$ particles for use in a wide variety of magnetophoretic applications.

## 7.4 Magnetically Recyclable Photocatalysts

### *7.4.1 Introduction*

Photocatalysis over nanoscale titanium dioxide is a highly promising technique for the deactivation or mineralization of a broad spectrum of waterborne contaminants, due to the low cost, abundance and minimal toxicity of $TiO_2$, as well as the central concept that the $TiO_2$ can theoretically be reused indefinitely for cost-effective purification of large volumes of water. Despite its potential, $TiO_2$ photocatalysis is widely understood to exhibit two critical failings limiting realistic implementation, namely difficult recovery of $TiO_2$ nanomaterials from treated water for reuse, as well as general low efficiency under natural lighting [36, 37]. Therefore substantial efforts have been devoted to improving the photocatalytic efficiency of $TiO_2$, as well as towards designing mechanisms for easier catalyst recovery and recycling [38–40]. However, few solutions have proposed to address both concerns simultaneously. Herein, we present a nanotechnology platform for highly efficient photocatalytic water treatment and simple $TiO_2$ recovery, in the form of mesoporous $TiO_2$ deposited on a magnetic particulate substrate. These composite particles are comprised of only cheap and earth-abundant oxides, while the mesoporous $TiO_2$ shell is developed through a simple and benign hydrothermal treatment, representing a practical engineering solution to the challenges of advanced water treatment. Furthermore, these recyclable photocatalysts are demonstrated to exhibit similar activity to P25, a benchmark $TiO_2$ formulation. This section is reproduced from [41].

### *7.4.2 Results and Discussion*

To adapt the magnetic separation paradigm to practical $TiO_2$ recycling, we started with a larger polycrystalline magnetite core to increase the total magnetic moment of each particle for faster separation, yet where the individual magnetite nanocrystals in the core are each below the ferromagnetic threshold size and thus collectively preserve the crucial property of superparamagnetism. These superparamagnetic magnetite nanospheres were formed through a one-pot hydrothermal reaction involving the controlled precipitation of ferric chloride and slow transformation to magnetite at 473 K (see Sect. 7.3 above). TEM images of these particles are presented in Fig. 7.10a. The magnetite spheres ("M" particles) presented good monodispersity with a mean particle diameter of ~180 nm, and their polycrystalline nature is clearly evident.

These magnetite spheres were coated with a silica shell through a modified Stöber method prior to deposition of $TiO_2$. The function of the silica interlayer is twofold: to protect the magnetite core from oxidation, dissolution and leaching during photocatalysis, as well as to prevent a detrimental electrical heterojunction

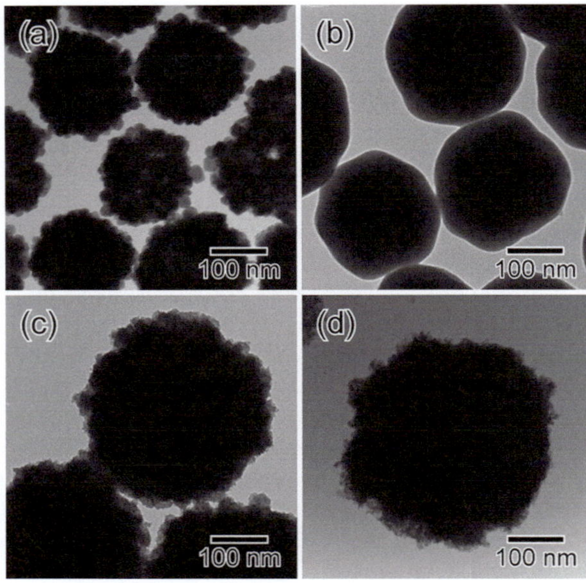


**Fig. 7.10** TEM images of (**a**) M, (**b**) MS, (**c**) MST, and (**d**) MSTh particles

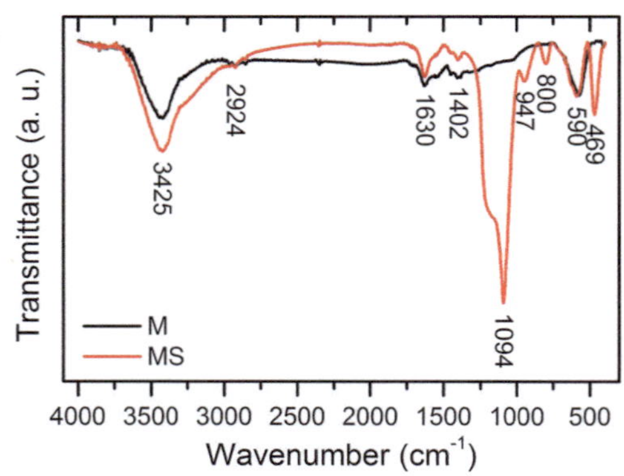


**Fig. 7.11** FTIR spectra of M and MS particles, as labeled

from forming between the $TiO_2$ shell and magnetite core (also a semiconducting oxide), which has been previously demonstrated to significantly impair photocatalysis [42, 43]. The silica-coated spheres ("MS" particles, Fig. 7.10b) were measured to be ~260 nm in size, indicating a silica shell thickness of ~40 nm. FTIR positively identified the shell material as silica due to the appearance of the characteristic Si–O–Si peaks at 1,094 and 800 $cm^{-1}$, while the peak at ~950 $cm^{-1}$ has been attributed to either Si–O–Fe bonds or silanol groups (Fig. 7.11) [44–46]. The vibrations at 1,630 and 1,402 $cm^{-1}$, present before the silica coating reaction, are attributed to residual PAM from the hydrothermal synthesis of the M cores. Indeed, as magnetite typically precipitates in less polar solvents such as ethanol [47], we propose that this residual PAM acts as a steric

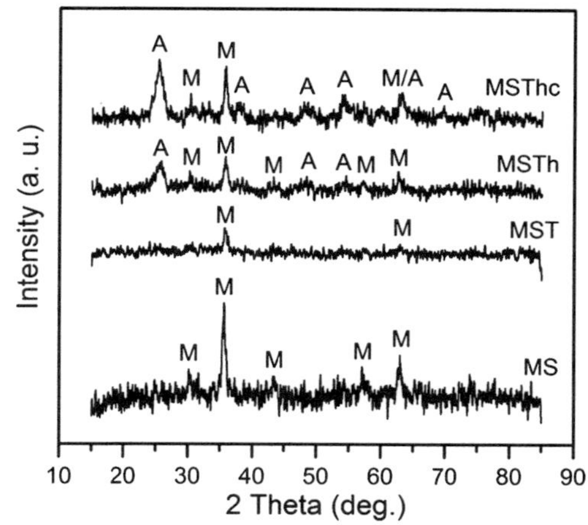


**Fig. 7.12** Powder XRD spectra of the indicated samples. Peaks labeled M are indexed to the magnetite/maghemite phase of iron oxide, and peaks labeled A are indexed to the anatase phase of $TiO_2$

stabilizer to prevent particle agglomeration during the sol-gel silica coating reaction, which allowed for a uniform and well-controlled shell to be deposited around each sphere [24].

A two-step sol-gel coating reaction was used to deposit a ~70 nm thick amorphous $TiO_2$ shell on the superparamagnetic MS particle substrates ("MST" particles, Fig. 7.10c), increasing the overall sphere diameter to ~400 nm. As the condensation reaction begins with heterogeneous nucleation at the silanol groups on the silica surface, the $TiO_2$ shell is chemically bonded to the core particle, unlike some previous designs [48, 49]. Furthermore, the thickness of the $TiO_2$ shell, while minimizing particle aggregation during the colloidal coating step, also represents a substantial improvement. Powder X-ray diffraction (XRD) revealed that the observed $TiO_2$ shell was indeed amorphous (Fig. 7.12), which, although photocatalytically inactive, offers a malleable substrate for nanostructuring.

Mesoporous structures have been extensively explored for environmental remediation, and for improving the photocatalytic efficiency of $TiO_2$, through increasing its specific surface area, while still allowing for the preservation of shaped or monolithic bulk morphologies [50]. However, the deposition of controlled mesoporous oxide shells on spherical colloidal substrates still remains challenging from a chemical engineering perspective. Here we employed a low-temperature hydrothermal treatment to simultaneously porosify and pre-crystallize the amorphous $TiO_2$ shell, transforming it to a disordered mesoporous layer. The mechanism of this transformation can be explained as the permeation of the boiling water into the incompletely condensed amorphous shell, wherein titanate monomers are rapidly dissolved only to reprecipitate as more thermodynamically stable nanocrystals [51, 52]. Due to the complete hydrolysis of the Ti monomers, these $TiO_2$ nanocrystals are enriched in surface hydroxyl groups, which bind the reprecipitated nanocrystals together through hydrogen bonding, maintaining the bulk geometry of the shell, while mesopores are generated through the densification process ("MSTh" particles, Fig. 7.10d). Indeed, the mean size of the MSTh particles was found to be essentially unchanged by the hydrothermal treatment (Table 7.2).

**Table 7.2** Measured TEM size, Brunauer-Emmett-Teller (BET) surface area, and XRD crystallite sizes of the indicated samples

| | M | MS | MST | MSTh | MSThc |
|---|---|---|---|---|---|
| TEM mean diameter (nm) | 180 | 260 | 395 | 400 | ~ |
| BET surface area ($m^2\ g^{-1}$) | ~ | $13.3 \pm 0.2$ | $95.9 \pm 0.3$ | $151.6 \pm 0.4$ | $98.7 \pm 0.3$ |
| Magnetite crystal size (nm)[a] | ~ | 14.1 | 13.2 | 12.3 | 15.9 |
| Anatase crystal size (nm)[b] | ~ | ~ | ~ | 5.7 | 6.3 |

[a]As calculated from the magnetite (311) peak using the Scherrer formula
[b]As calculated from the anatase (101) peak using the Scherrer formula

The crystallization of the shell into ~6 nm anatase phase $TiO_2$ nanocrystals was confirmed with XRD (Fig. 7.12, Table 7.2). $N_2$ adsorption was used to measure the specific surface area of the hydrothermally treated particles, which was found to have increased by ~60 % relative to the MST sample, to a value of 152 $m^2\ g^{-1}$ (Table 7.2), presumably due to the mesoporous shell structure. Thus although the spheres as a whole possessed a relatively large size of ~400 nm (primarily to enable faster magnetic separation), an exceptionally high specific surface area could still be created within the shell.

Prior to photocatalytic testing, the composite MSTh particles were calcined in air to further crystallize the shell and remove residual crystalline defects ("MSThc" particles). The improved crystallinity of the $TiO_2$ shell is evidenced by the stronger anatase peaks in the XRD spectrum (Fig. 7.12). Some loss in the specific surface area of the spheres was measured as well (Table 7.2), likely due to sintering of the fine nanocrystals within the disordered mesoporous shell. However, it should be noted that some degree of intercrystal sintering is desirable, as this would be expected to bind the shell structure more stably than simply hydrogen bonding. The final surface area of the particles at ~100 $m^2\ g^{-1}$ is still exceptionally high for these larger submicron spheres, indicating that substantial porosity is still present. While the surface area of the amorphous-shell MST particles would have been expected to substantially collapse upon calcination, previous studies have shown that a hydrothermal pre-crystallization step can serve to limit the amount of surface area loss due to sintering, for the preservation of porous nanoarchitectures [53, 54].

Another important feature of this synthesis approach is that the crystallinity of the magnetic core of the particles, along with their superparamagnetism, was well preserved even after calcination, without the appearance of non-magnetic phases of iron oxide such as hematite (Fig. 7.12). The cores were, however, likely oxidized to maghemite ($\gamma$-$Fe_2O_3$, indistinguishable on XRD) during calcination, due to their color change from black to brown, although this transformation hardly affects their magnetic properties. Previous studies on $Fe_3O_4/TiO_2$ composites have typically calcined their samples under anaerobic or reducing atmospheres in deference to protecting the oxidative loss of the magnetic phase [55, 56]. However, calcination under reducing conditions has been known to seed oxygen vacancy defects in the $TiO_2$ lattice [57], which can lead to lower-than-expected photocatalytic activities [58]. This perceived trade-off between magnetic and photocatalytic properties was avoided in the oxygenated calcination step in our study, which we hypothesize

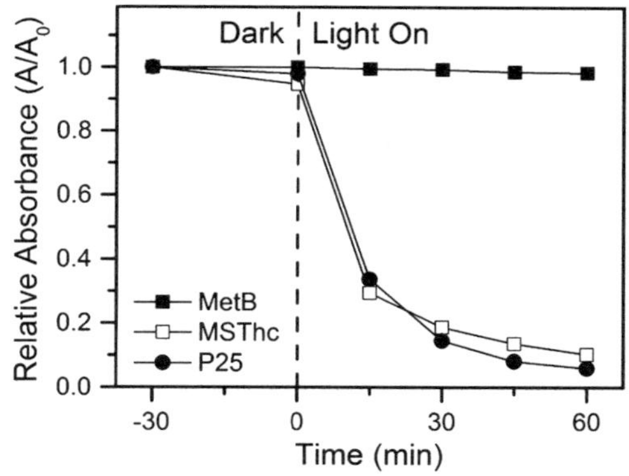


**Fig. 7.13** Photocatalytic degradation of methylene blue (MetB) under UV light

could be due to the presence of the silica shell constraining the recrystallization of the magnetic core, where silica has previously been shown to impede crystallization during calcination [59, 60].

Finally, the photocatalytic activity of the composite MSThc particles was tested in the decoloration of methylene blue dye, a common model organic contaminant, under UV light (Fig. 7.13). Ninety percent of the dye was found to be degraded within ~1 h, which compares very favorably with the rate of P25, a highly efficient $TiO_2$ formulation. Despite the much larger size of the MSThc spheres as compared to P25 (~21 nm nanocrystals), excellent photocatalytic activity was maintained on an easily recoverable platform, presumably due to the high surface area of the MSThc particles. After the photocatalytic dye degradation test, the MSThc particles could be rapidly and completely collected from solution by the simple application of a magnet to the side of the reaction vessel.

## *7.4.3 Experimental Details*

### 7.4.3.1 Materials

$FeCl_3{\cdot}6H_2O$ (99 %, Sigma), sodium citrate dihydrate (Fisher), polyacrylamide (PAM, $M_W$ ~5–6 MDa, Polysciences), tetraethyl orthosilicate (TEOS, 99 %, Sigma), titanium(IV) butoxide (TBOT, 97 %, Sigma), ammonium hydroxide ($NH_4OH$, 28 % solution $NH_3$ basis, Sigma), ethanol (EtOH, 99 %, ACS reagent grade), methylene blue (MetB, >96 %, Sigma), and hydroxypropyl cellulose (HPC, $10^5$ Da $M_W$, Sigma) were purchased and used without further purification.

### 7.4.3.2 Magnetite Sphere Synthesis (M Particles)

Magnetite spheres were prepared according to a hydrothermal reaction as previously (Sect. 7.3.2). Briefly, sodium citrate dihydrate, PAM, and $FeCl_3{\cdot}6H_2O$ (in a pre-prepared 1 M aqueous solution) were mixed and dissolved in Millipore deionized water through stirring for 24 h at room temperature. A small amount of ammonium $NH_4OH$ was then added to the solution under vigorous stirring, which was then allowed to continue to stir for 30 min. After addition of the ammonium hydroxide, the final concentrations of reagents in solution were 50 mM $FeCl_3 \cdot 6H_2O$, 100 mM sodium citrate, 7.5 g $L^{-1}$ PAM, and 0.1 M $NH_3$. This mixture was then poured into a 125 mL PTFE-lined stainless steel pressure vessel (Parr) at a 66 % fill fraction, which was then heated in an oven (Binder) at 2 °C $min^{-1}$ to 200 °C, and held at this temperature for 12 h, before removal from the oven to cool naturally in room temperature air. Once cool, the product was recovered magnetically and washed thoroughly with deionized water and ethanol by magnetic decantation, and then dried under $N_2$ and in a vacuum desiccator at room temperature.

#### 7.4.3.3 Silica Coating Reaction (MS Particles)

The M particle powder was dispersed into a solution of EtOH and Millipore deionized water by probe sonication (Branson, 100 W, 10 min, pulsed). $NH_4OH$ was then added to the dispersion, followed by the slow dropwise addition of 1 M TEOS in EtOH solution over 1 h, under vigorous mechanical stirring. After the TEOS addition, the final concentrations of reagents in solution were 25 mM TEOS, 0.3 M $NH_3$, 12 M $H_2O$, and 1.5 g $L^{-1}$ M particles. This mixture was then stirred at room temperature for 18 h, after which the product was recovered magnetically and washed thoroughly with EtOH by magnetic decantation, and then dried under $N_2$ and in a vacuum desiccator at room temperature.

#### 7.4.3.4 Titania Coating Reaction (MST Particles)

The titania coating reaction was performed similarly to a previously reported method [61]. The MS particle powder was dispersed into a solution of EtOH and Millipore deionized water by probe sonication (Branson, 160 W, 10 min, pulsed). This dispersion was then added to a solution of HPC in EtOH of equal volume. A 1.2 M TBOT in EtOH solution was then added dropwise to this mixture over 1 h with under mechanical stirring. After the TBOT addition, the final concentrations of reagents in solution were 0.12 M TBOT, 0.45 M $H_2O$, 3 g $L^{-1}$ HPC and 2.5 g $L^{-1}$ MS particles. The solution was then refluxed in an 85 °C oil bath for 90 min while stirring. After cooling to room temperature, the product was recovered magnetically and washed thoroughly with EtOH by magnetic decantation. The above titania coating process was then repeated using the same particles (without drying), in order to increase the titania shell thickness.

#### 7.4.3.5 Hydrothermal Treatment (MSTh Particles)

After the second titania coating step above, the particles were washed thoroughly with deionized water by magnetic decantation, and then dispersed into 30 mL deionized water by probe sonication (Branson, 160 W, 10 min, pulsed) at a concentration of 13.3 g $L^{-1}$ (MS particle basis, as the particles were not dried after the titania coating reaction step). This dispersion was then added to a 45 mL PTFE-lined stainless steel pressure vessel (Parr), which was then heated in an oven (Binder) at 2 °C $min^{-1}$ to 100 °C, and held at this temperature for 1.5 h, before removal from the oven to cool naturally in room temperature air. The product was then recovered magnetically and washed thoroughly with deionized water by magnetic decantation, before drying in air at 70 °C overnight.

#### 7.4.3.6 Calcination (MSThc Particles)

The MSTh particle powder was heated in air at 5 °C $min^{-1}$ to 500 °C, held at this temperature for 2 h, and then cooled to room temperature at 5 °C $min^{-1}$.

#### 7.4.3.7 Photocatalytic Tests

Particles were dispersed into deionized water by sonication (Branson, 100 W, 10 min, pulsed) at a concentration of 1 g $L^{-1}$. This particle dispersion was then added to an aqueous MetB stock solution in a 250 mL PTFE beaker such that the final concentrations of particles and MetB were 0.05 g $L^{-1}$ and 5 mg $L^{-1}$, respectively, and the total solution volume was 125 mL. This solution was mixed well and stored in the dark for 30 min to achieve adsorption-desorption equilibrium. The beaker was then placed under a UVC light source (UVP CL-1000 254 nm, ~6.67 mW $cm^{-2}$) and 1 mL aliquots were withdrawn at regular intervals and stored in the dark. After the completion of the test, these aliquots were centrifuged at 15,000 rpm for 10 min to remove suspended particles, and the concentration of MetB was determined by measuring the absorbance of the supernatant at 668 nm with a spectrophotometer (BioTek Epoch). A 5 mg $L^{-1}$ MetB solution without added particles was subjected to the same test, as a control.

#### 7.4.3.8 Characterization

The size and surface morphology of the particles were analyzed with a transmission electron microscope (TEM, Philips CM-10, 60 keV). The crystallinity of the prepared samples was assessed through X-ray diffraction (XRD, Bruker D8 Focus, 1.54 Å Cu Kα radiation) analysis. The surface area of the materials (pre-dried at 300 °C in $N_2$) was calculated from the BET equation with data from

$N_2$ adsorption isotherms obtained at 77 K (Micrometrics Gemini VII 2390 Surface Area Analyzer). The presence of silica was detected through use of Fourier-transform infrared spectroscopy (FTIR, Bruker Tensor 27) of powdered samples in KBr pellets.

### 7.4.4 Conclusion

We have demonstrated that the nanoscale phenomena of superparamagnetism and mesoporosity can be effectively combined in a functional hierarchical particle structure designed both to improve the photocatalytic efficiency of $TiO_2$ and allow for simple recovery of the photocatalyst from treated water. The careful control of each coating step to prevent interparticle aggregation resulted in high photocatalytic activity. Furthermore, due to the use of simple, scalable and environmentally friendly synthesis protocols, especially hydrothermal processing, in combination with cheap and earth-abundant oxide materials, we believe that these magnetic photocatalytic core-shell particles demonstrate excellent potential for commercialization and implementation as a practical water treatment solution.

## References

1. R. Ambashta, M. Sillanpaa, Water purification using magnetic assistance: a review. J. Hazard. Mater. **180**(1–3), 38–49 (2010). doi:10.1016/j.jhazmat.2010.04.105
2. C. Yavuz, A. Prakash, J. Mayo, V. Colvin, Magnetic separations: from steel plants to biotechnology. Chem. Eng. Sci. **64**(10), 2510–2521 (2009). doi:10.1016/j.ces.2008.11.018
3. S. Shylesh, V. Schunemann, W. Thiel, Magnetically separable nanocatalysts: bridges between homogeneous and heterogeneous catalysis. Angew. Chem. Int. Ed. **49**(20), 3428–3459 (2010). doi:10.1002/anie.200905684
4. V. Polshettiwar, R. Luque, A. Fihri, H. Zhu, M. Bouhrara, J.-M. Bassett, Magnetically recoverable nanocatalysts. Chem. Rev. **111**(5), 3036–3075 (2011). doi:10.1021/cr100230z
5. A.S. Teja, P.-Y. Koh, Synthesis, properties, and applications of magnetic iron oxide nanoparticles. Prog. Cryst. Growth Charact. Mater. **55**(1–2), 22–45 (2009)
6. P. Tartaj, Superparamagnetic composites: magnetism with no memory. Eur. J. Inorg. Chem. **2009**(3), 333–343 (2009). doi:10.1002/ejic.200800875
7. B.D. Cullity, C.D. Graham, *Introduction to Magnetic Materials* (IEEE/Wiley, Hoboken, NJ, 2009)
8. J.P. Jakubovics, *Magnetism and Magnetic Materials* (Institute of Materials, London, 1994)
9. T. Phenrat, N. Saleh, K. Sirk, R.D. Tilton, G.V. Lowry, Aggregation and sedimentation of aqueous nanoscale zerovalent iron dispersions. Environ. Sci. Technol. **41**(1), 284–290 (2007). doi:10.1021/es061349a
10. E. Climent, M.R. Maxey, G.E. Karniadakis, Dynamics of self-assembled chaining in magnetorheological fluids. Langmuir **20**(2), 507–513 (2004). doi:10.1021/la035540z
11. C. Yavuz, J. Mayo, W. Yu, A. Prakash, J. Falkner, S. Yean, L. Cong, H. Shipley, A. Kan, M. Tomson, D. Natelson, V. Colvin, Low-field magnetic separation of monodisperse Fe3O4 nanocrystals. Science **314**(5801), 964–967 (2006). doi:10.1126/science.1131475

12. H.N.W. Lekkerkerker, R. Tuinier, *Colloids and the Depletion Interaction* (Springer, Dordrecht, NY, 2011)
13. G. De Las Cuevas, J. Faraudo, J. Camacho, Low-gradient magnetophoresis through field-induced reversible aggregation. J. Phys. Chem. C **112**(4), 945–950 (2008). doi:10.1021/jp0755286
14. J. Faraudo, J. Camacho, Cooperative magnetophoresis of superparamagnetic colloids: theoretical aspects. Colloid Polym. Sci. **288**(2), 207–215 (2010). doi:10.1007/s00396-009-2107-z
15. J. Faraudo, J.S. Andreu, J. Camacho, Understanding diluted dispersions of superparamagnetic particles under strong magnetic fields: a review of concepts, theory and simulations. Soft Matt. **9**(29), 6654–6664 (2013). doi:10.1039/C3SM00132F
16. H.-J. Butt, *Physics and Chemistry of Interfaces*, 2nd., rev. and enl. edn. (Wiley-VCH, Weinheim, 2006)
17. B. Faure, G. Salazar-Alvarez, L. Bergström, Hamaker constants of iron oxide nanoparticles. Langmuir **27**(14), 8659–8664 (2011)
18. S. Bhattacharjee, M. Elimelech, M. Borkovec, DLVO interaction between colloidal particles: beyond Derjaguin's approximation. Croat. Chem. Acta **71**(4), 883–903 (1998)
19. C. Tsouris, T.C. Scott, Flocculation of paramagnetic particles in a magnetic field. J. Colloid Interface Sci. **171**(2), 319–330 (1995). doi:10.1006/jcis.1995.1186
20. K. Mandel, F. Hutter, The magnetic nanoparticle separation problem. Nano Today **7**(6), 485–487 (2012). doi:10.1016/j.nantod.2012.05.001
21. G. Liu, Q. Deng, H. Wang, S. Kang, Y. Yang, D.H.L. Ng, W. Cai, G. Wang, Synthesis and characterization of nanostructured Fe3O4 micron-spheres and their application in removing toxic Cr ions from polluted water. Chem. Eur. J. **18**(42), 13418–13426 (2012). doi:10.1002/chem.201200864
22. Y. Liu, Y. Wang, S. Zhou, S. Lou, L. Yuan, T. Gao, X. Wu, X. Shi, K. Wang, Synthesis of high saturation magnetization superparamagnetic Fe3O4 hollow microspheres for swift chromium removal. ACS Appl. Mater. Interfaces **4**(9), 4913–4920 (2012). doi:10.1021/am301239u
23. J.-S. Xu, Y.-J. Zhu, Γ-Fe2O3 and Fe3O4 magnetic hierarchically nanostructured hollow microspheres: preparation, formation mechanism, magnetic property, and application in water treatment. J. Colloid Interface Sci. **385**(1), 58–65 (2012). doi:10.1016/j.jcis.2012.06.082
24. H. Xu, L. Cui, N. Tong, H. Gu, Development of high magnetization Fe3O4/polystyrene/silica nanospheres via combined miniemulsion/emulsion polymerization. J. Am. Chem. Soc. **128**(49), 15582–15583 (2006). doi:10.1021/ja066165a
25. L. Cui, Developing a hybrid emulsion polymerization system to synthesize Fe3O4/polystyrene latexes with narrow size distribution and high magnetite content. J. Polym. Sci. A Polym. Chem. **45**(22), 5285–5295 (2007)
26. P. Qiu, C. Jensen, N. Charity, R. Towner, C. Mao, Oil phase evaporation-induced self-assembly of hydrophobic nanoparticles into spherical clusters with controlled surface chemistry in an oil-in-water dispersion and comparison of behaviors of individual and clustered iron oxide nanoparticles. J. Am. Chem. Soc. **132**(50), 17724–17732 (2010). doi:10.1021/ja102138a
27. J. Zhang, S. Xu, E. Kumacheva, Polymer microgels: reactors for semiconductor, metal, and magnetic nanoparticles. J. Am. Chem. Soc. **126**(25), 7908–7914 (2004). doi:10.1021/ja031523k
28. J. Ge, Y. Hu, M. Biasini, W. Beyermann, Y. Yin, Superparamagnetic magnetite colloidal nanocrystal clusters. Angew. Chem. Int. Ed. **46**(23), 4342–4345 (2007). doi:10.1002/anie.200700197
29. H. Deng, X. Li, Q. Peng, X. Wang, J. Chen, Y. Li, Monodisperse magnetic single-crystal ferrite microspheres. Angew. Chem. Int. Ed. **44**(18), 2782–2785 (2005). doi:10.1002/anie.200462551
30. W. Cheng, K. Tang, Y. Qi, J. Sheng, Z. Liu, One-step synthesis of superparamagnetic monodisperse porous Fe3O4 hollow and core-shell spheres. J. Mater. Chem. **20**(9), 1799–1805 (2010). doi:10.1039/b919164j

31. T. Leshuk, H. Krishnakumar, F. Gu, Size-tunable Fe3O4 spherical nanoclusters through a one-pot hydrothermal synthesis. J. Nanosci. Nanotechnol. (accepted Sept. 27, 2013, in press)
32. L. Wang, J. Li, Q. Jiang, L. Zhao, Water-soluble Fe3O4 nanoparticles with high solubility for removal of heavy-metal ions from waste water. Dalton Trans. **41**(15), 4544–4551 (2012). doi:10.1039/c2dt11827k
33. W. Cheng, K. Tang, J. Sheng, Highly water-soluble superparamagnetic ferrite colloidal spheres with tunable composition and size. Chem. Eur. J. **16**(12), 3608–3612 (2010). doi:10.1002/chem.201000014
34. V.K. LaMer, R.H. Dinegar, Theory, production and mechanism of formation of monodispersed hydrosols. J. Am. Chem. Soc. **72**(11), 4847–4854 (1950). doi:10.1021/ja01167a001
35. E. Matijevic, Preparation and properties of uniform size colloids. Chem. Mater. **5**(4), 412–426 (1993)
36. M.N. Chong, B. Jin, C.W.K. Chow, C. Saint, Recent developments in photocatalytic water treatment technology: a review. Water Res. **44**(10), 2997–3027 (2010). doi:10.1016/j.watres.2010.02.039
37. S. Malato, P. Fernandez-Ibanez, M. Maldonado, J. Blanco, W. Gernjak, Decontamination and disinfection of water by solar photocatalysis: recent overview and trends. Catal. Today **147**(1), 1–59 (2009). doi:10.1016/j.cattod.2009.06.018
38. G. Liu, L. Wang, H. Yang, H. Cheng, G. Lu, Titania-based photocatalysts-crystal growth, doping and heterostructuring. J. Mater. Chem. **20**(5), 831–843 (2010). doi:10.1039/B909930a
39. H. Han, R. Bai, Buoyant photocatalyst with greatly enhanced visible-light activity prepared through a low temperature hydrothermal method. Ind. Eng. Chem. Res. **48**(6), 2891–2898 (2009). doi:10.1021/ie801362a
40. X. Yu-Tang, X. Shuang-Shuang, D. Yong-Chao, F.Q. Shiang, Progress of novel TiO2 photocatalytic separation membrane. J. Inorg. Mater. **26**(4), 337–346 (2011). doi:10.3724/SP.J.1077.2011.00337
41. T. Leshuk, P. Everett, H. Krishnakumar, K. Wong, S. Linley, F. Gu, Mesoporous magnetically recyclable photocatalysts for water treatment. J. Nanosci. Nanotechnol. **13**(4), 3127–3132 (2013). doi:10.1166/jnn.2013.7396
42. D. Beydoun, R. Amal, G. Low, S. McEvoy, Occurrence and prevention of photodissolution at the phase junction of magnetite and titanium dioxide. J. Mol. Catal. A **180**(1–2), 193–200 (2002)
43. D. Beydoun, R. Amal, G. Low, S. McEvoy, Novel photocatalyst: titania-coated magnetite. Activity and photodissolution. J. Phys. Chem. B **104**(18), 4387–4396 (2000)
44. P. Innocenzi, Infrared spectroscopy of sol-gel derived silica-based films: a spectra-microstructure overview. J. Non-Cryst. Solids **316**(2–3), 309–319 (2003)
45. P. Innocenzi, P. Falcaro, D. Grosso, F. Babonneau, Order-disorder transitions and evolution of silica structure in self-assembled mesostructured silica films studied through FTIR spectroscopy. J. Phys. Chem. B **107**(20), 4711–4717 (2003). doi:10.1021/jp026609z
46. C. Parler, J. Ritter, M. Amiridis, Infrared spectroscopic study of sol-gel derived mixed-metal oxides. J. Non-Cryst. Solids **279**(2–3), 119–125 (2001)
47. Y. Deng, C. Wang, J. Hu, W. Yang, S. Fu, Investigation of formation of silica-coated magnetite nanoparticles via sol–gel approach. Colloids Surf. A **262**(1–3), 87–93 (2005). doi:10.1016/j.colsurfa.2005.04.009
48. S. Watson, J. Scott, D. Beydoun, R. Amal, Studies on the preparation of magnetic photocatalysts. J. Nanopart. Res. **7**(6), 691–705 (2005). doi:10.1007/s11051-005-7520-8
49. F. Chen, J. Zhao, Preparation and photocatalytic properties of a novel kind of loaded photocatalyst of TiO2/SiO2/gamma-Fe2O3. Catal. Lett. **58**(4), 245–247 (1999)
50. A.A. Ismail, D.W. Bahnemann, Mesoporous titania photocatalysts: preparation, characterization and reaction mechanisms. J. Mater. Chem. **21**(32), 11686 (2011). doi:10.1039/c1jm10407a

51. S. Li, Q. Shen, J. Zong, H. Yang, Simple preparation of sub-micron mesoporous TiO2 spheres consisting of anatase nanocrystals. J. Alloys Compd. **508**(1), 99–105 (2010). doi:10.1016/j.jallcom.2010.04.246
52. K. Yanagisawa, Y. Yamamoto, Q. Feng, N. Yamasaki, Formation mechanism of fine anatase crystals from amorphous titania under hydrothermal conditions. J. Mater. Res. **13**(4), 825–829 (1998). doi:10.1557/JMR.1998.0106
53. D. Kim, S. Kwak, The hydrothermal synthesis of mesoporous TiO2 with high crystallinity, thermal stability, large surface area, and enhanced photocatalytic activity. Appl. Catal. A **323**, 110–118 (2007). doi:10.1016/j.apcata.2007.02.010
54. T. Peng, D. Zhao, K. Dai, W. Shi, K. Hirao, Synthesis of titanium dioxide nanoparticles with mesoporous anatase wall and high photocatalytic activity. J. Phys. Chem. B **109**(11), 4947–4952 (2005). doi:10.1021/jp044771r
55. J. Chen, C. Chen, J. Liu, R. Xu, S. Qiao, X. Lou, Ellipsoidal hollow nanostructures assembled from anatase TiO2 nanosheets as a magnetically separable photocatalyst. Chem. Commun. **47** (9), 2631–2633 (2011). doi:10.1039/c0cc04471g
56. X. Lou, L. Archer, A general route to nonspherical anatase TiO2 hollow colloids and magnetic multifunctional particles. Adv. Mater. **20**(10), 1853–1858 (2008). doi:10.1002/adma.200702379
57. G. Wang, H. Wang, Y. Ling, Y. Tang, X. Yang, R.C. Fitzmorris, C. Wang, J.Z. Zhang, Y. Li, Hydrogen-treated TiO2 nanowire arrays for photoelectrochemical water splitting. Nano Lett. **11**(7), 3026–3033 (2011). doi:10.1021/nl201766h
58. M. Kong, Y. Li, X. Chen, T. Tian, P. Fang, F. Zheng, X. Zhao, Tuning the relative concentration ratio of bulk defects to surface defects in TiO2 nanocrystals leads to high photocatalytic efficiency. J. Am. Chem. Soc. **133**(41), 16414–7 (2011). doi:10.1021/ja207826q
59. J.B. Joo, Q. Zhang, I. Lee, M. Dahl, F. Zaera, Y. Yin, Mesoporous anatase titania hollow nanostructures though silica-protected calcination. Adv. Funct. Mater. **22**(1), 166–174 (2012). doi:10.1002/adfm.201101927
60. J. Liu, M. Li, J. Wang, Y. Song, L. Jiang, T. Murakami, A. Fujishima, Hierarchically macro-/mesoporous Ti-Si oxides photonic crystal with highly efficient photocatalytic capability. Environ. Sci. Technol. **43**(24), 9425–9431 (2009). doi:10.1021/es902462c
61. J. Lee, S. Kong, W. Kim, J. Kim, Preparation and characterization of SiO2/TiO2 core-shell particles with controlled shell thickness. Mater. Chem. Phys. **106**(1), 39–44 (2007). doi:10.1016/j.matchemphys.2007.05.019

# Chapter 8
# Removal of 4,6-Dinitro-*o*-Cresol, Congo Red Dye, and Decane from Water Using Magnetic-Activated Carbons

**Tarek Trad and Allen Apblett**

**Abstract** This chapter aims at developing magnetic extractants that can be utilized in conjunction with magnetic filtration devices to efficiently and economically remove a number of pollutants from aqueous solutions and mixtures. The material used for this study was based on activated carbon, where raw materials used for the development of high surface area-activated carbon were modified to produce novel magnetically active activated carbons (MACs). The unique properties and adsorption capacity of these materials allowed their application in the extraction of hydrocarbons from water and in breaking oil in water emulsions. An investigation on the ability of MACs for the removal of 4,6-dinitro-*o*-cresol pesticide, Congo red, and decane is presented.

## 8.1 Introduction

The widespread contamination of surface and ground waters by organic compounds can profoundly impact wildlife and drinking water quality. Nitroaromatic and phenolic compounds are among the most common water contaminants because they are widely used as pesticides, herbicides, chemical cleaning agents, and solvents. Due to their poor biodegradability, toxicity, and accumulation potential in animal and plant tissues, they represent an increasing environmental problem [1, 2]. As a result, these

T. Trad
Department of Chemistry and Environmental Sciences, University of Texas at Brownsville, 80 Fort Brown, Brownsville, TX 78526, USA

A. Apblett (✉)
Department of Chemistry, Oklahoma State University, 302 Physical Science 1, Stillwater, OK 74078, USA
e-mail: apblett@okstate.edu

A. Hu and A. Apblett (eds.), *Nanotechnology for Water Treatment and Purification*, Lecture Notes in Nanoscale Science and Technology 22,
DOI 10.1007/978-3-319-06578-6_8, 

materials have been classified as hazardous pollutants because of their harmful effects on human health via many possible routes of entry into the food chain [3]. Dyes and pigments, another major group of organic water pollutants, are generally found in the discharged effluents of textile, dyeing, leather, pulp and paper, and printing industries [4]. Their annual worldwide production comprises 70,000 t and 10,000 types, of which about 20–30 % are released into the environment through wastewaters and industrial effluents [5]. Many dyestuffs and pigments are structurally stable and difficult to biodegrade and many are considered to be carcinogenic and highly toxic to living beings [6]. Medical reports have shown that some dyes can cause dermatitis, skin irritation, allergy, and cancer in humans [7]. Thus, early removal of organic pollutants from industrial effluents before they get introduced into unpolluted natural waters is of significant interest to environmental agencies and researchers all over the world.

Basic water treatment methods including screening, aeration, sedimentation, flocculation, filtration (using sand beds), chlorination (disinfection), and hardness reduction are not particularly engineered to remove organic compounds and trace metals [8]. Moreover, using these processes that are important for the improvement of drinking water quality can create additional pollution problems through the incorporation of potentially toxic manufactured chemicals such as organic flocculants (acrylamide and epichlorohydrin) and disinfectants (chlorine, chloramine, or bromine). Therefore, the attainment of high quality drinking water is usually accomplished when advanced water treatment technologies are used. The most common of these technologies are the following [8]:

- Granular-activated carbon (GAC), used since the 1900s for the removal of dissolved organic compounds from water.
- Ion exchange resins, used for decades mainly to remove inorganic ions.
- Ultraviolet light, recently being evaluated for the treatment of aquatic viruses and bacteria, and oxidation of organics using photochemically generated hydroxyl radicals.
- Ozonation, used on a large scale in Europe to oxidize some organic compounds and as an alternative water disinfecting process (instead of bromine and chlorine).
- Membrane separation (reverse osmosis), a method developed in the early eighties for the removal of organic and inorganic contaminants from water.
- Organic polymers, a number of advanced polymers are introduced in the flocculation step in order to provide enhanced coagulation.

Magnetic filtration has emerged in the last few decades as a modern water treatment technology that can provide rapid, efficient removal of pollutants from aqueous waste streams. The electronic control over magnetic filters eliminates the need for mechanical contact, thus making the technology safer for water treatment plant workers by minimizing their exposure to harmful agents [9]. However, since most common pollutants are generally nonmagnetic, materials that can bind to environmental contaminants and be magnetically separated from aqueous solutions must be developed [9].

Activated carbon is considered to be one of the best technologies implemented in water purification systems and is the most used [10]. This material can be produced from a variety of carbonaceous solid precursors. Activated carbon's large surface areas, variable surface chemistry, and highly developed porosity account for their versatility in adsorbing pollutants [11]. These unique properties in combination with appropriate magnetic activity can provide the basis for developing highly effective magnetic adsorbents. In this study, the successful demonstration of the feasibility of synthesizing such adsorbents from low-cost carbonaceous materials is described, and the removal of a number of water pollutants using magnetic-activated carbons as single component magnetic extractants for magnetic filtration is discussed.

## 8.2 Experimental

### *8.2.1 Materials and Methods*

All ACS grade reagents used in this study were purchased either from Strem Chemicals or Aldrich Chemical Company. Along with HPLC grade hexane and other solvents, all of these chemicals were used as received without any further purification. Reverse osmosis and deionization methods were used to purify water.

### *8.2.2 Synthesis of Magnetic-Activated Carbons MAC-1 and MAC-2 Using Paper Towels as a Starting Material*

*MAC-1*: Iron(II) gluconate solution was prepared by dissolving 4.820 g of iron (II) gluconate in approximately 50 mL of distilled water. The solution was left to settle for 6 h at room temperature. Using a pipette, the solution was added dropwise to 10.10 g of paper towels to incipient wetness. The wet towels were left to dry under air for 24 h. After drying, the towels were shredded and placed in an open pyrex tube, then thermally treated under a nitrogen atmosphere at 500 °C in a tube furnace. Twelve hours later, the sample was left to cool under air at room temperature for a few hours. During the cooling step, the resulting powder was exposed to air at which point it became quite hot due to rapid oxidation of iron(II) oxide to magnetite. This phenomenon was observed for all iron oxide containing activated carbons. Fine grinding of the product and thorough washing with methylene chloride yielded a black powder that weighed 3.100 g and was magnetic in nature.

*MAC-2*: 15.00 g of commercial paper towels were treated with an iron(II) and nickel(II) gluconate solution. The solution was made by dissolving 4.821 g of iron (II) gluconate in 50 mL of water followed by mixing with 2.245 g of nickel (II) gluconate dissolved in 25 mL of water. Paper towels treated with the gluconate solution via the incipient wetness impregnation method were dried in air for 24 h then pyrolized under nitrogen at 500 °C for 12 h. The product from this reaction weighed 5.620 g and was crushed and finely ground to give a magnetic black powder. Finally the powder was washed several times with hexane and methylene chloride then dried under vacuum overnight.

### 8.2.3 Synthesis of MAC from Treated Sawdust

*MAC-3*: 100.00 g of coarse fir sawdust was sieved and treated with 50 mL of 10 % sulfuric acid solution. An iron(II) sulfate solution made by dissolving 5.00 g of $FeSO_4.7H_2O$ in 100 mL of water was added to the sawdust and the mixture was stirred over a hot plate for 30 min. After evaporating about 50 mL of the solvent, 20.32 g of the treated sawdust was packed in an open pyrex tube and fired at 500 °C under nitrogen for 4 h. This resulted in a yield of 6.41 g of black magnetic charcoal.

*MAC-4*: For the preparation of this sample, the same amounts of ferrous sulfate and sulfuric acid were used to impregnate the sawdust as was used for MAC-3. In this case, however, the treated sawdust was dried under air overnight. 15.70 g of it was initially pyrolized at 500 °C for 4 h under nitrogen. The temperature was decreased to 200 °C and the sample was maintained at this temperature for 5 h under air. Further cooling to room temperature allowed for the sample to be collected, ground, and weighed. The overall yield of this reaction was 4.74 g.

*MAC-5*: Preparation of this magnetic charcoal was accomplished according to the procedure mentioned above (MAC-4). However, a solution of 1:2 mol ratio Ni(II) gluconate to Fe(II) gluconate was used for treatment instead of the ferrous sulfate. Incipient wetness impregnation of the sawdust with the gluconate solution followed by its thermal treatment yielded the desired magnetic product. The initial weight of the treated sawdust prior to pyrolysis was 20.67 g. The final weight of the product was 5.85 g.

Similarly, *MAC-6* was prepared using nickelous sulfate and ferrous sulfate solution as a source for magnetic character.

Synthesis of activated carbon from sawdust without the addition of any iron or nickel salts was performed for comparative reasons. The reaction conditions were similar to those used in the preparation of magnetic charcoals derived from sawdust. This sample of activated carbon was designated as *SBAC*.

### 8.2.4 Synthesis of Cellulose-Based Magnetic-Activated Carbons

The matrix of this family of magnetic carbons was derived from microcrystalline cellulose. The following section describes the preparation of two magnetic powders, MAC-7 and MAC-8, along with an unmodified activated charcoal CBAC.

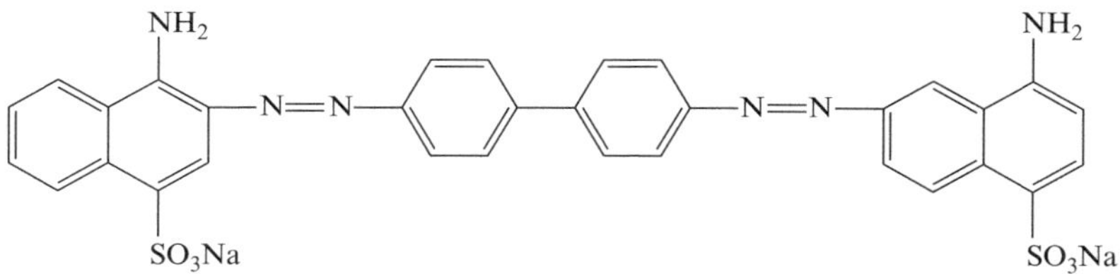

**Fig. 8.1** The molecular structure of Congo red

*MAC-7*: By incipient wetness impregnation, 30 mL of a concentrated iron (II) gluconate solution were added to 15.01 g of cellulose. 8 mL of 10 % $H_2SO_4$ solution were introduced to the wet cellulose by dropwise addition. A 21.58 g portion of the dried, treated cellulose was pyrolyzed and cooled according to the method for MAC-4 preparation. The reaction yielded 5.51 g of fine black magnetic powder.

*MAC-8*: In this case, a concentrated solution of $FeSO_4.7H_2O$ made by dissolving 7.50 g of the iron salt in 16 mL of water was used as an iron source instead of iron (II) gluconate. The amount of modified cellulose used for heat treatment was 23.96 g, and the yield of the reaction was 6.02 g.

CBAC is a cellulose-based activated charcoal synthesized using the technique mentioned in this section. The cellulose used for this activated carbon was only treated with a 10 % sulfuric acid solution, and dried before pyrolysis.

The samples selected for the adsorption of 6-dinitro-*o*-cresol (DNOC) were sawdust-based magnetic-activated carbon MAC-3, cellulose-based activated carbon CBAC, and cellulose-based magnetic-activated carbons MAC-7 and MAC-8. Commercial-activated carbon Darco G-60 (with a specific surface area of 651 $m^2/g$) purchased from Fisher Scientific was also used as received. Adsorption experiments were performed by reacting a fixed amount of the adsorbent (0.1 g) with a series of DNOC concentrations. Each mixture was left to react on a rotor mill for approximately 5 h. After calculating the equilibrium concentrations, adsorption capacities were determined using the Langmuir isotherm.

A magnetic-activated carbon made from sawdust (MAC-9, SSA = 258 $m^2/g$) was physically activated and tested for the adsorption of Congo red (Fig. 8.1) from an aqueous solution. An accurately weighed quantity of the dye was dissolved in the purified distilled water for the preparation of a stock solution (133.14 mg/L). Calibration curves between absorbance and concentration were constructed using a UV/Vis spectrophotometer. The concentrations of Congo red were determined using the absorbance at its characteristic wavelength of maximum absorption at 497 nm.

Low-cost magnetic extractants MAC-1 and MAC-2 were tested for the removal of decane from water. The adsorbents were directly mixed with a decane solution with known concentration, left to react for a few hours then magnetically filtered. The decane in the filtrate was extracted by liquid–liquid extraction using hexane, and the final decane concentration was determined by GC/MS as described below.

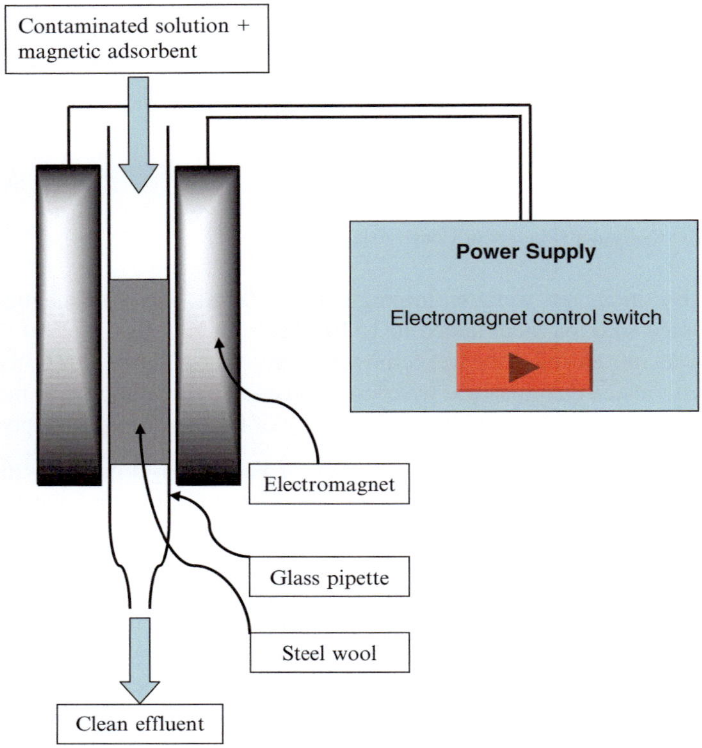


**Fig. 8.2** Sketch of the magnetic filtration device used in the water treatment procedure

### 8.2.5 *Magnetic Filter*

The lab scale magnetic filtration device was designed using a 24 V extended-reach-electromagnet with a 170 lb pull and dimensions 3″ × 1″ × 1.4″ and filtration columns. The electromagnet was powered by a 30 W direct current power supply. A filtration column constructed by packing glass pipettes loosely with fine grade steel wool attached to the latter. An illustration of this device is shown in Fig. 8.2.

### 8.2.6 *Analytical Measurements*

UV/Vis spectra were recorded on an Agilent 8453 UV-visible spectrophotometer that uses two light sources, a deuterium and a tungsten lamp. Agilent ChemStation software was used to analyze the generated data. The UV-visible spectrophotometer was used mainly with DNOC and Congo red dye extraction experiments.

Analysis of decane in treated solutions was performed using a Hewlett Packard G1800A gas chromatographic/mass spectroscopic (GC/MS) analysis instrument.

The temperature program used was an initial hold of 2 min at 35 °C, a ramp of 5 °C/min to 170 °C, and a final hold of 5 min. The instrument was equipped with a HP5 column (cross-linked 5 % PhME silicone) with dimensions of 30 m × 0.25 mm. The injection port was set at 250 °C and the helium flow rate was 1 mL/min. The mass spectrometer was used in the selective ion monitoring mode with the parent ion of decane ($m/e = 142$ a.m.u).

## 8.3 Results and Discussion

### 8.3.1 *Adsorption of DNOC on Sawdust and Cellulose-Based Magnetic-Activated Carbons*

For this application, all experiments were performed at room temperature using 20 mL glass vials. The procedure consisted of maintaining a fixed amount of the adsorbent and varying the initial concentrations of the DNOC aqueous solutions. In each case, approximately 0.1 g of the adsorbent was added directly to the reaction vessel containing 10 mL of the DNOC solution. The series of DNOC concentrations in the aqueous solutions ranged from 20 to 40 ppm. The capped vials were mixed for 5 h at around 70 rpm on a rotor mill in order to reach equilibrium. After equilibrium was attained, the adsorbent was magnetically retrieved using the previously illustrated magnetic filter. Equilibrium concentrations were determined using UV/Vis spectrophotometry, where the parent peak at maximum wavelength $\lambda_{max} = 369$ nm presented by the DNOC spectrum was used to quantify its concentration. The adsorbed amount (mg) of DNOC was calculated from the difference between initial and equilibrium concentrations multiplied by the volume of solution used in liters.

A defined distribution of solute between solid and liquid phases exists at equilibrium [12]. At a given temperature, the relationship between the solute equilibrium concentration in a solution and the amount of solute adsorbed on the solid is normally expressed in an adsorption isotherm. A variety of different adsorption equations describing liquid phase adsorption for single-component systems have been proposed [13]. Basically, some of these models rely on theoretical foundations and some are of an empirical nature [12]. For the purpose of describing adsorption equilibria, the Langmuir isotherm equation is one of the simplest and most widely used. The derivation of the Langmuir equation is based on both statistical and thermodynamic considerations [14, 15]. The basic assumption in the Langmuir model is localized adsorption, with only one molecule per adsorption site, no interaction between adsorbed molecules, and with equal adsorption energies for all active sites. Moreover, only one layer of adsorbed molecules can be formed on the surface [15]. The shape of the adsorption isotherms can be approximated by the Langmuir equation generally expressed as

$$X = \frac{\theta^{o} K C_{eq}}{1 + K C_{eq}} \tag{8.1}$$

where $X$ is the amount of DNOC adsorbed per unit weight of adsorbent (mg/g), it can also be referred as the adsorption density at the equilibrium solute concentration ($C_{eq}$), $C_{eq}$ is the equilibrium concentration of DNOC in aqueous solutions (mg/L), $\theta^{\circ}$ (mg/g) is the maximum adsorption capacity corresponding to the complete adsorbate coverage of an adsorbent monolayer, and it's one of Langmuir constants, $K$ (l/mg) is the Langmuir constant related to the energy of adsorption by an Arrhenius-type equation expressed in the form:

$$K = K_{o} e^{\frac{-E}{RT}} \tag{8.2}$$

where $K_o$ is a constant containing the entropy, $E$ is the adsorption energy, $R$ is the universal gas constant, and $T$ is the absolute temperature.

The rearrangement of Eq. (8.1) gives the following linear form:

$$\frac{C_{eq}}{X} = \frac{C_{eq}}{\theta^{o}} + \frac{1}{\theta^{o} K} \tag{8.3}$$

A linear plot of $C_{eq}/X$ versus $C_{eq}$ using this equation implies that the Langmuir model fits the generated data. Figures 8.3 and 8.4 show the Langmuir isotherm plots for the adsorption of DNOC using various adsorbents. The monolayer adsorption capacities of the magnetic extractants in comparison to the nonmagnetic commercial charcoal (CAC) along with the correlation coefficients are presented in Table 8.1. The coefficient of determination, $R^2$, values of the linear relationships are very close to unity, which demonstrates the applicability of the Langmuir model. The monolayer adsorption capacities of the different extractants are of the following order: CBAC < MAC-3 < MAC-7 < MAC-8 < CAC. As observed, cellulose and sawdust-based magnetic adsorbents have comparable adsorption capacities as the Darco G-60-activated carbon. However, the latter's surface area is more than double that of the magnetic extractants. Thus the surfaces of the magnetic-activated carbons have higher capacity for DNOC per unit area than the commercial-activated carbon. Consequently, it is necessary to discuss the possible factors involved in this adsorption process. One of the basic facts related to adsorption on activated carbon surfaces is the presence of oxygen surface complexes such as carbonyl, carboxyl, phenol, anhydride, lactone, quinone, hydroquinone, and ether. The dissociation of these oxygen-containing functional groups if they possess an acidic or basic character will result in the presence of negative and positive surface charges needed for adsorption [11]. Additionally, the participation of graphene layers in π–π dispersive interactions with aromatic molecules will lead to their adsorption on the surface. The adsorption capacity of CBAC, an activated carbon made from pure cellulose, was found to be less than that of its magnetic derivatives MAC-7 and MAC-8. Hence, it can be concluded that in addition to the previously mentioned factors, the presence of nanoparticulate iron oxides and ferrites throughout the carbon matrix of the MACs

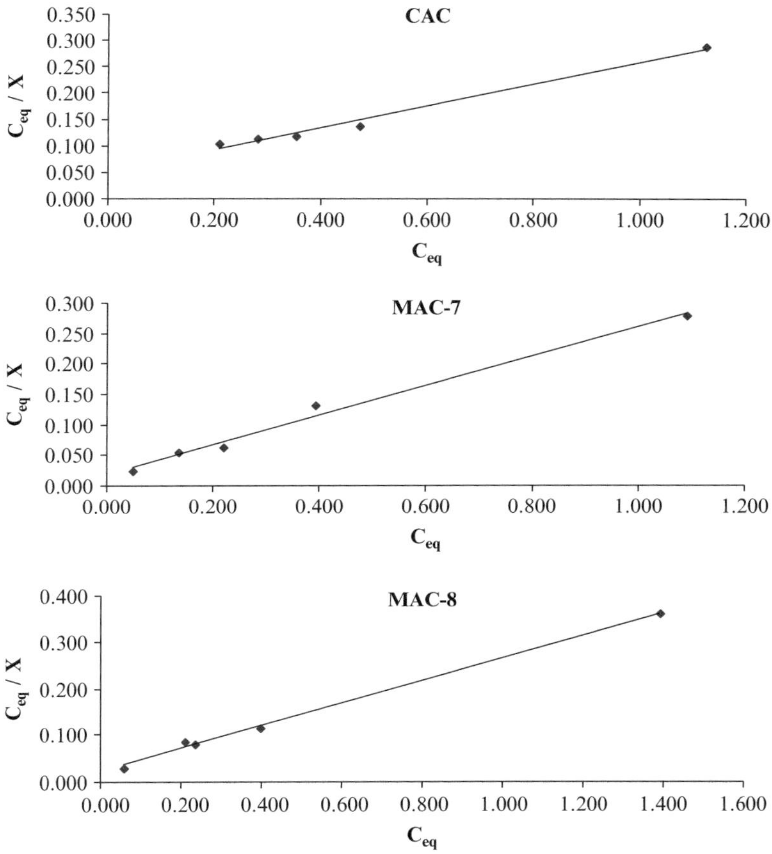


**Fig. 8.3** Langmuir isotherms for the adsorption of DNOC using commercial-activated carbon, and cellulose-derived activated carbon/iron oxide composites, MAC-7 and MAC-8

plays an important role in enhancing the materials' adsorption capacity perhaps by also adsorbing the DNOC. In regards to the difference in performance between sawdust (MAC-3) and cellulose (MAC-7 and MAC-8)-based magnetic adsorbents, the slight advantage held by the latter might be due to their higher organic character.

### *8.3.2 Adsorption Kinetics of DNOC Using MAC-7*

Adsorption of 40 ppm DNOC solutions onto cellulose-based magnetic-activated carbon of highest surface area, MAC-7, was monitored for several hours. Plotting the measured concentrations as a function of time gives the graph shown in Fig. 8.5.

In order to investigate the adsorption process of DNOC on the magnetic adsorbent, first order and second order kinetic models represented in Eqs. (8.4) and (8.5) respectively were applied [16].

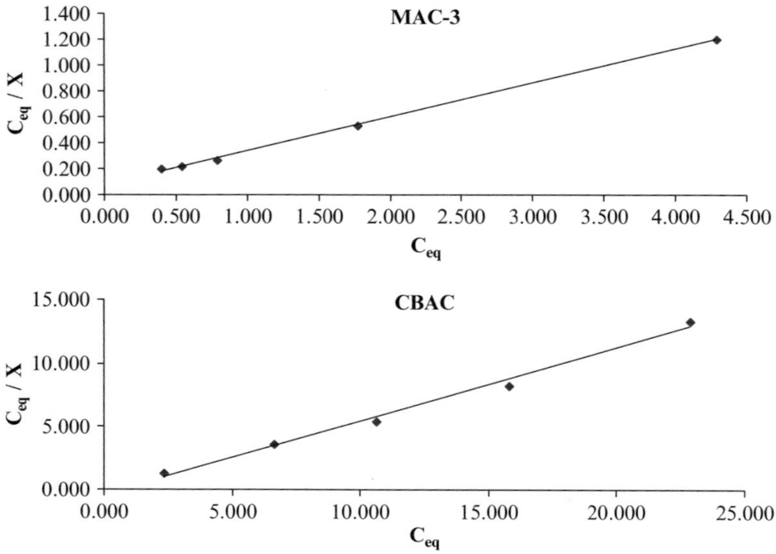


**Fig. 8.4** Langmuir isotherm for the adsorption of DNOC using sawdust-derived activated carbon/iron oxide composite (MAC-3), and cellulose-derived activated carbon (CBAC)

**Table 8.1** Adsorption capacities, coefficients of determination, and surface areas of magnetic and nonmagnetic-activated charcoals used in the adsorption of DNOC

| Adsorbent | SSA $m^2/g$ | $R^2$ | $\theta°$ (maximum adsorption mg\g) |
|---|---|---|---|
| Commercial-activated carbon | 651 | 0.98 | 4.83 |
| MAC-7 | 288 | 0.98 | 4.10 |
| MAC-8 | 279 | 0.99 | 4.11 |
| MAC-3 | 273 | 0.99 | 3.84 |
| CBAC | 237 | 0.99 | 1.73 |

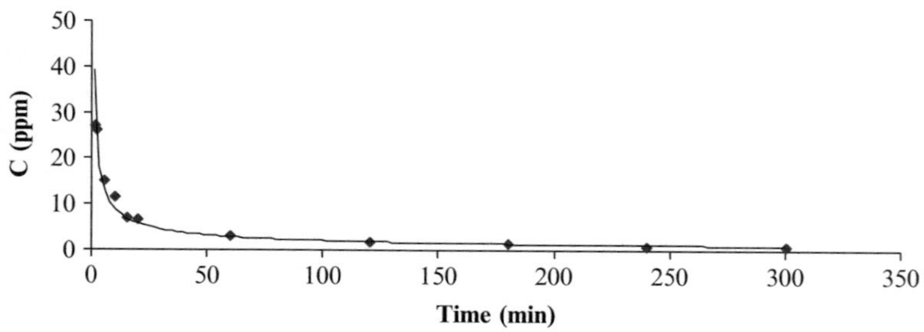


**Fig. 8.5** Adsorption behavior of DNOC from aqueous solutions of 40 ppm initial concentration onto MAC-7 as a function of time

**Table 8.2** Rate constants and coefficients of determination obtained from linear plots of the two kinetic models

| Kinetic model | Rate constants | $R^2$ |
|---|---|---|
| First order (1–15 min) | $k_1 = 0.0939$ | 0.965 |
| First order (60–240 min) | $k_2 = 0.0067$ | 0.997 |
| Second order | $k_3 = 0.0038$ | 0.993 |

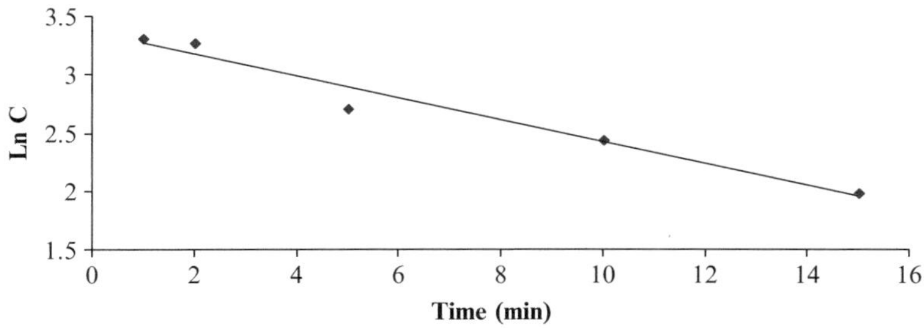


**Fig. 8.6** First order plot of lnC versus *t* (1–15 min) for DNOC adsorption using cellulose-derived magnetic-activated carbon (MAC-7)

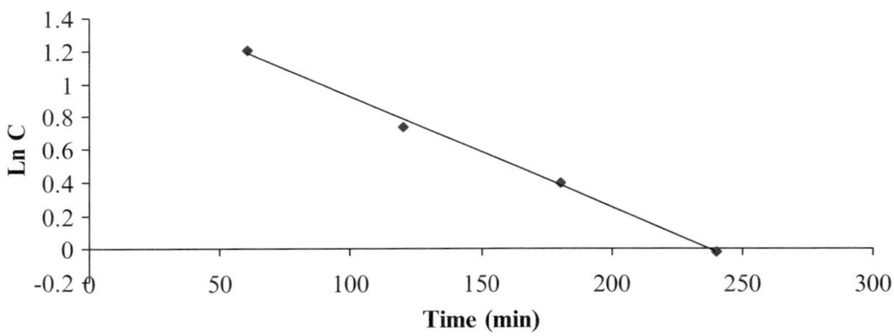


**Fig. 8.7** First order plot of lnC versus *t* (60–240 min) for DNOC adsorption using cellulose-derived magnetic-activated carbon (MAC-7)

$$\text{Ln } C - \text{Ln } C_i = -k_1 t \tag{8.4}$$

$$\frac{1}{C} - \frac{1}{C_i} = k_2 t \tag{8.5}$$

where $C_i$ is the initial concentration of adsorbate, $C$ is the final concentration of the treated solution at time $t$. $k_1$ and $k_2$ are the rate constants for first order and second order models, respectively.

The applicability of the kinetic models was tested by constructing linear plots of Ln $C$ versus $t$ for first order model and $1/C$ versus $t$ for second order model. Table 8.2 gives the rate constants determined from the slopes of the linear plots along with the coefficients of determination $R^2$. When these $R^2$ values are compared, it can be predicted that the first order model plotted in Figs. 8.6 and 8.7 seems to be

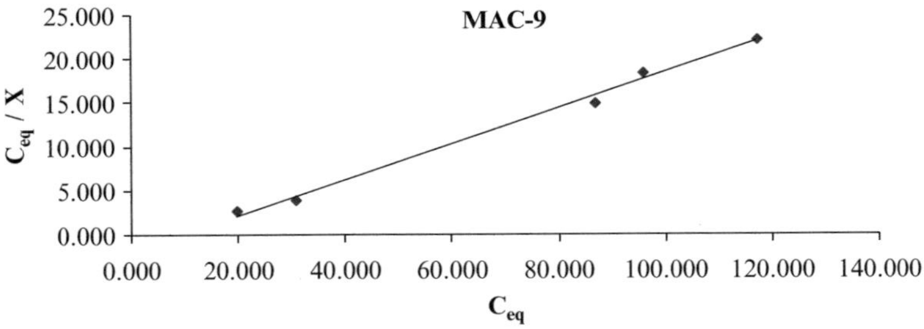


**Fig. 8.8** Langmuir isotherm for the adsorption of CR-dye using sawdust-derived activated carbon/nickel ferrite composite (MAC-9)

successfully applicable in this case. The time needed to reach a definite fraction of equilibrium adsorption was previously reported to be independent of initial concentrations [7]. Similar observations indicating a first order process of Congo red adsorption on various activated carbons were also reported [6]. Due to the similarity in nature for the different adsorbents used in this project, the first order sorption kinetics can be extrapolated for the rest of the magnetic extractants.

### *8.3.3 Removal of Congo Red Dye from Aqueous Solution Using MAC-9*

After the Congo red dye is dissolved in an aqueous solution, the dissociation of the sulfonate groups generates the anionic dye ions.

$$\text{R-SO}_3\text{Na} \xrightarrow{\text{H}_2\text{O}} \text{R-SO}_3^- + \text{Na}^+$$

Congo red ions exhibit a tendency to aggregate in aqueous solutions [17]. Therefore due to this aggregation and the large size of the Congo red molecule, the adsorption is more likely to take place in the mesopores [18]. It was efficiently adsorbed by the magnetic-activated carbon MAC-9. This successful adsorption can be attributed to a large number of mesopores within the adsorbent's structure in addition to the surface's partial occupation with ferrite particles. Upon water contact, ferrite particles develop a number of positively charged sites on the surface of the magnetic-activated carbons. Consequently an increase in electrostatic attraction between the adsorbent's positively charged surface and the anionic dye is believed to enhance the adsorption capacity of the magnetic adsorbent.

Figure 8.8 show the Langmuir adsorption isotherm of Congo red dye by MAC-9. The adsorption capacity is equal to 4.92 mg/g which exceeds that of commercial carbons recently used for the same purpose (0.636 and 1.875 mg/g) [18].

The effect of adsorbent dosage on the removal of Congo red was studied by plotting the mass of the adsorbent against the equilibrium concentration. An increase in the dye removal was observed to a certain limit. Increasing the

**Table 8.3** Results from treatment of aqueous solutions of decane with paper towel-derived magnetic-activated carbons (MAC-1 and MAC-2)

| Adsorbent | Initial concentration (ppm) | Final concentration (ppb) | Percent decane removal (%) |
|---|---|---|---|
| MAC-1 | 713 | 56.0 | 99.99 |
| MAC-2 | 713 | 62.4 | 99.99 |

adsorbent dosage will lead to higher adsorbent surface area which results in more availability of adsorption sites. Hence, the increase in adsorption can be directly attributed to the increase in adsorbent dosage. However, this increase in adsorbent dosage is accompanied by a decrease in unit adsorption. It is of importance to note that at lower concentrations (30 ppm), percent dye removal as high as 99.63 % was reached using 0.2 g of the magnetic adsorbent in 7 mL CR solution.

### *8.3.4 Removal of Decane from Water Using MACs Synthesized from Impregnated Paper Towels*

The testing of the magnetic extractants (MAC-1 and MAC-2) made from iron and nickel gluconate impregnated paper towels for the extraction of decane from aqueous solutions with a concentration of 713 ppm was performed. This concentration falls below the solubility limit of decane in water, which provides a stringent separation challenge for the MAC since the adsorption of decane must be from solution and not from a nonaqueous phase that is much easier to separate from water. The low ratio of adsorbent to solution was 2.5 % and the contact time with aqueous solutions was kept short, a matter of only a few hours. After the magnetic filtration step, liquid–liquid extraction was performed on the aqueous solution using hexane and the extracts were analyzed for decane by GC/MS. Both extractants removed significant quantities of decane from the aqueous solutions. The results for these experiments are outlined in Table 8.3.

A stable emulsion was prepared according to the method of Shin and Kim and the ability to break it was tested using the nickel ferrite/activated carbon extractant (MAC-2) [19]. With such an emulsion, MAC-2 was briefly mixed then separated by magnetic filtration. Determining the effectiveness of treatment was attained by measuring the transmittance of the emulsion at 500 nm before and after treatment. A significant increase in transmission from 2 to 62 % was observed. In terms of absorbance, a value of 1.66 was recorded before treatment while after it was found to be 0.21. Thus, magnetic extractants are capable of breaking emulsions and these initial results suggest that complete breaking of an oil in water emulsion could be accomplished by optimizing the magnetic extractants. Notably, it is likely that real world samples may not be as challenging since the emulsion used in this investigation was extremely stable and very high in surfactant [20].

## 8.4 Conclusions

It was found that low-cost magnetic-activated carbons are effective adsorbents for the removal of DNOC, Congo red dye, and decane from aqueous solutions via magnetic filtration. Additionally, successful separation of oil from aqueous emulsions was also possible using the same technique. The experimental adsorption equilibrium data for DNOC and Congo red on all the magnetic adsorbents studied show very good fit with the Langmuir isotherm. From a practical point of view, further optimization of the MACs is expected to insure their status as promising magnetic adsorbents for the magnetic separation of organic species from water.

## References

1. A. Ribeiro, M.H. Neves, M.F. Almeida, A. Alves, L. Santos, J. Chromatogr. A **975**, 267–274 (2002)
2. F. Salvador, M.D. Merchan, Carbon **34**, 1543–1551 (1996)
3. Calace, N.; Nardi, E.; Petronio, B. M.; Pietroletti, M. *Environmental Pollution (Oxford, United Kingdom)* **2002**, *118*, 315–319.
4. J. Easton, *Colour in dye house effluent* (Alden Press, Oxford, 1995)
5. B. Khan, P. Gupta, S. Jala, D. Goyal, Ind. J. Environ. Protect **23**, 415–418 (2003)
6. N. Kannan, Meenakshisundaram, M. Water Air Soil Pollut. **138**, 289–305 (2002)
7. A. Bhatnagar, A.K. Jain, M.K. Mukul, Environ. Chem. Lett. **2**, 199–202 (2005)
8. P.J. Sullivan, F.J. Agardy, J.J.J. Clark, *The environmental science of drinking water; 1st ed* (Elsevier Butterworth-Heinemann, Burlington, MA, 2005)
9. A. Apblett, S.M. Al-Fadul, T.M. Trad, Ceram Trans. **143**, 15–22 (2003)
10. W.E. Rashwan, B.S. Girgis, Adsorpt. Sci. Tech **22**, 181–194 (2004)
11. C. Moreno-Castilla, J. Rivera-Ultrilla, MRS Bull **26**, 890–894 (2001)
12. A.D. Cota-espericueta, *New Mexico State University*, 2003
13. S.A. Dastgheib, *New Mexico State University*, 2001
14. A. Clark, *The theory of adsorption and catalysis* (Academic, New York, 1970)
15. I. Langmuir, J. Am. Chem. Soc. **40**, 1361–1402 (1918)
16. E. Ayranci, N. Hoda, Chemosphere **60**, 1600–1607 (2005)
17. Coates, E. *J. Soc. Dyers Colour.* **1969**, 355-365.
18. I.D. Mall, V.C. Srivastava, N.K. Agarwal, I.M. Mishra, Chemosphere **61**, 492–501 (2005)
19. S.H. Shin, D.S. Kim, Environ. Sci. Tech. **35**, 3040–3047 (2001)
20. Allen W. Apblett, S. M. A., and Tarek Trad In *Integrated Petroleum Environmental Consortium* Tulsa, Oklahoma, 2001.

# Chapter 9
# Effects of Nanotechnologies on Disinfection By-product Formation

**Aleksandra Sokolowski, Stephanie Gora, and Susan Andrews**

**Abstract** Nanotechnologies such as $TiO_2$ photocatalysis (UV/$TiO_2$ or solar/$TiO_2$), silver nanoparticles, carbon nanotubes, and chitosan are currently being evaluated for use in drinking water and wastewater treatment applications. $TiO_2$ photocatalysis, in particular, is an attractive option for drinking water treatment because it can provide concurrent disinfection, disinfection by-product (DBP) precursor reduction, and degradation of recalcitrant compounds. Like any oxidation-based process, however, it has the potential to contribute to the formation of by-products, whether by directly oxidizing precursor compounds to form degradation products with potential health effects or indirectly by breaking down innocuous water constituents to form DBP precursors that can later be converted to DBPs when chemicals such as chlorine and chloramines are applied for disinfection. Research into DBP formation in water treatment processes employing $TiO_2$ photocatalysis is limited, but the studies that have been conducted to date suggest that these technologies create few DBPs directly and have site-specific effects on DBP precursor formation. The latter are also highly dependent on system design parameters, in particular UV dose and $TiO_2$ configuration or concentration. UV/$TiO_2$ compares favorably to existing disinfection and DBP precursor removal technologies in terms of both DBP formation and DBP precursor removal. Hybrid systems incorporating UV/$TiO_2$ along with another DBP precursor removal technology such as coagulation, adsorption, or membrane filtration are also promising for both applications.

A. Sokolowski • S. Gora • S. Andrews (✉)
Drinking Water Research Group, Department of Civil Engineering, University of Toronto, 35 St. George Street, Toronto, ON, Canada M5S 1A4
e-mail: sandrews@civ.utoronto.ca

A. Hu and A. Apblett (eds.), *Nanotechnology for Water Treatment and Purification*, Lecture Notes in Nanoscale Science and Technology 22,
DOI 10.1007/978-3-319-06578-6_9, 

## 9.1 Introduction

Since the early 1970s, research has demonstrated that chlorination, a process commonly used to disinfect drinking water, has the unintended consequence of forming disinfection by-products (DBPs). Approximately 600 DBPs have been identified over the decades but many more remain unidentified. Both identified and unidentified DBPs are suspected of having human health implications including carcinogenicity [1]. The formation of DBPs by alternative disinfection processes, such as chloramination, ozone, UV, advanced oxidation processes (AOPs), and nanotechnologies such as $TiO_2$ photocatalysis, is an active area of research [1–4]. Some of these nanotechnologies have been shown to reduce DBP formation by removing DBP precursors or by replacing a disinfection process that creates DBPs. However, they can also contribute new DBPs either by directly forming DBPs or by creating DBP precursors.

This chapter reviews the current body of knowledge and practices used in the drinking water industry to control the formation of DBPs and the possible impacts that nanotechnologies such as $TiO_2$ photocatalysis will have on these processes. A short summary of existing disinfection practices and processes presented in Sects. 9.2 and 9.3 provides a description of the human health concerns associated with DBPs, their precursors and formation pathways, and the standards and guidelines that are currently used to regulate them. An analysis of the direct and indirect DBPs formed in processes incorporating $TiO_2$ nanomaterials is provided in Sect. 9.4, focusing on $TiO_2$ photocatalysis because currently it is the only nanotechnology that has been researched in significant detail with respect to DBP formation. $TiO_2$ photocatalysis has been shown to lower DBP formation by removing DBP precursors ahead of disinfection and by providing an alternative source of disinfection. Finally, in Sect. 9.5, $TiO_2$ photocatalysis is compared to existing drinking water treatment processes in terms of the direct formation of DBPs and its ability to form or degrade DBP precursors.

## 9.2 Disinfection

### *9.2.1 Primary and Secondary Disinfection*

Drinking water treatment disinfection processes are generally classified as either primary or secondary. Primary disinfection kills or inactivates microorganisms including pathogenic viruses, bacteria, and protozoa. Secondary disinfection is the application of a long lasting disinfectant that remains active after water leaves a treatment plant and enters a distribution system, protecting the distribution system and ensuring that water quality objectives are met at the point of use. Nanotechnologies are more likely to be applied for primary disinfection because of concerns related to the potential toxicity of the nanomaterials themselves [5, 6].

### 9.2.2 Existing Disinfection Processes

Chlorine was first employed for drinking water disinfection approximately 100 years ago and continues to be incorporated into engineered water treatment systems worldwide. It is a strong oxidant that inactivates microorganisms by attacking cell walls, causing lysis and death. Despite its implication in the formation of DBPs, it continues to be popular as both a primary and a secondary disinfectant because it is well understood, has fast kinetics, can be safely stored and handled, and is generally considered to be economical. Chlorine is added to water as sodium hypochlorite (NaOCl) or chlorine gas ($Cl_2$), forming hypochlorous acid (HOCl), and hypochlorite ion ($OCl^-$) upon equilibrium with water according to Eqs. (9.1) and (9.2):

$$Cl_2 + H_2O \rightarrow HOCl + H^+ + Cl^- \quad (9.1)$$

$$HOCl \leftrightarrow OCl^- + H^+ \quad (9.2)$$

As society has become more aware of the potential human health effects of DBPs from chlorination, alternative disinfection processes have been developed in an attempt to decrease DBP formation [7]. Common alternative disinfection processes include chloramination, chlorine dioxide ($ClO_2$), ozone ($O_3$), and low or medium pressure ultraviolet light (UV) for primary disinfection, and chloramination and chlorine dioxide for secondary disinfection. Although AOPs are typically employed for contaminant removal, they also provide some disinfection capability and can contribute to primary disinfection.

Chloramination has become common for secondary disinfection because it lasts longer than chlorine in the distribution system [8] and forms lower levels of most regulated DBPs. The term "chloramines" refers to a group of three similar compounds: monochloramine ($NH_2Cl$), dichloramine ($NHCl_2$), and trichloramine ($NCl_3$). They are created when chlorine reacts with ammonia as described as shown in Eqs. (9.3–9.5) below [9].

$$HOCl + NH_3 \rightarrow NH_2Cl + H_2O \quad (9.3)$$

$$HOCl + NH_2Cl \rightarrow NHCl_2 + H_2O \quad (9.4)$$

$$HOCl + NHCl_2 \rightarrow NCl_3 + H_2O \quad (9.5)$$

Monochloramine is most often used as a secondary disinfectant since dichloramine and trichloramine can impart an unpleasant taste and odor to the water. Chloramine residuals have also been linked to negative health effects in dialysis patients and on aquatic organisms [10].

Chlorine dioxide is popular in Europe as a primary and secondary water disinfectant and has also been applied for taste and odor control [11]. It produces a longer lasting residual in the distribution system than free chlorine [8] and is comparable in terms of oxidative strength [9].

Ozone is a very strong oxidant and is so reactive that it is not able to provide any residual effect. It is used as a primary disinfectant and either directly oxidizes cell components or decays to hydroxyl ($HO^\bullet$) radicals that subsequently oxidize cell components, effectively killing most microorganisms [9].

UV light has been used for water disinfection for many decades but has only recently become popular in municipal drinking water and wastewater applications. Commercially available UV lamps emit light at wavelengths between 200 and 300 nm, which is commonly referred to as the "germicidal range." Light at these wavelengths can be absorbed by cell components, particularly DNA, RNA, and, to a lesser extent, proteins [12]. UV light absorption disrupts the DNA and RNA components, preventing cell replication and eventually resulting in cell inactivation. Like ozone, UV is used exclusively for primary disinfection as it does not provide a measurable disinfectant residual to prevent bacterial regrowth in the distribution system. Two types of UV lamps are commonly used in drinking water and wastewater applications, low pressure (LP) mercury vapor lamps and medium pressure (MP) mercury vapor lamps. LP lamps emit almost all of their light at approximately 254 nm and have a relatively low power output. MP lamps emit light over a wider spectrum and have a higher power output. A fluence (UV dose) of approximately 40 $mJ/cm^2$ is usually assumed to be adequate for disinfection, though fluences of up to 186 $mJ/cm^2$ are required to inactivate some organisms [12].

AOPs that combine two or more traditional disinfection processes have recently been introduced to some water treatment plants to degrade recalcitrant contaminants as well as to provide some concurrent disinfection. They have been shown to be effective in the removal of emerging water contaminants including synthetic organic compounds (SOCs) such as pesticides, pharmaceuticals, and personal care products. They rely on the powerful $HO^\bullet$ radical to oxidize contaminants and include $UV/H_2O_2$, $UV/O_3$, and $H_2O_2/O_3$. UV fluences of as much as 50 times that required for disinfection are required for some AOP applications [12].

### *9.2.3 Nanotechnologies for Drinking Water Disinfection*

Several nanotechnologies are emerging as viable and economical process elements for treating water. The term nanotechnology encompasses a wide variety of processes that employ nanometer-sized materials as disinfectants, catalysts, oxidants, or to create nano-sized filters for physical separation. For example, nano-sized particles of titanium dioxide ($TiO_2$), zinc oxide (ZnO), and fullerol can be used as photocatalysts to produce reactive oxygen species (ROS) that can either disinfect water or degrade contaminants [5]. Aqueous iron acts as a catalyst for oxidation/reduction reactions and has been used in conjunction with hydrogen peroxide ($H_2O_2$) to create $HO^\bullet$ radicals and ROS [13]. Nano-sized iron may prove to be an effective disinfectant [5]. Peptides, chitosan, carboxyfullerene, carbon nanotubes, nano-sized silver (nAg), and aqueous fullerene ($nC_{60}$) have been shown to effectively destroy waterborne microorganisms by compromising the cell membrane and/or wall, interrupting energy transduction, and/or inhibiting enzyme activity and DNA synthesis [5]. Finally, nanoporous membranes made of various materials can be used to remove microorganisms and other contaminants through size exclusion [14].

## 9.3 Disinfection By-products

DBPs are formed when naturally occurring or anthropogenic water contaminants are oxidized or otherwise transformed by a disinfection process. It has been suggested that DBPs will form any time an oxidant is used in water treatment [5]. Oxidants are also used for taste, odor, contaminant, hydrogen sulfide, color, iron, and manganese removal [9] and these processes may also form what are typically considered disinfection by-products. In addition, some unit treatment processes will indirectly affect DBP formation by either adding or removing DBP precursors. By knowing which water contaminants are DBP precursors, treatment processes may be optimized to degrade or remove them and reduce or prevent DBP formation. DBP precursor removal has the additional benefit of decreasing membrane fouling; biofilm growth in distribution systems; and taste, color, and odor issues. Although disinfectants and oxidants are used in drinking, sanitary, industrial, and recreational water treatment, most research on DBPs has been in the context of drinking water disinfection. No currently regulated DBPs have been definitively linked to the nanotechnologies that have been proposed for drinking water treatment.

To date, approximately 500–600 DBPs have been positively identified in treated drinking water; however, researchers suspect that many more DBPs remain unidentified and effectively unknown [1, 2, 4]. Many classes of halogenated DBPs are known because the identification of DBPs started from chlorination. Known halogenated DBP classes include halomethanes, haloacids, halonitriles, haloketones, haloaldehydes, halonitromethanes, haloamides, halofuranones, haloacetamides, haloacetonitriles, and oxyhalides. Known nonhalogenated classes of DBPs include nitrosamines, ketones, aldehydes, and carboxylic acids. A listing of the most well-known DBPs is provided in Table 9.1. Potential unidentified DBPs may include compounds from the chemical classes of bromoamines, nitrosamines, chloroketamines, chloroaldamines, polycarboxylic acids and polyamines, organic peroxides, and reaction products of disinfectants with carbohydrates or alkylphenol ethoxylates (used as surfactants) [2]. The types and concentrations of DBPs formed will vary from utility to utility depending on the type of disinfectant applied, available precursors, and overall water quality. For example, the DBPs formed at a chlorination demonstration plant were 2.8 % bromochloroacetic acid, 10 % haloacetic acids ($HAA_5$), 1.5 % chloral hydrate, 2 % haloacetonitriles, 20.1 % trihalomethanes (THMs), 1 % cyanogen chloride, and 62.4 % unidentified organic halides [2]. However, the DBPs formed at one ozone treatment plant were 26 %

**Table 9.1** Disinfection by-products identified in drinking water (Adapted from [9])

| Byproduct | Chemical agent | Molecular formula |
|---|---|---|
| *Trihalomethanes* (THMs) | | |
| Chloroform | Chlorine | $CHCl_3$ |
| Bromodichloromethane | Chlorine | $CHBrCl_2$ |
| Dibromochloromethane | Chlorine | $CHBr_2Cl$ |
| Bromoform | Chlorine, ozone | $CHBr_3$ |
| Dichloroiodomethane | Chlorine | $CHICl_2$ |
| Chlorodiiodomethane | Chlorine | $CHI_2Cl$ |
| Bromochloroiodomethane | Chlorine | $CHBrICl$ |
| Dibromoiodomethane | Chlorine | $CHBr_2I$ |
| Bromodiiodomethane | Chlorine | $CHBrI_2$ |
| Triiodomethane | Chlorine | $CHI_3$ |
| *Haloacetic acids* (HAAs) | | |
| Monochloroacetic acid | Chlorine | $CH_2ClCOOH$ |
| Dichloroacetic acid | Chlorine | $CHCl_2COOH$ |
| Trichloroacetic acid | Chlorine | $CCl_3COOH$ |
| Bromochloroacetic acid | Chlorine | $CHBrClCOOH$ |
| Bromodichloroacetic acid | Chlorine | $CBrCl_2COOH$ |
| Dibromochloroacetic acid | Chlorine | $CBr_2ClCOOH$ |
| Monobromoacetic acid | Chlorine | $CH_2BrCOOH$ |
| Dibromoacetic acid | Chlorine | $CHBr_2COOH$ |
| Tribromoacetic acid | Chlorine | $CBr_3COOH$ |
| *Haloacetonitriles* (*HAN*) | | |
| Trichloroacetonitrile | Chlorine | $CCl_3CN$ |
| Dichloroacetonitrile | Chlorine | $CHCl_2CN$ |
| Bromochloroacetonitrile | Chlorine | $CHBrClCN$ |
| Dibromoacetonitrile | Chlorine | $CHBr_2CN$ |
| *Haloketones* | | |
| 1,1-Dichloroacetone | Chlorine | $CHCl_2COCH_3$ |
| 1,1,1-Trichloroacetone | Chlorine | $CCl_3COCH_3$ |
| *Aldehydes* | | |
| Formaldehyde | Ozone, chlorine | $HCHO$ |
| Acetaldehyde | Ozone, chlorine | $CH_3CHO$ |
| Glyoxal | Ozone, chlorine | $OHCCHO$ |
| Methyl glyoxal | Ozone, chlorine | $CH_3COCHO$ |
| *Aldoketoacids* | | |
| Glyoxylic acid | Ozone | $OHCCOOH$ |
| Pyruvic acid | Ozone | $CH_3COCOOH$ |
| Ketomalonic acid | Ozone | $HOOCCOCOOH$ |
| *Carboxylic acids* | | |
| Formate | Ozone | $HCOO^-$ |
| Acetate | Ozone | $CH_3COO^-$ |
| Oxalate | Ozone | $OOCCOO^{2-}$ |

(continued)

**Table 9.1** (continued)

| Byproduct | Chemical agent | Molecular formula |
|---|---|---|
| *Oxyhalides* | | |
| Chlorite | Chlorine dioxide | $ClO_2^-$ |
| Chlorate[a] | Chlorine dioxide | $ClO_3^-$ |
| Bromate | Ozone | $BrO_3^-$ |
| *Nitrosamines* | | |
| *N*-nitrosodimethylamine (NDMA) | Chloramines | $(CH_3)_2N_2O$ |
| *N*-nitrosodiethylamine (NDEA) | Chloramines | $(C_2H_5)_2N_2O$ |
| *N*-nitrosomethylethylamine (NMEA) | Chloramines | $C_3H_8N_2O$ |
| *N*-nitrosodibutylamine (NDBA) | Chloramines | $(C_4H_9)_2N_2O$ |
| *N*-nitrosodi-*n*-propylamine (NPRO) | Chloramines | $(C_3H_7)_2N_2O$ |
| *N*-nitrosopiperidine (NPIP) | Chloramines | $C_5H_{10}N_2O$ |
| *N*-nitrosomorpholine (NMOR) | Chloramines | $C_4H_8N_2O_2$ |
| *N*-nitrosopyrrolidine (NPYR) | Chloramines | $C_4H_8N_2O$ |
| *Cyanogen halides* | | |
| Cyanogen chloride | Chloramines | ClCN |
| Cyanogen bromide | Chloramines | BrCN |
| *Trihalonitromethanes* | | |
| Trichloronitromethane (Chloropicrin) | Chlorine | $CCl_3NO_2$ |
| Bromodichloronitromethane | Chlorine | $CBrCl_2NO_2$ |
| Dibromochloronitromethane | Chlorine | $CBr_2ClNO_2$ |
| Tribromonitromethane | Chlorine | $CBr_3NO_2$ |
| *Halofuranones* | | |
| 3-Chloro-4-(dichloromethyl)-5-hydroxy-2(5H)-furanone (MX) | Chlorine | $C_5H_3Cl_3O_3$ |
| Mucochloric acid (MCA) | Chlorine | $C_4H_2Cl_2O_3$ |
| Mucobromic acid (MBA) | Chlorine | $C_4H_2Br_2O_3$ |
| *Miscellaneous* | | |
| Chloral hydrate | Chlorine | $CCl_3CH(OH)_2$ |

[a]Chlorate may also form from the degradation of liquid hypochlorite solutions during storage [9]

carboxylic acids, 7 % aldoketoacids, 4 % aldehydes, and 63 % unidentified AOC [2].

DBPs are typically present in drinking water at μg/L or ng/L levels, so their detection and measurement requires rigorous analytical methods. Many organic DBPs including THMs, HAAs, halofuranones, and nitrosamines are analyzed by gas chromatography with electron capture detection (GC-ECD) [7, 15–17] or GC with mass spectrometry (GC–MS) [15, 18–20]. The early targeted DBPs were thermally and chemically stable, volatile/semivolatile, neutral organic halides such as THMs and haloacetonitriles (HANs); this may not be the case for unidentified DBPs and may explain their escape from detection [21]. New analytical techniques being employed for DBP identification include GC with atomic emission detection (AED), membrane introduction MS (MIMS), membrane pre-concentration-capillary electrophoresis MS, high volume injection GC–MS, and MS–MS [2].

**Table 9.2** DBPs of concern; regulated and unregulated (Adapted from [23])

| Disinfection by-product | Health Effects | Classification |
|---|---|---|
| $THM_4$<br>Chloroform, bromoform, chlorodibromomethane, and bromodichloromethane | Liver effects (fatty cysts), kidney, and colorectal cancers | Chloroform—possible carcinogen |
| $HAA_5$<br>Monochloroacetic acid (MCA), dibromoacetic acid DBA), dichloroacetic acid (DCA), trichloroacetic acid (TCA), and monobromoacetic acid (MBA) | Liver cancer (DCA); organ cancers (DCA, DBA, TCA); body, kidney, and testes weights (MCA) | Probable carcinogen (DCA) |
| NDMA | Liver cancer | Probable carcinogen |
| Bromate | Renal cell tumors | Probable carcinogen |
| Chlorate | Thyroid gland effects (colloid depletion) | |
| Chlorite | Neurobehavioral effects (lowered auditory startle amplitude, decreased exploratory activity), decreased absolute brain weight, altered liver weight | |
| 3-Chloro-4-(dichloromethyl)-5-hydroxy-2(5H)-furanone (MX) | | Possible carcinogen [24] |

## 9.3.1 Human Health Concerns

DBPs have been linked to numerous negative human health outcomes including cancer and reproductive and developmental effects [2]. The suspected or confirmed human health effects related to the ingestion of a selection of regulated and unregulated disinfection by-products are summarized in Table 9.2. The effects of only a small percentage of known DBPs have been researched in detail, and the potential toxicity of many known DBPs is still unclear. Also, exposure routes such as inhalation or dermal absorption during showering may be significant [2, 22].

A 2009 in-depth review of the research conducted on DBPs over the previous 40 years concluded that the levels of THMs and many HAAs found in most drinking water distribution systems pose a low risk to human health compared to other unregulated DBPs [25]. For example, one study showed that 3-chloro-4-(dichloromethyl)-5-hydroxy-2(5H)-furanone, referred to as Mutagen X or MX, was the cause of 20–50 % toxicity of drinking water as measured by the Ames test method [26]. MX is typically present at concentrations 2–3 orders of magnitude lower than THMs and HAAs [22]. Other high priority compounds include bromo, chloro, and iodo halomethanes; bromo and chloro haloacetonitriles; haloketones; haloacids; halonitromethanes; haloamides; and MX analogues [2, 7]. Iodinated and brominated organics are more toxic than their chlorinated counterparts [2, 7].

For example, brominated nitromethanes have been found to be an order of magnitude more genotoxic than MX and more genotoxic than all HAAs regulated in USA except monobromoacetic acid [2].

The use of alternative disinfectants to decrease the formation of regulated DBPs such as THMs and HAAs may have the unintended consequence of increasing the levels of other DBPs with toxicological effects [2, 22, 27]. For example, one study found that the concentration of *N*-nitrosodimethylamine (NDMA) in water disinfected with chloramines was an order of magnitude higher than in water disinfected with chlorine [9]. A US EPA survey of DBP occurrence at drinking water treatment plants in the United States [7] showed that high levels of iodinated-trihalomethanes (iodo-THMs) were formed in waters treated with chloramines, that dichloroacetaldehyde formation was highest in systems employing both ozone and chloramines, and that bromo-trihalonitromethanes occurred at elevated levels in some cases where ozone was used [7]. The survey also found that the concentrations of MX and brominated MX analogues were higher when chlorine dioxide was used as a primary disinfectant before secondary chlorination.

The full effects of DBPs are not yet fully understood but as new analytical methods allow us to detect more of them, and new toxicity tests allow us to evaluate them, we will be better able to control them and manage any health concerns from drinking water disinfection.

### 9.3.2 DBP Precursors

DBPs are formed when naturally occurring or anthropogenic water constituents are oxidized or otherwise transformed by a disinfection process. Many of these constituents, commonly referred to as DBP precursors, are innocuous until they react with a disinfectant. There are three major categories of DBP precursors:

- Naturally occurring organic matter (NOM)
- Anthropogenic organic matter
- Inorganic halides (bromide, iodide)

A thorough understanding of the sources and characteristics of DBP precursors allows utilities to minimize the formation of DBPs by designing water treatment processes to remove the precursors or choosing disinfectants that are less likely to react with them.

All surface water supplies contain a heterogeneous mix of naturally occurring organic compounds that are often referred to as NOM [28]. NOM is generally thought to be composed of humic substances, carboxylic acids, proteins, amino acids, hydrocarbons, and polysaccharides [20]; however, its composition often varies seasonally and between different source waters [9].

Humic substances (humic and fulvic acids) account for the majority of NOM in most surface waters. These large, hydrophobic compounds are formed from the degradation of plant, algal, and microbial material and range from yellow to black in color [29]. They are hydrophobic with high carbon content and both aliphatic and aromatic structures and are rich in oxygen-containing functional groups such as carboxyl, phenol, alcohol, and quinoid [30]. Fulvic acids are lower in molecular weight and higher in oxygen content compared to humic acids and are categorized with humic acids as hydrophobic high molecular weight (HMW) humic compounds. Humic acids are DBP precursors [15, 18, 20, 30–32], and their concentration tends to trend with DBP formation potential (DBPfp) [15]. The functionalized aromatic structures in humic acids are considered to have a higher probability of chlorine attack compared to the aliphatic structures because of higher electron densities [18, 32]. THM and HAA formation has also been correlated with the $UV_{254}$ absorbance of NOM, which is a measure of the aromaticity of NOM [32]. THM and total organic halides (TOXs) are thought to be more strongly correlated to hydrophobic NOM levels while HAA formation is influenced by both hydrophobic and hydrophilic NOM [20, 32, 33].

Nitrogenous DBPs have also been extensively researched, particularly in the past decade. They are formed when dissolved organic nitrogen (DON) compounds interact with chemical disinfectants. DON, which includes compounds such as amino acids, proteins, and amino sugars, occurs naturally in surface water supplies as a result of microbial activities in the water and in the surrounding soils. DON is mostly found in the hydrophilic neutral, hydrophilic base, and colloidal fractions of NOM [34]. The two hydrophilic fractions are poorly removed in most traditional drinking water treatment processes (e.g., coagulation), making it more likely that DON will pass through the water treatment train to the primary and secondary disinfection steps [34, 35]. Wastewater also contains high levels of DON and the disposal of wastewater effluent into water supplies has been identified as a source of precursors for nitrogenous DBPs [36, 37].

Not all DBP precursors originate in the natural environment. Anthropogenic compounds such as pharmaceuticals and personal care products (PPCPs) are poorly removed by most wastewater treatment plants and are often released into the environment, including into existing source water [36]. Traditional water treatment processes are not designed or optimized to remove PPCPs and as a result, they frequently pass through to the primary disinfection process or into the distribution system with the secondary disinfectant. There, like NOM, they can be oxidized or otherwise transformed to form DBPs [35].

Inorganic DBP precursors are also of concern. Naturally occurring bromide can be oxidized by ozone, chlorine, $TiO_2$, and most AOPs to form bromate and brominated DBPs [9, 38, 39]. A similar process occurs with iodide, which can occur naturally in coastal areas experiencing salt water intrusion [2, 40]. The first iodo-acids found from drinking water treatment plants (using chloramination) were reported during a US nationwide DBP occurrence study [7].

Although research on DBPs has historically focused on drinking water applications, they can be formed during wastewater treatment and when water is disinfected for recreational purposes. Wastewater is typically disinfected as it leaves a wastewater treatment plant. Sanitary and industrial wastewater effluent may be relatively higher in organic and inorganic matter compared to drinking water raw water and thus DBP formation may be higher. For example, NDMA is typically present at 10 ng/L in chlorinated drinking water versus 100 ng/L in chlorinated sanitary wastewater [2]. In swimming pools, the concentration of secondary disinfectant and anthropogenic organic contamination is higher and thus it is possible that DBP formation can be higher than in drinking water.

## 9.3.3 Formation of DBPs in Water Treatment Processes

The exact mechanisms by which DBPs are formed are mostly unknown and involve many chain reactions with complex organic and inorganic constituents in water [4, 28]. DBPs form quickly during primary disinfection and also more slowly with time in the distribution system while in contact with the secondary disinfectant. Higher concentrations of disinfectant, organic precursors, reaction time (in the clearwell during primary disinfectant and in the distribution system during secondary disinfectant), or bromine or iodine will increase DBP formation [9]. pH adjustment during disinfection may help minimize DBP formation [2, 9]. Also, while the effect of a partial treatment process may not be to produce DBPs, it may alter DBP precursors and thereby change the types and concentrations of DBPs formed in later process steps. General mechanisms of formation for the various disinfectants are described below.

### 9.3.3.1 Nanotechnologies

Although their role in the treatment process may not be disinfection per se, since nanotechnologies are generally used in conjunction with disinfectants, their by-products can be considered DBPs. Nanotechnologies may be able to lower DBP formation by removing DBP precursors or by replacing a disinfection process that creates DBPs. Similarly they can contribute new DBPs by either creating new precursors or new DBPs.

Direct and indirect DBP formation in $TiO_2$ photocatalytic systems will be discussed in greater detail in Sect. 9.4 of this chapter, but in general it has been observed that DBP formation in $TiO_2$ photocatalytic systems is similar to that in AOP systems [20]. Large aromatics are oxidized to short chain aldehydes and ketones (including formaldehyde, acetaldehyde, acetone, *n*-propanal, and *n*-butanal), which are then oxidized to carboxylic acids before mineralization. This can be a cyclic process, resulting in the formation of smaller and smaller molecules, where aldehydes and ketones are oxidized to carboxylic acid, and then the carboxylic acid is mineralized to $CO_2$ and a carbon-centered radical. The carbon-centered radical with (UV) photogenerated holes or $HO^{\bullet}$ radical can form alcohol by-products that further degrade to aldehydes. Then the cycle repeats, yielding increasingly shorter chain carbonyls such as formaldehyde and acetone. Reaction intermediates and recalcitrant compounds can then react with other disinfectants to form DBPs.

Chitosan, nanostructured silver, fullerol, aqueous fullerene nanoparticles, and carbon nanotubes are not strong oxidants on their own and are relatively inert in water. As a result, they are not expected to produce significant DBPs and few studies have been conducted to investigate their application as drinking water disinfectants, particularly with respect to potential DBPs. Some of them may be toxic to humans and other living organisms, raising concerns about their presence in treated drinking water distributed to users or treated wastewater returned to the natural environment. For example, zinc oxide has been shown to act as a disinfectant but the mechanisms are not significantly understood. It dissolves in water and is toxic to aquatic organisms so its use in water treatment will likely be limited [5].

#### 9.3.3.2 Advanced Oxidation Processes

AOPs, which include $UV/H_2O_2$, $UV/O_3$, and $H_2O_2/O_3$, rely on the powerful $HO^{\bullet}$ radical to oxidize contaminants. They are currently being introduced to many drinking water treatment plants to improve the removal of organic compounds, especially recalcitrant contaminants [16].

The $HO^{\bullet}$ radical can oxidize NOM by electron transfer, hydrogen abstraction, or $HO^{\bullet}$ addition to double bonds [16]. Refer to Eqs. (9.6–9.8), respectively [41]. The second-order reaction rate constants for reactions of $HO^{\bullet}$ radicals with contaminants in aqueous solution are usually in the order of $10^8$–$10^{10}$ LM/s and are 3–4 orders higher than other oxidants [9, 13]. Inorganic ions such as bicarbonate, carbonate, phosphate, and sulfate will compete with NOM and scavenge $HO^{\bullet}$ radicals, prolonging reaction times for NOM degradation [9].

$$I^- + {}^{\bullet}OH \rightarrow {}^{\bullet}I + OH^- \tag{9.6}$$

$$CH_3OH + {}^{\bullet}OH \rightarrow CH_2{}^{\bullet}OH + H_2O \tag{9.7}$$

$$CHCl = CCl_2 + {}^{\bullet}OH \rightarrow {}^{\bullet}CHCl - C(OH)Cl_2 \tag{9.8}$$

The general degradation pathways observed, from initial to final are: organic pollutants; aldehydes; carboxylic acids; and finally carbon dioxide and mineral acids [9]. In theory, incomplete mineralization may increase or decrease the DBP formation potential (DBPfp) of the feed water. A research team at the University of British Columbia found that $UV/H_2O_2$ was able to reduce THM formation upon subsequent chlorination [42]. Other researchers have found that $UV/H_2O_2$ followed by chlorination increased the HAAfp of some reference compounds while that of others decreased [43]. The addition of $UV/H_2O_2$ to the treatment train can also impact the amount of NDMA formed during secondary disinfection with chloramines, both directly and by breaking down water constituents to form or destroy NDMA precursors [44]. When halogenated species such as bromide or trichloroethylene are present in the water, the use of AOPs can lead to the formation of halogenated DBPs such as bromate or HAAs, respectively [9].

#### 9.3.3.3 Chlorination

During the disinfection process chlorine can undergo several reactions with contaminants in water to create DBPs. These include the oxidation of organic and inorganic constituents, ammonia substitution, chlorine substitution of hydrogen, decomposition, and chlorine addition to double bonds.

Chlorine oxidizes organic constituents, such as ketones, aldehydes, and carboxylic acids, and is reduced to chloride salts (refer to Eqs. (9.9) and (9.10), the oxidation of a carbonyl to carboxylic acid and the subsequent decarboxylation, respectively). When complete oxidation to $CO_2$ does not occur, which is normally the case for the $Cl_2$ concentrations used in drinking water disinfection, DBPs can include organics such as carboxylic acids [2].

$$RCHO + HOCl \rightarrow RCOOH + H^+ + Cl^- \tag{9.9}$$

$$RCOOH + HOCl \rightarrow CO_2 + R + H^+ + Cl^- \tag{9.10}$$

Chlorine can also undergo substitution and addition reactions with organic molecules, creating halogenated carbon molecules. In a substitution reaction, chlorine replaces hydrogen in a hydrocarbon molecule. In Eq. (9.11) for example, free chlorine attacks the ketone at the carbonyl carbon–carbon bond, forming a carboxylic acid and trichloromethane. The carboxylic acid can then be further oxidized to form progressively smaller carboxylic acids, as described in Eqs. (9.9) and (9.10). However, complete mineralization to $CO_2$, mineral acids, and water does not typically occur because of the relatively low oxidation potential of chlorine and/or the presence of recalcitrant original or intermediate compounds.

$$RCOCH_3 + 3HOCl \rightarrow HCCl_3 + RCOOH + H^+ + 2H_2O \tag{9.11}$$

In an addition reaction, a carbon–carbon double bond is broken and chlorine binds to a carbon. With free chlorine this can also create a hydroxyl (HO) functional group as shown in Eq. (9.12).

$$RC = CR' + HOCl \rightarrow RC(OH)C(Cl)R' \tag{9.12}$$

The reaction kinetics of addition and substitution reactions between chlorine with NOM are in the order of hours and days [9] and thus concentrations of the products of these reactions will be highest at locations along a distribution system where the hydraulic retention time is the greatest.

Chlorine will oxidize inorganic compounds that are less electronegative than it is. Of particular concern is the oxidation of reduced halogen species such as bromide and iodide. The reaction of free chlorine with bromide ions occurs according to Eq. (9.13). A similar reaction occurs when chlorine comes into contact with iodide ions. The oxidized products of these reactions can in turn oxidize other contaminants and create DBPs.

$$Br^- + HOCl \rightarrow HOBr + Cl^- \tag{9.13}$$

In ammonia substitution, chlorine replaces the hydrogen atoms bonded to nitrogen in ammonia ($NH_3$) to create chloramines. Because of the favorable reaction of chlorine with ammonia, nitrogen species, and other readily oxidizable species, this initial chlorine demand of water must be accounted for when providing free chlorine residual for secondary disinfection [9].

Finally, the decomposition of free chlorine can yield chlorate according to reaction Eq. (9.14) [9].

$$3HOCl \rightarrow ClO_3^- + 3H^+ + 2Cl^- \tag{9.14}$$

#### 9.3.3.4 Chloramination

As mentioned previously, the use of free chlorine as a primary disinfectant with subsequent addition of ammonia to form monochloramine for secondary disinfection is a widely employed alternative secondary disinfection strategy because it results in less formation of THMs and HAAs [9]. Unfortunately, chloramination has been linked to the formation of other DBPs, most notably nitrosamines. Nitrosoamines such as NDMA are of particular interest to scientists and regulators because they are highly carcinogenic and have been detected in drinking water systems, particularly those that use chloramination for secondary disinfection [35]. Recent research indicates that many anthropogenic compounds, including some PPCPs and pesticides, can act as precursors to NDMA [45, 46]. Compounds that are likely to cause NDMA formation include quaternary, tertiary, and secondary amines [34]. Numerous reaction pathways have been suggested to explain the reactions between these compounds and chloramines. A number of these have been summarized by Bond et al. [35] and Mitch et al. [34, 35].

#### 9.3.3.5 Chlorine Dioxide

Chlorine dioxide does not produce the halogenated organic DBPs that free and combined chlorine do because it does not break carbon–carbon bonds [9]; however, its use has been limited because it produces the inorganic DBPs chlorite ($ClO_2^-$) and chlorate ($ClO_3^-$) at levels that warrant a health concern. Chlorite in particular carries serious health concerns for humans [9]. A 2002 survey of American water treatment plants also found the use of chlorine dioxide was linked with increased formation of Mutagen X (MX) and brominated MX analogues [7].

#### 9.3.3.6 Ozone

Ozone disinfection is also effective in breaking down NOM and creates organic DBPs such as aldehydes (e.g., formaldehyde) and ketoacids. Although it does not directly contribute to halogenated DBPs like the chlorine-based disinfectants [9], ozone does play a significant role in bromate ($BrO_3^-$) formation. The reaction kinetics of Eqs. (9.15–9.17) are in the order of minutes [9].

$$Br^- + O_3 \rightarrow OBr^- \quad (9.15)$$

$$OBr^- + O_3 \rightarrow BrO_3^- \quad (9.16)$$

$$OBr^- + OH^* \rightarrow BrO_3^- \quad (9.17)$$

Hypobromous acid ($HOBr/OBr^-$) described in Eq. (9.15) can also react further to create brominated organic compounds. Bromonitromethanes are sometimes found in larger concentrations when ozone is used as a disinfectant in waters with naturally occurring bromide [2]. Similarly, hypoiodous acid (HOI) may form from ozonation and create iodinated organic compounds.

#### 9.3.3.7 Ultraviolet Light

It was long believed that, unlike the various chemical oxidants traditionally used for disinfection, UV disinfection did not give rise to DBPs or DBP precursors. Recent work suggests that, under certain circumstances, DBP precursors can indeed be formed by UV [43, 47, 48]. Various studies have shown that THMfp [47], HAAfp [43], NDMAfp [44], and the formation potentials of a number of unregulated DBPs [48] can all be impacted by high UV doses ($\geq$186 mJ/cm$^2$), although the effect varies from parameter to parameter and study to study. For example, although Buchanan et al. observed a general decrease in THMfp upon application of UV light, bromoform formation potential actually increased [47]. It is also possible to use UV to remove DBPs formed during pretreatment oxidation and primary disinfection. For example, UV can breakdown NDMA through photolysis [9].

#### 9.3.3.8 Regulation of DBPs

Regulators around the world have set limits on the levels of specific DBPs permitted in drinking water. These limits must be met regardless of the process used for primary or secondary disinfection (including the use of any nanomaterials). The most commonly regulated organic DBPs are THMs, HAAs, and NDMA [23, 49–54]. THMs and HAAs make up a large percentage of the total amount of DBPs measured; however, the distribution of DBPs is site specific and dependent on water quality and disinfectant type [22]. Inorganic DBPs such as bromate, chlorate, and chlorite are also regulated in many jurisdictions. Summaries of permitted or recommended maximum levels of organic and inorganic DBPs in a selection of jurisdictions are provided in Tables 9.3 and 9.4, respectively. In some countries, provincial or local jurisdictions are permitted or encouraged to set DBP limits that are even more stringent than the national limits. Jurisdictions such as Ontario and California have set lower guidelines or standards for NDMA based on health concerns as well as on the availability of technologies that can achieve lower DBP concentrations in the finished water and/or analytical methods that can reliably confirm treatment goals. These progressive jurisdictions can be an example for others.

## 9.4 $TiO_2$ Photocatalysis and DBPs

Of the nanotechnologies used in water treatment listed previously, processes that incorporate $TiO_2$ photocatalysis are the only ones that are well researched, particularly with respect to their influence on DBP formation and prevention. As a result, $TiO_2$ photocatalysis is the focus of this section.

### *9.4.1 Background on $TiO_2$ Photocatalysis*

In the water treatment industry, the term photocatalysis refers to the process of using light to activate a catalyst that is able to produce hydroxyl radicals ($HO^{\bullet}$) and other reactive oxygen species (ROS) in water [57]. The effects of $TiO_2$ photocatalysis on DBP formation are similar to those of AOPs. Unlike other AOPs, however, $TiO_2$ has the added advantage that it does not require ongoing chemical inputs. ROS are capable of oxidizing organic and inorganic water contaminants and also of destroying many types of microorganisms. Hydroxyl radicals are the second best oxidizing agents [58] (after fluorine) with a reduction potential of 2.8 V, making $TiO_2$ photocatalysis a particularly effective oxidation process. When oxygen is present, the super oxide molecule ($O_2^{\bullet -}$) can also be formed. Although ROS can potentially mineralize organic compounds, incomplete decomposition often occurs due to insufficient treatment or the presence of recalcitrant compounds.

**Table 9.3** Organic DBP regulations and guidelines

| Disinfection By-product | Ontario (μg/L) | Canada (μg/L) | California (μg/L) | US EPA (μg/L) | WHO (μg/L) | China (μg/L) | EU (μg/L) | Singapore (μg/L) |
|---|---|---|---|---|---|---|---|---|
| Total THMs (TTHMs) | 100 | 100 | 80 | 80 | [(e)] | [(e)] | 100 | [(e)] |
| Bromodichloromethane | – | – | | – | 60 | 60 | – | 60 |
| Bromoform | – | – | | – | 100 | 100 | – | 100 |
| Chloroform | – | – | | – | 300 | 60 | – | 300 |
| Dibromochloromethane | – | – | | – | 100 | 100 | – | 100 |
| $HAA_5$ | [(a)] | 80[b] | 60 | 60 | – | – | – | – |
| Dichloroacetate (dichloroacetic acid) | – | – | | – | 50[c] | 50 | – | 50 |
| Monochloroacetate (monochloroacetic acid) | – | – | | – | 20 | – | – | 20 |
| Trichloroacetate (trichloroacetic acid) | – | – | | – | 200 | 100 | – | 200 |
| NDMA | 0.009[g] | 0.04 | 0.01/0.3[h] | – | 0.1 | – | – | – |
| Dibromoacetonitrile | – | – | | – | 70 | – | – | 70 |
| Dichloroacetonitrile | – | – | | – | 20[d] | – | – | 20 |
| Formaldehyde | – | – | | - | – | 900[f] | – | – |
| 2,4,6-Trichlorophenol | | | | | 200 | | | |
| Trichloroacetaldehyde | – | – | | – | – | 10 | – | – |

[a]A guideline value of 60 μg/L is currently under review
[b]As low as reasonably achievable
[c]Provisional guideline because disinfection is likely to create higher values
[d]Provisional guideline because health database has uncertainties
[e]The sum of the ratio of the concentration of each to its respective guideline value should not exceed 1
[f]When ozone is used as a disinfectant
[g]Provisional guideline is above calculated guideline because unable to achieve by practical treatment methods and unable to quantify [2, 9]
[h]Notification level/response level [55]
*Source*: [23, 49–54, 56]

**Table 9.4** Inorganic DBP regulations and guidelines

| Disinfection By-product | Ontario (μg/L) | Canada (μg/L) | California (μg/L) | US EPA (μg/L) | WHO (μg/L) | China (μg/L) | European Union (μg/L) | Singapore (μg/L) |
|---|---|---|---|---|---|---|---|---|
| Bromate ($BrO_3^-$) | 10 | 10 | 10 | 10 | 10[a] | 10 | 10 | 10 |
| Chlorate ($ClO_3^-$) | – | 1,000 | – | – | 700[b] | 700 | – | 700 |
| Chlorite ($ClO_2^-$) | – | 1,000 | 1,000 | 1,000 | 700[b] | 700 | – | 700 |
| Cyanogen chloride (as cyanide $CN^-$) | – | – | – | – | – | 70 | – | 70 |

[a] Provisional guideline is above calculated guideline because the calculated guideline is below current detection limits or cannot be achieved using practical treatment methods

[b] Provisional guideline because disinfection is likely to create higher values

*Source*: [23, 49–54, 56]

$TiO_2$ can be formed or manufactured in two phases, anatase and rutile. The anatase phase of $TiO_2$ is more photoreactive and absorbs UVA light at wavelengths less than 380 nm (3.2 eV) [59]. The outer shell electron in the valence band (VB) absorbs the energy and moves into the higher energy conduction band (CB) [30]. This results in the release of a negatively charged electron ($e^-$) and leaves behind a positively charged hole ($h^+$, refer to Eq. 9.18) [60]. At a pH of 7, the hole has a reduction potential of approximately 2.9 V while the conduction band electron has a reduction potential of approximately −0.3 V ([9]). Often, these electron–hole pairs recombine and produce heat and/or light [9], but some pairs will also migrate to the $TiO_2$ surface [61]. Molecules with high electropotentials such as oxygen can then pick up the CB electrons to form various ROS (as shown in Eqs. 9.19–9.21) [60]. The VB holes are filled either through the oxidation of organic compounds or water molecules, the latter of which results in the formation of ROS such as the $HO^•$ radical (Eqs. 9.22–9.24) [9, 60]. The standard electrode potential for the formation of $HO^•$ radical is 2.177 V at a typical drinking water pH of 7 [9]. As described for AOP processes, the $HO^•$ radical can oxidize NOM by electron transfer, hydrogen abstraction, or $HO^•$ addition to double bonds [16]. Unfortunately, the quantum yield for the reaction described in Eq. (9.22) is quite low at 0.04–0.05 [41].

$$TiO_2 + hv \leftrightarrow e^- + h^+ \tag{9.18}$$

$$e^- + O_2 \leftrightarrow +O_2^{*-} \tag{9.19}$$

$$e^- + O_2^{*-} + 2H^+ \leftrightarrow H_2O_2 \tag{9.20}$$

$$e^- + H_2O_2 + H^+ \leftrightarrow OH^* + H_2O \tag{9.21}$$

$$h^+ + H_2O \leftrightarrow +OH^* + H^+ \tag{9.22}$$

$$2h^+ + H_2O \leftrightarrow H_2O_2 \tag{9.23}$$

$$OH^- + h^+ \leftrightarrow OH^* \tag{9.24}$$

The formation of the superoxide radical ($O_2^{•-}$) described in Eq. (9.19) will only occur if dissolved oxygen is present in the water. $O_2^{•-}$ can react with microorganisms or water contaminants, or it can form hydrogen peroxide [60], which can then form $HO^•$ radicals as described in Eq. (9.21). Many other complex reactions and reactive species occur, including direct oxidation and reduction of contaminants by the $TiO_2$ electron–hole pairs [60].

In practice, $TiO_2$ is usually activated using a lamp that emits light in the UVA range (usually 365 nm). Light at this wavelength is sufficiently energetic to activate the photocatalyst. Sunlight can also be used to activate $TiO_2$, but the development of full-scale solar/$TiO_2$ applications remains limited by low efficiency. This is because the sun's energy is mostly made up of visible light energy (about 40–50 %) with only a small contribution from UVA light (about 3 %) [13]. Therefore, although the sun puts out approximately 986 $W/m^2$ (irradiance of one "standard sun"), only 30 $W/m^2$ of this is available to activate the photocatalyst in solar/$TiO_2$ applications.

#### 9.4.1.1 Nanoparticles, Nanorods, Nanobelts, and Nanowire Membranes

Nano-sized $TiO_2$ for water treatment has been traditionally prepared as nanoparticles, which are nano-scale $TiO_2$ crystals with relatively constant dimensions. Degussa P25 is a common $TiO_2$ nanoparticle commercial standard [20]. The powder can be sprayed onto surfaces as a fixed film or mixed with contaminated water to form a suspension. Recent research has resulted in the development of $TiO_2$ structures including nanowires, nanowhiskers, nanobelts, and nanorods. These are created by inducing nanoparticles to grow along an axis. They generally have length to width ratios of approximately 10:1 to 1,000:1 [61]. $TiO_2$ nanowires have the benefit of being able to form interlaced networks that may be applicable to membrane fabrication and also may help inhibit electron–hole recombination. Refer to Chap. 2 and 4 of this book for more detail on nanowire membranes.

Since photocatalysis occurs at or near the $TiO_2$ surface, either by direct oxidation of contaminants or reaction between oxidative species and contaminants, a large surface area is desirable. Their small size makes nanoparticles more reactive than nanowires and other novel nanostructures [61].

#### 9.4.1.2 Photoelectrocatalysis

Photoelectrocatalysis (PEC) is another technique that is used to increase the quantum efficiency of $TiO_2$. The catalyst is fixed on a conductive substrate and the photogenerated electrons are driven to a cathode by an external voltage. For example, $TiO_2$/Ti and $RuO_2$/Ti can be employed as the anode and cathode, respectively [62]. Recent research shows that the overall differences/increases in efficiencies of PEC as compared to photocatalysis are not significant [63].

#### 9.4.1.3 $TiO_2$ Photocatalysis with $H_2O_2$

$TiO_2$ photocatalysis can also be paired with $H_2O_2$ addition to increase the formation of $HO^•$ radicals. The kinetics of degradation increase with $H_2O_2$ addition, but recalcitrant compounds may nonetheless remain [18, 20]. The addition of $H_2O_2$ may also prevent electron–hole pair recombination, increasing the availability of holes for oxidation, for $HO^•$ radical production, and for direct NOM oxidation. $H_2O_2$ has a higher reduction potential than oxygen [9] and may be more efficient to add than oxygen, when added at an optimized concentration. Under certain circumstances, however, $H_2O_2$ may compete with NOM for $TiO_2$ adsorption sites [18, 20].

#### 9.4.1.4 Photosensitized $TiO_2$ with Iodine and Nitrogen

$TiO_2$ can be made to absorb longer wavelengths or make better use of the light that it does absorb by doping it with other atoms, for example, iodine and nitrogen [30]. It can also be sensitized to visible light wavelengths with ruthenium-based complexes, various dyes, and even humic acid [30]. Although humic acid can sensitize $TiO_2$, this requires a low pH so that the humic acid becomes adsorbed to the $TiO_2$ surface, which is not generally practical in drinking water treatment applications [30]. Refer to Chap. 2 for more information on visible light-sensitized $TiO_2$ nanomaterials for water treatment..

#### 9.4.1.5 $TiO_2$ with Silver

Metallic ions such as silver can be used to extend the range of wavelengths that can be absorbed by $TiO_2$ to the visible spectrum. It can also increase the quantum yield of UVA activation by preventing electron–hole recombination [5]. Silver is a disinfectant and can increase the overall disinfection efficiency of $TiO_2$ systems, but at the time of writing, the effects of silver on DBPs were only beginning to be investigated and preliminary results were inconclusive with respect to DBPs [64].

#### 9.4.1.6 Direct Effects of $TiO_2$ Photocatalysis on DBP Formation

In theory, all $TiO_2$ photocatalysis degradation products can be considered DBPs if the process is being applied primarily for disinfection. To date, there have been few studies to show significant DBP formation when $TiO_2$ is used alone (i.e., without a prior or subsequent additional disinfectant). A few studies have detected genotoxic degradation products without identifying the products themselves in UV/$TiO_2$ systems used to treat pesticides such as paraquat [65] and *p*-nitrophenol [66] when these systems were not optimized to achieve full mineralization of the original contaminant. These studies were conducted at contaminant concentrations well above those expected in drinking water supplies; however, they underline the risks associated with the formation of undesirable degradation products under nonideal treatment conditions.

A small number of studies have investigated the effects of UV/$TiO_2$ on toxicity or DBP formation in drinking water applications. A study by Richardson et al. [67] found one organic DBP tentatively identified as 3-methyl-2, 4-hexanedione in unchlorinated filtrate from an ultrafiltration membrane treated with $TiO_2$ photocatalysis. A more general study by Bekbolet et al. [50] measured the toxicity (percent immobilization of *Daphnia magnum*) before chlorination of water samples from three surface water supplies treated with ozonation, coagulation, or ozonation followed by coagulation and UV/$TiO_2$. Their results were largely inconclusive, but did suggest that the influence of UV/$TiO_2$ on toxicity was dependent on initial raw water quality and that, on occasion, toxicity might increase as a result of UV/$TiO_2$ treatment. The hydroxyl radicals formed through $TiO_2$ photocatalysis can also oxidize bromide and iodide to form HOBr and HOI, which may then react with NOM to create brominated and iodinated organics [38].

PEC has also been evaluated for DBP formation. It has been observed to form a number of DBPs directly, including bromate [39] when bromide is present in the feedwater and THMs if chloride ions are present in the photoanodic chamber of a PEC system [68].

#### 9.4.1.7 Indirect Effects of $TiO_2$ Photocatalysis on DBP Formation

Since $TiO_2$ photocatalysis has more often been used in water treatment for the removal of specific contaminants, it has usually been followed by a more conventional disinfection step. Pretreatment with $TiO_2$ photocatalysis has been shown to alter the types and concentrations of DBPs formed during subsequent disinfection.

$TiO_2$ photocatalysis is one of many emerging processes that has been proposed for DBP precursor reduction. Often the oxidation products from $TiO_2$ photocatalysis are less likely to form DBPs upon chlorination [16] and in general the DBPfp decreases because of the significant decrease in reactive NOM [67], particularly of the aromatic variety. Experimental results suggest, however, that under most conditions NOM oxidation by photocatalysis does not result in full NOM mineralization and can result in the formation of DBP precursors and refractory compounds [15] with low molecular weights [16, 69] and low aromaticity [19]. These recalcitrant compounds include acids and neutral compounds such as aliphatic acids, aldehydes, ketones, and alcohols [19, 20].

Some recent studies have suggested that UV/$TiO_2$ can decrease both trihalomethane formation potential (THMfp) and haloacetic acid formation potential (HAAfp) [18–20, 70]. The extent of these reductions is specific to the water supply and configuration of the treatment system under consideration. For example, a study conducted by Gerrity et al. [71] found that the THMfp of water treated using a pilot scale suspended UV/$TiO_2$ system was dependent on the amount of energy input into the process. Limited treatment ($<5$ kW/m$^3$ energy applied) increased overall THMfp, but extended treatment ($\geq 80$ kW/m$^3$ energy applied) decreased it dramatically.

Other studies have shown that the initial degradation of larger compounds can result in increased THMfp and HAAfp even while the overall concentration of DOC decreases [17, 19, 20, 47, 72]. Some researchers have hypothesized that this occurs because the fragmentation of aromatic structures within NOM exposes sites for chlorine attack and DBP formation. The degradation of humic acids (HAs) by photocatalysis leads to the formation of hydrophilic intermediates that have been shown to be HAA precursors [20]. With continued $TiO_2$ photocatalysis, these intermediate degradation products of photocatalysis are further broken down to recalcitrant compounds, which have a lower DBPfp [19].

$TiO_2$ photocatalysis has been shown to preferentially degrade larger molecules such as polysaccharides and HA and to attack aromatic rings and conjugated structures [15, 18–20, 69]. The high electron densities of aromatic structures give them a higher probability of attack by hydroxyl radicals. To some degree, this preference can also be explained by the fact that large molecules have more functional groups with which to interact with $TiO_2$ and undergo direct oxidation or reduction, or react with released ROS. A higher affinity for the $TiO_2$ surface will bring the contaminant in closer proximity to any $HO^{\bullet}$ radicals, $O_2^{\bullet -}$, and other ROS and make it more likely that a reaction will occur irrespective of the degradation mechanism.

Liu et al. [18] found that $TiO_2$ has a strong affinity for HA at low pH but that the adsorption of NOM onto $TiO_2$ decreases with increasing pH [18, 20]. At an alkaline pH of 9, there is a repulsion between HA and the $TiO_2$ surface [18]. Despite this, one study found that degradation kinetics increased with increased pH [18]. The researchers hypothesized that this increase could be attributed to the transformation of hydroxyl ions ($HO^-$) to $HO^{\bullet}$ radicals at higher pH, leading to increased NOM degradation via the $HO^{\bullet}$ radical oxidation mechanism. It was also hypothesized that at pH 9 the large HA molecules uncoiled due to deprotonation, loss of hydrogen bonding, and an increase in repelling negative charges, leaving them more susceptible to $HO^{\bullet}$ radical attack.

A drinking water study by Richardson et al. [67] found that when UV/$TiO_2$ replaced chlorine for primary disinfection ahead of secondary disinfection with chlorine a similar array of DBPs formed but at reduced levels. The researchers also found that MCA, an analogue of MX, was formed when $TiO_2$ photocatalysis was used ahead of chlorination but not when raw or ultrafiltered water was chlorinated.

As with AOP (Sect. 9.3.3.2), inorganic ions such as bicarbonate, carbonate, phosphate, and sulfate will compete with NOM and scavenge $HO^{\bullet}$ radicals in $TiO_2$ photocatalytic systems, prolonging the amount of time required for NOM degradation [9].

## 9.5 Comparing $TiO_2$ Photocatalysis to Existing Disinfection Processes for DBP Formation and DBP Precursor Reduction

### *9.5.1 Direct Effects of $TiO_2$ Photocatalysis on DBP Formation*

If $TiO_2$ photocatalysis is to be adopted for drinking water disinfection, its potential for DBP formation must be understood relative to other technologies and minimized. A potential benefit of $TiO_2$ photocatalysis is that it could result in the formation of fewer DBPs than other traditional and emerging disinfection processes such as chlorination, chloramination, chlorine dioxide, ozone, UV, and other AOPs.

**Table 9.5** DBPs formed in various disinfection processes (Adapted from [9])

| Disinfection process | Carbonaceous DBPs | Nitrogenous DBPs | Inorganic DBPs |
|---|---|---|---|
| Chlorination | THMs, HAAs, haloketones, aldehydes, halofuranones, chloral hydrate | HANs, trihalonitromethanes | – |
| Chloramination | THMs, HAAs[a] | Nitrosamines (e.g., NDMA), cyanogen halides, HANs | – |
| Chlorine dioxide | Limited | Limited | Chlorite/chlorate |
| Ozone | Aldehydes, ketones, aldoketoacids, carboxylic acids | – | Bromate, iodate |
| UV disinfection | No DBPs have been identified at UV doses/fluences used for disinfection | | |
| $UV/H_2O_2$ | Aldehydes, ketones, carboxylic acids | | Bromate, iodate |
| $TiO_2$ photocatalysis | Low molecular weight and low aromaticity organic molecules | – | Bromate, iodate |

[a]At lower concentrations compared to chlorination

Common DBPs formed directly by various processes used for primary and/or secondary disinfection of drinking water are summarized in Table 9.5.

### 9.5.2 *Indirect Effects of $TiO_2$ Photocatalysis on DBP Formation*

$TiO_2$ photocatalysis has also been proposed for DBP precursor control because it is capable of oxidizing NOM and many anthropogenic compounds. Studies to date have reported conflicting results [43, 44, 70, 71] with the only true consensus being that DBPfp reduction is only assured when the system is designed to achieve maximum mineralization of DBP precursors.

Other methods used for NOM removal and DBPfp reduction include:

- Enhanced coagulation
- Ozone
- Ozone with biofiltration
- Nanofiltration
- AOPs (e.g., UV/$H_2O_2$)

Enhanced coagulation is the addition of aluminum or iron salts to water with the express goal of removing NOM [73]. It works by inducing dissolved NOM compounds to agglomerate together into flocs that can be removed through clarification (e.g., sedimentation, flotation) or size exclusion (e.g., media or membrane filtration). Enhanced coagulation is particularly effective at removing large hydrophobic acid compounds but less effective at removing hydrophilic NOM [74, 75] and DON [35].

Ozone, which has been discussed in the context of disinfection, is also used for NOM removal and DBPfp reduction. It is a strong oxidant that is able to quickly break down hydrophobic NOM compounds into smaller, more hydrophilic compounds [75]. Ozone is often paired with biofiltration, which has been shown to remove the degradation products formed when ozone comes into contact with NOM [73].

Nanofiltration, which is similar to RO membrane filtration, has recently become a more common choice for NOM removal in drinking water applications. Unlike enhanced coagulation and ozone, nanofiltration removes NOM indiscriminately via size exclusion. It is an energy intensive process that relies on high pressure to force water through a semipermeable membrane while retaining dissolved organic compounds and divalent ions [73].

The water treatment processes described earlier each have advantages and limitations. For example, although nanofiltration is particularly effective for NOM removal, it must be preceded by a preliminary filtration step (e.g., microfiltration) and is highly energy intensive. Enhanced coagulation is well established in the industry and effective at removing the hydrophobic NOM that often gives rise to THMs. It is, however, less effective at removing hydrophilic NOM and results in the formation of a chemical waste product that must be treated and/or disposed of.

$TiO_2$ photocatalysis has its own advantages and disadvantages. Advantages include:

- Concurrent disinfection, DBP precursor reduction, and degradation of recalcitrant compounds
- No chemical inputs required (except $TiO_2$)
- No waste products formed
- Can potentially be made to operate using sunlight

There are numerous disadvantages as well. Although $TiO_2$ photocatalysis can potentially degrade the majority of DBP precursors, including recalcitrant organic anthropogenic compounds, full mineralization often requires lengthy contact times [71]. $TiO_2$ photocatalysis has fewer chemical inputs than other AOPs, but it also has more complex mass transfer limitations. Finally, at present, solar/$TiO_2$ is not feasible for full-scale drinking water treatment applications. As a result, the photocatalyst must be activated using a UV lamp, which represents a significant energy input.

It is difficult to compare DBP precursor removal methods to one another based on a review of literature because treatment efficiency is often site specific. A few recent studies have compared the ability of UV/$TiO_2$ and other treatment processes to remove DBP precursors from water samples gathered from actual water supplies. One study compared ozonation, coagulation, ozonation followed by coagulation, and UV/$TiO_2$ for DBPfp formation ahead of chlorination. Although in most cases the treatments were comparable in terms of treatment effectiveness, the researchers found that the concentration of one HAA, trichloroacetic acid was higher in water treated with UV/$TiO_2$, and bromochloroacetonitrile was only found in water treated with UV/$TiO_2$. A significant decrease in chloral hydrate was observed after UV/$TiO_2$ treatment [76]. Studies by Gerrity et al. [71] and Philippe et al. [16] have also compared bench-scale enhanced coagulation to UV/$TiO_2$ and have found that photocatalysis is superior for THMfp removal. These preliminary studies suggest that $TiO_2$ photocatalysis has the potential to be as effective or more effective than other treatment processes for DBP precursor removal. Table 9.6 compares $TiO_2$ photocatalysis to other common DBP precursor removal technologies in terms of treatment effectiveness, chemical and energy inputs, and other concerns.

DBP precursor reduction processes are often used in combination to ensure adequate treatment. Though $TiO_2$ photocatalysis may not always be practical as a stand-alone system, it could be a valuable addition to a more complex treatment train. Numerous bench- and pilot-scale studies have been conducted to evaluate the feasibility of coupling $TiO_2$ nanomaterials with conventional drinking water treatment processes to improve disinfection, DBP precursor degradation, or the removal

**Table 9.6** Comparison of $TiO_2$ to other DBP precursor control processes

| Treatment process | DOC removal | THMfp reduction | Chemical inputs | Energy inputs | Other concerns |
|---|---|---|---|---|---|
| Enhanced Coagulation | 22–72 % [9, 76, 77] | 37–84 % [77] | Iron or aluminum salts and pH adjustment | Mixing | Sludge production |
| Ozone | 0–13 % [77] | −36 to 29 % [77] | $O_2$ | Ozone generation | Operator safety |
| Ozone and biofiltration | 3–25 % [77] | −36 to 10 % [77] | $O_2$ | Ozone generation | Operator safety, cold weather operation |
| Nanofiltration | 70–95 % [78, 79] | 96–99 % [80]<br>60–100 % [79] | | Pumping | Pretreatment required |
| UV/$TiO_2$ | 75 %[a] [20]<br>60 %[b] [16]<br>94 %[c] [31]<br>90 %[d] [71] | 96 %[a] [20]<br>53 %[b] [16]<br>95 %[d] [71] | $O_2$ | UV lamp | Long contact time required |

[a]120 min contact time, 100 mg/L $TiO_2$
[b]10 min contact time, 100 mg/L $TiO_2$
[c]Nanowire membrane
[d]320 $kWh/m^3$ UV dose, 400 mg/L $TiO_2$

of recalcitrant organic contaminants. These hybrid processes have yet to be widely applied for drinking water treatment, however, partly due to concerns about the impact of $TiO_2$ on the formation of DBPs or DBP precursors [38].

Though research on DBP formation in hybrid systems is sparse, a number of studies have addressed their ability to remove DBP precursors. For example, the addition of UV/$TiO_2$ after coagulation has been shown to improve DOC and $UV_{254}$ removal compared to coagulation alone [81]. $TiO_2$ photocatalysis has also been shown to increase the biodegradability of NOM, making it an ideal pretreatment ahead of biofiltration for the removal of DBP precursors [16, 18]. Finally, numerous configurations have been proposed for hybrid UV/$TiO_2$ membrane-based treatment systems [38].

## 9.6 Conclusions

Chlorine remains popular for primary and secondary disinfection of drinking water because it is reliable, safe to handle, and cost effective. It does, however, react with naturally occurring water constituents to form disinfection by-products (DBPs), some of which have suspected or proven human health implications. Alternative disinfectants such as chloramines and ozone also create DBPs, albeit different ones than chlorine. As a result, there is demand for new water treatment processes that can provide adequate disinfection without forming disinfection by-products. Nanotechnologies, in particular $TiO_2$ photocatalysis (UV/$TiO_2$ and solar/$TiO_2$), are emerging as at least partial solutions to this problem.

A growing body of research is exploring the effects of $TiO_2$ photocatalysis on the formation of disinfection by-products and disinfection by-product precursors. Studies comparing UV/$TiO_2$ to common primary disinfectants suggest that UV/$TiO_2$ produces fewer regulated DBPs upon subsequent secondary chlorination, in particular carbonaceous DBPs such as THMs and HAAs. Under certain circumstances, however, $TiO_2$ photocatalysis can produce by-products of potential health concern and has been linked to the formation of regulated inorganic DBPs such as bromate and iodate.

Other researchers have compared UV/$TiO_2$ to DBP precursor reduction processes placed prior to primary disinfection including enhanced coagulation and ozonation and found that the effects of UV/$TiO_2$ on DBP precursor levels are site specific and highly dependent on process parameters such as irradiation and contact time (UV dose). Future research into the effects of these and other process parameters on the direct and indirect formation of DBPs in water treatment processes employing $TiO_2$ photocatalysis and other nanomaterials is required before these exciting new technologies can be adopted in drinking water applications.

## References


1. J.G. Pressman, S.D. Richardson, T.F. Speth et al., Concentration, chlorination, and chemical analysis of drinking water for disinfection byproduct mixtures health effects research: U.S. EPAs four lab study. Environ. Sci. Technol. **44**(19), 7184–7192 (2010)
2. S.D. Richardson, Disinfection by-products and other emerging contaminants in drinking water. TrAC **22**(10), 666–684 (2003)
3. S.W. Krasner, H.S. Weinberg, S.D. Richardson, S.J. Pastor, R. Chinn, M.J. Sclimenti, G.D. Onstad, A.D. Thruston Jr., Occurrence of a new generation of disinfection byproducts. Environ. Sci. Technol. **40**(23), 7175–7185 (2006)
4. C.M. Villanueva, G. Castano-Vinyals, V. Moreno et al., Concentrations and correlations of disinfection by-products in municipal drinking water from an exposure assessment perspective. Environ. Res. **114**, 1–11 (2012)
5. Q. Li, S. Mahendra, D.Y. Lyon, L. Brunet, M.V. Liga, D. Li, P.J.J. Alvarez, Antimicrobial nanomaterials for water disinfection and microbial control: Potential applications and implications. Water Res. **42**(18), 4591–4602 (2008)
6. S.A. Love, M.A. Maurer-Jones, J.W. Thompson, Y. Lin, C.L. Haynes, Assessing Nanoparticle Toxicity. Annu. Rev. Anal. Chem. **5**, 181–205 (2012)
7. H.S. Weinberg, S.W. Krasner, S.D. Richardson, A.D. Thruston Jr., and National Exposure Research Laboratory, *The Occurrence of Disinfection By-Products (DBPs) of Health Concern in Drinking Water: Results of a Nationwide DBP Occurrence Study, EPA/600/R-02/068* (USA, US Environmental Protection Agency, National Exposure Research Laboratory, Athens, GA, 2002)
8. E. Aieta, J. Berg, A Review of Chlorine Dioxide in Drinking-Water Treatment. J. AWWA **78** (6), 62–72 (1986)
9. J.C. Crittenden, R.R. Trussel, D.W. Hand, K.J. Howe, G. Tchobanoglous, *MWH's Water Treatment—Principles and Design*, 2nd edn. (Wiley, Hoboken, NJ, 2005)
10. H. Shorney-Darby, L.L. Harms, Chlorination of potable water, in *White's Handbook of Chlorination and Alternative Disinfectants*, ed. by G.C. White, Black & Veatch Corporation, 5th edn. (Wiley, Hoboken, NJ, 2010), p. 230
11. R.C. Hoehn, H. Shorney-Darby, J. Neemann, Chlorine dioxide, in *White's Handbook of Chlorination and Alternative Disinfectants*, ed. by G.C. White, Black & Veatch Corporation, 5th edn. (Wiley, Hoboken, NJ, 2010), p. 700
12. J.R. Bolton, C. Cotton, *The ultraviolet disinfection handbook* 1st edn. (American Water Works Association, Denver, CO, 2008)
13. S. Malato, P. Fernandez-Ibanez, M.I. Maldonado, J. Blanco, W. Gernjak, Decontamination and disinfection of water by solar photocatalysis: Recent overview and trends. Catal. Today **147**(1), 1–59 (2009)
14. R. Narayan, Use of nanomaterials in water purification. Mater. Today **13**(6), 44–46 (2010)
15. M. Mori, T. Sugita, A. Mase, T. Funatogawa, M. Kikuchi, K. Aizawa, S. Kato, Y. Saito, T. Ito, H. Itabashi, Photodecomposition of humic acid and natural organic matter in swamp water using a $TiO_2$-coated ceramic foam filter: Potential for the formation of disinfection byproducts. Chemosphere **90**(4), 1359–1365 (2013)
16. K.K. Philippe, C. Hans, J. MacAdam, B. Jefferson, J. Hart, S.A. Parsons, Photocatalytic oxidation of natural organic matter surrogates and the impact on trihalomethane formation potential. Chemosphere **81**(11), 1509–1516 (2010)
17. L. Wen Jun, Z. Yu, M. Gao, L. Zhang, X. Cai, F. Chao, Effect of ultraviolet irradiation on the characteristics and trihalomethanes formation potential of humic acid. Water Res. **30**(2), 347–347 (1996)
18. S. Liu, M. Lim, R. Fabris, C. Chow, K. Chiang, M. Drikas, R. Amal, Removal of humic acid using $TiO_2$ photocatalytic process - Fractionation and molecular weight characterisation studies. Chemosphere **72**(2), 263–271 (2008)

19. S. Liu, M. Lim, R. Fabris, C.W.K. Chow, M. Drikas, G. Korshin, R. Amal, Multi-wavelength spectroscopic and chromatography study on the photocatalytic oxidation of natural organic matter. Water Res. **44**(8), 2525–2532 (2010)
20. S. Liu, M. Lim, R. Fabris, C. Chow, M. Drikas, R. Amal, $TiO_2$ photocatalysis of natural organic matter in surface water: Impact on trihalomethane and haloacetic acid formation potential. Environ. Sci. Technol. **42**(16), 6218–6223 (2008)
21. H. Weinberg, Disinfection byproducts in drinking water: the analytical challenge. Analytical Chemistry News & Features **71**(23), 801A–808A (1999)
22. S.D. Richardson, M.J. Plewa, E.D. Wagner, R. Schoeny, D.M. DeMarini, Occurrence, genotoxicity, and carcinogenicity of regulated and emerging disinfection by-products in drinking water: A review and roadmap for research. Mutat. Res. - Rev. Mut. Res. **636**(1–3), 178–242 (2007)
23. H. Canada, *Guidelines for Canadian Drinking Water Quality - Summary Table (Water* (Air and Climate Change Bureau, Healthy Environments and Consumer Safety Branch, Health Canada, Ottawa, Ontario, 2012)
24. IARC Working Group on the Evaluation of Carcinogenic Risks to Humans, *IARC Monographs on the Evaluation of Carcinogenic Risks to Humans: Some Drinking-water Disinfectants and Contaminants, Including Arsenic* (International Agency for Research on Cancer, Lyon, France, 2004), 84, p. 526
25. S. Hrudey, Chlorination disinfection by-products, public health risk tradeoffs, and me. Water Res. **43**(8), 2057–2092 (2009)
26. L. Kronberg, B. Holmbom, M. Reunanen, L. Tikkanen, Identification and quantification of the Ames mutagenic compound 3-chloro-4-(dichloromethyl)-5-hydroxy-2(5H)-furanone and of its geometric isomer (E)-2-chloro-3-(dichloromethyl)-4-oxobutenoic acid in chlorine-treated humic water and drinking water extracts. Environ. Sci. Technol. **22**(9), 1097–1103 (1988)
27. S.W. Krasner, M.J. McGuire, J.G. Jacangelo, N.L. Patania, K.M. Reagan, E. Marco Aieta, Occurrence of disinfection by-products in US drinking water. J. AWWA. **81**(8), 41–53 (1989)
28. D.L. Norwood, R.F. Christman, ACID Structural Characterization of Aquatic Humic Material. 2. Phenolic Content and its Relationship to Chlorination Mechanism in an Isolated Aquatic Fulvic. Environ. Sci. Technol. **21**(8), 791–798 (1987)
29. F.J. Stevenson, *Humus chemistry : genesis, composition, reactions* 2nd edn. (Wiley, New York, 1994)
30. Y. Cho, W. Choi, Visible light-induced reactions of humic acids on $TiO_2$. J. Photochem. Photobiol. Chem **148**(1–3), 129–135 (2002)
31. X. Zhang, A.J. Du, P. Lee, D.D. Sun, J.O. Leckie, $TiO_2$ nanowire membrane for concurrent filtration and photocatalytic oxidation of humic acid in water. J. Membr. Sci. **313**(1–2), 44–51 (2008)
32. H. Kim, M. Yu, Characterization of natural organic matter in conventional water treatment processes for selection of treatment processes focused on DBPs control. Water Res. **39**(19), 4779–4789 (2005)
33. J. Croue, D. Violleau, L. Labouyrie, Disinfection by-product formation potentials of hydrophobic and hydrophilic natural organic matter fractions: A comparison between a low- and a high-humic water. ACS Symp. Ser. **761**, 139–153 (2000)
34. W.A. Mitch, S.W. Krasner, P. Westerhoff, A. Dotson, *Occurrence and formation of nitrogenous disinfection by-products* (Water Research Foundation, Denver, CO, 2009)
35. T. Bond, J. Huang, M.R. Templeton, N. Graham, Occurrence and control of nitrogenous disinfection byproducts in drinking water—A review. Water Res. **45**, 4341–4354 (2011)
36. S.W. Krasner, P. Westerhoff, B. Chen, G. Amy, S. Nam, Z.K. Chowdhury, S. Sinha, B.E. Rittmann, *Contribution of Wastewater to DBP Formation* (AWWA Research Foundation, Denver, CO, 2008)
37. E. Pehlivanoglu-Mentas, D.L. Sedlak, The fate of wastewater-derived NDMA precursors in the aquatic environment. Water Res. **40**, 1287–1292 (2006)

38. S. Liu, M. Lim, R. Amal, *Photocatalysis of Natural Organic Matter in Water: Characterization and Treatment Integration, inPhotocatalysis and Water Purification: From Fundamentals to Recent Applications*, 1st edn., ed. by P. Pichat (Wiley, Weinheim, 2013), p. 273
39. H. Selcuk, H.Z. Sarikaya, M. Bekbolet, M.A. Anderson, Bromate formation on the non-porous $TiO_2$ photoanode in the photoelectrocatalytic system. Chemosphere **62**, 715–721 (2006)
40. B. Cancho, F. Ventura, M. Galceran, A. Diaz, S. Ricart, Determination, synthesis and survey of iodinated trihalomethanes in water treatment processes. Water Res. **34**(13), 3380–3390 (2000)
41. J.R. Bolton, *Ultraviolet Applications Handbook* (Photosciences Inc., Ayr, 2001), p. 30
42. R. Toor, M. Mohseni, UV-$H2O_2$ based AOP and its integration with biological activated carbon treatment for DBP reduction in drinking water. Chemosphere **66**, 2087–2095 (2006)
43. T. Bond, E.H. Goslan, B. Jefferson, F. Roddick, L. Fan, S.A. Parsons, Chemical and biological oxidation of NOM surrogates and effect on HAA formation. Water Res. **43**, 2615–2622 (2009)
44. Y. Zhao, J.M. Boyd, M. Woodbeck, R.C. Andrews, F. Qin, S.E. Hrudey, X. Li, Formation of N-nitrosamines from eleven disinfection treatments of seven different surface waters. Environ. Sci. Technol. **42**, 4857–4862 (2008)
45. E. Shen, S. Andrews, Demonstration of 20 pharmaceuticals and personal care products (PPCPs) as nitrosamine precursors during chloramine disinfection. Water Res. **45**, 944–952 (2011)
46. W.H. Chen, T.M. Young, NDMA formation during chlorination and chloramination of aqueous diuron solutions. Environ. Sci. Technol. **42**(4), 1072–1077 (2008)
47. W. Buchanan, F. Roddick, N. Porter, Formation of hazardous by-products resulting from the irradiation of natural organic matter: Comparison between UV and VUV irradiation. Chemosphere **63**(7), 1130–1141 (2006)
48. B.A. Lyon, A.D. Dotson, K.G. Linden, H.S. Weinberg, The effect of inorganic precursors on disinfection byproduct formation during UV-chlorine/chloramine drinking water treatment. Water Res. **45**, 4653–4664 (2012)
49. European Commission, *Drinking Water Directive*, Council Directive 98/83/EC, (1998).
50. Government of Singapore, *Environmental Public Health Act (Chapter 95) Environmental Public Health Quality of Piped Drinking Water, Regulations 2008*, (2008).
51. Ontario Ministry of the Environment, *Ontario Drinking Water Quality Standards*, O.Reg. 327/08 (2008)
52. US EPA, *National Primary Drinking Water Regulations*, EPA 816-F-09-004, (2009)
53. World Health Organization, *Guidelines for Drinking-water Quality4th* (Gutenberg, Malta, 2011)
54. Ministry of Environmental Protection of the Government of the People's Republic of China, *Standards for Drinking Water Quality*, GB 5749-2006 (2006).
55. CDPH, NDMA and Other Nitrosamines—Drinking Water Issues. (State of California, 2011), http://www.cdph.ca.gov/certlic/drinkingwater/pages/NDMA.aspx. Accessed 2011
56. State of California, *California Code of Regulations, Title 22, Division 4, Chapter 15.5 Disinfectant Residuals, Disinfecation Byproducts, and disinfection byproduct precursors*, (2013)
57. K. Hashimoto, H. Irie, A. Fujishima, $TiO_2$ photocatalysis: a historical overview and future prospects. Jpn. J. Appl. Phys. **44**(12), 8269–8285 (2005)
58. J. Herrmann, Photocatalysis fundamentals revisited to avoid several misconceptions. Appl. Catal. B Environ. **99**, 461–468 (2010)
59. A. Fujishima, X. Zhang, Titanium dioxide photocatalysis: present situation and future approaches. C. R. Chimie **9**, 750–760 (2006)
60. Y. Nosaka, A.Y. Nosaka, *Identification and Roles of the Active Species Generated on Various Photocatalysts, inPhotocatalysis and Water Purification: From Fundamentals to Recent Applications*, 1st edn., ed. by P. Pichat (Wiley-VCH Verlag GmbH & Co. KGaA, Weinheim, Germany, 2013), p. 3
61. A. Hu, X. Zhang, K.D. Oakes, P. Peng, Y.N. Zhou, M.R. Servos, Hydrothermal growth of free standing $TiO_2$ nanowire membranes for photocatalytic degradation of pharmaceuticals. J. Hazard. Mater. **189**(1–2), 278–285 (2011)

62. A. Li, X. Zhao, H. Liu, J. Qu, Characteristic transformation of humic acid during photoelectrocatalysis process and its subsequent disinfection byproduct formation potential. Water Res. **45**(18), 6131–6140 (2011)
63. T.A. Egerton, Does photoelectrocatalysis by $TiO_2$ work? J. Chem. Technol. Biot. **86**(8), 1024–1031 (2011)
64. S. Dobrovic, H. Juretic, D. Ljubas, I.V. Vrcek, M.Z. Avdicevic, M. Milic, N. Kopjar, I. Nemet, Genotoxicity and effects of nanosilver contamination in drinking water disinfection. Wa. Sci. Technol. **12**(6), 829–836 (2012)
65. M.J. Cantavenera, I. Catanzaro, V. Loddo, L. Palmisano, G. Sciandrello, Photocatalytic degradation of paraquat and genotoxicity of its intermediate products. J. Photochem. Photobiol. A. **185**, 277–282 (2007)
66. M.S. Sekler, Y. Levi, B. Polyak, A. Novoa, P.S.M. Dunlop, J.A. Byrne, R.S. Marks, Monitoring genotoxicity during the photocatalytic degradation of p-nitrophenol. J. Appl. Toxicol. **24**, 395–400 (2004)
67. S. Richardson, A. Thruston, T. Collette, K. Patterson, B. Lykins, J. Ireland, Identification of $TiO_2$/UV disinfection byproducts in drinking water. Environ. Sci. Technol. **30**(11), 3327–3334 (1996)
68. H. Selcuk, Disinfection and formation of disinfection by-products in a photoelectrocatalytic system. Water Res. **44**, 3966–3972 (2010)
69. X. Huang, M. Leal, Q. Li, Degradation of natural organic matter by $TiO_2$ photocatalytic oxidation and its effect on fouling of low-pressure membranes. Water Res. **42**(4–5), 1142–1150 (2008)
70. F.C. Kent, K.R. Montreuil, R.M. Brookman, R. Sanderson, J.R. Dahn, G.A. Gagnon, Photocatalytic oxidation of DBP precursors using UV with suspended and fixed $TiO_2$. Water Res. **45**(18), 6173–6180 (2011)
71. D. Gerrity, B. Mayer, H. Ryu, J. Crittenden, M. Abbaszadegan, A comparison of pilot-scale photocatalysis and enhanced coagulation for disinfection byproduct mitigation. Water Res. **43**, 1597–1610 (2009)
72. A. Parkinson, F.A. Roddick, M.D. Hobday, UV photooxidation of NOM: Issues related to drinking water treatment. J. Water Supply Res. T. **52**(8), 577–586 (2003)
73. J.C. Crittenden, R.R. Trussel, D.W. Hand, K.J. Howe, G. Tchobanoglous, *MWH's Water Treatment - Principles and*, Design3rd edn. (John Wiley & Sons, Hoboken, N.J., 2012)
74. T.F. Marhaba, N.S. Pipada, Coagulation: Effectiveness in Removing Dissolved Organic Matter Fractions. Environ. Eng. Sci. **17**(2), 107–115 (2000)
75. J. Sohn, A. Gary, Y. Yoon, Process-train profiles of NOM through a drinking water treatment plant. J. AWWA. **99**(6), 145–153 (2007)
76. M. Bekbolet, C.S. Uyguner, H. Selcuk, L. Rizzo, A.D. Nikolaou, S. Meric, V. Belgiorno, Application of oxidative removal of NOM to drinking water and formation of disinfection by-products. Desalination **176**(1–3), 155–166 (2005)
77. S.W. Krasner, W.A. Mitch, P. Westerhoff, A. Dotson, Formation and control of emerging C- and N-DBPs in drinking water. J. AWWA. **104**, 582–595 (2012)
78. G. Amy, B.C. Alleman, C.B. Cluff, Removal of dissolved organic matter by nanofiltration. J. Environ. Eng. **116**, 200–205 (1990)
79. A. De la Rubia, M. Rodriguez, V.M. Leon, D. Prats, Removal of natural organic matter and THM formation potential by ultra- and nanofiltration of surface water. Water Res. **42**, 714–722 (2008)
80. M. Itoh, S. Kunikane, Y. Magara, Evaluation of nanofiltration for disinfection by-products control of drinking water treatment. Wa. Sci. Technol. **1**(5/6), 233–243 (2001)
81. C.S. Uyguner, M. Bekbolet, H. Selcuk, A comparative approach to the application of a physico-chemical and advanced oxidation combined system to natural water samples. Separ. Sci. Technol. **42**, 1405–1419 (2007)

# Chapter 10
# Nanotechnology in Contemporary Mine Water Issues

**Ken Oakes, Zhi Shan, Rajendran Kaliaperumal, Shine Xu Zhang, and Martin Mkandawire**

**Abstract** Nanotechnology holds great promise in advancing affordable mine water treatment and remediation technologies by improving treatment efficiency and performance. Further, it promises to overcome major challenges faced by existing treatment technologies through provision of highly efficient, modular, and multifunctional processes facilitating new treatment capabilities. Such technologies could further allow economic utilization of mine water as a source of commercially viable products. This chapter provides an overview of recent and future developments in nanotechnology that would benefit contemporary mine water treatment regimes, while investigating novel abate approaches possible through the use of nanotechnology. The chapter discusses candidate nanomaterials, their advantages and limitations relative to existing processes, and the unique properties and surface-active mechanisms that enable their adoption in mine water treatment applications.

## 10.1 Introduction

Natural resources derived from the Earth's lithosphere, hydrosphere, biosphere, and atmosphere are the key building blocks of a sustainable human society. Historically, mining affects the surrounding environment during both active operation and

K. Oakes • Z. Shan • R. Kaliaperumal • M. Mkandawire (✉)
Verschuren Centre for Sustainability in Energy and the Environment, Cape Breton University, 1250 Grand Lake Road, Sydney, NS, Canada B1P 6 L2
e-mail: martin_mkandawire@cbu.ca

S.X. Zhang
Verschuren Centre for Sustainability in Energy and the Environment, Cape Breton University, 1250 Grand Lake Road, Sydney, NS, Canada B1P 6 L2

Waterloo Institute of Nanotechnology and Department of Biology, University of Waterloo, 200 University Avenue West, Waterloo, ON, Canada N2L 3G1

A. Hu and A. Apblett (eds.), *Nanotechnology for Water Treatment and Purification*, Lecture Notes in Nanoscale Science and Technology 22,
DOI 10.1007/978-3-319-06578-6_10, 

for extended periods following mine closure. Among the myriad of environmental impacts associated with mining, including disturbed landscapes, erosion, formation of sinkholes, depletion and contamination of water sources, and soil contamination, the long-term production of mine water poses one of the greatest concerns. Mine water refers to any surface or groundwater present at the mine site, which has been altered due to mining activities or influenced by mining-related processes such as milling, extraction, or percolation over mine waste. Mine water may be produced by contact with materials from long-closed mine sites, or it may refer to water produced during active mining by dewatering processes.

Contemporary mine water issues range from depletion and/or contamination of water resources during active mining, to acid mine drainage (AMD) and leachate from abandoned workings and waste mine rock heaps, to management of tailings post-mine closure. Notably, most closed mines are sources of multiple pollutants including inorganic metals and organic chemicals and acids used to extract the minerals. Despite the high risks they pose, many mine water discharges are insufficiently treated, if treated at all, due to the substantial number of sites on properties where mines closed without sufficient funds set aside during operation to address the high (and often on-going) cost of remediation. However, recent advances in nanotechnology offer opportunities to develop next-generation mine water remediation and management technologies, which promise affordable treatments offering improved performance. Nanotechnology may further overcome traditional challenges faced by existing treatment technologies through provision of highly efficient, modular, and multifunctional processes with the potential for economic utilization of mine water as source of commercially viable products.

This chapter provides an overview of recent and future developments in nanotechnology that would address contemporary mine water issues, while raising potential concerns that may accompany the non-precautious use of nanotechnology. The discussion covers candidate nanomaterials, properties, and mechanisms enabling the applications, advantages, and limitations relative to existing processes, and barriers and research needs for commercialization. By tracing these technological advances to the physicochemical properties of nanomaterials, the present chapter outlines the opportunities and limitations to further capitalize on these unique materials for sustainable mine water remediation and management.

The chapter is presented through several sentinel topics; beginning with a perspective on contemporary mine water issues through the lens of challenges faced both during active mining and post-closure. Types of mine water, as well as issues associated with mine water discharge, are next addressed, together with current technologies used to treat the contaminated waters. This is followed by an analysis of the current state of nanotechnology as applied to contemporary mine water issues, potential future applications, and an overview of potential issues associated with the application of nanotechnology to mine waters. The chapter will explore these topics from a broad-based, multidisciplinary perspective relevant for both engineers and scientists outside the field, and for students at both the undergraduate and graduate level.

## 10.2 Mining and the Mine Life Cycle

To understand contemporary mine water issues, an understanding of a mines' life cycle, including the mining techniques employed, are critical. While exact processes differ by mineral, depth, region, and available resources, the fundamentals are similar. They begin with the discovery of a commercially viable ore body, leading to development and extraction of minerals during the life of the mine, concluding with the closure of the mine, after which the site is either restored to its natural state, or reclaimed to a condition suitable for other land uses. While the discovery phase is a mixture of geological knowledge, exploratory drilling, and luck, the extraction of mineral resources is constrained to two common mining techniques, namely excavation and in situ leaching [1].

### *10.2.1 Excavation and Mining Techniques*

Specific excavation techniques are used in surface or sub-surface mining applications, depending on the depth and quantity at which the mineral resource is found [2, 3]. Surface mining involves removing surface vegetation and overlying layers of bedrock to reach the ore deposits [4, 3]. Specific surface mining techniques include: (1) open pit mining, where relatively shallow ore is recovered from expansive open pits excavated into the ground; (2) quarrying of aggregate materials, from a usually shallow open pit mine; (3) strip mining which consists of stripping surface layers to reveal ore/seams underneath, and (4) mountaintop removal, commonly associated with coal mining, which involves removing the top of a mountain to access ore deposits at depth [3–6]. Conversely, sub-surface mining involves the digging of tunnels or shafts into the earth, often but not necessarily at significant depths, to access ore deposits. The targeted mineral-bearing ore and non-mineral-bearing waste rock are brought to the surface through the tunnels and shafts. Sub-surface mining can be further classified by the type of access shafts used, as well as the extraction method or technique used to reach the mineral deposit. For instance, drift mining utilizes horizontal access tunnels, while slope mining uses diagonally sloping access shafts, and shaft mining consists of vertical access shafts. Following excavation, the ore-bearing rocks are crushed and pulverized to facilitate mineral extraction. The target resource of interest is then recovered singularly, or with other minerals through combinations of several mechanical and/or chemical procedures [7–10]. Most metals are present in ores as oxides or sulphides, necessitating their reduction to their metallic forms via chemical means such as smelting or electrolytic reduction.

### 10.2.2 In Situ Leaching Technique

In situ leaching involves the drilling of holes into the ore deposit, with explosive or hydraulic fracturing potentially used to create openings in the deposit for a leaching solution to penetrate [1]. As the leaching solution is pumped into the deposit, it dissolves the target minerals and the solution carrying the dissolved ore content is pumped to the surface and processed [11]. In situ leaching allows for the extraction of metals and salts from an ore body without blasting, removing overburden and waste rock, or construction of underground mine infrastructure.

## 10.3 Mine Water and Contemporary Issues

For both excavative and in situ techniques, water is a major concern during both the active mining and post-closure periods. During active mining, both excavative and in situ techniques utilize dewatering processes, but excavative approaches typically produce more waste water as ore processing requires considerable quantities of water during washing and mineral extraction [5, 6]. Waste rock and tailing piles, if exposed to water, can leach contaminants. Mine flooding is common once the dewatering processes conclude with the cessation of active mining, leading to formation of underground mine water pools which remain a significant contemporary environmental issue [12]. With open pit mining, pit lakes may form by precipitation, surface runoff, or groundwater filling the open pit upon the completion of excavation activities [13, 14].

### 10.3.1 Classification of Mine Waters

Mine waters can be classified into numerous categories, varying in their quality, pH, and potential for environmental impacts. The definition of what actually constitutes mine water is still a major point of contention within scientific and engineering conversations and literature. Consequently, the term "mine water" has been used interchangeably with other terms like "mine drainage" and "mine pools" [15, 16]. In part, this confusion stems from the absence of a recognized body sanctioning terminology; so, the definition of mine water, as used in this chapter, will broadly constitute any water altered due to mining activities, or influenced by mining-related processes such as milling, extraction, or percolation over mine waste, as well as groundwater flooding active mine workings and subsequently de-watered [16, 17]. Consequently, "mine water" will generally refer to any surface or groundwater present at a mine site, while "mining water" will specifically refer to water that has come into contact with any mine workings. Within this framework, mine water is the all-encompassing term that includes mining water [2, 17]. Mining water includes mill water, which refers to water used during the crushing and grinding of ore, and may also contain dissolved minerals or metals. Process water, another mining water used in the chemical extraction of metals, similarly frequently contains high burdens of dissolved metals and other chemicals. Leachate, water that has trickled through solid mine wastes or tailings, process chemicals, or metals. When the mining, mill, or process water is discharged to the surface, often after being treated, it is known as effluent. Ideally, all mine water should be collected and stored in tailings ponds before being treated and released to surface water, if necessary [15]. Mine drainage includes water contaminated by interactions with waste rock, within impoundments, or through interactions with geological formations in water-filled abandoned mine shafts. Mine drainage can be classified by its pH as either acid or neutral mine drainage, both containing metals/constituents at relatively high concentrations [15, 16].

### *10.3.2 Mine Water Issues*

There are five outstanding mine water issues, mainly related to pollution monitoring and control that may be successfully addressed by the application of nanotechnology.

#### 10.3.2.1 Acid Mine Drainage

AMD usually forms when sulphide minerals in rocks are exposed to oxidizing conditions during or after mining activities, leading to the production of sulphuric acid contaminated water known as acid rock drainage (ARD), which contributes to the acidity of interacting waters [18]. While ARD formation is a naturally occur process, it is markedly accelerated when sulphide-containing rocks are exposed to air and water through construction of shafts, tunnels, and collieries that maintain oxidizing conditions over greater geolithic surfaces leading to production of AMD [18, 19]. In most cases, AMD formation requires the presence of iron sulphides such as pyrite and marcasite ($FeS_2$), which are common in coal regions, but may also be formed by chalcopyrite ($CuFeS_2$), covellite ($CuS$), and arsenopyrite ($FeAsS$) (also see Table 10.1).

The concentration and composition of metals in AMD is a function of both the type and amount of sulphide mineral, and the presence or absence of buffering alkaline materials, which regulate acidity. The oxidation of Fe disulphides and subsequent conversion to acid occur through several reactions. The first chemical process, iron sulphide oxidation, releases ferrous iron ($Fe^{2+}$), sulphate ($SO_4^{2-}$), and protons, as summarized in Eq. (10.1):

**Table 10.1** Some important metal-sulphide minerals responsible for AMD at mining sites

| Chemical formula | Mineral | Chemical formula | Mineral |
|---|---|---|---|
| $FeS_2$ | Pyrite | $MoS_2$ | Molybdenite |
| $FeS_2$ | Marcasite | NiS | Millerite |
| $Fe_xS_x$ | Pyrrhotite | PbS | Galena |
| $Cu_2S$ | Chalcocite | ZnS | Sphalerite |
| CuS | Covellite | FeAsS | Arsenopyrite |
| $CuFeS_2$ | Chalcopyrite | $Cu_5FeS_4$ | Bornite |

*Source*: Adapted from Ref. [20]

$$FeS_2 + {}^7\!/_2O_2 + H_2O = Fe^{2+} + 2SO_4{}^{2-} + 2H^+ \qquad (10.1)$$

The resulting ferrous iron can further oxidize to ferric iron ($Fe^{3+}$), according to Eq. (10.2):

$$2Fe^{2+} + {}^1\!/_4O_2 + H^+ = Fe^{3+} + {}^1\!/_2H_2O \qquad (10.2)$$

Ferric iron can then either be hydrolyzed to ferric hydroxide, $Fe(OH)_3$, increasing $H^+$ acidity (Eq. 10.3), or it can directly attack pyrite as a catalyst, generating much greater amounts of ferrous iron, sulphate, and acidity (Eq. 10.4).

$$Fe^{3+} + 3H_2O = Fe(OH)_3 + 3H^+ \qquad (10.3)$$

$$FeS_2 + 14Fe^{3+} + 8H_2O = 15Fe^{2+} + 2SO_4{}^{2-} + 16H^+ \qquad (10.4)$$

The generation of AMD would slow or cease when any of the constituent processes are inhibited as these steps are sequential and interdependent [21, 22]. Obviously, with removal of two of the three principal reactants, air and water from the system, pyrite would not be oxidized. The rate of pyrite oxidation depends on numerous variables such as the reactive surface area of pyrite, form of pyritic sulphur, oxygen concentrations, solution pH, catalytic agents, flushing frequencies, and the presence of Fe-oxidizing bacteria (FeOB). Under most conditions, the oxidation of $Fe^{2+}$ to $Fe^{3+}$ (Eq. 10.2) is the rate-limiting pyrite oxidation step as the conversion of ferrous iron to ferric iron is slow at pH values below 5 under abiotic conditions [21, 23]. However, FeOB, principally *Thiobacillus* sp., greatly accelerate the reaction, and are consequently critical regulators of AMD generation. Availability of oxygen may be the rate-limiting step in AMD formation within geological materials of low porosity and permeability, while in porous and permeable formations, especially those composed of coarse sandstone, air convection driven by the heat generated by pyrite oxidation provide high amounts of oxygen deep into the rocks, ensuring an oxidizing environment.

#### 10.3.2.2 Alkaline Mine Drainage

Of perhaps less concern, but closely related to AMD is neutral or alkaline mine drainage (NAMD). NAMD has a pH of 6.0 or above, as a reflection of local alkalinity or proton buffering capacity attributable to the local presence of hydroxyl ion ($OH^-$), carbonate, silicate, borate, organic ligands, phosphate, and ammonia within the mine drainage [16, 24]. The principal source of alkalinity in mine water is dissolved carbonaceous compounds, which exist in solution as bicarbonate ($HCO_3^-$) or carbonate ($CO_3^{2-}$). The quality of mine water emanating from underground mines or backfills of surface mines is consequently dependent on both the acidic (i.e. sulphide) and alkaline (i.e. carbonate) materials contained in the disturbed geological material. It follows that NAMD may result from any combination of relatively low levels of sulphide minerals, a deoxygenated but sulphide-rich environment, the presence of monosulphides rather than pyrite or marcasite, a large pyrite grain-size limiting oxidation rate, the presence of highly reducing or alkaline influent water, neutralization of acid by carbonate or basic silicate minerals, and/or the neutralization of acid by naturally highly alkaline groundwater [25]. Other reactions that contribute to the formation of NAMD include the oxidation of other sulphides, dissolution of host rock minerals, precipitation and dissolution of iron hydroxide and hydroxysulphate, co-precipitation and adsorption of dissolved metals, precipitation and dissolution of iron sulphate, gypsum precipitation and dissolution, and $CO_2$ degassing [26]. Mine waters that contain alkalinity can also be contaminated with divalent metal ions including ferrous iron and manganese. None are contaminated with significant levels of dissolved ferric iron or aluminium because the solubility of these metal hydroxides is low in mine waters with pH values greater than 5.0.

#### 10.3.2.3 Toxic Metal Contamination and Leaching

Both surface and underground as well as in situ leach mining approaches vary greatly but share risks for environmental contamination with toxic metals, metalloids, and radionuclides. Waste rock, common to the first two extraction techniques, accumulate in large quantities under fairly oxidizing conditions associated with surface storage in large piles, and often contain acid-generating sulphides, metals, and other contaminants. The waste rock and the exposed bedrock walls from which ore is excavated are the sources of most of the metal pollution. Toxic metal and metalloids such as arsenic, cobalt, copper, cadmium, lead, silver, iron, chromium, manganese, mercury, molybdenum, nickel, lead, selenium, vanadium and zinc as well as some radionuclides are liberated and move into mine water when excavated rock exposed in an underground mine come in contact with water [2, 27–31]. In open pit mines, new metals are in continuous contact with the surface air and water, and after mine closure, the metals continue to dissolve in the pit lake waters. With conventional extraction techniques, considerable quantities of tailings are generated, as the percentage of extractable material in ore rocks is usually very low.

For instance, high-grade uranium ore contains a maximum of 0.07 % extractable uranium [32]. As a result, uranium mining and processing inexorably leads to masses of geological materials, which are often surface stored long-term as mountains of waste. From these heaps, metals are leached out and carried downstream as water washes over the rock surface or seep through waste rock piles and tailings. Although metals can be mobilized at neutral pH, leaching is accelerated under low pH conditions, including those created by AMD.

Tailings also represent a major source of metal contamination of waterways. After the waste rock is removed and the ore is excavated, it is then processed to separate the target mineral from the valueless portion; the remaining ore becoming tailings. Mine tailings often contain the same toxic metals and acid-forming minerals that waste rock does. Tailings can also contain chemical agents used to process the ores, such as cyanide or sulphuric acid. With the in situ leaching technique, it is usually impossible to recover all the leaching solution from the ground. In most cases considerable volumes remain underground and continue dissolving rocks and releasing toxic metals and radionuclides into the ground that can have impacts for many generations after the active mining ceased.

#### 10.3.2.4 Radioactive Contamination

A common misconception is that radioactive contamination is unique to uranium producing mines, when in fact, the potential to generate radioactive mine water is equally high at any abandoned mine as natural radioactivity is common in the geological materials of our planet. Radioactivity indicates that matter, often known as radionuclides, is decaying in order to reach, according to the law of physics, a more energetically stable state. As materials decay, they emit radiation—emission of gamma ray(s) or subatomic particles such as alpha or beta particles—which disintegrate entirely over short or long intervals (measured in seconds, years, or millennia). Although most mine water emits insignificant levels of gamma radiation [33, 34], the presence of short-lived alpha particles or radon emitted as a decay product of readily soluble isotopes of uranium and radium can pose a significant risk.

There are over 1,500 different radionuclides, with potassium ($^{40}K$) responsible for the highest radioactivity and being the most predominant radionuclide in soil and rock material [35]. However, more radioactive elements such as radium and radon in the ore arise from the radioactive decay of uranium over hundreds of millions of years. Most radioactive mine water, from both ground and open pit mining, results from the solubilization of radionuclide-containing minerals. Several studies have shown radionuclide contamination in coal mining is usually attributed to secondary mineralization of uranium with coal development [36]. Similarly, radioactivity has been observed in iron and phosphate due to mineralization processes started with localization of uranium and related nuclides co-precipitated with either iron and/or phosphorus [10, 37–40].

#### 10.3.2.5 Processing Chemicals Contamination

Process chemical contamination occurs when chemical agents (such as cyanide or sulphuric acid) used by mining companies to separate the target mineral from the ore spill, leak, or leach from the mine and processing site.

## 10.4 State of Nanotechnology as Applied to Mine Water Issues

Advances in nanotechnology are enabling development of more efficient and cost-effective waste treatment and environmental remediation technologies. However, application of nanotechnology to mine water remediation, just as to mineral discovery, mining, extraction, and processing, has so far received little attention. To date, most applications of nanotechnology in mine water remediation have essentially followed the general application of nanomaterials in wastewater treatment and water purification. Most research in this area, especially that pertaining to mine water issues, has been performed in a small-scale laboratory context, with large-scale implementation of nanotechnology by the mining industry expected to take a few more decades.

### *10.4.1 Strategies in Water Treatment and Remediation*

Some nanotechnology applications designed for wastewater treatment and water purification may be used in mine water treatment and remediation including nanofiltration, nanocatalysis, nanomagnetism, nanosensors, nanoextraction, and disinfection. Nanofiltration involves development of nanomaterial-enhanced membranes that mechanically exclude contaminating substances, thereby removing pollutants from water. Such filters and membranes are made from a variety of nanomaterials including carbon nanotubes, nanoporous ceramics (clays), dendrimers, zeolites, nanofibres, and nanosponges. Multi-tasking filtration systems can detect, separate out, and/or detoxify mixtures of contaminants. A key advancement is the development of nanowire membranes with tuneable wettability ranging from superhydrophobic to superhydrophilic, which have high potential as a clean-up media [41–43]. Further, dendritic nanomaterials, highly branched nanoscale polymers, have been successfully employed to "encapsulate" environmental pollutants. Dendritic nanomaterials are often recyclable, water-soluble, and have demonstrated great potential for removing inorganic pollutants, heavy metals, biological, and radiological compounds [44, 45]. During the last decade, significant advances have been made in the development of nanoscale supramolecular hosts that can serve as high capacity, selective, and recyclable ligands and sorbents for extracting valuable metal ions from artificially synthesized mine water based on solutions and mixtures [44, 46–52]. For instance, dendrimer-based chelating agents and separation systems have been developed to recover valuable metal ions from mine water, such as Cu(II), Ni(II), Zn(II), Fe(III), Co(II), Pd(II), Pt(II), Ag(I), Au(I), Gd(III), and U(VI). Similarly, there have been advances in development of nanosorbents based on self-assembled monolayers on mesoporous supports for recovering a number metal ions, including most of those mentioned above.

Nanocatalysis, which utilizes nanoparticles with catalytic properties to chemically degrade pollutants, is a promising approach for treating waterborne contaminants present at very low levels. Magnetic nanoparticles have large surface areas to mass ratios, and easily bind with contaminants such as arsenic, AMD constituents, and oils, and following complexation, are easily removed from solution using a magnet, making them an appealing solution for mine water treatment [53–56]. Another example is nanoscale zero-valent iron (NZVI) particles, which are very efficient redox-active media for the degradation of organic contaminants, and can be applied to the treatment of AMD and other mine processing contaminants [53–57].

### *10.4.2 Detection and Monitoring*

There have also been recent advances utilizing nanotechnology to develop small, portable sensors with enhanced capabilities for detecting trace biological and chemical contaminants in environmental matrices, including water. Nanotechnology-enabled sensors are enabling rapid and accurate detection of harmful compounds by exploiting a variety of sensing and detection modalities, such as chemical (e.g. molecular recognition), optical (e.g. fluorescence), and mechanical (e.g. resonance). Potential applications of engineered nanomaterials for mine water monitoring include detection of various compounds in gaseous, aqueous, or soil matrices; sampling and detecting mine water pollutants accumulated within biological systems (cells, organs, tissues, etc.), and monitoring of physical parameters (pressure, temperature, distance, etc.). To date, many excellent publications have been produced to address such topics as "Environmental sensors" or "Nanotechnology for environmental monitoring" [58–60], of which most emphasized the design, construction, and operational mechanisms of the nanostructure-based sensor, but very few focused on specific target analytes, especially those heavy metals species such as (Hg, As, Mn, Pb, Cd, Ni, Zn, Cu, Cr, U, Ra) and polycyclic aromatic hydrocarbons (PAHs) found in mine water.

## 10.5 Applicable Nanotechniques in Mine Water Issues

### *10.5.1 Detection and Monitoring Techniques*

Monitoring mine water pollution with expensive and sophisticated instrumentation such as atomic absorption spectroscopy (AAS), inductively coupled plasma-mass spectrometry (ICP-MS), mass spectroscopy (MS), and X-ray fluorescence

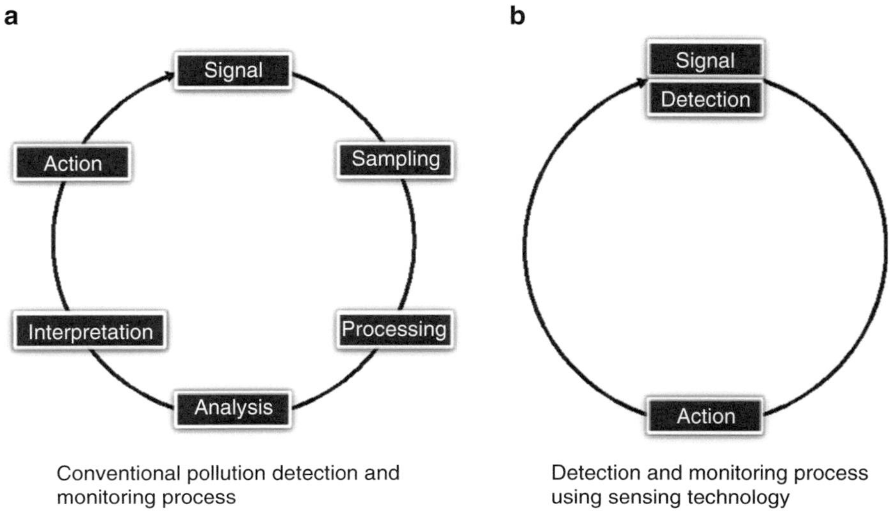


**Fig. 10.1** Comparison of conventional (**a**) and sensing technology (**b**) processes in pollution detection and monitoring. In (**a**), the process has five steps, each having potential sources of error due to human and instrumentation factors; while in (**b**) the process is greatly simplified, reducing sources of errors while being relatively inexpensive

spectroscopy (R-FS) is still the standard approach. However, these instrumentally intensive methods only measure total analyte concentrations in environmental samples, but provide no information on speciation. Further, these instruments often require intense sample preparation efforts. In short, the complexity of current instrumental approaches introduces many potential sources of error, and the expenses associated with monitoring mine water pollution preclude its application in many developing economies. Consequently, the availability of simple, portable, and affordable devices capable of on-site monitoring would be a critical advance in the monitoring of mine water pollution (Fig. 10.1). There are two approaches to develop on-site detection and monitoring apparatus; the first is simply to miniaturize current laboratory instrumentation (i.e. developing miniaturized gas- or liquid-chromatography/mass spectrometry instrumentation), the second is to develop devices such as micro- or nanosensors or sensor arrays. The idea underlying the former approach can be described as "top-down", while the latter is a "bottom-up" approach. Nanotechnology and nano-/micro-engineering methodologies are essential for both approaches, but in this chapter, we will focus on the development of nano-/micro-sensor technology for environmental monitoring.

The merit of the nanotechnology sensor-based approach and device development to monitor mine water pollutants can be demonstrated through recent progress made in this field. The following is not an exhaustive summary of all publications generated in this field; rather, we selected some published data since 2010 to exemplify the trend in the development of nanotech-based environmental sensing, updating the excellent review by Namour et al. [61].

The nanosensor technology normally involves nanostructured materials in either the recognition element or the signal-transducing element, so as to improve the performance of the sensor in terms of sensitivity, selectivity, reproducibility, portability (miniaturization), or throughput. When biomacromolecules or structures are incorporated into the recognition element for target recognition, the sensor is called a biosensor. Depending on the signal transduction mechanism [62], nanosensors can be categorized as optical sensors (i.e. fluorescent, colorimetric, SERS) or electrochemical sensors. Based on their configuration, there are simple nanoparticle-based sensors, nanomaterial-modified sensors/devices, and microfluidic sensors. We will briefly introduce the mechanisms of several representative nano-sensors and sensing devices in the following sections.

#### 10.5.1.1 Optical Nanosensors

Optical nanosensors convert the recognition and binding of target analytes to an optical change in intensity, lifetime, fluorescence anisotropy, colour based on shift of light absorbance, or Raman Spectrum. Of these, colorimetric sensors are most desirable for on-site environmental monitoring due to their operational simplicity and convenience, as normally no power supply or other instruments are needed. While colorimetric sensors can detect analyte presence with the naked eye, instruments may be necessary to obtain more accurate quantification. In many cases, smartphones incorporating a digital camera could collect the data (colour or colour change) and send to the lab via wireless communication for accurate quantification. To construct colorimetric nanosensors, gold nanoparticles (AuNPs) serve as an excellent signal-transducing element owning to their unique optical properties. Due to the plasmonic effect, AuNPs ($d > 3.5$ nm) show specific absorbance dependent on inter-particle proximity. For example, 13 nm-sized monodispersed AuNPs in aqueous solution appear red in colour, but change to purple or blue due to shifts in surface plasmon resonance (SPR) peak in the case of AuNP aggregation. Herein, the aggregation of AuNPs could be induced by different targets when AuNPs are functionalized with the specific receptor (ligand) for a given analyte of interest, which forms the basis for the sensor design.

For heavy metal detection, the analyte recognition element immobilized on the AuNP surface is normally a chelating ligand that can specifically bind to the heavy metal ions. The red-to-blue colour change results from metal ion-induced AuNP aggregation by forming multidentate inter-particle complexes (Fig. 10.2) [63]. To date, small molecules [such as 11-mercaptoundecanoic acid (MUA) (for $Pb^{2+}$, $Cd^{2+}$, and $Hg^{2+}$), carboxylate, 15-crown-5 (for $Pb^{2+}$), cysteine (for $Cu^{2+}$), 5,5′-dithiobis (2-nitrobenzoic acid) (for $Cr^{3+}$), quaternary ammonium (for $Hg^{2+}$), tetramethylmalonamide (for $Ln^{3+}$)], and biopolymer molecules [such as peptide (for $Hg^{2+}$), pentapeptide (CALNN, for $Al^{3+}$), single-stranded DNA] have been used as analyte recognizers on AuNP.

It is worth noting that single-stranded DNA (ssDNA) has been established as a powerful and versatile functional material serving as a recognition element for biosensor design [64]. The discovery of DNA aptamers [65], synthesized ssDNA

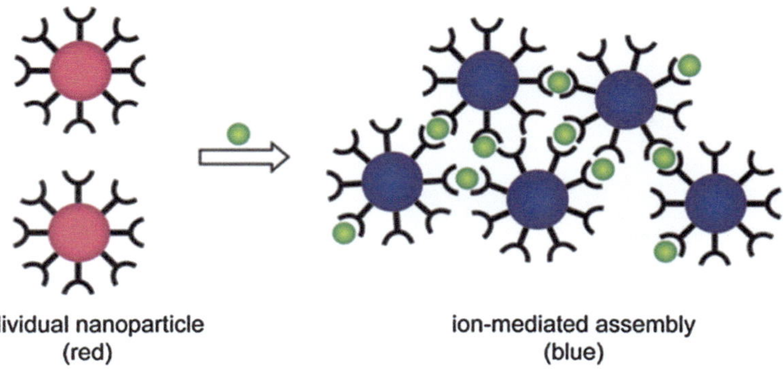


**Fig. 10.2** Schematic of colorimetric sensing of metal ions (*red-to-blue*) using AuNPs functionalized with ligands. The ligand may be chelating molecules, peptides, or DNA strands. Adapted from Ref. [63]

that can specifically bind to a wide variety of metal ions or organic molecules or structures (similar to an antibody, but more chemically stable), paves the way for widespread use of ssDNA as a binding ligand for various targets including heavy metal ions. For example, the T-$Hg^{2+}$-T coordination chemistry has been widely employed to design DNA probes for sensor design [66]. In addition, DNAzyme, the catalytic DNA, has been incorporated into sensor design for signal-transducing [67–69]. For example, $Pb^{2+}$ and $UO_2^{2+}$ have been detected with the DNAzyme-based colorimetric sensors (blue-to-red transition), with excellent specificity against other metal ions such as $Ca^{2+}$, $Co^{2+}$, $Ni^{2+}$, and $Cd^{2+}$.

Nanomaterials such as AuNP, carbon nanotube, graphene, and graphene oxide are efficient fluorescence quenchers, while other nanoparticles such as inorganic quantum dots (QDs), carbon dots, nanoclusters, and up-conversion luminescent nanoparticles have been emerging as alternative fluorophores due to their improved optical properties, low background, multiple colours, and high photostability and brightness (quantum efficiency). As such, these nanomaterials are widely employed to construct fluorescent sensors, even for in vivo analysis. When designing fluorescent sensors with nanomaterials, the main relevant mechanisms are either "Förster resonance energy transfer (FRET)" or nanometal surface energy transfer (NSET) [70–72].

For example, Li et al. reported a DNA-directed QD–DNA–AuNP sensor for $Hg^{2+}$ detection [72]. As shown in Fig. 10.3, QDs and AuNPs were functionalized with two complimentary oligonucleotide strands, each containing TTT as the recognition element for $Hg^{2+}$ (based on T-$Hg^{2+}$-T coordination chemistry). The mixture of these two functionalized nanoparticles is fluorescent, as the proximity of AuNPs to QDs is not close enough to quench the fluorescence of the QDs. However, when the water sample containing $Hg^{2+}$ ions were introduced into the fluorescent mixture, $Hg^{2+}$ ions can link together the DNA strands on the AuNPs and QDs to form a stable duplex linking the AuNPs and QDs, which results in an energy transfer

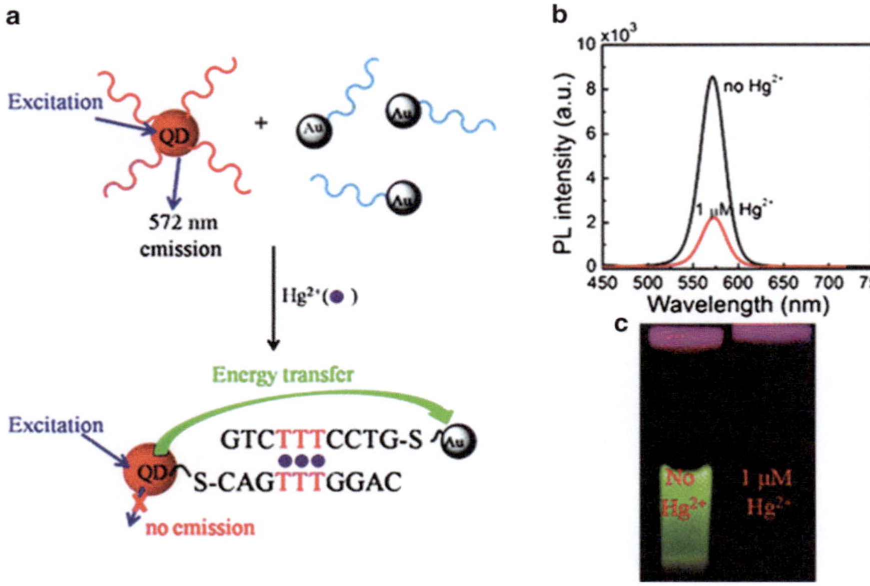


**Fig. 10.3** The scheme of $Hg^{2+}$ detection based on QD-DNA-AuNP sensor [72]

that quenches the QDs fluorescence. This sensor showed excellent sensitivity (LOD of 0.4 and 1.2 ppb $Hg^{2+}$ in buffer solution and in river water, respectively) and specificity (Fig. 10.3).

In another example, graphene oxide (GO) was used as the fluorescence quencher [73]. In this case, the underlying sensor mechanism is the much lower affinity double-stranded DNA (duplex) demonstrates for GO than is displayed by ssDNAs. As shown in Fig. 10.4, a sensing system containing salt, fluorescein-labelled A15 (FAM-A15), T15, and GO, is fluorescent since FAM-A15 will bind to FAM-T15 to form a duplex, and is thus not likely to be adsorbed and quenched by GO. However, when a $Hg^{2+}$-containing sample was introduced, the $Hg^{2+}$ induced T15 strands to form a T15/T15 duplex based on T-$Hg^{2+}$-T coordination. This leaves FAM-A15 single stranded, which was then adsorbed onto GO to quench the fluorescence. This sensor also showed good sensitivity (LOD in buffer: 0.5 nM) and selectivity.

Optical sensors, based on surface-enhanced Raman scattering (SERS), have been developed for heavy metal analysis, though by an indirect means. By exploiting the aggregation of silver nanoparticles (AgNPs), we can significantly enhance the SERS signal of the Raman reporter molecules (such as negatively charged 4-mercaptopyridine (4-MPY) on AgNPs) as Chen's group has done with its development of a SERS sensor for $As^{3+}$ ions [74]. First, AgNPs were functionalized with both 4-MPY and glutathione (GSH), where 4-MPY not only serves as the Raman reporter but also provides negative charge to electrostatically stabilize

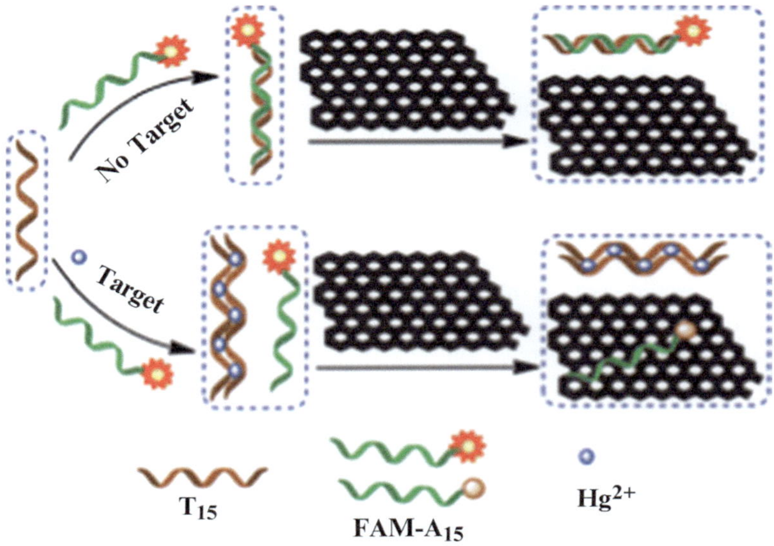


**Fig. 10.4** The scheme of GO-DNA sensor for $Hg^{2+}$ detection [73]

AgNPs, while GSH functions as a specific binding element for $As^{3+}$ ions. When $As^{3+}$ ions bind to GSH, their positive charges destabilize the AgNPs and induce aggregation, resulting in an enhanced SERS signal for 4-MPY molecules. The sensor showed good sensitivity (LOD: 0.76 ppb) and selectivity. Similarly, a SERS sensor for $Cd^{2+}$ with $Cd^{2+}$-chelating polymer brush functionalized on AuNPs as the binding element was developed with good sensitivity and selectivity. More importantly, SERS provides structural information on the molecules, suggesting application potential for understanding heavy metal speciation, which is deserving of additional future research attention.

#### 10.5.1.2 Electrochemical Nanosensors

For on-site detection of heavy metal ions and other charged pollutants in water samples, electrochemical sensing, such as potentiometry, voltammetry, coulometry, and amperometry, shows great potential. Owning to the rapid development of microelectronics, electrochemical sensors might be devised and fabricated commercially with high uniformity (no significant batch-to-batch variations). These sensors could be easily miniaturized and integrated into state-of-art wireless mobile technologies (such as smart phones) for fast detection and signal processing. As a result, long-term continuous on-line and remote monitoring may be practical, which could be very important for environmental or medical applications, for example, monitoring glucose or hormone levels in blood with an implanted electrochemical sensor, or continuous monitoring of mine water discharges. The fundamental mechanism is the highly specific electrochemical detection of metal ions, suggesting a molecule recognition element might not be required to construct the electrochemical sensors, improving the sensor design simplicity by allowing for constant re-use, making these devices suitable for continuous on-line measurement over long lifespans. Nanomaterials have been widely exploited to upgrade traditional electrochemical sensing, with many review articles published over the last few years providing in-depth expertise on detailed aspects of this field. Herein, we provide only some simple examples to briefly overview the electrochemical sensing field.

One powerful technique is ion selective electrode (ISE), belonging to the potentiometric membrane sensors family, which detects specific target ions in water based on the selectivity of the ionophore membrane. Recently, significant improvements in the performance of carbon pastes were achieved by incorporating multi-walled carbon nanotubes (MWCNT), nanosilica, and functional polymer to form synergistic functional composites which serve as the potentiometric membrane for the detection of $Yb^{3+}$, $Er^{3+}$, $Ho^{3+}$, $Pb^{2+}$, and $Cu^{2+}$, so that the sensitivity and response time were enhanced [75–78].

For voltammetric sensors, which record current as a function of systematically varied potential, nanomaterials were widely adopted for electrode modification to achieve better performance. For example, by using carbon-based nanomaterials such as carbon nanotubes and graphene sheets with unique electric and chemical properties to modify the working electrode, the response time, sensitivity (signal to noise ratio), and stability could be significantly improved, presumably due to improved electron transfer rates, lower solution resistance, increased mass transfer kinetics, and enhanced chemical stability and mechanical strength. Some intriguing review articles may provide in-depth understanding in this regard [79–81].

When utilizing the nanomaterial-modified electrodes for stripping analysis, such as anodic stripping voltammetry, nanomaterials can serve as a sorbent, thus further improving the sensitivity of the assay. Compared to unmodified approaches, the carbon nanomaterial (graphene or CNTs)-modified electrode showed higher efficiency for stripping analysis (i.e. sensitivity was improved by thousands of times), which accumulates the heavy metal ions from the sample matrix prior to the reduction of those ions to the zero-valent metal. In this case, CNTs or graphene are often functionalized with various molecules to improve their affinity to target analyte ions. For example, when using cysteine-modified CNTs to modify the electrode surface, the detection limit for $Pb^{2+}$ and $Cu^{2+}$ reached 1 and 15 ppb, respectively [82]. Similarly, thiacalixarene-functionalized CNTs showed enhanced sensitivity towards $Pb^{2+}$, with the detection limit of 40 pM [83]. Likewise, a super-sensitive and selective $Hg^{2+}$ sensor (LOD: 3 ppt) was developed using layered titanate nanosheets, a typical cation-exchange material [84].

Field effect transistor (FET) has been adopted as a promising technological platform for real-time and label-free chemical- and bio-sensing for decades. Similar to CNT-modified electrodes, SWCNT-FET demonstrates promise for heavy metal analysis. For example, SWCNT-FET was used to monitor $Hg^{2+}$ ions in water up to 10 nM, based on the change in conductance of the CNTs on exposure to $Hg^{2+}$ ions due to specific redox reactions between CNTs and $Hg^{2+}$ [85].

#### 10.5.1.3 Microfluidics-Based Sensor Devices

Microfluidics technology has demonstrated versatility in analytical chemistry, even for heavy metal detection. Based on the unique strengths of the miniaturized microfluidic platform, it is reasonable to expect on-site measurement and multiplexed detection of multiple analytes simultaneously. Currently, both optical (absorbance, SERS, and fluorescence) and electrochemical sensors have been incorporated into microfluidic platforms to achieve enhanced portability, energy- and time-efficiency. Date et al. [86] demonstrated a microfluidic immunoassay for heavy metal ($Cd^{2+}$, $Cr^{3+}$, and $Pb^{2+}$) absorbance measurement; herein, the bound and free AuNP-labelled antibody were separated by microparticles functionalized with the EDTA chelated metal ions, where the microparticles were entrapped in a dam-like structure in the microfluidic channel. When the heavy metal ion sample and AuNP-Abs mixture flow through the channel, the free AuNP-Abs (i.e. those in excess not binding to metal ions) are retained by the microparticles while the bound AuNP-Abs flow through the microparticles without retention. The retained AuNP-Abs are quantified by absorbance at 520 nm, demonstrating the simplicity and portability of the whole device. Other examples involve fabrication of on-chip quantitation devices such as micro-electrodes for electrochemical detection [86], fluorescence detection [87], and SERS [88]. To date, only simple chemical reactions have been realized on-chip and real environmental and biological samples were rarely tested against on-chip nanosensors.

### *10.5.2 Treatment and Remediation Techniques*

#### 10.5.2.1 Mine Water Remediation Strategies

To understand the potential of nanotechnology for mine water remediation and treatment, it is necessary to understand current techniques and strategies. Currently, there is a wide spectrum of mine water treatment and remediation technologies which have been developed, proven, and are being applied within different abandoned mine environments. The generic range of technologies is reflected in Fig. 10.5, and can be broadly classified into active, passive, and in situ treatment approaches [18, 19]. Nanotechnology has potential to enhance the capacity, as well as increase the efficiency, of all existing treatment technologies.

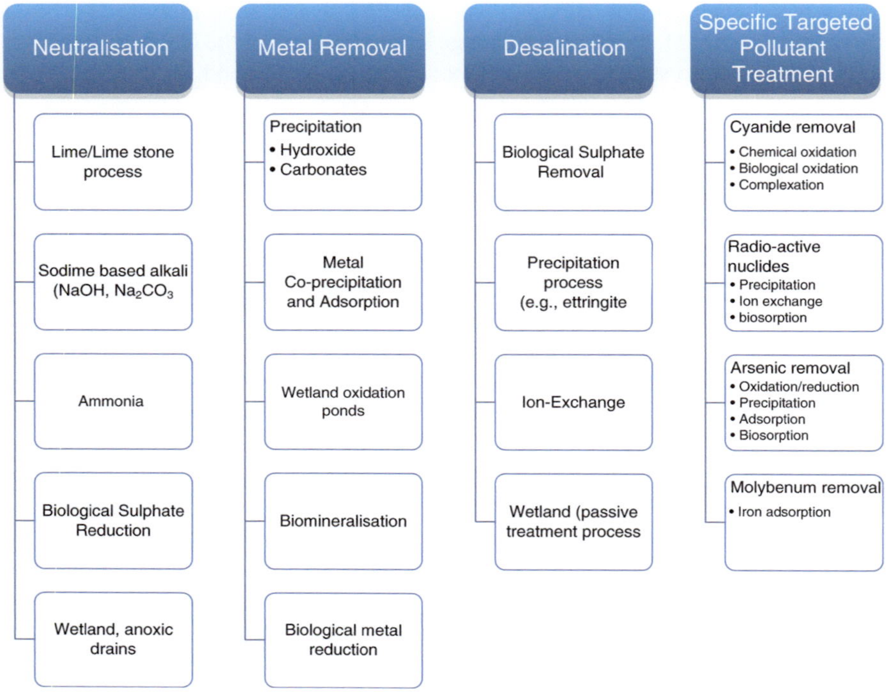


**Fig. 10.5** Generic strategies in mine water treatment technologies. Modified from Ref [19]

*Passive Treatment Systems*: Passive treatment systems utilize chemical, biological, and physical removal approaches, many of which mimic or enhance naturally occurring processes in the environment, to modify influent characteristics. The major types of passive treatment applicable to mine water include:

1. Anoxic alkaline drains—used mainly in the treatment of AMD, this system utilizes trenches filled with alkaline material (often crushed limestone), sealed under plastic and geotechnical fabric and covered by soil, through which an un-aerated, contaminated influent stream flows by gravity. As AMD flows through the system, it gradually dissolves the limestone, releasing carbonate, which along with bicarbonate, raises effluent pH.
2. Constructed wetlands—ecological systems designed to optimize a variety of natural physical, chemical, microbial, and plant-mediated processes [89–93]. In a constructed wetland, influent mine water drains by gravity through the wetland, progressively undergoing metal removal in addition to pH neutralization. Metals are removed by precipitation, chelation, and exchange reactions, while neutralization is primarily achieved by the activity of sulphate-reducing bacteria (SRB), or by increasing alkalinity via chemical and microbial reactions including limestone dissolution.

3. Microbial reactor system—reactor comprised of a shell containing an electron-donating substrate (i.e. carbon donors), which supports the growth of microorganisms. Aerobic bacteria, such as sulphate-reducing bacteria, treat mine water (including AMD) by reducing sulphate to hydrogen sulphide [94–97]. The sulphide in turn reacts with metal ions leading to their precipitation as metal sulphides [91, 98]. Such bacterial activity consumes hydrogen ions by the reduction of substrates, thereby increasing the pH of the solution while producing carbon dioxide which contributes to the $CO_2$-bicarbonate-carbonate buffer system.
4. Biosorption systems—utilize living biomass including bacteria, algae, fungi, and yeasts to adsorb heavy metals and radionuclide pollutants from mine water. Such biosorption systems are limited to treating influent with metal concentrations below toxicity thresholds for each microorganism. Such systems are also seasonally limited, as biosorption by living biomass during the colder temperatures of winter are limited when growth is negligible. Consequently, biosorption systems cannot be effective stand-alone treatment systems for AMD, but are a viable alternative form of treatment, or for incorporation as a final polishing step.

*Active Treatment Systems*: Active treatment is the approach to improving mine water quality by the addition of chemical reagents (e.g. hydrogen peroxide, caustic soda) in the context of extensive infrastructure, energetic inputs, and engineered systems. Active treatment technologies include aeration, neutralization, metal precipitation, metals removal, chemical precipitation, membrane processes, ion exchange, and biological sulphate removal.

1. Aeration—this critical step increases dissolved oxygen, which is necessary as mine water from flooded shafts is often hypoxic or anoxic; this step promotes the oxidation of iron and manganese, and increases chemical treatment efficiency. Aeration also drives off dissolved $CO_2$, which is in dynamic equilibrium with carbonic acid in solution, thereby increasing the pH and significantly reducing reagent use. As the principal contaminant in AMD is often dissolved ferrous iron, and at average water temperatures, only about 10 mg/L of oxygen can dissolve to achieve saturation [19, 99]; aeration is required if there is more than ~50 mg/L of $Fe^{2+}$ in the mine water. Aeration can be done before or during treatment, using gravity cascades or mechanical aeration mixing devices [19].
2. Neutralization and hydrolysis—key aspects of mine water treatment, which involves the addition of alkali materials such as limestone ($CaCO_3$), hydrated lime ($Ca(OH)_2$), un-hydrated (quick) lime (CaO), soda ash ($Na_2CO_3$), caustic soda (NaOH), magna lime (MgO), fly ash, kiln dust, or slag.
3. Metal removal—the classical approach to metals removal from mine drainage is based on chemical precipitation leading to the formation of solid particles containing metal precipitates. The process normally requires a pH adjustment to selectively remove metals of interest, as many metals are amphoteric.

**Table 10.2** Theoretical minimum metal hydroxide solubility pH ($pH_{MMH}$) for some selected metals common in mine water

| Metal | $pH_{MMH}$ | Metal | $pH_{MMH}$ |
|---|---|---|---|
| Ferric iron, $Fe^{3+}$ | ~3.5 | Ferrous iron, $Fe^{2+}$ | ~8.0 |
| Antimony, $Sb^{2+}$ | ~4.2 | Zinc, $Zn^{2+}$ | ~8.5 |
| Aluminium, $Al^{3+}$ | ~4.5 | Nickel, $Ni^{2+}$ | ~9.3 |
| Lead, $Pb^{2+}$ | ~6.5 | Cadmium, $Cd^{2+}$ | ~10.0 |
| Copper, $Cu^{2+}$ | ~7.0 | Manganese, $Mn^{2+}$ | ~10.6 |

Amphoteric metals (M) decrease their solubility until a threshold pH is reached, above which the metal solubility increases again because of the formation of soluble complexes [18, 100]. Dissolved metals can form a number of insoluble compounds with anions, such as

$$\text{Hydroxides}: \quad M^{x+} + xOH^- \rightarrow M(OH)_x \tag{10.5}$$

$$\text{Carbonates}: \quad 2M^{x+} + xCO_3{}^{2-} \rightarrow M_2(CO_3)_x \tag{10.6}$$

$$\text{Sulphides}: \quad 2M^{x+} + xS^{2-} \rightarrow M_2(S)_x \tag{10.7}$$

Table 10.2 presents pH values corresponding to the theoretical thermodynamic and minimum solubility of some selected metal hydroxides. Pre-oxidation of metals with multiple oxidation states prior to precipitation improves the process as the oxidized form of most metals have lower solubility with the exception of chromium, selenium, and uranium, which are more soluble in their oxidized forms [91, 101–103]. To enhance removal of targeted metals, chemical pre-treatment or co-precipitation strategies are often employed, such as aeration, iron addition to co-precipitate or adsorb certain metals onto ferric hydroxide precipitates, and chemical reduction or oxidation to alter the valence state of a target metal.

4. Chemical precipitation for sulphate removal—mine water may contain a wide range of anionic species, but sulphate is typical of many mine drainages and often represents the primary anion contaminant. Some sulphate is removed by gypsum precipitation during neutralization reactions if lime, limestone, or another calcium source is added during water treatment [26, 104]. In addition, a number of precipitation processes have been developed for specific application to high sulphate content mine waters, including: Barium sulphate process to re-precipitate sulphate [105, 106]; Ettringite ($Ca_6Al_2(SO_4)_3(OH)_{12}{\bullet}26H_2O$) precipitation process, which is based on the addition of aluminium hydroxide in a high pH environment resulting in precipitation of ettringite (a hydrated calcium aluminosulphate mineral) [107]:

$$6Ca^{2+} + 3SO_4{}^{2-} + 2Al(OH)_3 + 38H_2O = Ca_6Al_2(SO_4)_3(OH)_{12} \bullet 26H_2O + 6H_3O^+ \tag{10.8}$$

5. Membrane treatment—usually used to treat mine water that is brackish or saline. There are a wide range of existing membrane treatment technologies for mine drainage, but their application is challenging as mine water typically contains several compounds with fouling potential such as metals, sulphates, and carbonates [41, 108]. Thus, typical membrane treatments for mine water utilize spiral wound reverse osmosis or nanofiltration membranes, but other membrane processes such as electrodialysis reversal can also be applied to treat mine water, and are currently operational in large-scale desalination of mine drainage.
6. Ion exchange—involves a chemical reaction in which free mobile ions from a solid phase, the ion exchanger, are traded for different ions of similar charge from the solution. Like any ionic compound, the solid ion exchange material can dissociate and participate in ion exchange reactions, but without dissolving of the material. This property facilitates many specific phenomena and intriguing possibilities to design heterogeneous systems with ionic properties. The process has been traditionally used by mining companies in copper cementation or precipitation processes [53, 109–111]. Within these applications, metallic iron was placed in a copper-containing stream of leach solution from a waste or low-grade ore pile. Copper in solution would plate on the surface of the iron metal, and in the process, exchange electrons with the underlying iron, leading to irons oxidation and the reduction of copper to the metallic state. This process created a higher-value product from a waste product (e.g. precipitation of copper from waste) utilizing low-value scrap iron to remove dissolved copper from mine water. The reaction is

$$3Cu^{2+} + 2Fe^0 \rightarrow 3Cu^0 + 2Fe^{3+} \tag{10.9}$$

Recently, ion exchange processes using specialized resin materials have been developed for high sulphate mine water. There are both cation and anion exchange process; the cation resin exchanges $Ca^{2+}$, $Mg^{2+}$, and other cations (i.e. metal ions) by the following reaction:

$$2R - H + Ca^{2+} \rightarrow R_2 \bullet Ca + 2H^+ \tag{10.10}$$

During the process, water acidifies requiring the degassing of $CO_2$. The anion resin exchanges $SO_4^{2-}$, $Cl^-$, and other anions by the following reaction:

$$2R - OH + SO_4^{2-} \rightarrow R_2SO_4 + 2OH^- \tag{10.11}$$

Many natural mineral compounds, such as clays (e.g. bentonite, kaolinite, and illite), vermiculite, and zeolites (e.g. analcite, chabazite, sodalite, and clinoptilolite) exhibit ion exchange properties.

7. Biological sulphate removal—a process based on sulphate reduction in an anaerobic reactor, which converts sulphate to sulphide. The process is mediated by SRB, which preferentially utilize substrates such as fatty acids, alcohols, and $H_2$ gas. The SRB includes a consortium of fermenting bacteria and methanogens, some of which help to hydrolyze and ferment complex carbon compounds to readily accessible substrates for the SRBs [97, 112]. In mine water applications, an electron-donating carbon source such as alcohol, sugar, $H_2$ gas, or even complex substrates such as sewage sludge, as well as nutrients including nitrogen, phosphate, potassium, and trace minerals should be added to the bioreactor to stimulate bacterial growth. Using this process, sulphide can be partially oxidized to sulphur in a carefully controlled micro-aerobic environment, with the potential for the sulphur to be separated as a potentially commercial by-product. Stripping of the sulphide and converting to sulphur involves a substitution of $H_2S$ by $CO_2$, which results in an increase in carbonate alkalinity and potential precipitation of carbonates such as calcite [91, 98]. This can lead to biomineralization of a number of metallic contaminants present in the mine water.
8. Sulphide precipitation—this is a process used to remove metals such as lead, copper, chromium (VI), silver, cadmium, zinc, mercury, nickel, thallium, antimony, and vanadium from mine water, which works under the same basic principle as hydroxide precipitation [91, 98]. The precipitation process converts soluble metal compounds into relatively insoluble sulphide compounds through the addition of precipitating agents, such as sodium sulphide ($Na_2S$), sodium hydrosulphide (NaHS), ferrous sulphide (FeS), and calcium sulphide (CaS). Over a broad pH range, sulphides ($S^{2-}$, $HS^-$) are extremely reactive with metal ions but the precipitation occurs only near neutral conditions (pH 7.0–9.0) [91, 112]. The metal-sulphide precipitates are physically removed from the treated mine water through coagulation, flocculation, and clarification, or filtration thereby leaving a metal-sulphide sludge.
9. Biomineralization—a broad diversity of microorganisms influence and regulate geochemical processes such as the mineralization of inorganic materials [113, 114]. This process, known as biomineralization can be mimicked in mine water as a promising bionanotechnological method for contaminant removal. For instance, bacteria in natural systems can form calcium carbonate, influencing carbonate rock (i.e. limestone) and sediment formation. Such biominerals are hybrids of inorganic and organic components, and in living systems, are the basis for generating bones, teeth, or shells. The organic matrix contributing to these recalcitrant structures consists of proteins, lipids, or polysaccharides, and functions as a template or as nucleation sites for the mineralization of the inorganic phase. Minerals, synthesized by biomineralization processes include silica, iron oxides, hydroxyapatite, and calcium carbonate in various polymorphic orientations (e.g. calcite, aragonite, and vaterite) [115, 116]. Calcium carbonate mineralization by bacteria is regarded as a biologically induced and mediated process [113–115]. Microorganisms can influence most of the factors leading to precipitation, initially by bacterial cells acting as nucleation sites for metal accumulation. Positively charged ions

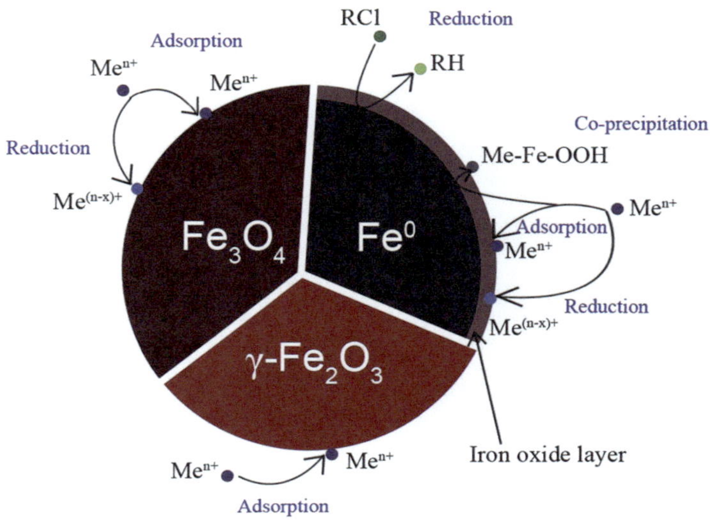


**Fig. 10.6** Schematic of contaminant removal mechanisms from water by nanoscale iron and derivatives, illustrating nZVI in the core mainly provides the reducing power for reactions with contaminants while the oxide shell provides sorption sites. However, adsorption also occurs on the iron oxide ($Fe_3O_4$ and $\gamma$-$Fe_2O_3$) surfaces, while $Fe_3O_4$ also possesses reducing power. *Source*: Ref. [118]

like $Ca^{2+}$ can be accumulated on negatively charged functional groups on the cell surface and subsequently react with anions to form insoluble inorganic solids like calcium carbonate [23, 113]. Metabolic pathways of heterotrophic bacteria, namely the nitrogen and the sulphur cycle, can be involved in biomineralization processes by the generation of (hydrogen-) carbonate ions as well as ammonia, influencing their surrounding medium [113, 114, 117].

#### 10.5.2.2 Potential of Nanoscale Iron and Derivatives

Iron-based nanomaterials are gaining prominence in water nanoremediation, with the most popular being magnetic nanoparticles including nZVI, reactive nanoscale iron product (RNIP), magnetite ($Fe_3O_4$), and ($\gamma$-$Fe_2O_3$) nanoparticles [116, 118, 119]. These magnetic nanoparticles not only have a large removal capacity, fast kinetics, and high reactivity for contaminant removal due to their extremely small particle size and high surface-area-to-volume ratios but also one more important property, magnetism. Tang and Lo [118] have summarized the nanoremediation mechanism and processes involving nanoscale iron species (Fig. 10.6). The advantage of nanoscale iron is its ability to be reused by desorbing the contaminants and regaining removal capacity over successive treatment cycles [120].

(a) *Nanoscale Zero-Valent Iron*

NZVI particles range from 10 to 100 nm in diameter. They are usually synthesized through reductive precipitation of $FeCl_3$ with $NaBH_4$, or reduction of

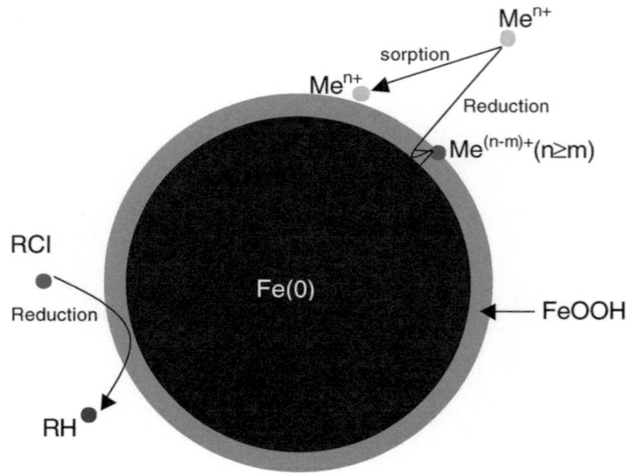


**Fig. 10.7** Schematic representation of zero-valent iron nanoparticles illustrating the zero-valent iron core which provides the reducing power for reactions with environmental contaminants, and the iron oxide/hydroxide shell formed largely from the oxidation of zero-valent iron. Adapted from Ref. [55, 121]

goethite and hematite particles with $H_2$ at high temperatures (200–600 °C). They exhibit a typical core-shell structure, which consists primarily of zero-valent or metallic iron whereas the mixed valent (i.e. Fe (II) and Fe (III)) oxide shell is formed by oxidation of the metallic iron. The nZVI are efficient agents of nanoremediation due to their large surface area and large number of reactive sites, and their dual properties of adsorption and reduction [57, 121]. The use of nZVI in mine water remediation would benefit from iron being a strong reducing agent, reducing inorganic pollutants and immobilizing AMD-associated metals such as uranium, arsenic, and chromium, while degrading organic ore processing chemicals. For instance, in excess of 99 % of arsenic in water can be removed using 12 nm diameter iron oxide nanoparticles [57, 122–125], and they are highly effective for the transformation and detoxification of chlorinated organic solvents and polychlorinated biphenyls (PCBs). Further, surface sorption and co-precipitation of contaminants also occur due to the formation of an iron oxide or oxy-hydroxide shell when nZVI contacts air or water. Consequently, nZVI can be used in wide range of in situ decontamination applications including sub-surface injection into reactive treatment zones [18, 125]. Furthermore, they can be used as reactive barriers to control plumes migrating from contaminated sites and in filter systems (Fig. 10.7).

(b) *Reactive Nanoscale Iron Product*

In the presence of oxygen, nZVI is readily oxidized to $Fe^{2+}$ and can be further oxidized to $Fe^{3+}$ ions. The conversion from $Fe^{2+}$ to $Fe^{3+}$ is dominant under acidic and oxygenated conditions. Further, the oxidation of nZVI in the

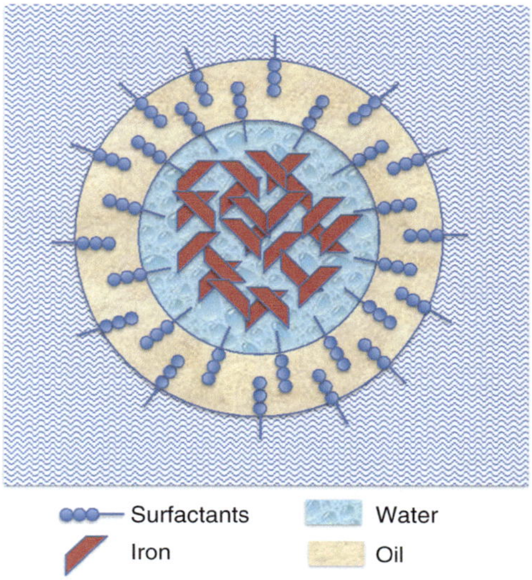


**Fig. 10.8** Schematic of an eZVI droplet

presence of water can also result in the production of hydrogen, causing the nanoparticles to "burn out"; a major constraint when using nZVI for remediation of AMD. Providing surface coatings such as a thin layer of either $Fe_3O_4$, silica, or polymers reduces oxygen contact with nZVI, retaining their reactivity [120, 125, 126]. Particles formed when nZVI is coated with a thin layer of magnetite ($Fe_3O_4$) are called Reactive Nanoscale Iron Products (RNIP) [44, 116]. The RNIP consists of approximately a 50:50 mixture by weight of elemental iron (α-Fe) as the core, surrounded by a $Fe_3O_4$ outer shell [116]. Unfortunately, the reaction kinetics of RNIP are sometimes slower than nZVI a lower accessibility of contaminants to $Fe^0$ due to the covering $Fe_3O_4$ shell.

(c) *Emulsified Zero-Valent Iron*

Zero-valent iron can be modified to comprise catalysts like palladium, coatings such as polyelectrolyte or triblock polymers, or they can be encased in emulsified vegetable oil micelles. A potential benefit of eZVI (Fig. 10.8) over nZVI for mine water applications is the hydrophobic membrane surrounding the nZVI, which protects it from other waterborne constituents such as inorganic compounds that might otherwise react with the nZVI, reducing its capacity or inactivating the iron [116]. As with the addition of metal catalysts to nZVI particles, the formation of emulsified zero valent iron (eZVI) also represents an enhancement to the existing nZVI technology. In fact, such modifications are necessary to treat ore extracting and processing chemicals in mine water, as well as treating petroleum by-products as the surface of the nZVI nanoparticle requires modification to contain an oil-liquid membrane. As the composition of this membrane is hydrophobic and forms an emulsion with ZVI, the resulting eZVI can interact with non-aqueous phase liquid petroleum contaminants [127]. As the eZVI emulsion is miscible with

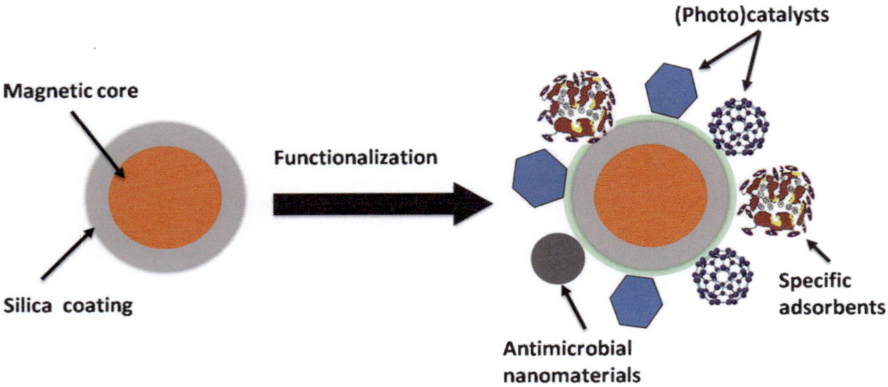


**Fig. 10.9** Multifunctional magnetic nanoparticle showing the core material in shell nanoparticle structure where the shell provides the desired specialized function while the magnetic core facilitates magnetic separation. Silica coating helps with functionalization due to the rich silica chemistry. Adapted from Ref. [41]

hydrophobic contaminant, there is broad contact between the contaminant and the ZVI present within the oil emulsion droplet. While the ZVI in the emulsion remains reactive, contaminants are continuously degraded within the aqueous emulsion droplet, which produces a concentration gradient within the oil membrane. This gradient drives additional contaminant migration into the membrane, where continual degradation can occur.

(d) *$Fe_3O_4$ and $\gamma$-$Fe_2O_3$ Nanoparticles*

$Fe_3O_4$ and $\gamma$-$Fe_2O_3$ magnetic nanoparticles have many potential uses within mine water remediation applications due to their ability to remove metallic contaminants using both physical and chemical adsorption. Further, $Fe_3O_4$ and $\gamma$-$Fe_2O_3$ magnetic nanoparticles can also remove contaminants using ion exchange mechanisms, and can serve as carriers of other nanoparticles with complimentary functions in water treatment (Fig. 10.9). $Fe_3O_4$ nanoparticles use both physical and chemical adsorption to remove metals from water. For instance, Hu et al. [128] demonstrated that the $FeCr_2O_4$ crystalline structure formed on the surface of $Fe_3O_4$ nanoparticles was generated by the removal of Cr(VI) from water. The mechanism underlying the removal of Cr(VI) is its reduction to Cr(III), and the subsequent surface precipitation of Cr(III) onto the magnetic nanoparticles [118, 128].

Generally, the adsorption of inorganic contaminants onto magnetic nanoparticles is low at high pH values indicating that adsorption is largely through physicosorption mechanisms but posssibly with limited chemosorption. Since electrostatic interaction is one of the major removal mechanisms of magnetic nanoparticles, background electrolytes have little effect on removal performance when the affinity of pollutants for the nanoparticles exceeds their affinity for the background electrolytes. Natural organic matter (NOM) such as humic and fulvic acids affect pollutant extraction performance of nanoparticles by slowing reaction kinetics through blocking sorption on the nanoparticle surface. pH dramatically affects the reduction of nZVI and adsorption performance of $Fe_3O_4$ and $\gamma$-$Fe_2O_3$ magnetic Nanoparticles the optimal pH for selective removal of Cr, Cu, and Ni determined to be 2.5, 6.5, and 8.5, respectively [116, 129, 130].

Similar to $Fe_3O_4$ nanoparticles, the dominant contaminant removal mechanism for $\gamma$-$Fe_2O_3$ nanoparticles is physical adsorption, with a limited amount of chemical adsorption contributing. The crystallite structure of $\gamma$-$Fe_2O_3$ nanoparticles does not change after the desorption, indicating that electrostatic interactions are the primary pollutant removal mechanism. Oxygen atoms on the surface of $Fe_3O_4$ and $\gamma$-$Fe_2O_3$ nanoparticles are polarized when ambient pH is below the zero point of charge (pHZPC), charging the iron oxides surface positive, which attracts negatively charged pollutants (e.g. Cr(VI) and As(V)) [41, 116, 131]. Consequently, the pollutant removal performance of $Fe_3O_4$ and $\gamma$-$Fe_2O_3$ nanoparticles is highly pH-dependent.

#### 10.5.2.3 Potential of Bimetallic Nanoparticles and Other Metals

Another type of nanoparticle with strong potential for mine water remediation applications is the bimetallic nanoparticle (BNP). Bimetallic nanoparticles consist of elemental iron or other metals in conjunction with a metal catalyst, such as platinum (Pt), gold (Au), nickel (Ni), or palladium (Pd) [132]. BNPs can also be made using other noble metal catalysts, such as plutonium, gold, and nickel; however palladium-iron BNPs are the only particles commercially available, and are consequently the most common [118, 128, 133]. Metals such as zinc and tin possess similar reduction capabilities as iron, with all types of BNP nanoparticles increasing the oxidation–reduction (redox) reaction kinetics, thus catalyzing the reaction. Other inert metals, such gold and aluminium can also used to make BNP nanoparticles, but unlike iron, are non-reactive. Subsequently, gold and aluminium can only serve as a base metal core in conjunction with a catalyst coating. Nevertheless, their inclusion still increases the catalytic rates of the coating metal [46]. Various base metals can be fused with different noble metals to form a multitude of metal combinations [134, 133]. When metal catalysts are coupled with reactive metals such as iron, they serve to enhance the reactive properties of the iron. In contrast, when metal catalysts are coupled with non-reactive metals such as gold, they act as the catalysts themselves. Lately, palladium-coated gold nanoparticles have emerged as one of the most promising alternatives to nZVI and palladium-coated nZVI. The enhancing properties of gold, while not yet clearly understood, are illustrated by Pd-on-Au particles generating 100-fold greater reaction rates than palladium [135–137] alone, with a sub-monolayer of palladium on gold producing the highest catalytic rates [135–137].

There are several ways BNP can be used in mine water treatment, the most prominent application being in situ passive treatment utilizing slurry injections where particles dissipate into the ground. However, such an approach, while

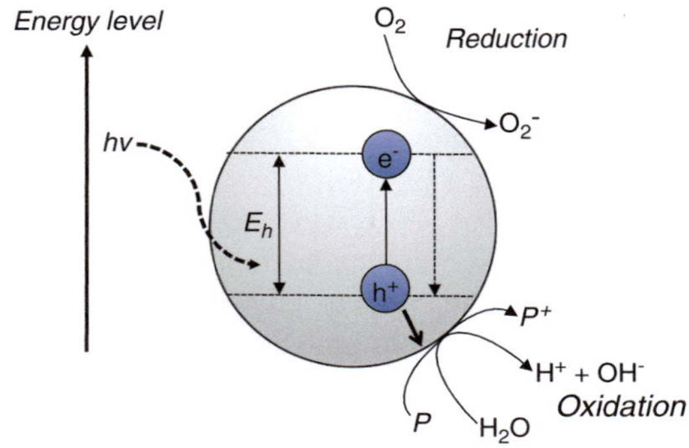


**Fig. 10.10** Mechanism for photo-induced formation of electron–hole pair in a semiconductor nanoparticle in the presence of a mine water pollutant, *P*

technically feasible, may not currently be cost-effective for Pd–Au nanoparticle applications. Interestingly, even expensive BNPs can be applied cost-effectively using two recent innovations. The first technology involves mounting the nanoparticles onto membranes, and then allowing contaminated mine water to be pushed through the membranes. The second technology binds the nanoparticles to a powder, which enables particle recovery from solution by filtration, a process unachievable with individual nanoparticles [138–140].

#### 10.5.2.4 Nanoscale Semiconductor Photocatalysts

A number of materials, such as titanium dioxide ($TiO_2$), zinc oxide (ZnO), iron oxide ($Fe_2O_3$), cadmium sulphide (CdS), zinc sulphide (ZnS), and tungsten oxide ($WO_3$) can obtain energy by absorbing light, thereby acting as photocatalysts [141, 142]. These materials have semiconducting properties, and can act as sensitizers for light-induced redox processes due to their electronic structure, which is characterized by a filled valence band and an empty conduction band [46]. When a photon with energy of $hv$ matches or exceeds the band gap energy, $E$, of the semiconductor, an electron, $e^-$, is promoted from the valence band, VB, into the conduction band, leaving a hole, $h^+$ behind (Fig. 10.10). Excited state conduction-band electrons and valence-band holes can recombine and dissipate the input energy as heat, get trapped in metastable surface states, or react with electron donors and electron acceptors adsorbed on the semiconductor surface or within the surrounding electrical double layer of the charged particles. In the absence of suitable electron and hole scavengers, the stored energy is dissipated within a few nanoseconds by recombination. If a suitable scavenger or surface-defect state is available to trap the electron or hole, recombination is prevented and subsequent redox reactions may occur. The valence-band holes are powerful oxidants (+1.0 to +3.5 V depending on the semiconductor and pH), while the conduction-band electrons are good reductants (+0.5 to −1.5 V). Most organic photodegradation reactions utilize the oxidizing power of the holes directly or indirectly; however, to prevent a build-up of charge one must also provide a reducible species to react with the electrons. In contrast, on bulk semiconductor electrodes, only one species, either the hole or electron, is available for reaction due to band bending.

Semiconductor photocatalysts have the ability to oxidize pollutants into non-toxic or immobile forms, and for the reductive deposition of heavy metals from aqueous solution such as mine water to adjacent surfaces. Traditionally, $TiO_2$ has been used in advanced photochemical oxidation (APO) processes for environmental remediation because of its low toxicity, high photoconductivity, high photostability, availability, low cost, and ability in many cases to completely mineralize organic compounds [142]. $TiO_2$, for example, exhibits photoconductivity when illuminated by light with an energy level exceeding 3.2 eV, the band gap for $TiO_2$ [143, 144]. This energy level translates to light with a wavelength shorter than 387.5 nm, which falls into the category of UV light [142–144]. Doping particles or modifying the surfaces of the photocatalysts with metals has become an increasingly popular enhancement. Metals such as platinum, copper, silver, and gold have been tested for their ability to improve $TiO_2$ decontamination rates. Coupling $TiO_2$ with these metals can also induce a sensitivity and subsequent response to visible, rather than ultraviolet light. Such doping overcomes the traditional problem faced when utilizing semiconductors as photocatalysts, the requirement for UV light, which is a small component of incident solar radiation [143, 144]. However, by coupling $TiO_2$ with copper, we can utilize visible light for photocatalysis, including the remediation of metal contaminants like Cr(VI). Coupling $TiO_2$ with gold and silver produces similar reductive capabilities as that of $TiO_2$ and copper [52]. Further, if $TiO_2$ is used to coat silver and gold particles with a $TiO_2$ shell, reversing the traditional shell:core relationship may enhance photocatalytic activity and increase light absorbing capabilities. Other recent advancements include photocatalytic $TiO_2$-based p–n junction nanotubes containing platinum on the inside and $TiO_2$ on the outside [141, 142]. The nature of the p–n junction allows the outside of the tube to act as an oxidizing surface, while the inside of the tube acts as a reductive surface.

Recently, ZnO has been proposed as a dual function photocatalytic material. ZnO films possess both sensing and remediating potential for organic contaminants in mine water, including a high degree of sensitivity and ability to degrade aromatic compounds under UV light [145, 146]. These two features can facilitate monitoring of the degradation process, since as decontamination progressed, a direct change in emission intensity followed [52].

#### 10.5.2.5 Nanocomposite Materials

(a) *Self-Assembled Monolayer on Mesoporous Supports*
Self-assembled monolayers on mesoporous silica (SAMMS) have a structure resembling a hexagonal honeycomb with monolayers formed within the porous surfaces that adsorb or bind molecules [147, 148]. Synthesis of SAMMS involves multiple steps, mostly reliant on molecular self-assembly [53, 54].

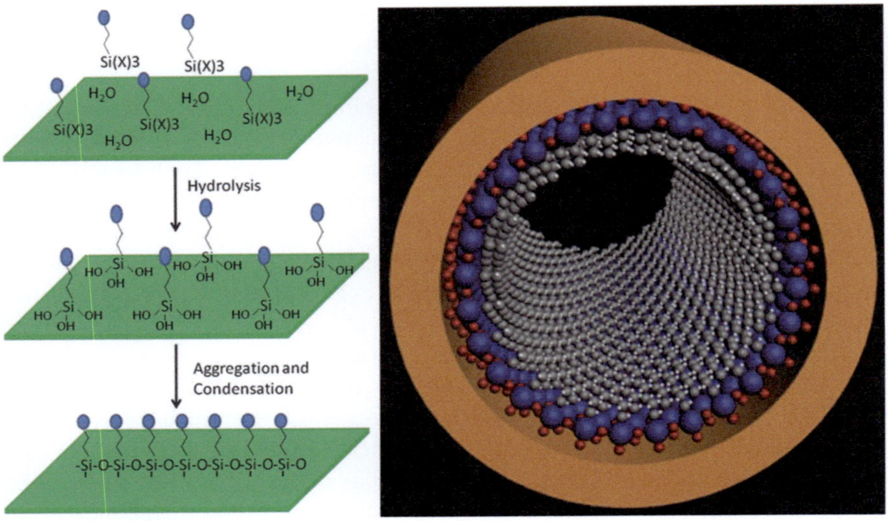


**Fig. 10.11** Schematic of organosilane self-assembly on ceramic oxide surface, showing production stages (*upper illustration, top schematic*). Hydrolysis of organosilanes on the oxide surfaces to create the corresponding hydroxylsilanes, which then absorb on the surface via hydrogen bond (*upper illustration, middle schematic*). Aggregation and condensation of these hydrogen-bound species lead to a closely packed monolayers on the oxide surface (*upper illustration, lower schematic*). Self-assembled monolayer structure within a mesopore (*right illustration*), where the *red, blue, grew*, and *white* spheres indicate the oxygen, silicon, carbon, and hydrogen atom, respectively. *Source*: Ref. [147]

Figure 10.11 illustrates the process and steps involved; the first, self-assembly, involves the aggregation of the starting surfactant molecules into micelle templates of ordered liquid crystalline structures [148, 149]. Then, oxide materials are precipitated onto the surface of the micelles in the presence of solvents under mild hydrothermal conditions [147, 148]. This leads to formation of the preliminary mesoporous backbone, the second phase of self-assembly. The final step in creating the mesoporous ceramic requires the calcination of the organic-oxide material to remove the surfactants. Functionalized silane molecules are mixed in excess with the mesoporous ceramics and self-assemble in an ordered monolayer on the pore surfaces of the ceramic (Fig. 10.11). The biofunctional silanes used in the process can be engineered to have hydrophilic head groups that bind the target materials, and hydrophobic tail groups, which covalently bind to the ceramic substrate.

The potential of using SAMMS in mine water remediation is based on their ability to alter the exposed functional group of the monolayer in the porous surface, allowing potential binding of a broad range of contaminants [148, 150]. One of the most common functional groups is thiol (thiol-SAMMS), designed mainly for the sequestration of mercury. However, thiol-SAMMS can also bind other metallic cations, such as silver, cadmium, lead, molybdenum, and thallium.

Currently, there are Anion-SAMMS or metal-capped ethylenediamine (EDA) SAMMS which absorb anions, such as chromate and arsenate [148, 150]. Their derivative, chelate-SAMMS, contains only the EDA functional group, and can bind metals, such as copper, nickel, cobalt, and zinc (SAMMS Technical Summary). Cu-EDA-SAMMS (a type of anion-SAMMS) has been further functionalized to bind cesium by incorporating ferrocyanide-forming Cu-Ferrocyanide-SAMMS [116, 148, 150]. Furthermore, a variety of phosphonate and hydroxypyridone (HOPO) functionalized SAMMS have been created for the sequestration of actinides and lanthanides [150]. The advantage of SAMMS is twofold: they can be custom functionalized towards target contaminants for on-site remediation, while able to be recycled and reused in consecutive treatments [116]. However, mercury-laden SAMMS are tightly bound and cannot be reused, but must be disposed of [116]. SAMMS can be mixed with aqueous solutions where, following sequestration of targeted molecules, they can be filtered from solution and recycled. Acid-stripping causes SAMMS to release bound compounds, at which point they can be separated out of solution and reused.

(b) *Sol–Gel and Biocers*

In the sol–gel process, simple molecular precursors are converted into nanometer-sized particles to form a colloidal suspension, or sol. The colloidal nanoparticles are then linked with one another in a three-dimensional, liquid-filled solid network. This transformation to a gel can be initiated in several ways, but the most convenient approach is to change the pH of the reaction solution [151]. In general, the major processes involved are hydrolysis and condensation of alkoxide-based precursors such as Si (OEt) 4 (tetraethyl orthosilicate (TEOS)). The reactions involved in the sol–gel process are based on the hydrolysis and condensation of metal alkoxides $M(OR)_z$ and can be described as follows:

$$MOR + H_2O \rightarrow MOH + ROH \text{ (hydrolysis)} \tag{10.12}$$

$$MOH + ROM + ROH \text{ (condensation)} \tag{10.13}$$

Sol–gel materials can be different forms, such as powders, films, fibres, and freestanding pieces of material called monoliths (Fig. 10.12). The sol–gel's properties are determined by the method used to remove liquid from a solid. For example, a gel can be dried in a sol–gel process to make aerogels, a special class of ultralow-density materials. Similarly, there are broad sets of sol–gel materials for applications ranging from optics, coatings, to energy storage, to ceramics, to nanoelectronics and to remediation. For example, specific forms of sol–gel material are designed to remove oil from water. Such sol–gel are designed porous and hydrophobic but absorb organics such as oil. A similar aerogel composite is used to remove contaminants such as uranium, chromium, copper, and arsenic from groundwater [152, 153]. Other new sol–gel materials include ultrathin films, which can be used to coat silicon wafers and protect

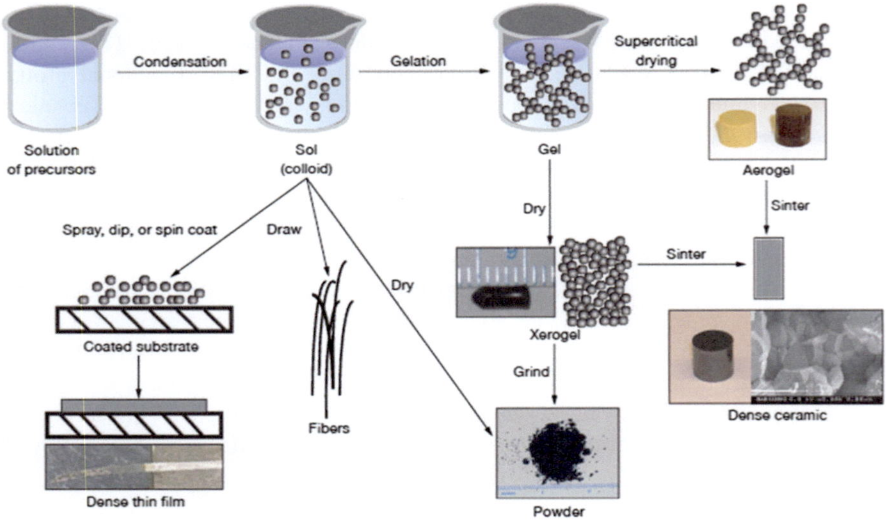


**Fig. 10.12** Sol–gel chemistry involves molecular precursors, which are converted to nanometer-sized particles to form a colloidal suspension, or sol. Adding epoxide to the sol produces a gel network. The gel can be processed by various drying methods (shown by the *arrows*) to develop materials with distinct properties [153]

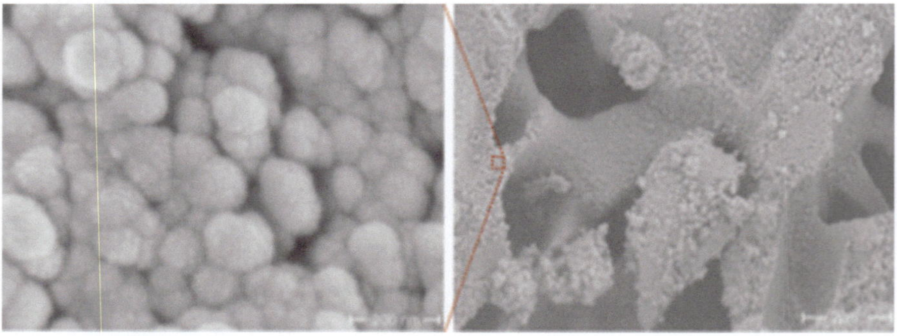

**Fig. 10.13** Sol–gel material showing different porosity levels at nano- and meso-scale (courtesy of Martin Mkandawire—Habilitation research work)

optics, and sol–gel-derived powders, which can be used to produce ceramics with various properties (Fig. 10.13).

Sol-gels can be further modified to be biologically active by either incorporating bioactive molecules like proteins and enzymes, or by embedding them as an immobilizing matrix for living microorganisms. These types of sol–gel composite materials are popularly known as biological ceramic composites (biocers), and are made according to aqueous sol–gel protocols and can be used as selective metal-binding filters. Biocers are prepared by dispersing

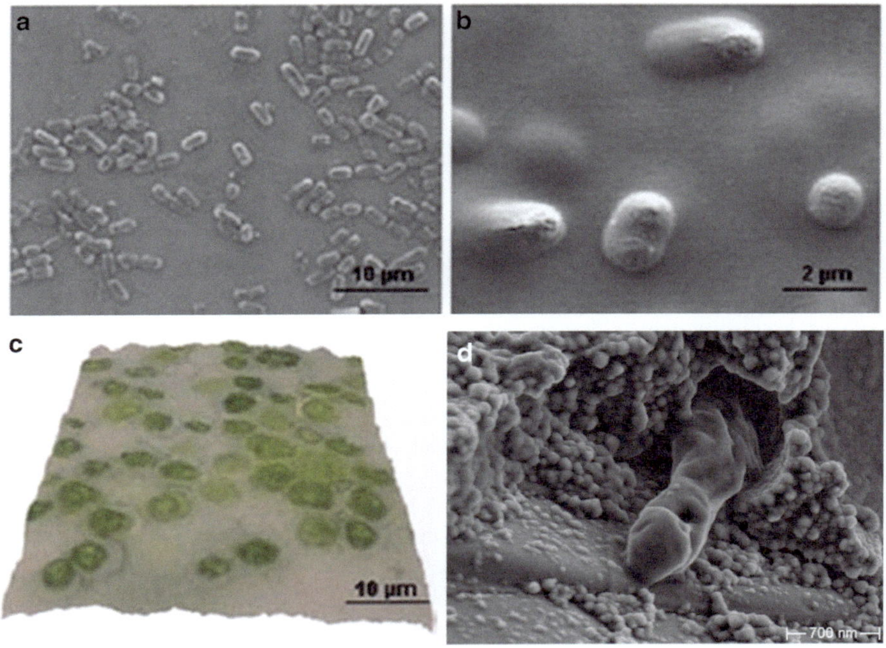


**Fig. 10.14** Biological sol-gels (biocers) demonstrating different means of immobilizing living microorganisms: (**a**) surface immobilization of *Rhodoccocus ruber*, (**b**) bacteria under a thin film coating of biocer, (**c**) surface immobilization of algae, and (**d**) embedded bacteria growing in biocer. *Sources*: (**a–c**) Ref. [154] and (**d**) from Ref. [155]

vegetative cells, spores, and stabilized surface-layer proteins (S-layer) within aqueous silica nanosols prior to geling and drying. Both freeze-drying of prepared biocers as well as the addition of water-soluble compounds such as sorbitol lead to higher porosity and faster metal-binding properties. The biological component of the biocers can be customized for desirable adaptations in bacteria by isolating and incorporating those strains acclimated to contaminated sites or mine water discharges. Thereby, vegetative cells of algae, bacteria, and some fungus and their spores (where applicable) can be added to the biocer to selectively bind U, Cu, Al, Cd, and Pb in large amounts (Fig. 10.14).

#### 10.5.2.6 Nanoscale Polymers and Nanosponges

(a) *Dendrimers*

Dendrimers are three-dimensional, highly branched, globular polymeric macromolecules usually comprised of three covalently bound components: a core, interior branch cells, and terminal branch cells (Fig. 10.15) [45, 156]. Flexibility during synthesis allows for customized molecular design, including size, shape, surface or interior chemistry, flexibility, and topology to be produced to

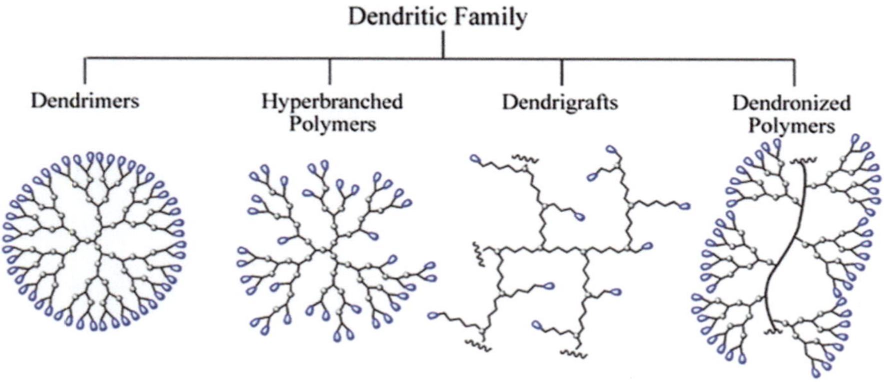


**Fig. 10.15** Different dendritic polymers: dendrimer, core-shell tecto (dendrimer), dendrigraft polymer, hyperbranched polymer. *Source*: Carlmark et al. [45]

specifications [157, 158]. The size of dendrimers ranges between 2 and 20 nm, with common shapes including cones, spheres, and disc-like structures able to be generated. Dendrimers can be broadly applied to a variety of remediation applications, particularly in water treatment. For instance, poly(amidoamine) (a.k.a. PAMAM) are a type of dendrimer which have been used to treat wastewater contaminated with a variety of transition metal ions such as copper (Cu(II)) [159–161]. Diverse cores and functional groups that can be incorporated into dendrimers translate into equally diverse water remedial and biodetection applications [58, 159, 162]; for metal-remediating dendrimers, an EDA core is employed [157, 163]. The high concentration of nitrogen ligands within the interior branches makes PAMAM dendrimers particularly useful as chelating agents for metal ions [163–165] and the ability to choose a multitude of functional groups as terminal cells also contributes to their functionality as metal chelators. Surface terminal groups, including primary amine, succinamic acid, gycidol, hydroxyl, and acetamide have been tested successfully [165].

Currently, dendrimer-enhanced ultrafiltration (DEUF) is a recent remediation tool, which has emerged as a promising technology for removing metal ions from mine water streams. With DEUF, metal ions bind to the dendrimers allowing for removal of contaminants though membrane filtration. While the current status of dendrimer research for remediation only includes EDA core PAMAM dendrimers for copper and other metal ion recovery, PAMAM dendrimers can be functionalized with redox active metal clusters of FeS for reductive decontamination of organic pollutants [162].

(b) *Polymeric Nanoparticles*

Polymeric nanoparticles are molecules or molecular aggregates that possess amphiphilic properties originating from distinct segments present in the particle [166]. Individual nanoparticles contain both a hydrophobic and hydrophilic region, which self-assemble in the presence of water to form polymer vesicles with diameters in the nanometer range, with hydrophobic segments oriented inwards, and hydrophilic segments forming the outer layer [166]. Polymeric nanoparticles offer a potential replacement for traditional surfactants commonly used to enhance the remediation of hydrophobic organic contaminants using pump-and-treat systems. They can be applied to treat mine process water as well as drilling oil contamination, which otherwise sorb strongly to soils or form non-aqueous phase liquids [166–168].

#### 10.5.2.7 Bionanomaterials

(a) *Ferritin Nanoparticles*

Constituents of biological systems, including proteins such as ferritin, can regulate the formation of mineral structures. Iron, while an essential element for living organisms, is also highly toxic in unregulated, so living organisms store iron as ferritin to regulate iron homeostasis. The need to sequester iron, and hence the presence of ferritin, is found across phyla including the animal, plant, and microbial kingdoms. The structure of ferritin consists of a spherical protein shell (apoferritin) composed of 24 polypeptide subunits chains, surrounding an aqueous cavity with internal and external diameters of about 8 and 12 nm. The multi-subunit construction of the apoferritin shell facilitates the generation of eight hydrophilic channels, each of about 4–5 Å and leading to the protein cavity. Water, metallic cations, and hydrophilic molecules of appropriate size diffuse through these channels from the external matrix to the cavity, or vice versa. Consequently, iron molecules can diffuse into the ferritin cavity through channels in the protein shell, where mineralization converts the molecules into a nanoparticle of ferrihydrite, a ferric oxy-hydroxide.

Ferritin nanoparticles offer multiple potential applications for mine water remediation, the most promising being the photoreduction of contaminants in the presence of visible light or solar radiation [169–171]. Iron oxides in general have received a significant amount of attention for their potential remedial abilities. While able to carry out significant photochemical processes, Fe(III)-bearing iron oxides quickly undergo photoreduction to Fe(II), rendering the catalyst inactive [116]. However, ferritin naturally converts Fe(II) to Fe(III), preventing photoreduction of the iron oxide. While providing stability, the apoferritin cage does not inhibit photoreduction of environmental contaminants, conferring ferritin a significant advantage over traditional freestanding particles. In addition to iron hydroxide, the apoferritin protein cage supports the synthesis of the other metal hydroxides, such as Mn(O)OH, Fe(O)OH, and Co(O)OH [116, 172]. Exploiting these capabilities could potentially expand this technology, thereby increasing remediation speed and effectiveness as well as expanding the list of contaminants that could be addressed. Further, ferritin can be employed in the production of both metallic and metal hydroxide nanoparticles like iron and cobalt metallic- and oxide-based nanoparticles.

(b) *Surface-Layer Proteins*

Surface layers (S-layers) form the outermost cell envelope of most archaea and many bacteria by serving as an interface between the bacterial cell and its environment. S-layers are macromolecular paracrystalline arrays of proteins or glycoproteins that self-assemble into two-dimensional semi-permeable meshworks with regularly distributed pores of about 2–8 nm [151, 173]. The self-assembly consists of a spontaneous entropy-driven process with the size of each individual cell ranging from 2.5 to 35 nm, whereas the thickness of the layer varies between 5 and 70 nm [174]. The S-layers are about 50 % hydrophobic amino acids, 15 % glutamic and aspartic acids, and 10 % lysines [175]. Surface layer proteins can be extracted with chaotropic agents such as guanidine hydrochloride from Gram-positive bacteria, or metal chelating agents (EDTA) from Gram-negative bacteria [174, 175]. Upon removal of the agent used for isolation, SLPs can be recrystallized to re-form meshworks identical to those on intact bacteria [175, 176]. The isolated S-layers can recrystallize into two-dimensional regular arrays in suspension or on various surfaces including silicon wafers, metal, glass, mica, or lipid, thus being an appropriate material for several bionanotechnological or potentially water treatment purposes.

Among the promising applications of S-layers is their use as biotemplates for capturing metal ions or for the synthesis of metal nanoclusters as they can be excellent bottom-up building blocks for metal immobilization structures and forming supramolecular scaffolding assemblies [176]. The high density of functional groups on the S-layer surface form matrices for controlled formation of inorganic nanocrystal superlattices (e.g. CdS, Au, Ni, Pt, or Pd), which are used for molecular electronics and non-linear optics [175, 177]. The S-layer's high capacity for binding metals is attributed to passive accumulation of metal on the surface of the bacteria cell, lending potential for applications removing toxic metals such as U, Cu, Pb, Al, Cd, Zn, and Ni [151, 174, 177, 178] from contaminated water, as well as for the recovery of precious metals (e.g. Au, Pt, Pd, Rh) from mine water [151, 174, 177, 178]. For instance, S-layers from different bacteria of *Lactobacillus kefir* bind $Pb^{2+}$, $Cd^{2+}$, $Zn^{2+}$, and $Ni^{2+}$ while S-layers of *Lysinibacillus sphaericus* have high binding capacities towards $UO_2^{2+}$ as well as $Cu^{2+}$, $Pd^{2+}$, $Pt^{2+}$, and $Au^{3+}$ [151, 174, 177]. S-layer-metal interactions are mediated through coordination with side chain carboxyl groups of Asp and Glu residues, and NH groups from the peptide backbone while others with uranium are coordinated through carboxyl and phosphate groups (Fig. 10.16). Isolated S-layers can be used to nucleate biomineralization as they intrinsically tend to self-assemble in suspension into two-dimensional arrays and onto various surfaces, exhibiting pores of identical size and morphology and presenting functional groups aligned in a well-defined order and orientation [179]. Mineralization begins within the holes of the S-layer, where cations such as $Ca^{2+}$ bind negative charges of the S-layer, followed by carbonate binding to initiate the formation of a mineral aggregate, dependant on the pH and ionic composition of the surrounding solution.

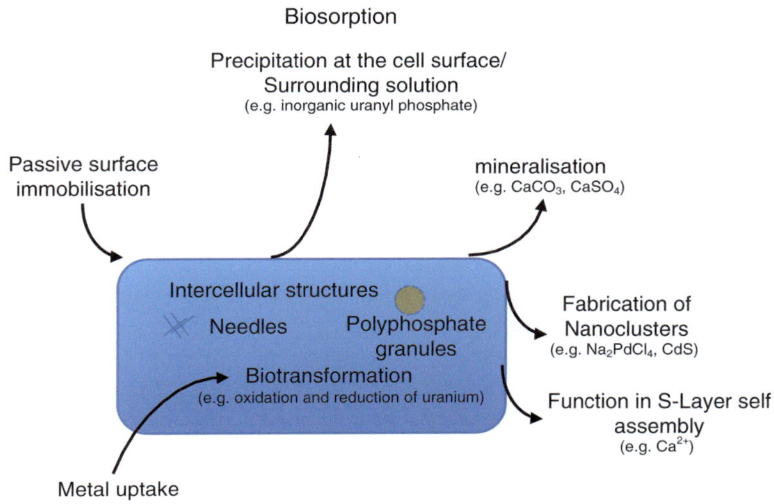


**Fig. 10.16** Modes of interaction between metals and bacteria cells bearing S-layers, which can be exploited in mine water remediation. Modified from Ref. [181]

S-layers can be utilized in mine water remediation through construction of filter matrices with high metal-binding capacities possessing the ability to selectively and reversibly bind high amounts of toxic metals from mine water [180]. The S-layer can be engineered by inserting amino acid residues like cysteine and histidine carrying functional groups with a high affinity for heavy metals to metal sequestration. In other words, engineered recombinant S-layers enlarge the possibilities of designing nanoclusters with a higher affinity for metals.

(c) *Single-Enzyme Nanoparticles*

To increase longevity and enhance versatility, single enzymes are "caged" or encapsulated to create a class of catalysts called single enzyme nanoparticles (SENs) [182]. The "cage" is actually a nanostructured silicate shell, linked to the surface of the enzyme to protect the enzymes' catalytic function by allowing it to remain active for months relative to the normal proteinaceous enzymes life span of just a few hours. Consequently, these stable novel enzyme-containing nanoparticles are capable of significantly more catalytic activity, offering vast capabilities for chemical conversions, bio-sensing, and bioremediation. Their specificity and targeted effectiveness make them much more effective than synthetic catalysts. Synthesis of SENs initially involves covalent modification of the enzyme surface, creating vinyl group functionality and solubilization of the enzyme in a non-polar hydrophobic solvent such as hexane. Then, the modified enzymes are mixed with silane monomers with both vinyl and trimethoxysilane groups. Vinyl group polymerization creates linear polymers with free trimethoxysilane groups attached to the enzyme surface. Lastly, the cross-linked silicate shell resembling armour is created by hydrolysis of the trimethoxysilane groups and the subsequent condensation of the silanols [183]. The SENs offer a variety of remediation advantages in being able to withstand more extreme conditions (pHs, high contaminant concentration, high salinity, and temperature) making them suitable for remediation of mine discharges and drainages with the contaminant of interest dictating the enzyme employed. Despite the benefits achievable through enzymatic remediation, concerns over enzyme stability and lifetime remain a cost-limiting factor for large-scale remedial applications.

(d) *Tunable Biopolymers*

Genetic and protein engineering have emerged as valuable tools for the construction of nanoscale materials that can be controlled precisely at the molecular level [184]. For instance, it is currently possible, using recombinant DNA techniques, to create "artificial" protein polymers with fundamentally new molecular organization. These nanoscale biopolymers can be specifically pre-programmed within a synthetic gene template and controlled precisely in terms of size, composition, and function at the molecular level [185]. In this manner, it is possible to design specific protein-based nano-biomaterials with both metal-binding and tuneable properties that can be used to selectively remove heavy metals from dilute solutions in a single process.

Elastin-like polypeptides (ELPs) are one such class of biopolymers that are of particular interest because of their "smart"-stimuli responsive-properties [185]. They are comprised of the repeating pentapeptide VPGVG and undergo a reversible phase transition from water-soluble forms or polymeric solutions into aggregates with increases in temperature [186]. The transition temperature of the ELP can be tuned/controlled by altering the chain length and sequence, but also with changes in pH, ionic strength, pressure, and covalent modifications [185]. The ELPs can be functionalized with other peptides or proteins, while still maintaining their temperature responsive characteristics. Tuneable biopolymers can be easily manipulated to selectively bind and remove heavy metals from water. For example, fusion of ELP with specific proteins creates biopolymers specifically targeting individual metals, such as cadmium, mercury, arsenic, and lead [184–186]. Currently, tuneable biopolymers that bind to cadmium and mercury are available and in use; those ELPs binding cadmium, for example, posses the repeating pentapeptide VPGVG, but can be tuned to incorporate a hexahistidine tail that serves as the cadmium-binding moiety [185, 186].

#### 10.5.2.8 Nanocrystalline Zeolites

Zeolites are crystalline hydrated aluminosilicate materials composed of three-dimensionally connected framework structures constructed from corner-sharing tetrahedral coordinated Si and Al cations and oxygen [187]. These framework structures are composed of rings with 4–6 atoms of Si and Al, and large pore openings of 8-, 10-, and 12-rings. The specific surface area of zeolites is attributed to their porous crystalline structure. There are over 40 different kinds of natural zeolites, and over 100 synthetics found in various part of the world. Conventional synthesis methods produce zeolites on the scale of 1–10 μm, however their pores are 0.4–1 nm, which are close to the molecular diameters of lighter hydrocarbons and consequently, zeolites are considered nanomaterials. The sizes of the intracrystalline pores and nanospaces depend on the type of zeolite. Zeolites have high cation-exchange capacities, high specific surface areas, and high hydrothermal stability [188]. The porosity of zeolite structure enables cations to migrate in and out freely, exchanging their cations for those of the surrounding fluid. The cation preference of a given zeolite can be due to ion sieving or competition between the zeolite phase and aqueous phase for the cations [189]. Strong acid sites exist on the nanopore surfaces, enabling the zeolites to be used as shape-selective catalysts. Zeolites are extremely active catalysts with selective catalysis-steric phenomena. They can be manipulated to catalyze reactions selectively by avoiding formation reactants or by preventing transition states from forming within the pores because of their size or shape.

Zeolite can act as a molecular sieve to produce sharp separations of molecules by size and shape while strong electrostatic fields within a zeolite cavity result in very strong interactions with polar molecules [189, 190]. However, they also effectively and strongly absorb non-polar molecules due to the polarizing power of the electric field. As a result, zeolites are excellent separators, even in the absence of steric hindrance. The unique properties of zeolites which make them effective sorbents and ion exchange media for metal ions, qualify them for use as a relatively inexpensive means of cleaning acidic mine water. Synthetic zeolites have been demonstrated to effectively remove Cr(III), Ni(II), Zn(II), Cu(II), and Cd(II) from electroplating waste water [190]. Regular-sized zeolites have the ability to remediate water containing cationic species, such as ammonium and heavy metals, as well as radioisotopes such as $^{137}$Cs and $^{90}$Sr [191, 192]. Zeolites can be used as permeable barriers to adsorb contaminants from groundwater and mine water as they pass to receiving environments. Overall, the separatory properties of zeolites based on their molecular sieve structure, electrostatic fields, and polarity enable their use for mine water treatment due to their contaminant separation and removal abilities.

## 10.6 Concerns with Nanomaterial Use in Mine Water

### *10.6.1 Perceived Nanotoxicological Risks*

#### 10.6.1.1 Toxicity Inherent to Nanoscale

The perceived risks of applying nanotechnology to contemporary mine water issues are inherent to characteristics and properties of materials at the nanoscale, and do not differ markedly from those associated with the use of nanomaterials in other environmental applications. While nanoscale particles can be produced naturally by forest fires, geochemical reactions or as biological or abiotic particulates in aquatic systems, their distribution is usually limited in time and space [193]. Consequently, concerns surrounding the application of engineered nanoparticles to mine water issues are fuelled by the greater quantities and scope of use, as well as the uncertainties of short- and long-term behaviour and effects of the nanomaterials in the environment [194]. Production of anthropogenic nanoparticles and their resultant availability in the biosphere are indeed increasing at a remarkable rate. As of March 2011, 580 companies in 30 countries produced over 1,317 nanoparticle-based products from textiles to electronics to wound dressings [195]. The unique physicochemical properties of nanoparticles have resulted in a rapid expansion of their use, with nanoproduct development representing one of the fastest growing sectors of the high-tech economy [196, 197]. A recent estimate is that more than 40,000 tons of titanium nanoproducts were produced in the USA alone, a figure which is projected to reach 2.5 million tons annually by 2025 [198]. With increasing production and application of nanoparticles into diverse consumer products, their ultimate release to natural environments is inevitable [199]. Concern regarding releases of nanomaterials to aquatic systems, both at the consumer-level and through large-scale water treatment applications such as mine water, is growing among the scientific, academic, industrial, and regulatory community. As aquatic environments are the primary long-term sinks for nanomaterials, uncertainties regarding their fate and ecological effects are considerable [196, 197, 200].

Although the long-term impacts of the rapid and substantial loading of engineered nanomaterials to aquatic environments are largely unknown and difficult to quantify, they concentrate on the enhanced electrical, mechanical, and optical properties for which nanoparticles are designed. These same properties, desirable under the circumstances for which they were engineered, may elicit unintended consequences upon release to the environment, as will be expanded upon in detail in the following sections, but primarily revolve around three concerns. The first focus on the unique surface-reactive properties of nanomaterials, which can lead to unexpected consequences upon interaction with both abiotic and biological systems, including the generation of reactive oxygen species [196, 193, 200]. Further, not only are the materials themselves of concern, but the toxicity of ions dissolving off metallic nanomaterials, depending on the environmental conditions, are sometimes more toxic than the nanomaterials themselves [201]. Finally, nanomaterials in aquatic environments may easily serve as conduits to increase the availability, and hence potential for bioconcentration, bioaccumulation, and biomagnification of trace contaminants within the environment and their transfer to higher trophic levels [202, 203].

Although bulk materials themselves are well understood, when made progressively smaller, their surface chemical reactivity increases, as does the complexity of the processes regulating the nanoparticle surface chemistry upon contact with living systems, making it challenging, but crucial, to understand and predict possible toxicological effects [199, 204]. Complicating our understanding of these reactions is the reality that nanoparticles are highly heterogeneous with none having identical size, shape, surface area or optical properties, although similar materials of similar size exhibit similar behaviours [202]. Toxicity of nanomaterials varies greatly with atomic composition, but to a large degree is contingent on whether the nanostructure is immobilized within or on the surface of a bulk material, or if they are free, unbound particles. For unbound particles with diameters in the nanometre range, nanomaterials and their degradation products are capable of great mobility within natural environments and organisms [204].

In surface waters, the consequences of unintentional exposure, particularly over the long term (as is the case with treated mine water discharges) is unknown. Most of the growing body of literature on nanocontaminants has focussed on short-term, acute toxicity assessments of nanoparticles to fish; however, it is likely that longer term exposure at lower concentrations is more ecologically relevant [205]. Unfortunately, few studies have investigated the consequences of long-term exposure to environmentally relevant concentrations of nanomaterials, so little is known of their chronic effects on aquatic organisms [196]. While much remains unknown, what seems clear is the distribution and potential for exposure to insoluble nanoscale particles of $<50$ nm is a new environmental phenomenon, which was non-existent in the pre-industrial biosphere of mere centuries ago [204]. Consequently, organisms have few, if any adaptive responses to accommodate exposure to nanomaterials [204]. If the widespread application and distribution of these relatively new products in consumer goods and potentially within water treatment systems is translated into widespread exposure and incorporation into organisms, potential nanotoxicity will be heightened by an inability to detoxify and/or remove these nanopollutants, the impacts of which are almost completely unknown [202].

As nanotoxicology is a relatively new field of study, a majority of toxicity investigations initially treated nanomaterials as if they behaved similarly to conventional chemicals or their bulk-sized counterparts rather than considering their distinctive physicochemical properties [202]. In not recognizing the unique surface chemistries of nanomaterials, and changes in their reactivities with changing environmental variables, many studies yielded inconclusive, controversial, and in many cases, conflicting results [202, 206, 220]. While we largely understand the properties of materials and/or chemicals at the molecular and bulk levels, there are novel and often distinct properties of materials in the zone between "molecule" and "bulk", the nanoscale. Specifically, the highly reactive surfaces of nanomaterials owing to their heightened surface area to volume ratios as well as wet biochemistry, both confer unique potential utility to nanoscale molecules, but also a means of interacting with biomaterials, which should be the focus of nanotoxicology [203, 206]. For example, the total surface area of nanoparticles may be a more sensitive metric for assessing the "dose" of nanomaterials than their mass, a paradigm-shifting perspective that causes no end of grief for classically trained toxicologists [204].

For example, Karlsson et al. [207] observed that CuO nanoparticles were much more toxic to mitochondria than micrometer-sized CuO particles; however, no size-related differences in toxicity were observed for iron oxides, with [208] also reporting no discernible correlation in vivo between primary particle size and toxicity for $TiO_2$. In contrast, Karlsson et al. [207] found micrometer particles of $TiO_2$ caused more damage than $TiO_2$ nanoparticles, leading these authors to conclude that nanoscale nanoparticles are not always more toxic than micrometer particles. Such conclusions are common enough that Donaldson and Poland [209], while recognizing there are changes in quantum size effects for some particles at certain size thresholds, question the toxicological relevance of size changes. Specifically, these authors ask what changes in toxicity endpoints must be linked to nano-relative to macro-sized materials before we describe them as "nano-specific effects"? In most cases, Donaldson and Poland [209] believe the toxicity of nanomaterials is unlikely to differ from those established for larger particles. With caveats, other authors agree, with Bae et al. [200] stating that "the paradigm that smaller nanoparticles are more readily taken up, and thereby capable of inducing toxicity relative to bulk forms, is too simplistic, and may not always be true under specific experimental conditions (i.e. increased agglomeration reducing bioavailability, thereby limiting toxicity)".

Further complicating nanotoxicity assessments is the revelation toxicity from ions dissolving off metallic nanomaterials, depending on the environmental conditions, may be more toxic than the nanomaterials themselves. For example, in assessments with the green algal species *Chlorella*, $Zn^{2+}$ was more toxic than nano-ZnO, which in turn was more toxic than bulk ZnO at concentration less than 50 mg/L. However, nanoscale ZnO was most toxic to *Chlorella* (then $Zn^{2+,}$ then bulk ZnO) at concentrations >50 mg/L [210]. However, in other phyla $TiO_2$ nanoparticles were more acutely toxic than their bulk counterparts to zebrafish (*Danio rerio*), with the toxicity of ZnO nanoparticles and bulk ZnO being very similar [211]. In studies conducted by Heinlaan et al. [212], both bulk and nano $TiO_2$ were not toxic to any evaluated organism up to very high concentrations (<20 g/L), while ZnO in bulk and nanoscales was toxic to bacteria and crustaceans, but again only at very high (0.18–8.8 g/L $LC_{50}$) concentrations. Such clearly different experimental results with ZnO exposure across phyla also highlight differences in species uptake and sensitivity. With greater surface area per unit weight than their bulk counterparts, nanometal oxides such as ZnO, while having superior performance in many engineered respects, may in fact differ in biological availability, and hence toxicity, to different organisms depending on potential uptake mechanisms [199].

While nanoparticle size and surface area are significant factors modifying organismal uptake (and hence toxicity), rates of distribution and particle reactivity in solution allow many nanoparticles to easily enter living organisms and produce adverse reactions [193, 203]. Sun et al. [213] report that smaller nanoparticles including AuNPs, $TiO_2$NPs, $SiO_2$, quantum dots, and carbon nanoparticles can cross the placental barrier easier than larger ones, suggesting unique mammalian susceptibilities potentially impacting early developmental stages relative to other phyla. Female vertebrate populations are particularly vulnerable to nanotoxicity as female hypothalamic-pituitary-ovarian axis, and foetal development appear particularly sensitive to nanoparticles [213].

Hydroxyl radical generation by nanoparticles are at least partially responsible, if not the dominant mechanism of nanoparticle toxicity relative to bulk counterparts [214]. However, malondialdehyde (MDA) generation, a metric of oxidative damage in vivo, further illustrates the confusion surrounding size-dependent toxicity. For example, MDA concentrations in liver were 204 and 286 % elevated in zebrafish exposed to 5 mg/L ZnO NP and bulk ZnO respectively, suggesting bulk ZnO was more bioavailable and exerted greater oxidative damage-mediated toxicity than the nanoscale form [215].

#### 10.6.1.2 Mechanism of Nanoscale Toxicity

(a) *ROS-Induced Toxicity*
A wide range of NPs such as iron oxides, fullerenes, and $TiO_2$ [197, 216], aluminium oxide [217], and even talc and silica nanoparticles [217] and chitosan nanoparticles [219] induce the generation of reactive oxygen species (ROS). Many of these ROS, including singlet oxygen, superoxide anion, and hydroxyl radical, are extremely reactive due to unpaired electrons in their outer valence shells, conferring great reactivity, which is largely responsible for their toxicity. There are four possible mechanisms postulated by which nanomaterials may interact with biological tissue [220], the first three of which are directly ROS-mediated, including:

1. UV activation of electron hole pairs leading to ROS generation.
2. Composition of the nanomaterial itself, whereby discontinuous crystal planes and defects could generate active electronic configurations that could form ROS radicals.
3. Metals or metal coatings such as Fe could redox cycle forming ROS.
4. Dissolution of metal ions from particles to surrounding medium, which both releases toxic chemicals, but will also potentially changes surface charge.

However, even the fourth mechanism, while not directly linked to ROS generation, may in fact also generate radicals indirectly in vivo. A prevalent suggestion is that free metal ions released from NPs exert their toxicity, at least in part, by crossing nuclear or mitochondrial membranes and reacting with endogenously generated hydrogen peroxide via Fenton-type processes, in turn inducing generation of highly reactive but short-lived ROS radicals such as hydroxyl radical and superoxide anion [220, 221]. However, some NPs as heterogeneous catalysts may have surface reactions that could generate ROS even in the absence of free ion dissolution, but more often than not, ion dissolution appears involved in nanomaterial toxicity, and it is the major postulated mechanism of metal oxide toxicity [221]. For instance, the antibacterial mechanism of AgNPs is related to the formation of free radicals and ROS-mediated membrane damage, likely as a function of Ag ion dissolution [200]. Silver nanoparticles and silver ions may in fact work synergistically to promote microbial membrane damage, as the ions move into the cells and produce ROS, which in turn can increase cell permeability leading to uncontrolled AgNP and cellular constituent transport across the cell membrane leading to cell death [201]. A more than tenfold increase in ROS was observed with 15 nm silver nanoparticles relative to 55 nm, suggesting silver cytotoxicity is mediated by oxidative stress and AgNP size is critical [193]. Iron oxide nanoparticles, releasing metal ions in vivo, have been demonstrated to induce hydroxyl radical generation via Fenton-type processes under condoning biological microenvironments (physiological pH, reducing agent presence, etc.), which again damage plasma membranes leading to cell death. As alluded to previously, naturally occurring talc and silica nanoparticles both generate ROS to mediate their toxicity through oxidative stress as evidenced by lactate dehydrogenase leakage, lipid hydroperoxide increases, and depletion of reduced glutathione, although it is unclear if dissolved ions mediate this toxicity in any way [218].

(b) *Non-ROS-Mediated Toxicity*

While ROS-mediated toxicity is a prominent concern associated with in vivo nanoparticle exposure, nanomaterials can exert toxic effects by other significant pathway as well. Numerous studies have found NPs can induce cellular autophagy or cell death by free ion leaching, but these effects appear to be highly dependent on the nature of the cells [222]. It has been experimentally determined that negatively charged nanoparticles bound to fluid areas of a plasma membrane-induced localized gelation, whereas positively charged nanoparticles turned gelled areas into a fluid state for easier penetration. This phenomenon explains why cationic particles are more toxic than neutral or anionic analogues of the same size as positively charged NP surfaces induce holes in plasma membranes, while neutrals or negatives do not [223]. Biophysical structural impairment induced by nanomaterials may be a significant mechanism of toxicity; for example, a plasma membrane may suffer rapid strain-induced damage when nanomaterials bind, ion channels are blocked, or holes introduced to membranes [223].

Silver nanoparticles and silver ions also work synergistically to promote membrane damage in plants, although it is unclear if ROS mediates this toxicity. What is clear is the dynamic equilibrium between dissolved and nanoparticle-bound Ag allows the silver ions to be continuously replaced in solution as algal cells assimilate free $Ag^+$ ions, with nanoparticles serving as a reservoir of $Ag^+$ [224]. The efficient assimilation of $Ag^+$ by algae promotes further Ag dissolution from nanoparticles into solution, likely mediated by oxidation. Ironically, $Ag^+$-induced toxicity may be accelerated by the presence of algae due to their production of $H_2O_2$ as a metabolic product, which may be converted to hydroxyl radical in the presence of $Ag^+$ [224]. Published rates of $H_2O_2$ release are in excess of estimated $Ag^+$ releases, lending support to this theory [224]. Although only about 1 % of the silver in the algal media were present as $Ag^+$ ions, these dissolved ions were 18 times more toxic than the Ag nanoparticles, suggesting long-term toxicity is attributable to both the nanoparticles presence, and that of their dissolved free ions [224]. Gene expression profiles reveal AgNPs and

$AgNO_3$ (i.e. $Ag^+$ in solution) induce distinct expression profiles, suggesting differing modes of toxicity [225, 226]. Other studies confirm nanoparticle metal oxides such as ZnO and $TiO_2$ also exerted their algal toxicity through soluble metal ions continuously released from these nanoparticles [200]. Similarly, it is the free $Cu^{2+}$ ions dissolved from CuO nanoparticles, rather than the NPs themselves, which inhibited cholinesterase activity in carp (*Cyprinus carpio*), suggesting the potential neurotoxicity is largely associated with exposure to the free metal species [205]. Aluminium oxide nanoparticles which leached aluminium ions after 48 h also crucially mediated toxicity in the freshwater crustacean *Ceriodaphnia dubia* [217]; however, other studies utilizing these and other metal oxide nanoparticles found negligible toxicity associated with ions dissolved from these nanoparticles to a suite of waterborne microorganisms [227].

Protein interactions with nanomaterials are another significant source of in vivo toxicity. As the dimensions of a nanoparticle are on par with that of a large plasma protein, several lines of evidence suggest the topographical features of NPs permit interactions with proteins in non-trivial ways [228]. For example, the formation of nanoparticle:protein complexes could enhance transport of nanoparticles, making them accessible to sites where larger particles could not reach. For many charged nanoparticles, especially those approximately the size of protein molecules, the interaction of proteins with their surface charges can alter the proteins' configuration, changing tertiary and quaternary structure, leading to loss of function or catalytic activity [204]. Conversely, smaller nanoparticles with their larger surface areas can enhance protein degradation, again leading to functional changes which do not occur with interactions on the relatively smaller surface areas of larger particles and bulk materials [193]. High and potentially deleterious local protein concentrations may result from proteins binding to nanoparticles by avidity effects, the sum total of the interactions between molecules at multiple binding sites [229]. Avidity effects are distinct from affinity, which is the strength of the interaction of one molecule with another at a single binding site, arising from close spatial repetition of the same protein. However, it is also possible proteins interacting with nanoparticles could produce co-operative effects such as promotion or inhibition of protein fibrillation or self-assembling of amino acid residues on a macromolecule serving as a template, such RNA. The large surface areas and high reactivity of nanoparticles also permit them to negatively interact with proteins and nucleic acids, thereby producing long-term adverse effects on living organisms [203]. Such interactions with DNA may be by direct intercalation or physical and/or electrochemical interactions, but recent evidence suggests nanoparticles may not necessarily need to permeate the cell nucleus to induce genotoxicity as this can be accomplished by ROS inducing oxidative stress within or adjacent to the nuclear membrane [204]. There are numerous and diverse means postulated whereby nanomaterials may exert toxicity together with, or independent of, ROS generation that do not fall cleanly into the preceding categories. Most nanoparticles stimulate the immune system, at least to some extent, and elicit an inflammatory response (often including ROS generation) through the abnormal secretion of cytokines, while other nanopartoicles, such as cerium oxide, significantly suppress inflammation and ROS generation [193]. Due to surface charges, nanoparticle uptake by immune cells (monocytes, macrophages, leukocytes) may lead to inflammatory responses [193]. Such immune activation by nanomaterials may pose a carcinogenic risk, triggered by ROS released by macrophages attempting to destroy foreign material at the inflammation site, leading to DNA damage [223]. Nanometals are further suspected of inducing sublethal effects including respiratory toxicity, disturbances to trace element equilibriums in tissues, and inhibition of $Na^{+}K^{+}$ATPase activities [196]. For example, exposure of zebrafish to $TiO_2$ NPs produced oedema and thickening of the gill lamellae, likely as a physical irritation response, but interestingly, increased Cu and Zn concentrations in the brain, clearly reflecting their uptake at the gill epithelium, but also their disruption of endogenous trace metal homeostasis [230]. $Na^{+}K^{+}$ATPase activity was suppressed in the gills and intestine, while 2-thiobabituric acid reactive substances (TBARS), a metric of plasma membrane peroxidation, were elevated in the gill, intestine, and brain with exposure to $TiO_2$ NPs, suggesting cell membrane damage may have been linked to the increase in trace metal (Cu and Zn) concentrations [230].

#### 10.6.1.3 Mine Water Conditions Mediating Nanotoxicity

The behaviour of NPs under real-world conditions has been under-investigated relative to laboratory simulations, but are known to be strongly influenced by the presence of NOM and salinity, which affect agglomeration, deposition, and adsorption [201]. The toxicity of nanomaterials may differ from their reactivity due to atomic composition, surface charge, particle agglomeration, and in natural systems, how these factors vary with environmental pH, presence of charged ions in solution, and through interactions with biomolecules in vivo [201]. For instance, ZnONPs and $TiO_2$NPs are more cytotoxic in the presence of sunlight, but cytotoxicity is ameliorated by the presence of humic acids [231]. Similarly, the potential toxicity of nanoparticles in mine waters is likely mediated by the interaction of the mine waters' and nanomaterials' physicochemical properties. Size, agglomeration rate (which decreases with increasing particle size), and adsorption properties towards dissolved and suspended organic material all determine the fate and ability of nanomaterials in mine water to interact with abiotic and biotic entities in receiving environments [201]. It is widely recognized that nanoparticles in a biological or environmental context never consist of "bare" particles as they rapidly contact atoms, atom clusters, single molecules, and macromolecules, which bind to the surface of the molecule and thereafter reflect the properties of the complex, rather than the pre-existing nanomaterial [204]. For example, in a biological environment where proteins are present, they will form a surface layer known as a corona that will ultimately define the surface properties of this new "particle + corona" compound [204]. Consequently, to understand a particle's nanotoxicity, it is necessary to understand both the particle and its interactions with its environment. Natural organic material (NOM) likely rapidly coats NPs discharged into aquatic environments, which influences their availability and relative toxicity to aquatic organisms. This is well illustrated by both dissolved and suspended humic acids, when bound to surfaces of nanoparticles, greatly reduced their bonding to algal cells by increasing electrostatic repulsion [206]. As humic acids are the main NOM in environments impacted with mine water, their tendency to absorb or chelate metals, hydrous metal oxides, and multivalent cations to both alkyl and aromatic units of their humic acid skeleton suggests their presence will significantly affect both nanoparticles and metal pollutants within mine water. However, this relationship is dependent on pH, as many nanoparticles are negatively charged, or nearly uncharged at higher pH values [201]. Mechanisms underlying the myriad of potential interactions between nanoparticles and living systems are not yet fully understood, and may well never be. This complexity is largely due to the particles' ability to bind and interact with many forms of biological matter, and to change their surface characteristics, depending on the environment they are in [204]. Even particles of the same material can show completely different behaviour, and often toxicities, with slight differences in surface coating, charge, size, and especially environment [204].

### *10.6.2 Risk of Nanoparticles as Pollution Vectors*

Untangling the mechanisms underlying the transport of contaminants in natural waters by colloidal particles is analogous to untying the mythical Gordian knot. Typical environmental colloids are nanoparticles consisting of silicates or oxides, organic colloids such as humic substances or polysaccharides, and biological particles (bacteria, viruses). Point source discharges such as municipal wastewater treatment plants, industry, and mine water are significant contributors of environmental colloids, which continually perfuse aquatic environments and expose biota with nanopollutants. However, the vast majority of nanomaterials in wastewater or in natural environments are retained among the biosolids and sediments [216]. Biosolids and sediments, in turn, can serve as a reservoir for non-point source, pulsatile nanomaterial release to surface waterways during rain events [195, 200, 216]. Considering how ubiquitous waterborne nanoparticles are in contemporary natural systems influenced by human activities, it is not surprising that they also influence the transport of other contaminants in the environment. However, the addition of significant quantities of engineered nanomaterials to aquatic systems may influence the migration of contaminants well beyond the role of natural nanoparticles or colloids, likely as highly efficient contaminant transport conduits. Traditionally, non-volatile contaminants within ground, surface, and mine waters are shuttled between the two-phase system of water and sediments by natural or anthropogenic nanoconduits linking the mobile aqueous phase and an immobile solid phase [232]. Sparingly soluble or strongly adsorbing contaminants partition primarily to the later phase, and should, therefore, be sparingly detected in mine water. However, when contaminants interact with colloids in the aquatic phase, there is the formation of "pseudocolloids" or "eigencolloids" ("real colloids", or "intrinsic colloids") [233]. This leads to even sparingly soluble and strongly adsorbing contaminants becoming mobile, seemingly against the dictates of their individual physicochemical properties [232]. Conversely, dissolved contaminants may be removed from water and immobilized by adsorption onto colloids, which may aggregate leading to their sedimentation, thereby serving as a form of water treatment.

## 10.7 Concluding Remarks

Adopting nanotechniques for the monitoring and remediation of mine water pollution can radically reduce complexity, as well as material and energetic requirements relative to traditional approaches. Nanomaterials are uniquely surface-active due to their extensive surface area to volume ratios, which confer expansive active sites for sensing and/or binding proximal elements and compounds. Utilizing these unique attributes, specific nano- and biosensors can be developed to rapidly identify biological and chemical substances through variations produced upon interaction that alter physicochemical properties. Proximal sensing and signal change abilities, long-term stability, high adsorption efficiencies, extraordinary sensitivities, tuneable selectivity, and reusability are some of the highly desirable and useful attributes of engineered nanomaterials which facilitate their incorporation into a large suite of mine water treatments and sensors. As a result, nanosensors are expected to contribute to novel, revolutionary mine water applications, facilitating early and accurate detection of environmental pollutants. It follows that nanosensors, and nanoremediation have the potential not only to reduce the overall costs of cleaning up large-scale abandoned mining and processing sites but also to reduce clean-up time, eliminate the need for treatment and disposal of contaminated sludge, and reduce some contaminant concentrations in situ. Substantial initial investment would be needed to incorporate or switch to nanotechnology-based mine water treatments. However, once adopted, these techniques could considerably lower mine water treatment costs over the long term. Unfortunately, large-scale application of nanotechnology within mine water management has been extremely limited, as techniques are currently largely constrained to the bench-scale. However, it is anticipated that within the next two decades, such real-world applications will be more commonly adopted, as has been the paradigm in other fields. Prior to the wholesale adoption of large-scale nanotechnology-based mine water projects, further toxicological studies, including assessments of potential environmental impacts, must be considered. Relative to bulk material toxicity evaluations for which much is known and which have been well standardized against mass-based dose metrics, nanotoxicity assessments suffer from a paucity of data, and are made more complicated by the need to consider the influences of material, size, shape, surface charge, coating, dispersion, agglomeration, aggregation, concentration, and matrix effects. Proper evaluation of materials used in nanoremediation, particularly full-scale ecosystem-wide studies, needs to be conducted to prevent potential adverse environmental impacts.

## References


1. M. Gavrilescu, L.V. Pavel, I. Cretescu, J. Hazard. Mater. **163**, 475 (2009)
2. I.G. Carvalho, R. Cidu, L. Fanfani, H. Pitsch, C. Beaucaire, P. Zuddas, Environ. Sci. Technol. **39**, 8646 (2005)
3. R.R. Tatiya, *Surface and Underground Excavations*. Methods, Techniques and Equipment, 2nd edn. (CRC Press, London, 2013)
4. H.B.L. Pettersson, J. Koperski, Health Phys. **60**, 681 (1991)
5. E.G. Kelly, D.J. Spottiswood, *Introduction to Mineral Processing* (Wiley, Oxford, 1982)
6. B.A. Wills, T. Napier-Munn, *Wills' Mineral Processing Technology: An Introduction to the Practical Aspects of Ore Treatment and Mineral Recovery* (Butterworth-Heinemann, Oxford, 2006)
7. J. Ahn, Y. Park, J.-Y. Kim, K.-W. Kim, Environ. Geochem. Health **27**, 147 (2005)
8. J.A. Davis, D.E. Meece, M. Kohler, G.P. Curtis, Geochim. Cosmochim. Acta **68**, 3621 (2004)
9. R. Donahue, M.J. Hendry, Appl. Geochem. **18**, 1733 (2003)
10. M.C. Duff, J.U. Coughlin, D.B. Hunter, Geochim. Cosmochim. Acta **66**, 3533 (2002)
11. IAEA, *Manual of acid in situ leach uranium mining technology* (International Atomic Energy Agency, Vienna, Austria, 2001), p. 283
12. C. Wolkersdorfer, *Water Management at Abandoned Flooded Underground Mines: Fundamentals, Tracer Tests, Modelling, Water Treatment* (Springer, Heidelberg, 2008)
13. IAEA, *Guidebook on good practice in the management of uranium mining and mill operations and the preparation for their closure* (International Atomic Energy Agency, Vienna, Austria, 1998), p. 81
14. D.N. Castendyk, L.E. Eary (eds.), *Mine Pit Lakes: Characteristics, Predictive Modeling, and Sustainability* (Society for Mining, Metallurgy, and Exploration, Littleton, 2009)
15. P.L. Younger, N.S. Robins (eds.), *Mine Water Hydrogeology and Geochemistry* (The Geological Society, Brassmill Lane, 2002)
16. P.L. Younger, S.A. Banwart, R.S. Hedin (eds.), *Mine Water: Hydrology, Pollution, Remediation* (Springer, Heidelberg, 2002)
17. G.G. Bowen, C. Dussek, R.M. Hamilton, Geological Society, London, Special Publications **128** 93 (1998)
18. D.B. Johnson, K.B. Hallberg, Sci. Total Environ. **338**, 3 (2005)
19. INAP, Global Acid Rock Drainage Guide (GARD Guide) Global Acid Rock Drainage Guide (GARD Guide), ICARD 8. INAP: The International Network for Acid Prevention, Skelleftea, Sweden, 2009
20. A.S. Sheoran, V. Sheoran, Miner. Eng. **19**, 105 (2006)
21. R.E. Mielke, D.L. Pace, T. Porter, G. Southam, Geobiology **1**, 81 (2003)
22. A.R. Elsetinow, M.J. Borda, M.A.A. Schoonen, D.R. Strongin, Adv. Environ. Res. **7**, 969 (2003)
23. K. Benzerara, G. Morin, T.H. Yoon, J. Miot, T. Tyliszczak, C. Casiot, O. Bruneel, F. Farges, G.E. Brown Jr., Geochim. Cosmochim. Acta **72**, 3949 (2008)
24. R.S. Hedin, Mine Water Environ.ment **27**, 200 (2008)
25. N. Shikazono, M. Utada, Miner. Deposita **32**, 596 (1997)
26. C. Woulds, B.T. Ngwenya, Appl. Geochem. **19**, 1773 (2004)
27. H. Shaoping, C. Xincai, S. Jiyan, C. Yingxu, L. Qi, Chemosphere **71**, 2091 (2008)
28. K. Leopold, B. Michalik, J. Wiegand, J. Environ. Radioact. **94**, 137 (2007)

29. M.K. Banks, A.P. Schwab, G.R. Fleming, B.A. Hetrick, Chemosphere **29**, 1691 (1994)
30. E.R. Landa, E.J.P. Phillips, D.R. Lovley, Appl. Geochem. **6**, 647 (1991)
31. M. Mkandawire, *Environ Sci Pollut Res 1*, 2013
32. P. Diehl, Uranium Mining in Eastern Germany: The WISMUT Legacy. WISE Uranium Project, Arnsdorf, Germany, 2003, p. http://www.antenna.nl/wise/uranium/uwis.html.
33. H.M. Abdel-Aziz, J. Appl. Polym. Sci. **120**, 1547 (2011)
34. A. Abdelouas, Elements **2**, 335 (2006)
35. D.W. Engelkemeir, K.F. Flynn, L.E. Glendenin, Phys. Rev. **126**, 1818 (1962)
36. IAEA, World Distribution of Uranium Deposits (UDEPO) with Uranium Deposit Classification, International Atomic Energy Agency, Vienna, 2009.
37. S. Yusan, S. Erenturk, Desalination **269**, 58 (2011)
38. L.J. Qian, J.N. Zhao, P.Z. Hu, Y.X. Geng, W.S. Wu, J Radioanal Nucl Chem **283**, 653 (2010)
39. D.L. Guerra, A.A. Pinto, C. Airoldi, R.R. Viana, Inorganic Chemistry Communications **11**, 539 (2008)
40. T. Allard, P. Ildefonse, C. Beaucaire, G. Calas, Chem. Geol. **158**, 81 (1999)
41. X. Qu, P.J.J. Alvarez, Q. Li, Water Res. **47**, 3931 (2013)
42. X.G. Li, H. Feng, M.R. Huang, Chem. Eur. J **15**, 4573 (2009)
43. T.B. Henry, F.M. Menn, J.T. Fleming, J. Wilgus, R.N. Compton, G.S. Sayler, Environ. Health Perspect. **115**, 1059 (2007)
44. D.K. Tiwari, J. Behari, P. Sen, World Appl. Sci. J. **3**, 417 (2008)
45. A. Carlmark, C. Hawker, A. Hult, M. Malkoch, Chem. Soc. Rev. **38**, 352 (2009)
46. Y.C. Sharma, V. Srivastava, J. Chem. Eng. Data **56**, 819 (2011)
47. X.X. Wang, CLEAN—Soil Air Water **39**, 13 (2011)
48. Y.F. Lin, H.W. Chen, K.L. Lin, B. Chen, C. Chiou, J. Environ. Sci. **23**, 44 (2011)
49. X.G. Li, H. Feng, M.R. Huang, Chem. Eur. J **16**, 10113 (2010)
50. N.N. Nassar, Sep. Sci. Technol. **45**, 1092 (2010)
51. A. Eslami, S. Nasseri, B. Yadollahi, A. Mesdaghinia, F. Vaezi, R. Nabizadeh, S. Nazmara, J. Chem. Tech. Biotechnol. **83**, 1447 (2008)
52. N.S. Wigginton, K.L. Haus, M.F.H. Jr, J. Environ. Monit. **9**, 1306 (2007)
53. A.J. Du, D.D. Sun, J.O. Leckie, J. Hazard. Mater. **187**, 96 (2011)
54. H.J. Shipley, S. Yean, A.T. Kan, M.B. Tomson, Environ. Sci. Pollut. Res **17**, 1053 (2010)
55. X.-q. Li, D.W. Elliott, W.-x. Zhang, Critical Reviews in Solid State and Materials Sciences **31** 111 (2006).
56. C.-H. Liao, S.-F. Kang, Y.-W. Hsu, Water Res. **37**, 4109 (2003)
57. L.J. Dong, P.V. Zinin, J.P. Cowen, L.C. Ming, J. Hazard. Mater. **168**, 626 (2009)
58. O.A. Sadik, A.L. Zhou, S. Kikandi, N. Du, Q. Wang, K. Varner, J. Environ. Monit. **11**, 1782 (2009)
59. I. Linkov, J. Steevens, G. Adlakha-Hutcheon, E. Bennett, M. Chappell, V. Colvin, J.M. Davis, T. Davis, A. Elder, S. Hansen, P.B. Hakkinen, S.M. Hussain, D. Karkan, R. Korenstein, I. Lynch, C. Metcalfe, A.B. Ramadan, F.K. Satterstrom, J. Nanopart. Res **11**, 513 (2009)
60. D.G. Rickerby, M. Morrison, Sci. Tech. Adv. Mater. **8**, 19 (2007)
61. P. Namour, M. Lepot, N. Jaffrezic-Renault, *Sensors (Basel, Switzerland)* ***10*** *7947*, 2010
62. M. Li, H. Gou, I. Al-Ogaidi, N. Wu, ACS Sustain. Chem. Eng. **1**, 713 (2013)
63. K. Saha, S.S. Agasti, C. Kim, X. Li, V.M. Rotello, Chem. Rev. **112**, 2739 (2012)
64. J. Liu, Z. Cao, Y. Lu, Chem. Rev. **109**, 1948 (2009)
65. D.L. Robertson, G.F. Joyce, Nature **344**, 467 (1990)
66. X. Xue, F. Wang, X. Liu, J. Am. Chem. Soc. **130**, 3244 (2008)
67. J. Liu, Y. Lu, J. Am. Chem. Soc. **125**, 6642 (2003)
68. J. Liu, Y. Lu, J. Am. Chem. Soc. **126**, 12298 (2004)
69. J. Liu, A.K. Brown, X. Meng, D.M. Cropek, J.D. Istok, D.B. Watson, Y. Lu, Proc. Natl. Acad. Sci. **104**, 2056 (2007)

70. E. Oh, M.-Y. Hong, D. Lee, S.-H. Nam, H.C. Yoon, H.-S. Kim, J. Am. Chem. Soc. **127**, 3270 (2005)
71. C.S. Yun, A. Javier, T. Jennings, M. Fisher, S. Hira, S. Peterson, B. Hopkins, N.O. Reich, G.F. Strouse, J. Am. Chem. Soc. **127**, 3115 (2005)
72. M. Li, Q. Wang, X. Shi, L.A. Hornak, N. Wu, Anal. Chem. **83**, 7061 (2011)
73. J.R. Zhang, W.T. Huang, W.Y. Xie, T. Wen, H.Q. Luo, N.B. Li, Analyst **137**, 3300 (2012)
74. J. Li, L. Chen, T. Lou, Y. Wang, ACS Appl. Mater. Interfaces **3**, 3936 (2011)
75. F. Faridbod, M.R. Ganjali, M. Pirali-Hamedani, P. Norouzi, Int. J. Electrochem. Sci. **5**, 1103 (2010)
76. P. Norouzi, Z. Rafiei-Sarmazdeh, F. Faridbod, M. Adibi, M.R. Ganjali, Int. J. Electrochem. Sci. **5**, 367 (2010)
77. F. Faridbod, M.R. Ganjali, B. Larijani, M. Hosseini, P. Norouzi, Mater. Sci. Eng. C **30**, 555 (2010)
78. M.R. Ganjali, T. Poursaberi, M. Khoobi, A. Shafiee, M. Adibi, M. Pirali-Hamedani, P. Norouzi, Int. J. Electrochem. Sci. **6**, 717 (2011)
79. Y. Shao, J. Wang, H. Wu, J. Liu, I.A. Aksay, Y. Lin, Electroanalysis **22**, 1027 (2010)
80. A.K. Wanekaya, Analyst **136**, 4383 (2011)
81. D.W. Kimmel, G. LeBlanc, M.E. Meschievitz, D.E. Cliffel, Anal. Chem. **84**, 685 (2011)
82. J. Morton, N. Havens, A. Mugweru, A.K. Wanekaya, Electroanalysis **21**, 1597 (2009)
83. L. Wang, X. Wang, G. Shi, C. Peng, Y. Ding, Anal. Chem. **84**, 10560 (2012)
84. S. Yuan, D. Peng, D. Song, J. Gong, Sens. Actuators B **181**, 432 (2013)
85. T.H. Kim, J. Lee, S. Hong, J. Phys. Chem. C **113**, 19393 (2009)
86. Z. Zou, A. Jang, E.T. MacKnight, P.M. Wu, J. Do, J.S. Shim, P.L. Bishop, C.H. Ahn, IEEE Sensors J **9**, 586 (2009)
87. J.P. Lafleur, S. Senkbeil, T.G. Jensen, J.P. Kutter, Lab Chip **12**, 4651 (2012)
88. S.J. Lee, M. Moskovits, Nano Lett. **11**, 145 (2010)
89. L.C. Batty, P.L. Younger, Environ. Pollut. **132**, 85 (2004)
90. G.E. Dudel, C. Brackhage, H. Dienemann, M. Mkandawire, A. Weiske, in *Mine Water—Process, Policy and Progress*, ed. by A.P. Jarvis, B.A. Dudgeon, P.L. Younger (University of Newcastle Upon Tyne Press, Newcastle, 2004)
91. K.B. Hallberg, D.B. Johnson, Land Contam. Reclam. **11**, 213 (2003)
92. C.E. Rostad, B.S. Martin, L.B. Barber, J.A. Leenheer, Environ. Sci. Technol. **34**, 2703 (2000)
93. A. Zayed, S. Gowthaman, N. Terry, J. Environ. Qual. **27**, 715 (1998)
94. M. Morita, N.S. Malvankar, A.E. Franks, Z.M. Summers, L. Giloteaux, A.E. Rotaru, C. Rotaru, D.R. Lovley, *Mbio* **2**, 2011
95. H. Daims, M.W. Taylor, M. Wagner, Trends Biotechnol. **24**, 483 (2006)
96. U. Soltmann, H. Wand, A. Müller, P. Kuschk, U. Stottmeister, Acta Biotechnol. **22**, 161 (2002)
97. S. Dave, M. Damani, D. Tipre, J. Hazard. Mater. **173**, 231 (2010)
98. C. White, S.C. Wilkinson, G.M. Gadd, Int. Biodeter. Biodegrad. **35**, 17 (1995)
99. M. Kilic, M.E. Keskin, S. Mazlum, N. Mazlum, Chem. Eng. Process **47**, 31 (2008)
100. Y.B. Zang, W.G. Hou, J. Xu, Chin. J. Chem. **29**, 847 (2011)
101. S.K. Hamdona, O.A. Al Hadad, Desalination **228**, 277 (2008)
102. M. Saeedi, M. Hosseinzadeh, M. Rajabzadeh, Environ. Earth Sci **62**, 519 (2011)
103. G.C.L. Araujo, S.G. Lemos, A.G. Ferreira, H. Freitas, A.R.A. Nogueira, Chemosphere **68**, 537 (2007)
104. M. Islam, R. Patel, Desalination **256**, 120 (2010)
105. H. Beddow, S. Black, D. Read, J. Environ. Radioact. **86**, 289 (2006)
106. R.A.P. Thomas, L.E. Macaskie, Appl. Microbiol. Biotechnol. **49**, 202 (1998)
107. C. Bobirica, L. Bobirica, R. Stanescu, I. Constantinescu, Environ. Eng. Manag. J **8**, 759 (2009)
108. X.X. Guo, F.Z. Zhang, Q. Peng, S.L. Xu, X.D. Lei, D.G. Evans, X. Duan, Chem. Eng. J. **166**, 81 (2011)

109. N.C. Feng, X.Y. Guo, S. Liang, Y.S. Zhu, J.P. Liu, J. Hazard. Mater. **185**, 49 (2011)
110. V.E. Copcia, C.M. Hristodor, N. Bilba, E. Popovici, Rev. Chem **62**, 195 (2011)
111. M.L. Zhang, H.Y. Zhang, D. Xu, L. Han, D.X. Niu, B.H. Tian, J.A. Zhang, L.Y. Zhang, W.S. Wu, Desalination **271**, 111 (2011)
112. M.L. Coleman, D.B. Hedrick, D.R. Lovley, D.C. White, K. Pye, Nature **361**, 436 (1993)
113. K. Benzerara, J. Miot, G. Morin, G. Ona-Nguema, F. Skouri-Panet, C. Férard, Compt. Rendus Geosci. **343**, 160 (2011)
114. S. Mann, J. Webb, R.J.P. Williams, *Biomineralisation - Chemical and Biological pespectives* (Weinhein, VCH Verlagsgesellschaft, 1989)
115. M. Epple, E. Bäuerlein, W. Pompe (Ed.)^(Eds.), Handbook of Biomineralization. Wiley-VCH, Weinheim, 2009.
116. K. Watlington, Emerging Nanotechnologies for Site Remediation and Wastewater Treatment. National Network for Environmental Management Studies (NNEMS), 2005, p. 40.
117. G.M. Gadd, Microbiology **156**, 609 (2010)
118. S.C.N. Tang, I.M.C. Lo, Water Res. **47**, 2613 (2013)
119. N. Singh, G.J. Jenkins, R. Asadi, S.H. Doak, *Nano reviews **1***, 2010
120. C. Noubactep, Environ. Technol. **29**, 909 (2008)
121. C.S. Rajan, Int. J. Environ. Sci. Dev **2**, 3 (2011)
122. S. Shariati, M. Faraji, Y. Yamini, A.A. Rajabi, Desalination **270**, 160 (2011)
123. H. Parham, B. Zargar, Z. Heidari, A. Hatamie, J. Iran. Chem. Soc. **8**, S9 (2011)
124. J.R. Peralta-Videa, L.J. Zhao, M.L. Lopez-Moreno, G. de la Rosa, J. Hong, J.L. Gardea-Torresdey, J. Hazard. Mater. **186**, 1 (2011)
125. C. Noubactep, J. Radioanal. Nucl. Chem **267**, 13 (2005)
126. C. Noubactep, G. Meinrath, P. Dietrich, M. Sauter, B.J. Merkel, Environ. Chem **2**, 71 (2005)
127. S.O. Hara, T. Krug, J. Quinn, C. Clausen, C. Geiger, Remed. J. **16**, 35 (2006)
128. J. Hu, I.M.C. Lo, G. Chen, *Water Science and Technology 50, 50 139*, 2004
129. H.K. Boparai, M. Joseph, D.M. O'Carroll, J. Hazard. Mater. **186**, 458 (2011)
130. W. Wu, Q. He, C. Jiang, Nanoscale Res. Lett. **3**, 397 (2008)
131. R.A. Root, S. Fathordoobadi, F. Alday, W. Ela, J. Chorover, *Environmental Science & Technology*, 2013
132. A. Knauer, A. Thete, S. Li, H. Romanus, A. Csaki, W. Fritzsche, J.M. Kohler, Chem. Eng. J. **166**, 1164 (2011)
133. N. Savage, M. Diallo, J. Nanopart. Res **7**, 331 (2005)
134. M.S. Diallo, N. Savage, J. Nanopart. Res **7**, 25 (2005)
135. S.B. Shen, T.L. Pan, X.Q. Liu, L. Yuan, Y.J. Zhang, J.C. Wang, Z.C. Guo, J. Colloid Interface Sci. **345**, 12 (2010)
136. W.G. Shim, J.W. Lee, S.C. Kim, Appl. Catal. B Environ. **84**, 133 (2008)
137. Y. Wang, J.H. Qu, H.J. Liu, C.Z. Hu, Catal. Today **126**, 476 (2007)
138. C. Boschi, H. Maldonado, M. Ly, E. Guibal, J. Colloid Interface Sci. **357**, 487 (2011)
139. V.K. Gupta, B. Gupta, A. Rastogi, S. Agarwal, A. Nayak, J. Hazard. Mater. **186**, 891 (2011)
140. K. Jainae, K. Sanuwong, J. Nuangjamnong, N. Sukpirom, F. Unob, Chem. Eng. J. **160**, 586 (2010)
141. Q.J. Geng, Q.J. Guo, C.Q. Cao, H.Q. Wang, Chem. Eng. Tech **31**, 1023 (2008)
142. M.R. Hoffmann, S.T. Martin, W. Choi, D.W. Bahnemannt, Chem. Rev. **95**, 69 (1995)
143. V. Rodríguez-González, R. Zanella, G. del Angel, R. Gómez, J. Mol. Catal. Chem **281**, 93 (2008)
144. V. Stengl, S. Bakardjieva, T.M. Grygar, J. Bludska, M. Kormunda, Chem. Cent. J **7**, 41 (2013)
145. J.J. Chen, K. Wang, R. Huang, T. Saito, Y.H. Ikuhara, T. Hirayama, W.L. Zhou, IEEE Trans. Nanotech **9**, 634 (2010)
146. J.E. Rossini, A.S. Huss, J.N. Bohnsack, D.A. Blank, K.R. Mann, W.L. Gladfelter, J. Phys. Chem. C **115**, 11 (2011)

147. L. Wei, D. Shi, Z. Zhou, P. Ye, J. Wang, J. Zhao, L. Liu, C. Chen, Y. Zhang, Nanoscale Res. Lett. **7**, 334 (2012)
148. G. Fryxell, S. Mattigod, Y. Lin, H. Wu, S. Fiskom, K. Parker, F. Zheng, W. Yantasee, T. Zemanian, R. Addleman, J. Liu, K. Kemner, S. Kelly, X. Feng, J. Mater. Chem. **17**, 2863 (2007)
149. E. Blanco, A. Pineiro, R. Miller, J.M. Ruso, G. Prieto, F. Sarmiento, Langmuir **25**, 8075 (2009)
150. T.S. Zemanian, G.E. Fryxell, J. Liu, S. Mattigod, J.A. Franz, Z. Nie, Langmuir **17**, 8172 (2001)
151. U. Soltmann, J. Raff, S. Selenska-Pobell, S. Matys, W. Pompe, H. Böttcher, J. Sol-Gel Sci. Tech **26**, 1209 (2003)
152. A. Bögershausen, Ruprecht - Karls - Universität Heidelberg, Mannheim, 2004
153. K. Rath, *Science and Technology Review 24*, 2005
154. U. Soltmann, H. Böttcher, J. Sol-Gel Sci. Tech **48**, 66 (2008)
155. M. Mkandawire, J. Förster, D. Fiedler, H. Böttcher, W. Pompe, Int. J. Environ. Anal. Chem. **89**, 529 (2009)
156. T.J. Barnes, I. Ametov, C.A. Prestidge, Asia Pac. J. Chem. Eng **3**, 13 (2008)
157. C.C. Chu, N. Ueno, T. Imae, Chem. Mater. **20**, 2669 (2008)
158. M. Kubeil, H. Stephan, H.-J. Pietzsch, G. Geipel, D. Appelhans, B. Voit, J. Hoffmann, B. Brutschy, Y.V. Mironov, K.A. Brylev, V.E. Fedorov, Chem. Asian J. **5**, 2507 (2010)
159. W.-D. Jang, K.M. Kamruzzaman Selim, C.-H. Lee, I.-K. Kang, Prog. Polym. Sci. **34**, 1 (2009)
160. M. Gaab, S. Bellemin-Laponnaz, L.H. Gade, Chemistry **15**, 5450 (2009)
161. A.N. Petit, P. Eullaffroy, T. Debenest, F. Gagne, Aquat. Toxicol. **100**, 187 (2010)
162. N.M. Mahmoodi, B. Hayati, M. Arami, F. Mazaheri, J. Chem. Eng. Data **55**, 4660 (2010)
163. F. Ma, R.J. Qu, C.M. Sun, C.H. Wang, C.N. Ji, Y. Zhang, P. Yin, J. Hazard. Mater. **172**, 792 (2009)
164. A. Shakhbazau, D. Shcharbin, I. Seviaryn, N. Goncharova, S. Kosmacheva, M. Potapnev, B. Gabara, M. Ionov, M. Bryszewska, Mol. Biol. Rep. **37**, 2003 (2010)
165. T. Pietsch, D. Appelhans, N. Gindy, B. Voit, A. Fahmi, Colloids Surf. A Physicochem. Eng. Asp. **341**, 93 (2009)
166. M.J. DeMarco, A.K. SenGupta, J.E. Greenleaf, Water Res. **37**, 164 (2003)
167. A.A. Agrawal, B.J. Nehilla, K.V. Reisig, T.R. Gaborski, D.Z. Fang, C.C. Striemer, P.M. Fauchet, J.L. McGrath, Biomaterials **31**, 5408 (2010)
168. A.G. Cattaneo, R. Gornati, E. Sabbioni, M. Chiriva-Internati, E. Cobos, M.R. Jenkins, G. Bernardini, J. Appl. Toxicol. **30**, 730 (2010)
169. P. Arosio, S. Levi, Free Radic. Biol. Med. **33**, 457 (2002)
170. J.-F. Briat, M. Lebrun, Compt. Rendus Acad. Sci. III Sci. Vie **322**, 43 (1999)
171. E. Pilon-Smits, M. Pilon, Crit. Rev. Plant Sci. **21**, 439 (2002)
172. M. Uchida, S. Kang, C. Reichhardt, K. Harlen, T. Douglas, *Biochim Biophys Acta **1800***, 2010
173. Y. Gorby, J. McLean, A. Korenevsky, K. Rosso, M.Y. El-Naggar, T.J. Beveridge, Geobiology **6**, 232 (2008)
174. K. Fahmy, M. Merroun, K. Pollmann, J. Raff, O. Savchuk, C. Hennig, S. Selenska-Pobell, Biophys. J. **91**, 996 (2006)
175. N. Ilk, E.M. Egelseer, U.B. Sleytr, Curr. Opin. Biotechnol. **22**, 824 (2011)
176. V. Saravia, S. Küpcü, M. Nolte, C. Huber, D. Pum, A. Fery, U.B. Sleytr, J.L. Toca-Herrera, J. Biotechnol. **130**, 247 (2007)
177. M.L. Merroun, J. Raff, A. Rossberg, C. Hennig, T. Reich, S. Selenska-Pobell, Appl. Environ. Microbiol. **71**, 5532 (2005)
178. P.J. Panak, J. Raff, S. Selenska-Pobell, G. Geipel, G. Bernhard, H. Nitsche, Radiochimica Acta **88**, 71 (2000)
179. A. Martın-Molina, S. Moreno-Flores, E. Perez, D. Pum, U.B. Sleytr, J.L. Toca-Herrera, Biophys. J. **90**, 1821 (2006)

180. U. Jankowski, M.L. Merroun, S. Selenska-Pobell, K. Fahmy, Spectroscopy **24**, 177 (2010)
181. P. Mobili, M.D.L.A. Serradell, C. Mayer, V. Arluison, A. Gomez-Zavaglia, Int. J. Biochem. Res. Rev **3**, 39 (2012)
182. J. Kim, J.W. Grate, Nano Lett. **3**, 1219 (2003)
183. I. Perez-Baena, F. Barroso-Bujans, U. Gasser, A. Arbe, A.J. Moreno, J. Colmenero, J.A. Pomposo, ACS Macro Lett. **2**, 775 (2013)
184. S.R. MacEwan, A. Chilkoti, Biopolymers **94**, 60 (2010)
185. J. Kostal, A. Mulchandani, W. Chen, Macromolecules **34**, 2257 (2001)
186. J. Pu, K. Fukushi, Sci. World J **2013**, 6 (2013)
187. S. Akbar, R. Shahnaz, T.H. Shah, M. Sami, J. Chem. Soc. Pakistan **31**, 543 (2009)
188. A.I.P. Cordoves, M.G. Valdes, J.C.T. Fernandez, G.P. Luis, J.A. Garcia-Calzon, M.E.D. Garcia, Micropor. Mesopor. Mater. **109**, 38 (2008)
189. O. Gok, A. Ozcan, B. Erdem, A.S. Ozcan, Colloids Surf. A Physicochem. Eng. Asp. **317**, 174 (2008)
190. R. Apiratikul, P. Pavasant, Chem. Eng. J. **144**, 245 (2008)
191. V.K. Jha, M. Nagae, M. Matsuda, M. Miyake, J. Environ. Manage. **90**, 2507 (2009)
192. P. Misaelides, A. Godelitsas, A. Filippidis, D. Charistos, I. Anousis, Sci. Total Environ. **173–174**, 237 (1995)
193. I.D. Rastogl, J. Toxicol. Environ. Health Sci **4**, 1 (2012)
194. I. Ahmad, J. Biomed. Toxicol **7**, 32 (2011)
195. R. Zhang, Y. Bai, B. Zhang, L. Chen, B. Yan, J. Hazard. Mater. **212**, 404 (2012)
196. B.J. Shaw, R.D. Handy, Environ. Int. **37**, 1083 (2011)
197. B. Wang, J.-J. Yin, X. Zhou, I. Kurash, Z. Chai, Y. Zhao, W. Feng, J. Phys. Chem. **C117**, 383 (2012)
198. A. Pourmand, Abdollahi, M., Daru J. Pharm. Sci. **20** 95 (2012)
199. Y. Wang, W.G. Aker, H.-M. Hwang, C.G. Yedjou, H. Yu, P.B. Tchounwou, Sci. Total Environ. **409** 4753 (2011)
200. E. Bae, H.-J. Park, J. Yoon, Y. Kim, K. Choi, J. Yi, Korean J. Chem. Eng. **28**, 267 (2011)
201. H.C. Poynton, J.M. Lazorchak, C.A. Impellitteri, B.J. Blalock, K. Rogers, H.J. Allen, A. Loguinov, J.L. Heckman, S. Govindasmawy, Toxicogenomic responses of nanotoxicity in Daphnia magna exposed to silver nitrate and coated silver nanoparticles. Environ. Sci. Technol. **46**, 6288–6296 (2012)
202. K.J. Lee, L.M. Browning, P.D. Nallathamby, T. Desai, P.K. Cherukuri, X.-H.N. Xu, Chem. Res. Toxicol. **25**, 1029 (2012)
203. K.J. Lee, L.M. Browning, P.D. Nallathamby, X.-H.N. Xu, Chem. Res. Toxicol. **26**, 904 (2012)
204. A. Elsaesser, C.V. Howard, Adv. Drug Deliv. Rev. **64**, 129 (2012)
205. J. Zhao, Z. Wang, X. Liu, X. Xie, K. Zhang, B. Xing, J. Hazard. Mater. **197**, 304 (2011)
206. D. Lin, J. Ji, Z. Long, K. Yang, F. Wu, Water Res. **46**, 4477 (2012)
207. H.L. Karlsson, J. Gustafsson, P. Cronholm, L. Möller, Toxicol. Lett. **188**, 112 (2009)
208. A. Menard, D. Drobne, A. Jemec, Environ. Pollut. **159**, 677 (2011)
209. K. Donaldson, C.A. Poland, Curr. Opin. Biotechnol. **24**, 724 (2013)
210. J. Ji, Z. Long, D. Lin, Chem. Eng. J. **170**, 525 (2011)
211. D. Xiong, T. Fang, L. Yu, X. Sima, W. Zhu, Sci. Total Environ. **409**, 1444 (2011)
212. M. Heinlaan, A. Ivask, I. Blinova, H.-C. Dubourguuier, A. Kahru, Chemosphere **71**, 1308 (2008)
213. J. Sun, Q. Zhang, Z. Wang, B. Yan, Int. J. Mol. Sci. **14**, 9319 (2013)
214. M. Pumera, Chem. Asian J. **6**, 340 (2011)
215. D.W. Xiong, T. Fang, X.D. Chen, X.F. Sima, W.T. Zhu, Huan Jing Ke Xue **31**, 1320 (2010)
216. J.F. Reeves, S.J. Davies, N.J.F. Dodd, A.N. Jha, Mutat. Res. **640**, 13 (2008)
217. S. Pakrashi, S. Dalal, A. Humayun, S. Chakravarty, N. Chandrasekaran, A. Mukherjuu, PLoS ONE **8**(9), e74003 (2013)

218. I. Ahmad, I. Mohmood, C. Mieiro, J. Coelho, M. Pacheco, M. Santos, A. Duarte, E. Pereira, *Aquatic Toxicology*, 2011
219. Y.-L. Hu, W. Qi, F. Han, J.-Z. Shao, J.-Q. Gao, Int. J. Nanomed. **6**, 3351 (2011)
220. S. Arora, J.M. Rajwade, K.M. Paknikar, Toxicol. Appl. Pharmacol. **258**, 151 (2012)
221. T.P. Dasari, H.-M. Hwang, J. Environ. Sci. **25**, 1925 (2013)
222. S.J. Soenen, J. Demeester, S.C. De Smedt, K. Braeckmans, Nano Today **8**, 121 (2013)
223. Y.-L. Wu, N. Putcha, K.W. Ng, D.L. Leong, C.T. Lim, S.C.J. Loo, X. Chen, Acc. Chem. Res. **46**, 782 (2013)
224. E. Navarro, F. Piccapietra, B. Wagner, F. Marconi, R. Kaegi, N. Odzak, L. Sigg, R. Behra, Environ. Sci. Technol. **42**, 8959 (2008)
225. J. Choi, S. Kim, J. Ahn, P. Youn, J. Kang, K. Park, J. Yi, D.-Y. Ryu, *Aquatic Toxicology*, 2009
226. Y. Wu, Q. Zhou, H. Li, W. Liu, T. Wang, G. Jiang, *Aquatic Toxicology*, 2009
227. Y.-W. Baek, Y.-J. An, Sci. Total Environ. **409**, 1603 (2011)
228. T. Vorup-Jenson, D. Peer, Adv. Drug Deliv. Rev. **64**, 1661 (2012)
229. A.A. Shemetov, Nabiev, I., Sukhanova, A., ACSNano **6**, 4585 (2012).
230. G. Federici, B.J. Shaw, R.D. Handy, Aquat. Toxicol. **84**, 415 (2007)
231. V. Aruoja, H.-C. Dubourguier, K. Kasemets, A. Kahru, Sci. Total Environ. **407**, 1461 (2009)
232. K.-U. Ulrich, A. Rossberg, H. Foerstendorf, H. Zänker, A.C. Scheinost, Geochim. Cosmochim. Acta **70**, 5469 (2006)
233. K.J. Wilkinson, J.R. Lead (eds.), *Environmental Colloids and Particles Behaviour, Separation and Characterisation* (Wiley, West Sussex, 2007)

# Index



A. Hu and A. Apblett (eds.), *Nanotechnology for Water Treatment and Purification*, Lecture Notes in Nanoscale Science and Technology 22,
DOI 10.1007/978-3-319-06578-6, 

**Q**



**R**

MIX
Papier aus verantwortungsvollen Quellen
Paper from responsible sources
FSC® C105338

If you have any concerns about our products,
you can contact us on
**ProductSafety@springernature.com**

In case Publisher is established outside the EU,
the EU authorized representative is:
**Springer Nature Customer Service Center GmbH**
**Europaplatz 3, 69115 Heidelberg, Germany**

Printed by Libri Plureos GmbH
in Hamburg, Germany